Principles of Cancer Biology

Lewis J. Kleinsmith
University of Michigan, Ann Arbor

San Francisco ▪ Boston ▪ New York ▪
Cape Town ▪ Hong Kong ▪ London ▪ Madrid ▪ Mexico City ▪
Montreal ▪ Munich ▪ Paris ▪ Singapore ▪ Sydney ▪ Tokyo ▪ Toronto

Sponsoring Editor: *Susan Winslow*
Project Editor: *Cinnamon Hearst*
Marketing Manager: *Jeff Hester*
Developmental Artists: *Ken Probst and Russell Chun*
Executive Managing Editor: *Erin Gregg*
Production Supervisors: *Vivian McDougal, Nancy Tabor*
Production Management and Art Coordination: *Crystal Clifton, Progressive Publishing Alternatives*
Composition: *Progressive Information Technologies*
Illustrator: *Corinna Dibble, Dartmouth Publishing*
Text Designer: *Jill Little, iDesign*
Cover Designer: *Jeff Puda*
Manufacturing Buyer: *Stacy Wong*
Cover Printer: *LSC Communications*
Printer: *LSC Communications.*

ISBN 0-8053-4003-3

Library of Congress Cataloging-in-Publication Data
Kleinsmith, Lewis J.
Principles of cancer biology / Lewis J. Kleinsmith.
p. ; cm.
Includes bibliographical references and index.
ISBN 0-8053-4003-3
1. Cancer. I. Title.
[DNLM: 1. Neoplasms–etiology. 2. Neoplasms–prevention & control. QZ 202 K637p 2006]
RC261.K54 2006
616.99′4–dc22

2005010378

34 2019

www.aw-bc.com

About the Author

LEWIS J. KLEINSMITH is Professor Emeritus of Molecular, Cellular, and Developmental Biology at the University of Michigan, where he has served on the faculty since receiving his Ph.D. from Rockefeller University in 1968. His teaching experiences have involved courses in introductory biology, cell biology, and cancer biology, and his research interests have included studies of growth control in cancer cells, the role of protein phosphorylation in eukaryotic gene regulation, and the control of gene expression during development. In addition to numerous research publications and review articles, he has written several award-winning educational software programs and is a co-author of *The World of the Cell*, a cell biology textbook currently in its 6th edition. His honors include a Guggenheim Fellowship, the Henry Russell Award, a Michigan Distinguished Service Award, citations for outstanding teaching from the Michigan Students Association, a Thurnau Professorship, a Best Curriculum Innovation Award from the EDUCOM Higher Education Software Awards Competition, and an NIH Plain Language Award for website materials he developed for the National Cancer Institute to educate the general public about cancer biology.

Preface

It is difficult to imagine anyone whose life has not been touched by the disease we call *cancer*. More than ten million new cases of cancer are being diagnosed globally each year, and if current trends continue, close to half the population of the United States will eventually develop some form of the disease. In the face of such statistics, it is not surprising that there is widespread interest in (and anxiety about) the topic of cancer.

For many years the study of cancer was restricted largely to the medical school curriculum. When I was first encouraged in 1983 by the chair of our biology department to develop an undergraduate course related to cancer, few such classes were being taught to undergraduates. Over the past two decades, however, the relevance of cancer biology to the undergraduate biology curriculum has become increasingly apparent as we uncover the roles of individual genes and their protein products in the abnormal behavior of cancer cells. As it turns out, studying the properties of cancer cells has deepened our understanding of normal cells and, conversely, our growing understanding of normal cells has provided numerous insights into the properties of cancer cells. The natural outcome of these trends is that courses dealing with cancer are now more common in the undergraduate curriculum, and even when a formal class about cancer is not offered, material related to cancer comes up more frequently when teaching courses in genetics or cell biology.

Principles of Cancer Biology is an outgrowth of twenty years of teaching about the biology of cancer to students at the University of Michigan, mostly undergraduate science majors but a significant number of nonscience majors and graduate students as well. The goals and content of this book have been heavily influenced by my experiences with these students and by the enthusiastic interest and numerous comments, suggestions, and questions they brought to the subject matter. I am also aware, however, that cancer biology is not taught in the same way or with the same objectives on every college campus. Unlike courses in introductory biology or genetics or cell biology, where the basic content and goals are similarly defined across a broad spectrum of institutions, the organization and subject matter covered in cancer biology courses exhibit a great deal of diversity. Such diversity presents a great challenge to anyone wishing to write an introductory cancer textbook that would be useful to a broad spectrum of students and faculty.

In confronting the challenge, I have been guided by three objectives. First, my primary goal has been to create a book that provides a broad introduction to the *basic principles* that underlie the field of cancer biology. The decision to name this book *Principles of Cancer Biology* was not made casually but rather to highlight its purpose as a "big picture" book—an instructional text rich in concepts rather than an encyclopedia of technical details. For those instructors who delve into particular aspects of cancer biology in greater depth, each chapter contains a list of suggested readings that can help expand the detail as needed for individual courses.

My second objective has been to emphasize the scientific evidence that underlies the basic concepts of cancer biology. Teaching students to think critically about scientific evidence is important regardless of the subject matter, but it is especially relevant to the field of cancer biology. Hardly a week goes by without another new "breakthrough" appearing in the newspapers about the causes, treatment, or prevention of cancer. It is not unusual, however, for an apparently definitive news report to be published one year only to be contradicted by another report the following year (or month or week). My second goal is therefore to help students learn how to *critically assess the scientific evidence* upon which our conclusions about cancer are built. In several places, such as the first half of Chapter 4, this objective is explicitly pursued by discussing the strengths and weaknesses of various scientific approaches. Further emphasis is provided throughout the text by discussing the evidence that supports various conclusions and by providing numerous tables and figures containing data from the scientific literature.

Finally, my third goal has been to produce an engaging book about cancer that can be *readily understood by students with diverse backgrounds* in the time allotted for a typical one semester course. Meeting this objective required selectivity in the examples chosen for discussion and meant that it was not possible (or desirable) to discuss every oncogene, or every tumor suppressor gene, or every hereditary cancer syndrome, or every drug used for chemotherapy. Accommodating a diverse group of readers also created the need to introduce or review a variety of basic biological concepts. While the book assumes that the reader will have taken at least one introductory biology course, familiarity with the basic concepts of cellular and molecular biology and genetics will naturally vary among such an audience. I have therefore included brief reviews of several topics related to DNA replication and repair, cell division, cell signaling, and inheritance patterns in the chapters where they become relevant. By including this background material and exercising selectivity in the examples chosen, I have tried to create a readable story that will be accessible and engaging to a broad audience of readers becoming acquainted with cancer biology for the first time.

ORGANIZATION AND COVERAGE

The reviewers of my initial outline and manuscript had a broad spectrum of opinions as to the order in which various topics should be presented. The sequence of chapters that I ended up selecting represents one common way of organizing the material, but given the diversity of opinion, I have tried to organize these chapters into clusters that can be read in several different sequences, using frequent page cross-references to provide appropriate background information if students read the chapters in a different order.

The first cluster, consisting of three chapters, begins with the general question "*What is cancer?*" in Chapter 1 followed by an introduction to the two defining properties of cancer cells: their ability to proliferate in an uncontrolled fashion (Chapter 2) and their ability to spread throughout the body (Chapter 3). The next cluster of five chapters deals with the causes of cancer, starting with a general consideration of the tools used by scientists to establish cause-and-effect relationships in Chapter 4 and then describing the role played by chemicals (Chapter 5), radiation (Chapter 6), infectious agents (Chapter 7), and heredity (Chapter 8) in the etiology of the disease. These various factors contribute to the development of cancer by triggering changes in the structure and expression of two classes of cancer-related genes that are covered in the following two chapters: Chapter 9 dealing with oncogenes and Chapter 10 dealing with tumor suppressor genes. The book then concludes with two final chapters dealing with the topics of cancer detection and treatment (Chapter 11) and cancer prevention (Chapter 12).

PEDAGOGICAL FEATURES

To enhance its effectiveness as a learning tool, *Principles of Cancer Biology* incorporates a variety of pedagogical devices:

- Every chapter is subdivided into a series of conceptual sections, each introduced by a Sentence Heading that summarizes the Principle being described in that section.
- A complete list of these basic Principles from the entire book is provided for reference purposes in a Detailed Table of Contents (p. xiii).
- To help students learn how to read and communicate about cancer, **boldface type** is used to highlight the most important technical terminology in each chapter.
- *Italics* are employed to identify additional technical terms that are less important than boldfaced terms but significant in their own right. In addition, I have adopted the standard convention of printing all gene names in italics, and italics are occasionally used to emphasize important phrases or sentences as well.
- Each chapter is concluded by a bulleted *Summary of Main Concepts* that can be used by the reader to review the major principles covered in that chapter.
- A list of **Key Terms** at the end of each chapter includes all of the boldfaced terms in the chapter and indicates the page on which each term appears in boldface and is defined or described.
- A *Suggested Reading* list is included at the end of each chapter with an emphasis on review articles and carefully selected research publications that students are likely to find especially relevant and understandable.
- A **Glossary** containing a brief definition of every boldfaced term is included as a reference tool at the end of the book. Each definition is followed by one or more page references indicating where in the book the meaning of the term is described.
- The artwork that accompanies the text is designed to accomplish several different objectives. Some line drawings are designed to illustrate or summarize important biological concepts and activities related to cancer, whereas others describe the ways in which particular experiments have been conceptualized and carried out. In addition, the importance of scientific evidence is highlighted by including more than 50 graphs, histograms, and pie charts containing data from the scientific literature, each accompanied by a citation to the original source of the data.

Acknowledgments

I am deeply indebted to the many students and reviewers whose comments have helped make this book clearer, more accurate, more encompassing, more useful, and better written than it would otherwise have been. In particular, I wish to thank the following individuals whose comments and suggestions were enormously helpful: Annemarie Bettica (Manhattanville College), Leonard Beuving (Western Michigan University), Elliott Blumenthal (Indiana University), Carol Burdsal (Tulane University), Katayoun Chamany (New School University), Thomas Fondy (Syracuse University), Paul Greenwood (Colby College), Alan Kelly (University of Oregon), Leila Koepp (Bloomfield College), Arlene Larson (University of Colorado, Denver), Carl Maki (Harvard University), Matthew Michael (Harvard University), Hao Nguyen (California State University, Sacramento), Carrie Rinker-Schaeffer (University of Chicago), Suzy Torti (Wake Forest University), Len Troncale (California State Polytechnic University, Pomona), Cathy Tugmon (Augusta State University), Jean Wang (University of California, San Diego), Robert Weaver (University of Kansas), and Edwin Wong (West Connecticut State University).

These reviewers provided numerous helpful criticisms and suggestions and their words of appraisal and counsel were gratefully received and greatly appreciated. Nonetheless, the final responsibility for what you read here remains my own, and you may confidently attribute any errors of omission or commission to me.

I am also deeply indebted to the many publishing professionals whose encouragement, ideas, and hard work contributed to the effectiveness of both the text and the art. Special recognition and sincere appreciation go to Jim Smith and Susan Winslow at Benjamin Cummings for their editorial support and encouragement for this project, and to Cinnamon Hearst, Vivian McDougal, and Nancy Tabor for the great care they took in guiding the final preparation of the manuscript and artwork for production. The outstanding efforts of Corinna Dibble and her colleagues at Dartmouth Publishing in creating the illustrations were greatly appreciated, and I am especially indebted to Crystal Clifton and her colleagues at Progressive Publishing Alternatives for the enormous efforts they made in tackling the myriad of details encountered in the final production phases of the project.

Finally, words cannot begin to capture the gratitude I feel for the support provided by my wife Cindy. Without her encouragement, good humor, sense of perspective, and never-ending love, this book would not have seen the light of day.

Lewis J. Kleinsmith
lewisk@umich.edu

Ann Arbor, Michigan
March 2005

Brief Contents

Detailed Contents

7 Infectious Agents and Cancer 119

8 Heredity and Cancer 140

11 Cancer Screening, Diagnosis, and Treatment 200

12 Preventing Cancer 234

1 What Is Cancer?

It is difficult to imagine anyone who has not heard of the disease we call "cancer." Ten million new cases are diagnosed annually worldwide, and the number is expected to increase to 20 million by the year 2020. If such trends continue, close to one out of every two people in the United States will eventually develop some form of the disease and cancer will be the most common cause of death. To put it another way, in a country of roughly 300 million people we are expecting more than 100 million new cancer cases! In the face of such numbers, it is not surprising that there is widespread interest in (and fear of) the topic of cancer biology.

The good news is that enormous progress has been made in the past few decades in unraveling the cellular and molecular mechanisms that underlie the development of cancer. The study of cancer biology, once restricted almost entirely to the medical school curriculum, is becoming a topic of broad relevance to biologists as they uncover the roles of various genes and their protein products in the abnormal behavior of cancer cells. As it turns out, investigating the properties of cancer cells has deepened our understanding of normal cells and, conversely, our rapidly expanding knowledge about the behavior of normal cells is providing numerous insights into the properties of cancer cells.

It is reasonable to expect that our growing understanding of the principles that govern the behavior of cancer cells will eventually lead to better approaches for cancer diagnosis, treatment, and prevention. The goal of this book is to provide a basic introduction to these underlying principles. In this first chapter we begin by considering the magnitude of the cancer problem, followed by an introduction to the biological nature of the disease and an overview of the topics to be discussed in subsequent chapters.

CANCER INCIDENCE AND MORTALITY

The term *cancer*, which means "crab" in Latin, was coined by Hippocrates in the fifth century BC to describe a family of diseases in which tissues grow and spread unrestrained throughout the body, eventually choking off life. Although the disease has therefore existed for at least several thousand years, its prevalence has been steadily increasing. In just the past 50 years, a person's chance of developing cancer within his or her lifetime has doubled, and doctors are now seeing more cases of the disease than ever before.

This increase has fostered the common misconception that the growing contamination of our environment by cancer-causing agents is creating a cancer epidemic. In fact, most of the increase in cancer rates has a somewhat different explanation: Cancer strikes older people more frequently than younger people, and more cancer cases are being seen simply because people are living longer than they did in the past. The increase in average lifespan—due mainly to the availability of vaccines and antibiotics that have lowered death rates from infectious diseases—means that more and more people are living long enough to develop cancer.

The Most Common Cancers Arise in the Skin, Prostate, Breast, Lung, and Colon

Cancer is a disease in which abnormal cells proliferate in an uncontrolled fashion and spread through the body. Such cells can arise in a variety of tissues and organs, and each of these sites contains different cell types that may be affected. The net result is more than 100 kinds of cancer distinguished from one another on the basis of where they originate and the cell type involved.

The various kinds of cancer differ significantly in terms of how frequently they arise (Figure 1-1, *left*). The most common type is skin cancer, which accounts for roughly half of all human cancers. The next group in terms of frequency in the United States includes cancers of the prostate, breast, lung, and colon (the latter is often combined with cancer of the rectum and designated *colorectal* cancer). Each of these four cancer types accounts for about 5% to 10% of all cancers. Of the dozens of other kinds of cancer routinely encountered, none accounts for more than a few percent of the total number of cancer cases.

Patterns of cancer incidence exhibit many similarities around the world, although some striking geographical differences have also been noted. When cancer rates are compared between different countries, it is crucial that the data be adjusted for two variables. First, the countries being compared will not have the same exact number of people, so cancer rates are generally expressed per 100,000 individuals to adjust for differences in population size. Second, statistics must also be adjusted for differences in age distribution because of the higher cancer rates observed in older individuals. As a consequence, cancer rates are usually expressed as *age-adjusted incidence per 100,000 people per year.*

When this type of statistic is used to compare cancer rates in different countries, a number of interesting patterns emerge. The incidence rates for most cancers tend to be similar in the United States, Canada, and Western Europe, but significant differences become apparent when comparisons are made with other regions of the world. For example, liver cancer is especially prominent in Africa and Southeast Asia, reaching age-adjusted rates in some regions of China and Thailand that are more than 25 times higher than in the United States. Likewise, stomach cancer is roughly 10 times more frequent in Japan than in the United States. Conversely, prostate cancer is 10 times more frequent in the United States than in Japan and 20 to 40 times more frequent than in northern Africa. Colon, breast, and lung cancers are also especially frequent in the

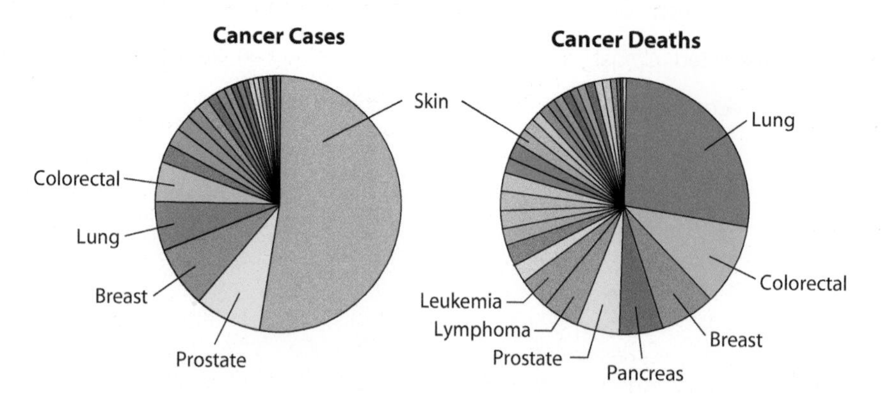

Figure 1-1 Relative Frequencies of Cancer Cases and Cancer Deaths in the United States. (*Left*) Relative frequency of various forms of cancer in the United States in 2004. Roughly 80% of all cancer cases are accounted for by cancers originating in only five different sites, with skin cancer by far the most prevalent. (*Right*) Relative frequency of cancer deaths in the same year. Because most forms of skin cancer are easy to cure, cancers that are less frequent than skin cancer end up killing more people. Heading this list is lung cancer, the cause of roughly one out of every three cancer deaths. [Based on data from *Cancer Facts & Figures 2005* (Atlanta, GA: American Cancer Society, 2005), p. 4.]

United States. We will return to a consideration of these geographical patterns in later chapters and examine what they tell us about the causes of cancer.

Lung Cancer Causes the Most Cancer Deaths

More than 500,000 people currently die from cancer in the United States each year, making it second only to heart disease as a cause of death. Although this statistic means that anyone who develops cancer is likely to react with feelings of fear and anxiety, it is important to understand that all forms of cancer are not equally dangerous. Most forms of skin cancer, for example, are easily cured and skin cancer is not a major source of cancer deaths, even though it is the most common type of cancer.

The number one cancer killer is not skin cancer but lung cancer, a disease that accounts for roughly one of every three cancer deaths in the United States (see Figure 1-1, *right*). Lung cancer is one of the five most frequent types of cancer, and it also has the worst prognosis of the five, killing roughly 85% of affected individuals within five years of diagnosis. Colorectal, breast, and prostate cancers are also among the top five in terms of total cancer fatalities, although their survival rates are significantly better than is seen with lung cancer. Cancer of the pancreas, which does not even fall among the top ten cancers in terms of overall frequency, is the fourth leading cause of cancer deaths in the United States because it is one of the most difficult cancers to treat successfully, killing an even higher percentage of its victims than does lung cancer.

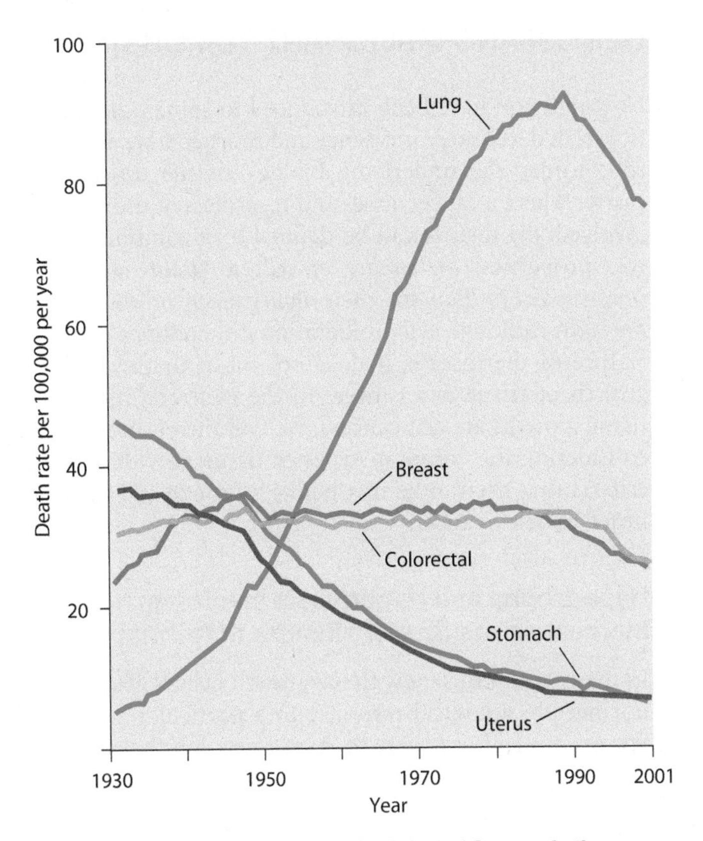

Figure 1-2 Trends in Death Rates for Selected Cancers in the United States. Age-adjusted data are shown starting in 1930 for lung, colorectal, and stomach cancer in men, and for breast and uterine cancer in women. Rates for most cancers, like colorectal and breast cancer illustrated here, have not changed much over the years. The dramatic increase in lung cancer rates was caused by increasing rates of cigarette smoking in the early part of the twentieth century. [Based on data from *Cancer Facts & Figures 2005* (Atlanta, GA: American Cancer Society, 2005), pp. 2–3.]

The Prevalence of Various Cancers Has Changed Over Time

Although lung cancer is currently the number one cancer killer in the United States, that has not always been the case. Lung cancer was one of the rarest forms of cancer 100 years ago, so rare that in a book written about lung cancer in the early 1900s, the author apologized for devoting so much time to such an esoteric subject! The past century, however, has seen a steady and dramatic increase in lung cancer rates, to the point where an obscure disease has been transformed into a major public health problem.

Recall that when we assess long-term cancer trends, the data need to be adjusted for age; otherwise, all forms of cancer will appear to be increasing in frequency because people are living longer and cancer is largely a disease of the elderly. Figure 1-2 provides age-adjusted data showing the long-term trends in cancer rates for several types of cancer. The data reveal that even when the statistics are adjusted for age, a dramatic increase in lung cancer rates has occurred since 1930. Yet most kinds of cancer (such as colorectal and breast cancer shown in Figure 1-2) have not changed much in age-adjusted incidence, and a few kinds of cancer are even decreasing in frequency. Stomach cancer is a particularly striking example. Since 1930, the annual death rate from stomach cancer has dropped about fivefold in the United States. While an astute observer might question whether such data reflect an improvement in cancer treatment rather than a decrease in stomach cancer rates, this explanation does not hold; survival rates for stomach cancer have improved only slightly, and the disease still kills about 80% of its victims within five years of diagnosis.

The discovery that stomach cancer rates were decreasing at the same time that lung cancer rates were increasing makes it fairly clear that different kinds of cancer have different causes. As we will see in Chapter 4, the increase in lung cancer rates is attributed to the fact that lung cancer is caused by smoking cigarettes, a practice that became increasingly popular during the twentieth century. The reason for the decrease in stomach cancer rates is not as well understood, but it is thought to be related to changes in diet and the growing use of antibiotics, which kill the bacteria involved in triggering stomach cancer. In Japan, where the traditional diet is quite unlike that in the United States, stomach cancer rates are almost ten times higher and stomach cancer kills almost as many Japanese as does lung cancer.

BENIGN AND MALIGNANT TUMORS

Now that you have been introduced to some basic statistics related to cancer incidence and mortality, we are ready to consider the underlying biology of the disease. No matter where a cancer arises and regardless of the cell type involved, the disease can be defined by a combination of two properties: *the ability of cells to proliferate in an uncontrolled fashion and their ability to spread throughout the body*. Although cell proliferation is therefore a defining feature of the disease, it does not mean that every new growth of tissue is a cancer. To the contrary, most new tissue growths are not cancers. We will therefore begin by considering the common types of tissue growth and the criteria that need to be met before a growth can be diagnosed as being cancer.

Hypertrophy and Hyperplasia Involve an Increase in the Size or Number of Normal Cells

In many situations, new tissue growth occurs as part of a normal physiological response to a particular stimulus. For example, if you were to start a new job that involves lifting heavy objects, your muscles would soon grow larger in response to the physical activity. Professional athletes and body builders adapt this principle when they use weight-lifting exercises to build muscle mass, yet the growing muscles on their arms and legs are not signs of cancer! Or perhaps you want to learn how to play the guitar. You will find that after a few days of practice, the skin on the tips of your fingers becomes thickened with calluses from pressing on the strings. Once again, the growing calluses on your fingertips are not signs of cancer.

The preceding examples illustrate two types of new tissue growth, neither of which is cancer. The increase in muscle mass triggered by exercise arises from the growth of individual muscle cells rather than an increase in the number of muscle cells. Such tissue growth based on an increase in cell size is called **hypertrophy** (Figure 1-3a). In contrast, calluses are produced by **hyperplasia**, a process in which cell division creates an increased number of normal cells (Figure 1-3b). In both cases, the process is potentially reversible. If you stop exercising or stop playing the guitar, the size of your muscles will decrease or your calluses will disappear.

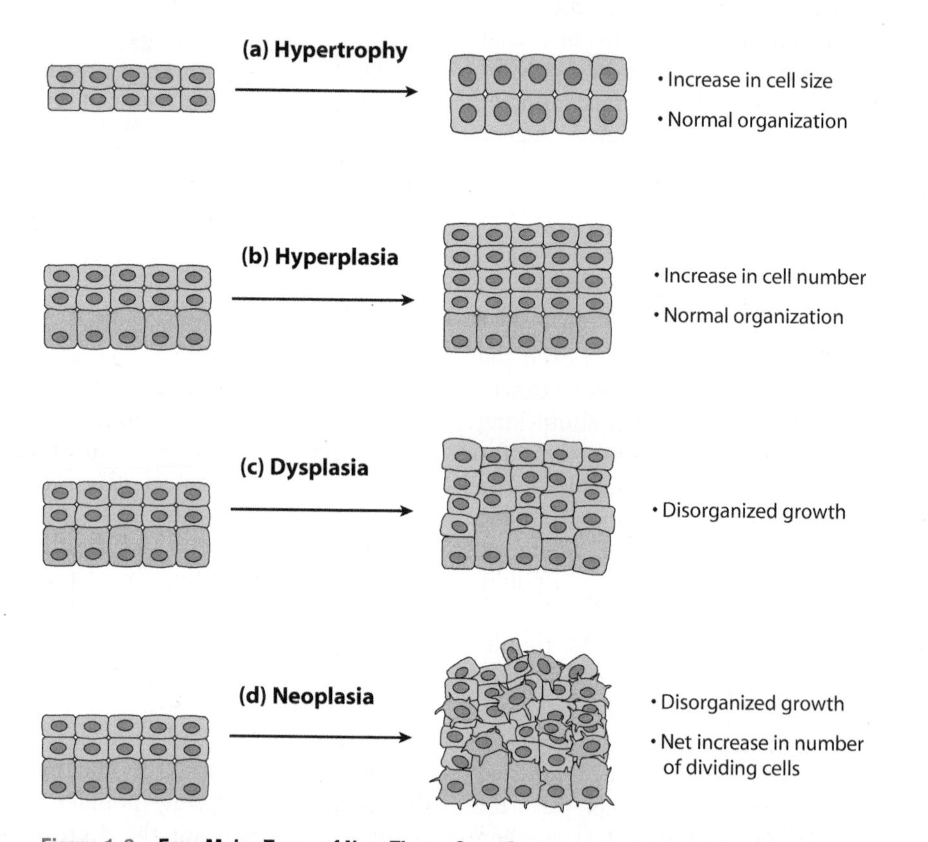

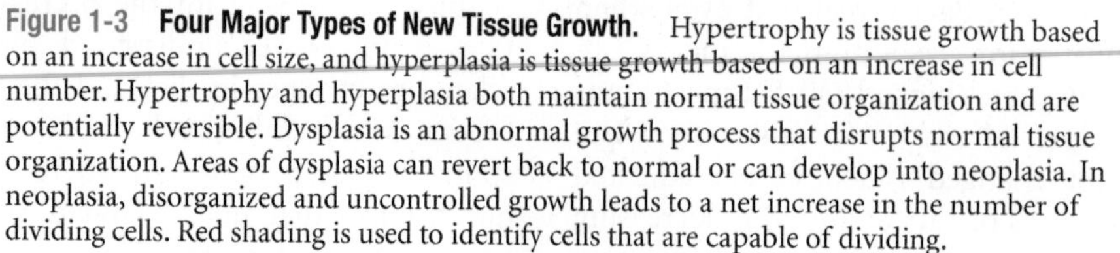

Figure 1-3 Four Major Types of New Tissue Growth. Hypertrophy is tissue growth based on an increase in cell size, and hyperplasia is tissue growth based on an increase in cell number. Hypertrophy and hyperplasia both maintain normal tissue organization and are potentially reversible. Dysplasia is an abnormal growth process that disrupts normal tissue organization. Areas of dysplasia can revert back to normal or can develop into neoplasia. In neoplasia, disorganized and uncontrolled growth leads to a net increase in the number of dividing cells. Red shading is used to identify cells that are capable of dividing.

Dysplasia and Neoplasia Are Characterized by Abnormalities in Cell Organization and Proliferation

When a piece of tissue in which hypertrophy or hyperplasia has occurred is examined with a microscope, the cell and tissue organization will look relatively normal. In contrast, the next type of tissue growth to be considered, called **dysplasia**, is an abnormal growth process that produces tissue in which proper cell and tissue organization have been disrupted (Figure 1-3c). The extent of the abnormalities varies across a broad range referred to as mild, moderate, or severe dysplasia. One site in which dysplasia commonly arises is the uterine cervix. As we will see in Chapter 11, the *Pap smear* is a routine screening procedure that can detect uterine dysplasia in its early stages before it has progressed to a more serious condition. Dysplasia has two possible outcomes. In some cases, especially in its less severe forms, dysplasia is reversible and the tissue will revert back to normal appearance and behavior. Alternately, dysplasia can become more severe, eventually progressing to a more dangerous form of tissue growth known as neoplasia.

Neoplasia is an abnormal type of tissue growth in which cells proliferate in an uncontrolled, relatively autonomous fashion, leading to a continual increase in the number of dividing cells (Figure 1-3d). This loss of growth control creates a proliferating mass of abnormal cells called a **neoplasm** or **tumor**. Uncontrolled proliferation does not mean that tumor cells always divide more rapidly than normal cells. The crucial issue is not the rate of cell division but rather the relationship between cell division and **cell differentiation**, the process by which cells acquire the specialized properties that distinguish different types of cells from one another. Cell differentiation is a trait of multicellular organisms, which are composed of complex mixtures of specialized or "differentiated" cell types—for example, nerve, muscle, bone, blood, cartilage, and fat—brought together in various combinations to form tissues and organs. The various types of differentiated cells are distinguished from one another by differences in their structural organization and in the products they manufacture. For example, red blood cells produce hemoglobin, nerve cells synthesize chemicals that transmit nerve impulses, pancreatic islet cells make insulin, and so forth. In addition to manufacturing such specialized products, differentiated cells often lose the capacity to divide.

In normal tissues, the rates of cell division and cell differentiation are kept in proper balance. To illustrate, let us briefly consider normal cell proliferation in the skin, where new cells are continually being produced to replace the cells being shed from the outer surfaces of the body. These replacement cells are generated through the division of undifferentiated cells located in the *basal layer* of the skin (Figure 1-4, *top*). Each time one of these **basal cells** divides, it gives rise to two cells with differing fates: One cell stays in the basal layer and retains the capacity to divide, whereas the other cell loses the capacity to divide and undergoes differentiation as it leaves the basal layer and moves toward the outer skin surface. As it differentiates, the migrating cell gradually acquires a flattened shape and begins to make *keratin*, the fibrous structural protein that imparts mechanical strength to the outer layers of the skin. The fully differentiated, thin flat cells that form the outer layer of the skin look almost like scales and are referred to as **squamous cells.**

Because only one of the two cells produced by each basal cell division normally retains the capacity to divide, there is no increase in the total number of dividing cells. Cell divisions are simply creating new differentiated cells to replace the ones that are being lost from the outer surface of the skin. A similar phenomenon takes place in the bone marrow, where new blood cells are produced to replace aging blood cells that are constantly being destroyed, and in the lining of the gastrointestinal tract, where new cells are produced to replace the cells that are being shed. In each of these situations, cell division is carefully balanced with cell differentiation so that no net accumulation of dividing cells takes place.

In tumors, this finely balanced arrangement is disrupted and some cell divisions give rise to two cells that both continue to divide, thereby feeding a progressive increase in the number of dividing cells (see Figure 1-4, *bottom right*). If the cells divide quickly, the tumor will rapidly increase in size; if the cells divide more slowly, tumor growth will be slower. But regardless of how slow or fast the cells divide, the tumor will continue to grow because new cells are being produced in greater numbers than needed. As the dividing cells accumulate, the normal organization and function of the tissue gradually become disrupted (Figure 1-5).

Malignant Tumors (Cancers) Are Capable of Spreading by Invasion and Metastasis

Now that the concept of a tumor has been introduced, we are finally ready to define the term *cancer*. The meaning of this word is based on differences in the growth patterns of tumors that allow them to be subdivided into two fundamentally different categories. One group consists of **benign tumors**, which grow in a confined local area. In contrast, **malignant tumors** can invade surrounding tissues, enter the bloodstream, and spread to distant parts of the body by a process called **metastasis.** The term **cancer** is a generic term that refers to any malignant tumor—that is, any tumor capable of spreading by invasion and metastasis.

Table 1-1 summarizes some of the properties of benign and malignant tumors. One distinguishing feature is that malignant tumors are frequently life threatening, whereas benign tumors usually are not. The reason for the

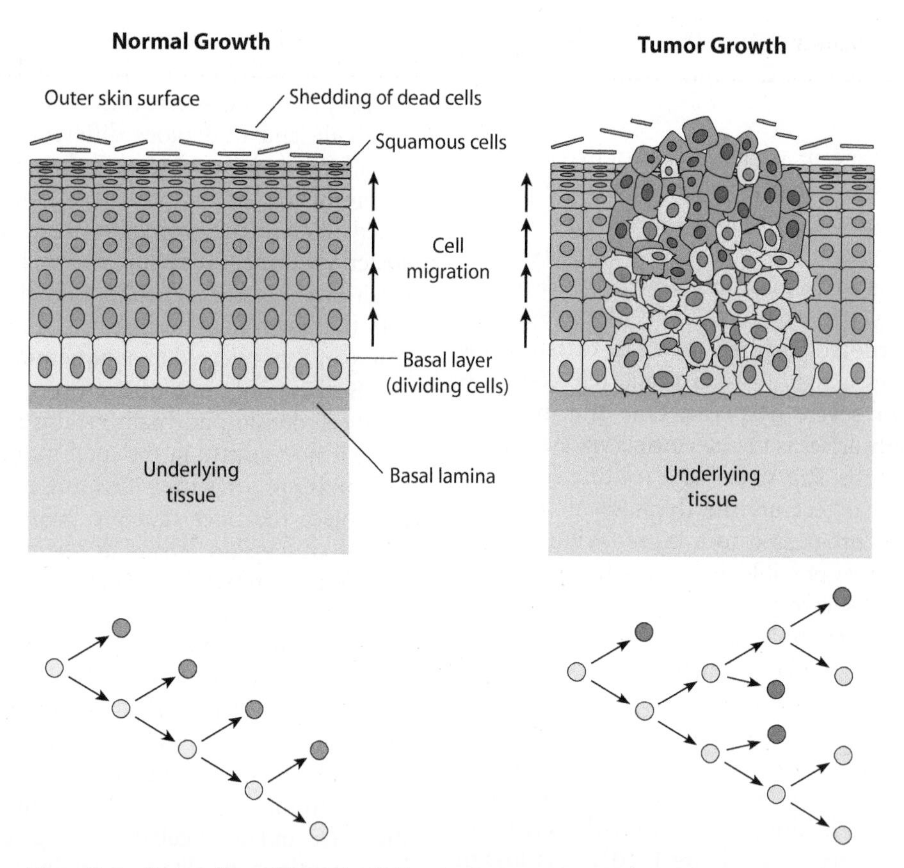

Figure 1-4 Comparison of Normal and Neoplastic Growth in the Epithelium of the Skin. (*Top left*) In normal epithelial growth, proliferation of cells located in the basal layer gives rise to new cells that migrate toward the outer surface of the skin, changing shape and losing the capacity to divide. (*Top right*) In neoplastic growth, this orderly process is disrupted and some of the cells that migrate toward the outer surface retain the capacity to divide. In both diagrams, lighter shading is used to distinguish cells that retain the capacity to divide. (*Bottom*) Schematic diagram illustrating the fate of dividing cells. In normal skin, each cell division in the basal layer gives rise to one cell that retains the capacity to divide (lighter color) and one that differentiates, thereby losing the capacity to divide. As a result, no net accumulation of dividing cells occurs. In neoplastic growth, the balance between cell division and cell differentiation is disrupted, thereby leading to a progressive increase in the number of dividing cells (lighter color).

difference is that the cells of a malignant tumor have often spread to other parts of the body by the time a person is diagnosed as having cancer. A surgeon will frequently be able to remove the original tumor, but cancer cells that have already spread through the body are difficult to locate and treat. Malignant tumors therefore tend to be more hazardous than benign ones, although some exceptions do occur. For example, the most common forms of skin cancer rarely metastasize and are easy to diagnose and remove, so these skin cancers are hardly ever fatal. Certain benign tumors, on the other hand, arise in surgically inaccessible locations, such as the brain, making them hazardous and potentially life threatening.

Differences in growth rate and state of differentiation are also common between benign and malignant tumors. Benign tumors generally grow rather slowly and are composed of *well-differentiated* cells, meaning that the cells bear a close structural and functional resemblance to the normal cells of the tissue in which the tumor has arisen. Malignant tumors, on the other hand, often (but not always) grow more rapidly and their state of differentiation is variable, ranging from relatively well-differentiated tumors to tumors whose cells are so poorly differentiated that they bear almost no resemblance to the original cells from which they were derived.

Benign and Malignant Tumors Are Named Using a Few Simple Rules

Because tumors can arise from a variety of cell types located in different tissues and organs, some basic conventions have been established to facilitate the naming of tumors. Depending on their site of origin, cancers are grouped into three main categories. (1) **Carcinomas** are cancers that arise from the **epithelial cells** that form covering layers over external and internal body surfaces. Carcinomas are by far the most common type of malignant tumor, accounting for roughly 90% of all human cancers. (2) **Sarcomas** are cancers that originate in supporting tissues such as bone, cartilage, blood vessels, fat, fibrous tissue, and muscle. They are the

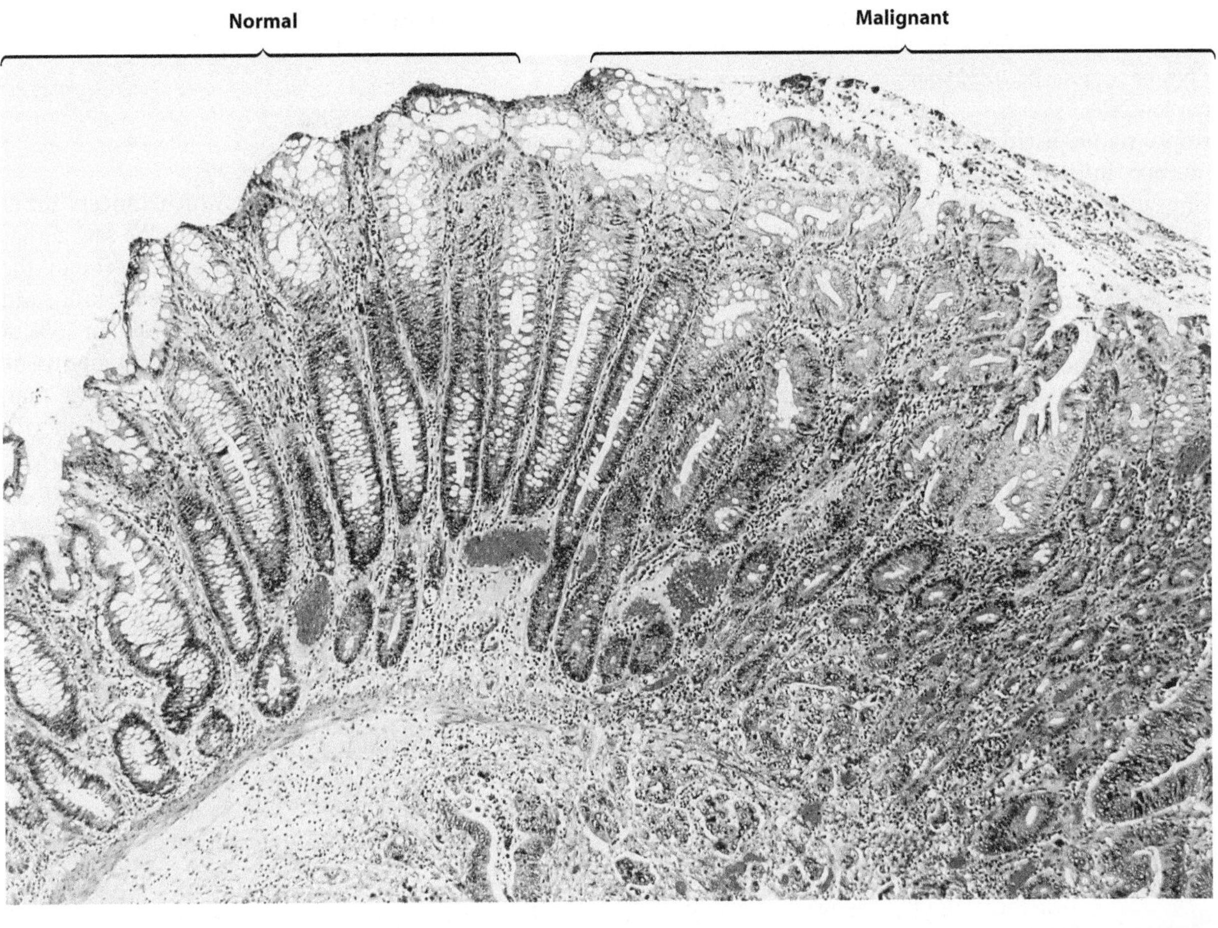

Figure 1-5 Colon Cancer Specimen Viewed by Light Microscopy. The left side shows normal colon tissue covered by an epithelium containing numerous tubular mucous glands. On the right side, proliferating cancer cells derived from the epithelium have disrupted the organized pattern of mucous glands and are invading the underlying tissue. [Courtesy of Gerald D. Abrams.]

rarest group of human cancers, accounting for about 1% of the total. (3) The remaining cancers are the **lymphomas** and **leukemias**, which arise from cells of lymphatic and blood origin. The term *lymphoma* refers to tumors of lymphocytes (white blood cells) that grow mainly as solid masses of tissue, whereas *leukemias* are cancers in which malignant blood cells proliferate mainly in the bloodstream.

Within each of the three groups, individual cancers are named using prefixes that identify the cell type involved. For example, consider a cancer arising from a gland cell, which is a specialized type of epithelial cell. Such a cancer is named by inserting the prefix *adeno-* (meaning "gland") in front of *carcinoma* (the term for epithelial cancers), yielding the name *adenocarcinoma*. Depending on the organ where it originated, the tumor might be called a lung adenocarcinoma, a colon adenocarcinoma, a breast adenocarcinoma, and so forth. What if the tumor were benign instead of malignant? In this case the suffix *-oma* is used instead of *-carcinoma*, yielding the name *adenoma*. Cancers of supporting tissue origin are named in a similar fashion, except the suffix *-sarcoma* is employed instead of *-carcinoma*. Thus a cancer of bone cells is called an *osteosarcoma* (the prefix *osteo-* means "bone"), and a benign tumor of bone is an *osteoma*.

Once you know the meanings of the prefixes that designate each of the common cell types, it is relatively straightforward to construct the proper technical names for a wide variety of tumors. The meanings of these prefixes, and the ways in which they are combined to create tumor names, are summarized in Table 1-2. Malignant tumors can usually be recognized by the presence of "carcinoma" or "sarcoma" in their name, but there are several exceptions. For example,

Table 1-1 Some Properties of Benign and Malignant Tumors

	Benign	Malignant
Growth pattern	Local growth only	Spreads by invasion and metastasis
Life threatening	Rarely	Often
Growth rate	Usually slow	May be rapid
State of differentiation	Well differentiated	Variable

despite their benign-sounding names, *melanomas* are malignant tumors of pigmented cells, *lymphomas* are malignant tumors of lymphocytes, and *myelomas* are malignant tumors of bone marrow cells.

The tumor names listed in Table 1-2 refer to the site at which a tumor initially arises—the so-called **primary tumor**. For example, a tumor discovered in a person's liver might be a liver adenocarcinoma, but it might also consist of stomach or colon cancer cells that had metastasized to the liver via the bloodstream and began growing there. In such cases, the tumor is *not* a liver cancer but rather a colon or stomach cancer that has metastasized to the liver.

WAYS IN WHICH CANCERS DIFFER

The need to have distinctive names for the various kinds of cancer is not just an academic exercise; it reflects the fact that the various types of cancer often behave like different diseases. For example, cancers grow at varying rates, look different when observed with a microscope, metastasize at varying frequencies, spread to different sites, and respond differently to various treatments. Differences are even exhibited by cancers of the same type. In other words, all melanomas do not behave the same, all lung adenocarcinomas do not behave the same, and so forth. In the following sections, we will briefly discuss some of the more important ways in which cancers differ.

Cancers Vary in Their Site and Cell Type of Origin

The first and most obvious way in which cancers differ from one another is in their site of origin. We saw earlier in the chapter that the most common cancers in the United States arise in the skin, prostate, breast, lung, and colon. While these five sites account for more than 75% of all cancers, the disease can originate in almost any tissue of the body. Table 1-3 provides a list of more than 40 locations where cancer may arise and the frequency of its occurrence at each site.

Even when cancers develop in the same tissue or organ, different forms of the disease can be distinguished based on the cell type involved. For example, a skin cancer might be a basal cell carcinoma, a squamous cell carcinoma, or a melanoma. The situation is even more complex for lung cancer, where at least seven main types have been distinguished (squamous cell carcinoma, small cell carcinoma, large cell carcinoma, adenocarcinoma, adenosquamous carcinoma, carcinoid tumor, and bronchial gland carcinoma). Subdivisions even occur within these seven groups; for example, large cell carcinoma is further divided into giant cell carcinoma and

Table 1-2 Naming Tumors

Prefix	Cell Type	Benign Tumor	Malignant Tumor
Tumors of epithelial origin			
Adeno-	Gland	Adenoma	Adenocarcinoma
Basal cell	Basal cell	Basal cell adenoma	Basal cell carcinoma
Squamous cell	Squamous cell	Keratoacanthoma	Squamous cell carcinoma
Melano-	Pigmented cell	Mole	Melanoma*
Terato-	Multipotential cell	Teratoma	Teratocarcinoma
Tumors of supporting tissue origin			
Chondro-	Cartilage	Chondroma	Chondrosarcoma
Fibro-	Fibroblast	Fibroma	Fibrosarcoma
Hemangio-	Blood vessels	Hemangioma	Hemangiosarcoma
Leiomyo-	Smooth muscle	Leiomyoma	Leiomyosarcoma
Lipo-	Fat	Lipoma	Liposarcoma
Meningio-	Meninges	Meningioma	Meningiosarcoma
Myo-	Muscle	Myoma	Myosarcoma
Osteo-	Bone	Osteoma	Osteosarcoma
Rhabdomyo-	Striated muscle	Rhabdomyoma	Rhabdomyosarcoma
Tumors of blood and lymphatic origin			
Lympho-	Lymphocyte		Lymphoma* or lymphocytic leukemia
Erythro-	Erythrocyte		Erythrocytic leukemia
Myelo-	Bone marrow		Myeloma* or myelogenous leukemia

*Note that certain tumors, such as melanoma, lymphoma, and myeloma, are malignant despite a benign-sounding name.

clear cell carcinoma. A similar level of complexity exists in many other organs, explaining why cancer ends up behaving like hundreds of different diseases.

Table 1-3 Estimated New Cancer Cases and Deaths for the Year 2005 in the United States

Primary Site	New Cases	Deaths
Skin (nonmelanoma)	> 1,000,000	2,000
Prostate	232,090	30,350
Breast	212,930	40,870
Lung (including bronchus)	172,570	163,510
Colon and rectum	145,290	56,290
Urinary bladder	63,210	13,180
Melanoma (skin)	59,580	7,770
Non-Hodgkin's lymphoma	56,390	19,200
Uterine endometrial	40,880	7,310
Kidney (and renal pelvis)	36,160	12,660
Pancreas	32,180	31,800
Thyroid	25,690	1,490
Ovary	22,220	16,210
Stomach	21,860	11,550
Brain, other nervous system	18,500	12,760
Liver	17,550	15,420
Myelogenous leukemias	16,560	9,850
Multiple myeloma	15,980	11,300
Esophagus	14,520	13,570
Lymphocytic leukemias	13,700	6,090
Uterine cervix	10,370	3,710
Mouth	10,070	1,890
Larynx	9,880	3,770
Soft tissues	9,420	3,490
Pharynx	8,590	2,130
Testis	8,010	390
Tongue	7,660	1,730
Gallbladder, other biliary	7,480	3,340
Hodgkin's disease (lymphoma)	7,350	1,410
Other nonepithelial skin	6,420	2,820
Small intestine	5,420	1,070
Other digestive	5,210	2,400
Other leukemia	4,550	6,630
Anus	3,990	620
Vulva	3,870	870
Other oral cavity	3,050	1,570
Bones and joints	2,570	1,210
Ureter, other urinary organs	2,510	750
Other respiratory	2,350	860
Vagina, other female genital	2,140	810
Eye and orbit	2,120	230
Other endocrine	1,960	880
Penis, other male genital	1,470	270
All other sites	28,590	46,250

Based on data from *Cancer Facts & Figures 2005* (Atlanta, GA: American Cancer Society, 2005), p. 4.

Cancers Vary in Their Survival Rates

It is common knowledge that cancers can kill people. The main threat comes from the ability of cancer cells to enter the bloodstream and spread to distant organs. If circulating cancer cells enter a vital organ such as the liver, brain, or kidneys, the growth of cancer cells in these locations can disrupt tissue organization and destroy normal cells, eventually leading to organ failure and death. In addition, some cancer cells produce substances that interfere with the function of the immune system, thereby making people more susceptible to potentially lethal infections. The drugs or radiation treatments that are commonly used to treat cancer patients also tend to inhibit immune function, again creating the risk of infections that can be lethal. Finally, the presence of cancer cells in the body sometimes triggers *cachexia*, a life-threatening condition characterized by extensive weight loss, weakness, and malnutrition.

Although most cancers are potentially life threatening for one or more of the preceding reasons, the dangers vary enormously because of differences in how fast cancers grow and how frequently they metastasize. One way of comparing the risks associated with various kinds of cancer is to examine the **five-year survival rate,** which is defined as the percentage of cancer patients who are still alive five years after initial diagnosis. As shown in Figure 1-6, five-year survival rates vary significantly, ranging from 99% five-year survival for skin cancer to less than 20% five-year survival for cancers of the lung, esophagus, liver, and pancreas.

The rates illustrated in Figure 1-6 are averages for all cancers arising in each organ. Because different types of

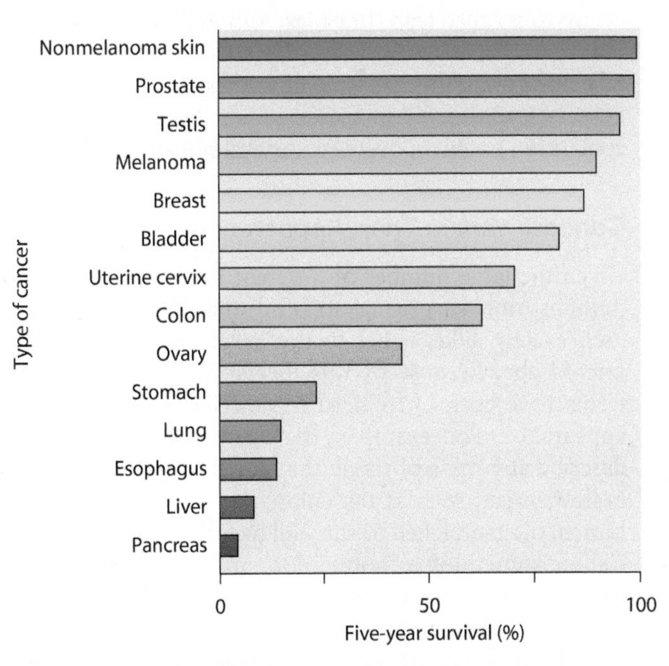

Figure 1-6 Five-Year Survival Rates for Selected Types of Cancer. Cancer survival rates are for the United States during the period 1995–2000. [Based on data from *Cancer Facts & Figures 2005* (Atlanta, GA: American Cancer Society, 2005), p. 17.]

cells can become cancerous in any given organ, such averages often mask important information. In the case of skin cancer, for example, the overall 99% rate does not tell you that the five-year survival rate for basal cell carcinoma of the skin is greater than 99.9%, which means that fewer than 1 person in 1000 dies from it. In contrast, melanoma—a skin cancer arising from pigmented cells—has a five-year survival rate of roughly 90%. While 90% is still relatively high, it means that melanoma kills about 1 person in 10, compared with 1 in 1000 for basal cell carcinoma. Melanoma is thus 100 times more dangerous than basal cell carcinoma, even though both are forms of skin cancer.

Despite the usefulness of five-year survival statistics in making such comparisons, a note of caution is required regarding their interpretation. Five-year survival is calculated starting from the *time of diagnosis*, which means that this statistic is influenced by differences in detection rates. To illustrate, let us consider two hypothetical women, named Carol and Diane, who both develop breast cancer at age 50. Suppose that Carol has her cancer detected that same year at an annual medical exam, whereas Diane does not have ready access to medical care and her cancer isn't detected until she discovers it three years later at age 53. If Carol and Diane were both to die of breast cancer at age 57, the five-year survival statistics would indicate that Carol was alive five years after diagnosis (diagnosis at 50, death at 57) whereas Diane was not (diagnosis at 53, death at 57). Yet both women developed cancer at the same time and died at the same time! Carol was simply diagnosed earlier and was therefore represented as being alive in the five-year survival statistics, whereas Diane was not. This tendency for survival rates to *appear to be improved* by early diagnosis when no improvement has actually occurred is called **lead time bias.** Although we will see later that early diagnosis really does improve a person's chances of being cured, this example illustrates that the phenomenon of lead time bias makes it impossible to reach such a conclusion based on five-year survival statistics alone.

Cancers Vary in Their Appearance

Given the large number of different kinds of cancer, it is perhaps not surprising to find that cancers seldom look exactly alike, either to the naked eye or under the microscope. Because of this diversity, various technical terms have been introduced to describe a given cancer's appearance. For example, the term **polyp** is used to describe any mass of tissue that arises from the wall of a hollow organ, such as the colon, and protrudes into the lumen, often attached to the wall by a relatively thin stalk. Calling something a "polyp" does no more than describe what the tissue growth looks like to the naked eye. A polyp might be a benign or a malignant tumor, or it might not even be a tumor at all (for example, nasal polyps are inflammatory swellings of the tissue lining the nose).

Cancers also have a distinctive appearance when they are viewed with a microscope, and this appearance usually provides the basis for cancer diagnosis. Thus when a person exhibits symptoms or signs that suggest the possible presence of cancer, a doctor will usually cut out a small piece of tissue for microscopic examination. The process of removing tissue for diagnostic purposes is called a **biopsy**, and examination of the tissue specimen is carried out by a specialist called a **pathologist**. By looking at a biopsy specimen under a microscope, a pathologist can determine whether a tumor is present, whether it is benign or malignant, and the cell type involved.

Although pathologists cannot rely on any single trait, cancer cells usually exhibit a number of distinctive features that together facilitate the diagnosis of a cancer from its microscopic appearance (Table 1-4). For example, cancer cells frequently have large, irregularly shaped nuclei, prominent nucleoli, and a high ratio of nuclear size to cytoplasmic volume; they also exhibit significant variations in cell size and shape accompanied by a loss of normal tissue organization. To varying extents, cancer cells lose the specialized structural and biochemical properties exhibited by normal cells residing in the tissue of origin; in other words, cancer cells undergo a loss of cell differentiation. In addition, the boundary separating cancer cells from the surrounding normal tissue will usually be poorly defined and tumor cells may be seen penetrating into adjacent tissues. Benign tumors, on the other hand, have a relatively distinct outer boundary that clearly separates the tumor cells from the surrounding tissue.

Cancers also tend to be characterized by an excessive number of dividing cells, so a large number of cells may be caught in the actual process of undergoing division

Table 1-4 Some Differences in the Microscopic Appearance of Benign and Malignant Tumors

Trait	Benign	Malignant
Nuclear size	Small	Large
N/C ratio (ratio of nuclear size to cytoplasmic volume)	Low	High
Nuclear shape	Regular	Pleomorphic (irregular shape)
Mitotic index (relative number of dividing cells	Low	High
Tissue organization	Normal	Disorganized
Differentiation	Well differentiated	Poorly differentiated (anaplastic)
Tumor boundary	Well defined ("encapsulated")	Poorly defined

(mitosis) when a tissue specimen is prepared for microscopic examination. The percentage of cells detected in the process of dividing, called the **mitotic index**, is measured by examining a biopsy specimen with a microscope, counting the number of cells undergoing mitosis in a given area of tissue, and dividing it by the total number of cells in that area. Cancers generally have a higher mitotic index than benign tumors, and faster-growing cancers will have a higher mitotic index than slower-growing cancers.

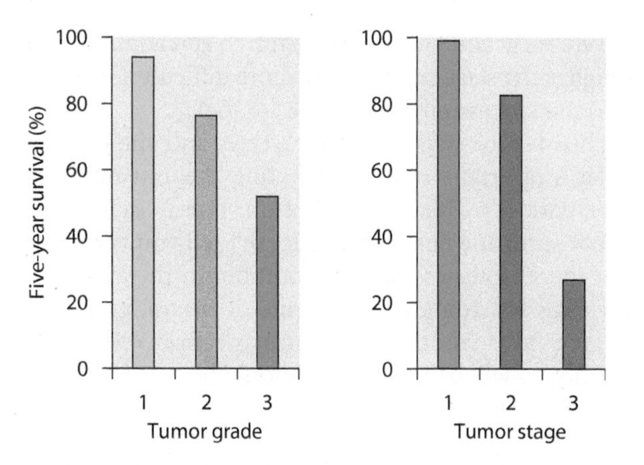

Figure 1-7 Effect of Tumor Grade and Stage on Survival Rates. The data are five-year survival statistics for breast cancer. (*Left*) Lower-grade cancers, whose cells are better differentiated and more closely resemble normal cells, have better survival rates than higher-grade cancers, which are poorly differentiated and bear little resemblance to normal cells. (*Right*) In this figure, stage 1 refers to localized breast cancer, stage 2 is breast cancer that has spread to regional lymph nodes, and stage 3 is breast cancer that has metastasized to distant sites. The data show that survival is much better for cancers detected at earlier stages. [Based on data from C. W. Elston and I. O. Ellis, *Histopathology* 19 (1991): 403 (Figure 3); and *Cancer Facts & Figures 2005* (Atlanta, GA: American Cancer Society, 2005), p. 17.]

Tumor Grading Is Based on Differences in Microscopic Appearance

When a sufficient number of the preceding traits are seen upon microscopic examination of a biopsy specimen, a pathologist can conclude that cancer is present, *even if invasion and metastasis have not yet occurred.* In other words, these microscopic traits indicate the presence of a tumor that, if left untreated, will eventually spread by invasion and metastasis. The severity of the observed microscopic abnormalities can vary widely among cancers, even for cancers involving the same cell type and organ. This variability has led pathologists to devise systems for **tumor grading** in which cancers of the same type are assigned different numerical grades based on their microscopic appearance. Lower numerical grades (e.g., grade 1) are assigned to tumors whose cells are well differentiated (resemble normal tissue in cell structure and organization), divide slowly (low mitotic index), and exhibit only modest abnormalities in the traits listed in Table 1-4. Higher numbers (e.g., grade 4) are assigned to tumors containing rapidly dividing, poorly differentiated cells that exhibit severe abnormalities in the traits listed in Table 1-4.

The highest-grade cancers contain cells that are said to be **anaplastic**, which means that they are so poorly differentiated and abnormal in appearance and organization that they bear little resemblance to the cells of the tissue in which the tumor arose. Such anaplastic, high-grade cancers tend to grow and spread more aggressively and be less responsive to therapy than lower-grade cancers whose cells have a more normal appearance. As a result, patients with lower-grade cancers often have a better prognosis for long-term survival than those with higher-grade cancers (Figure 1-7, *left*). Tumor grades tend to remain fairly constant for each person's cancer, although a low-grade cancer may occasionally evolve over time into a higher-grade cancer.

While tumor grading provides some general information regarding the likely behavior of a given cancer, tumors of the same grade do not always behave in the same way. It has been known for many years that patients with cancers that cannot be distinguished from one another by any traditional criteria, including cell type and grade, may nonetheless exhibit different outcomes when patients receive the same treatments. The reason for this differing behavior appears to be related to the molecular and genetic properties of different tumors. One powerful approach for assessing these properties is *DNA microarray analysis* (p. 191), a technique that allows the activity of thousands of genes to be measured simultaneously. The use of DNA microarray techniques has allowed researchers to take tumors that appear to be identical by other criteria and subdivide them into groups based on differing patterns of gene expression. In Chapter 11, we will see how such approaches permit cancers to be grouped into categories that respond differently to various treatments, thereby facilitating the development of treatment strategies that are tailor-made for each patient.

Cancers Vary in Their Clinical Stage

Another means of categorizing cancers, called **tumor staging**, employs a series of criteria to estimate how far a person's cancer has progressed *at the time of diagnosis.* Unlike tumor grading, which is a description of microscopic appearance that tends to be relatively consistent over time, tumor staging is a description of how large a tumor has grown and how far it has spread, which are traits that by definition change with time.

The most widely used system for tumor staging is the **TNM system** ("T" for Tumor size, "N" for lymph Node, "M" for Metastasis). In this system, the following three questions are asked: (1) How large is the tumor and how far has it invaded into surrounding tissues? (2) Are lymph nodes "positive" for cancer cells; that is, have cancer cells spread to regional lymph nodes? (3) To what extent have cancer cells metastasized to other organs? Based on the answers to these questions, cancers are assigned a stage number. A lower stage number means that a cancer has been detected relatively early and has not yet begun to spread. In general, the higher the stage number, the more difficult it is to treat the disease successfully (see Figure 1-7, *right*).

Thus tumor stage, grade, cell type, and site of origin all play important roles in affecting the outcome of a cancer patient's disease. The bottom line is that finding out that someone has "cancer" doesn't tell you very much about the seriousness of the condition. In the absence of knowledge regarding cell type, site of origin, grade, and stage, the mere fact that a person has cancer reveals little about how it will behave, how it should be treated, and the prospects for long-term survival.

CANCER: AN INTRODUCTORY OVERVIEW

Now that you understand what the term *cancer* means, we are ready to consider the underlying biology of the disease. From a biological point of view, several broad questions quickly come to mind: What kinds of cellular abnormalities allow cancers to grow and spread in an unrestrained fashion? What causes cancer? What role do genes play in the development of cancer? How is cancer detected and treated? Can cancer be prevented? In the remainder of the chapter, we will examine how this textbook is organized to address each of these questions in detail.

What Kinds of Cellular Abnormalities Allow Cancers to Grow and Spread in an Unrestrained Fashion?

Cancer is a disease of abnormal cells. We have already seen that a central malfunction is the ability of cancer cells to proliferate in an uncontrolled fashion, producing an ever-increasing number of cells without regard to the needs of the rest of the body. To understand the mechanisms responsible for such behavior, you need to know how the proliferation and survival of normal cells are controlled. When these control mechanisms are compared in normal cells and cancer cells, a number of important differences become apparent. These differences will be described in Chapter 2, "Profile of a Cancer Cell."

Although uncontrolled proliferation is a defining feature of cancer cells, it is not the property that makes the disease so dangerous. After all, the cells of benign tumors also proliferate in an uncontrolled fashion, but benign tumors are rarely life threatening because the cells remain in their original location. The hazards posed by cancer cells come from uncontrolled proliferation *combined with* the ability to spread throughout the body. Spreading of cancer is a complex process involving multiple steps. First, tumors must trigger the development of blood vessels that supply nutrients and oxygen to the tumor and remove waste products. Without this step, tumors cannot grow beyond a tiny size. After tumors have triggered the formation of a blood supply, cancer cells invade through surrounding tissues, enter the circulatory system, travel throughout the body, and establish new tumors at distant sites. The cellular traits and mechanisms that make each of these steps possible will be described in Chapter 3, "How Cancers Spread."

What Causes Cancer?

The uncontrolled proliferation of cancer cells, combined with their ability to spread throughout the body, makes cancer a potentially life-threatening disease. What causes the emergence of such cells that have the ability to destroy the organism of which they are a part? The conversion of normal cells into cancer cells is a complex, multistep process that typically takes many years to unfold. Despite the complexity of this process, however, many of its initiating causes are known. In Chapter 4 ("Identifying the Causes of Cancer"), we will see how scientists have gone about uncovering the causes of cancer, followed by an introduction to the main classes of cancer-causing agents that have been identified.

In the following four chapters, the agents that cause cancer will be examined in detail. Chapter 5 ("Chemicals and Cancer") describes how chemicals trigger the development of cancer; Chapter 6 ("Radiation and Cancer") examines the ability of radiation to cause the disease; Chapter 7 ("Infectious Agents and Cancer") deals with the ability of infectious agents to cause cancer; and finally, Chapter 8 ("Heredity and Cancer") discusses the influence of heredity on cancer risk.

What Role Do Genes Play in the Development of Cancer?

The causes of cancer described in Chapters 5 through 8 are quite diverse, but they often lead to the same outcome, namely gene mutations. A large body of evidence points to the pivotal role played by mutations in the development of cancer. These cancer-causing mutations can be triggered by chemicals, radiation, or infectious agents, or they may arise spontaneously, arise from errors in DNA replication, or be inherited. But regardless of these differences in how they arise, cancer-related mutations affect the same two classes of genes: *oncogenes* and *tumor suppressor genes*.

Oncogenes are defined as genes whose presence can lead to cancer. They arise by mutation from normal genes that code for proteins involved in stimulating cell proliferation and survival. By producing abnormal forms or excessive quantities of these proteins, oncogenes contribute to the uncontrolled proliferation and survival of cancer cells. In Chapter 9 ("Oncogenes"), we will see how normal genes are converted into oncogenes and how the proteins produced by oncogenes contribute to the development of cancer.

In contrast to oncogenes, which are *abnormal genes* whose activity can lead to cancer, **tumor suppressor genes** are *normal genes* whose deletion or loss of function can likewise lead to cancer. Tumor suppressor genes produce proteins that either directly or indirectly exert a restraining influence on cell proliferation and survival. The loss of such proteins can therefore allow cell proliferation and survival to evade normal restraints and controls. In Chapter 10 ("Tumor Suppressor Genes and Cancer Overview"), we will examine the pathways affected by tumor suppressor genes and will see how defects in such pathways contribute to cancer development.

Understanding the behavior of cancer-related genes requires some familiarity with DNA structure and function. This book assumes that readers have a basic understanding of the relationship between DNA and genes that is at least equivalent to that provided by a typical introductory biology course. For review purposes, Figure 1-8 illustrates the building blocks of DNA. As shown in this figure, DNA chains are constructed from varying sequences of four building blocks called **nucleotides.** Each nucleotide contains a sugar, a phosphate group, and a nitrogen-containing base that may be **adenine (A), guanine (G), cytosine (C),** or **thymine (T).** An intact DNA molecule consists of two intertwined DNA chains wound into a *double helix* that is held together by hydrogen bonds between the base adenine (A) in one chain and thymine (T) in the other, or between the base guanine (G) in one chain and cytosine (C) in the other (see Figure 2-14). The base sequence of one chain therefore determines the base sequence of the opposing chain, and the two chains of the DNA double helix are said to be held together by **complementary base pairing.** The way in which this complementary relationship makes DNA replication possible will be reviewed in Chapter 2.

A **gene** is any nucleotide sequence in DNA that codes for a functional product, in most cases a protein chain. The flow of information from a DNA gene to a protein chain occurs in a two-step process called transcription and translation (Figure 1-9). During **transcription**, the base sequence in one strand of the DNA double helix serves as the template for the synthesis of a complementary molecule of RNA. The base-pairing rules are similar to those used in making DNA except that RNA utilizes the base **uracil (U)** where DNA would use thymine (T). The RNA molecules produced by protein-coding genes function as **messenger RNAs (mRNAs)** that in turn guide the synthesis of protein chains in a process known as **translation.** During translation, mRNA associates with ribosomes and the genetic information in the mRNA is read in units of three bases called **codons.** Most codons specify an amino acid, but a few function as stop signals that mark the end of a protein chain. The net effect of the two-step process is that the nucleotide sequence of DNA molecules is used to guide the production of protein molecules, which in turn perform most cellular functions. As we will see in Chapters 9 and 10, disruptions in these proteins—caused by either DNA mutations or changes in the way DNA is expressed—lie at the heart of cancer cell behavior.

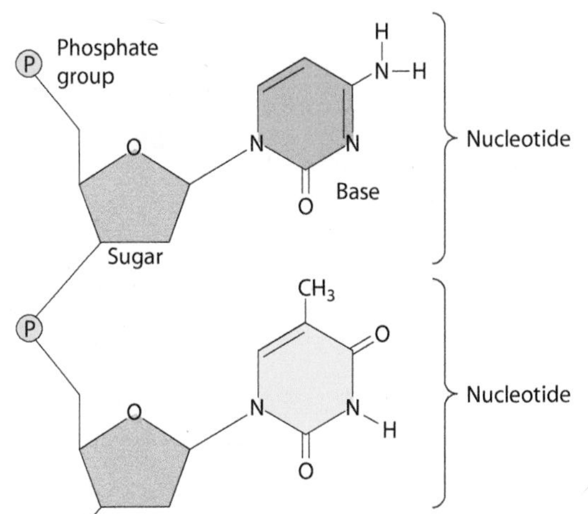

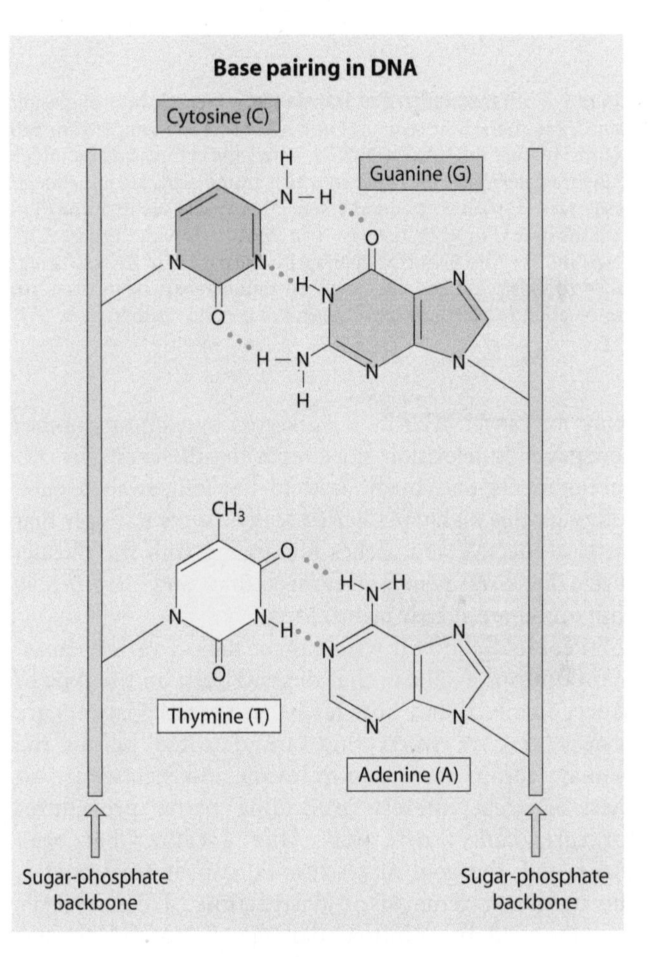

Figure 1-8 Building Blocks of DNA. (*Top*) DNA is built from chains of nucleotides, which are building blocks composed of a sugar, a phosphate group, and a nitrogen-containing base. (*Bottom*) A DNA molecule consists of two chains held together by hydrogen bonds between the bases cytosine (C) and guanine (G) or between the bases thymine (T) and adenine (A).

How Is Cancer Detected and Treated?

During the past several decades, great strides have been made in unraveling the molecular and genetic abnormalities exhibited by cancer cells. One of the hopes for such research is that our growing understanding of the mechanisms responsible for the development of cancer will eventually lead to improved strategies for diagnosing and treating the disease. The bottom line, of course, is the urgent desire for a

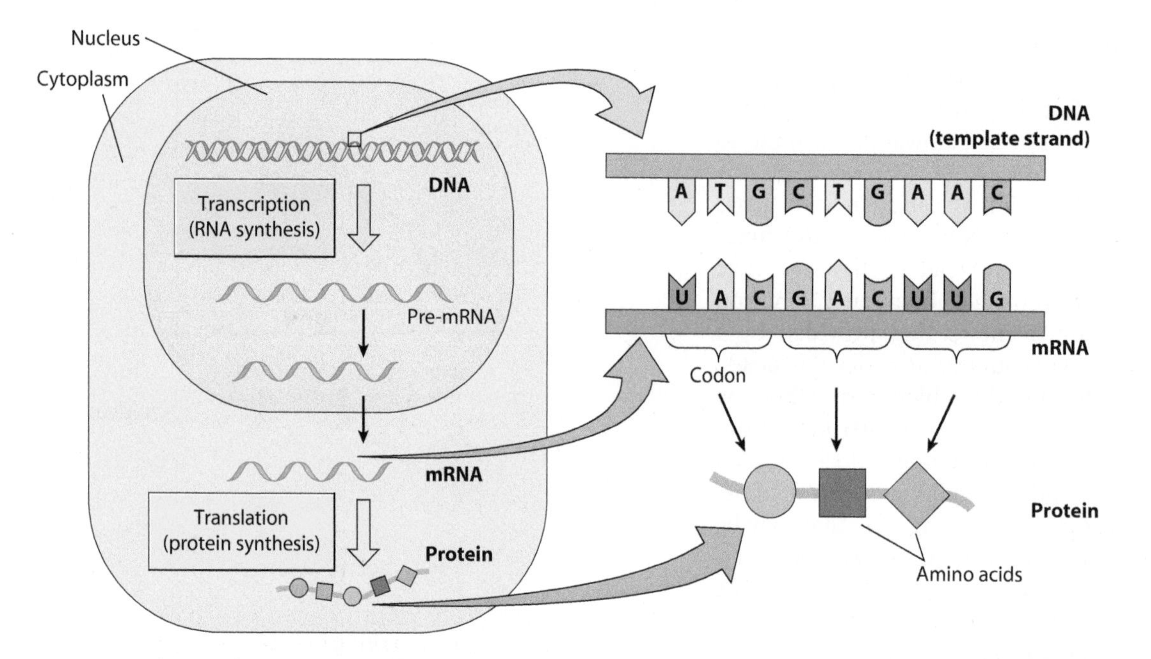

Figure 1-9 Transcription and Translation. Genetic information flows from DNA to RNA to protein. (*Left*) During transcription, the base sequence in DNA serves as a template for producing pre-mRNA that is processed in various ways to form mature mRNA. The mRNA enters the cytoplasm and undergoes translation, a process in which the mRNA base sequence determines the order in which amino acids are joined together to form a protein chain. (*Right*) In RNA, the base uracil (U) is incorporated where DNA would use thymine (T). Thus during transcription, the base A in DNA pairs with the base U in RNA, the base T in DNA pairs with the base A in RNA, the base G in DNA pairs with the base C in RNA, and the base C in DNA pairs with the base G in RNA. During translation, the mRNA is read in units of three bases called codons. Most codons specify an amino acid to be incorporated into a growing protein chain, but a few function as punctuation marks that signal the end of protein synthesis.

"cure for cancer." Hardly a week goes by without another newspaper or television story reporting the latest research developments that might lead to the long-awaited cure. Although this highly publicized search seems to imply that we lack effective approaches for dealing with the disease, such a dire assessment is far from accurate and many people with cancer are already being cured.

People diagnosed with cancer have a variety treatment options available that depend both on the type of cancer involved and how far it has spread. Cancers are easiest to cure when they are detected before the primary tumor has begun to invade and metastasize, so there is great interest in finding better procedures for early cancer detection. After a cancer has been diagnosed, the goal of traditional cancer treatment is the complete removal or destruction of cancer cells accompanied by minimal damage to normal tissues. This goal is usually pursued through a combination of surgery to remove the primary tumor, followed if necessary by the use of radiation or chemotherapy (drugs) or both to destroy any remaining cancer cells. Newer approaches to cancer treatment involve using the immune system to fight cancer and developing drugs that target proteins known to be required for the proliferation and spread of cancer cells. Examples of these various approaches and the mechanisms involved will be discussed in Chapter 11, "Cancer Screening, Diagnosis, and Treatment."

Can Cancer Be Prevented?

The practical importance of understanding the events that lead to cancer is not restricted to guiding the development of new treatment strategies. It also provides crucial information as to how cancer might be prevented, and as the old adage states, "An ounce of prevention is worth a pound of cure." One tactic for preventing the disease is based on the fact that many of the environmental agents that cause cancer have been identified. Avoiding these known risk factors can significantly decrease a person's chances of developing the disease. For example, more than 30% of all cancer deaths could be prevented if people simply did not smoke cigarettes.

A second approach to cancer prevention is based on the discovery that cancers arise through a multistep process that unfolds over several decades rather than occurring as a single discrete event. So even after a person has been exposed to agents that "cause" cancer, there are opportunities to intervene and block one of the subsequent steps required for the progression to malignancy. Current statistics suggest that if people adopt both prevention approaches—that is, if they reduce their exposure to cancer-causing agents and take steps to protect against the development of cancer after such agents have been encountered—more than half of all cancer deaths could be prevented. A detailed examination of these two strategies for cancer prevention is provided in the final chapter of the book (Chapter 12, "Preventing Cancer").

Summary of Main Concepts

- The total number of cancer cases has been steadily increasing over the years, mainly because people are living longer today than they did in the past and cancer risk increases dramatically with age.
- More than 100 types of cancer can be distinguished from one another based on differences in their sites of origin and the cell types involved. Some cancers have been increasing or decreasing in frequency on an age-adjusted basis, but the rates for most cancers have not changed much during the past century.
- Hypertrophy and hyperplasia are normal types of tissue growth based on an increase in cell size or number, respectively. Dysplasia is an abnormal type of cell proliferation involving the loss of normal cell and tissue organization. Although it is potentially reversible, dysplasia can also evolve into neoplasia. In neoplasia, cells proliferate in an uncontrolled, relatively autonomous fashion that produces tumors in which there is a progressive increase in the number of dividing cells.
- Benign tumors grow in a confined local area and so are rarely dangerous, whereas malignant tumors are potentially life threatening because they can spread to distant sites by invasion and metastasis. Malignant tumors are referred to as cancers.
- Survival rates differ significantly for various kinds of cancer, ranging from better than 99% five-year survival for certain kinds of skin cancer to less than 20% five-year survival for cancers of the lung, esophagus, liver, and pancreas.
- In tumor grading, cancers of the same type are assigned numerical grades based on differences in microscopic appearance. In tumor staging, cancers of the same type are assigned numerical stages based on tumor size and how much the tumor has invaded and metastasized at the time of initial diagnosis.
- The remainder of this book is organized around the following questions: What kinds of cellular abnormalities allow cancers to grow and spread in an unrestrained fashion? What causes cancer? What role do genes play in the development of cancer? How is cancer detected and treated? Can cancer be prevented?

Key Terms for Self-Testing

Benign and Malignant Tumors

hypertrophy (p. 4)
hyperplasia (p. 4)
dysplasia (p. 5)
neoplasia (p. 5)
neoplasm (tumor) (p. 5)
cell differentiation (p. 5)
basal cell (p. 5)
squamous cell (p. 5)
benign tumor (p. 5)
malignant tumor (p. 5)
metastasis (p. 5)
cancer (p. 5)
carcinoma (p. 6)
epithelial cell (p. 6)
sarcoma (p. 6)
lymphoma (p. 7)
leukemia (p. 7)
primary tumor (p. 8)

Ways in Which Cancers Differ

five-year survival rate (p. 9)
lead time bias (p. 10)
polyp (p. 10)
biopsy (p. 10)
pathologist (p. 10)
mitotic index (p. 11)
tumor grading (p. 11)
anaplastic (p. 11)
tumor staging (p. 11)
TNM system (p. 11)

Cancer: An Introductory Overview

oncogene (p. 12)
tumor suppressor gene (p. 12)
nucleotide (p. 13)
adenine (A) (p. 13)
guanine (G) (p. 13)
cytosine (C) (p. 13)
thymine (T) (p. 13)
complementary base pairing (p. 13)
gene (p. 13)
transcription (p. 13)
uracil (U) (p. 13)
messenger RNA (mRNA) (p. 13)
translation (p. 13)
codon (p. 13)

Suggested Reading

Cancer Incidence and Mortality

American Cancer Society. *Cancer Facts & Figures 2005*. Atlanta, GA: American Cancer Society, 2005.

Howe, H. L., et al. A vision for cancer incidence surveillance in the United States. *Cancer Causes Control* 14 (2003): 663.

Jemal, A., et al. Annual report to the nation on the status of cancer, 1975–2001, with a special feature regarding survival. *Cancer* 101 (2004): 3.

Quinn, M. J. Cancer trends in the United States—a view from Europe. *J. Natl. Cancer Inst.* 95 (2003): 1258.

Quinn, M. J., et al. Cancer mortality trends in the EU and acceding countries up to 2015. *Annals Oncol.* 14 (2003): 1148.

Ways in Which Cancers Differ

Chung, C. H., P. S. Bernard, and C. M. Perou. Molecular portraits and the family tree of cancer. *Nature Genetics* 32 Suppl. (2002): 533.

Shipp, M. A., et al. Diffuse large B-cell lymphoma outcome prediction by gene-expression profiling and supervised machine learning. *Nature Medicine* 8 (2002): 68.

Welch, H. G., L. M. Schwartz, and S. Woloshin. Are increasing 5-year survival rates evidence of success against cancer? *JAMA* 283 (2000): 2975.

Profile of a Cancer Cell

Cancer cells exhibit a number of unusual properties that distinguish them from normal cells. One key property, which was introduced in Chapter 1, is the ability of cancer cells to proliferate in an uncontrolled fashion, thereby leading to a progressive accumulation of dividing cells without regard to the needs of the body as a whole. In reality, uncontrolled proliferation is not a single discrete property but rather a collection of traits that together allow cancer cells to escape from the usual restraints on cell proliferation. The net result is that cancer cells circumvent the mechanisms designed to ensure that cells divide only when (and where) new cells are needed.

In this chapter we will examine the normal mechanisms for controlling cell proliferation and see how they behave in cancer cells. Through such a discussion it will become apparent that cancer cells exhibit a distinctive collection of abnormal properties, and while no single property is necessarily seen in every cancer cell, as a group these traits contribute to the "profile" of a typical cancer cell.

FEATURES OF CANCER CELL PROLIFERATION

The concept of *uncontrolled* proliferation is unique to multicellular organisms. In most single-celled organisms, such as bacteria or yeast, the presence of sufficient nutrients in the surrounding environment is the main factor that determines whether cells will grow and divide. The situation is reversed in multicellular organisms; cells are usually surrounded by nutrient-rich extracellular fluids, but the organism as a whole would be quickly destroyed if each cell were to continually grow and divide just because it had access to adequate nutrients. Cancer is a potentially lethal reminder of what happens when cell proliferation continues unabated without being coordinated with the needs of the organism as a whole.

Cancer Cells Produce Tumors When Injected into Laboratory Animals

It is the loss of normal growth control that causes cancer cells to produce a continually growing mass of tissue—in other words, a tumor—but uncontrolled growth does not mean that tumor cells always divide more rapidly than normal cells. Tumors can grow slowly, quickly, or somewhere in between. The distinctive feature of tumor growth is not the speed of cell division but its uncontrolled nature. In contrast to the proliferation of normal cells, where cell division and cell differentiation are kept in proper balance, this finely tuned arrangement is disrupted in tumors and cell division is uncoupled from cell differentiation, thereby leading to a progressive increase in the number of dividing cells (see Figure 1-4, *bottom*).

To determine experimentally whether a particular cell behaves in this way, the cell must be injected into an appropriate host organism to see if a tumor will develop. Experiments involving animal cells are fairly straightforward because the cells can simply be injected into animals of the same genetic type. The situation with human cells is more complicated. Injecting human cancer cells into other humans for testing purposes would be unethical, and using standard laboratory animals is not reliable: An animal's immune system is likely to reject human cells because they are of foreign origin. One way around this obstacle is to inject human cells into mutant strains of mice whose immune systems are unable to attack and destroy foreign cells. When human cancer cells are injected into such immunologically deficient animals, the cells will usually grow into tumors without being rejected.

Cancer Cells Exhibit Decreased Density-Dependent Inhibition of Growth

Although studying cancer cells in intact organisms is useful for investigating some of the properties of malignant tumors, issues related to the control of cell proliferation are often easier to investigate in cells grown under artificial laboratory conditions. In such *cell culture* studies, cancer cells are isolated from a tumor and placed in a defined *growth medium* containing nutrients, salts, and other molecules required for cell growth. The main reason for studying cells in culture is that this method allows the cells to be observed under carefully controlled conditions where their behavior can be assessed without the complicating effects of secondary factors present in an intact organism.

When most types of normal cells are placed in a culture vessel (test tube, bottle, flask, or dish) and then covered with an appropriate growth medium, they divide until the surface of the container is covered by a single layer of cells. When this *monolayer stage* is reached, cell movements and cell division both tend to stop. In the early 1950s, Michael Abercrombie and Joan Heaysman introduced the term "contact inhibition" to refer to the decrease in cell motility that occurs when cells make contact with one another in culture. The same term has also been used to refer to the inhibition of cell division that takes place when culture conditions become crowded. Because of the confusion that can result from the double meaning of this term, the phrase **density-dependent inhibition of growth** is now routinely used when referring to the inhibition of cell division that occurs in crowded cultures.

In contrast to normal cells, cancer cells do not stop dividing when they reach the monolayer stage. Instead, they continue to divide and gradually pile up on top of one another, forming multilayered aggregates (Figure 2-1). In other words, cancer cells are less susceptible to density-dependent inhibition of growth than are their normal counterparts. The relationship between the tendency of cancer cells to grow to high population densities in culture and their ability to form tumors has been investigated using cancer cells that differ in their susceptibility to density-dependent inhibition of growth. Cells that are very sensitive to density-dependent inhibition can be produced by growing cells under uncrowded conditions; every time the population density increases and crowding is imminent, the cells are simply diluted and transferred to a new culture flask. Cells obtained in this way will not grow to high population densities in culture. Alternatively, cell populations that are insensitive to density-dependent inhibition of growth can be produced by consistently growing cells in overcrowded conditions. Such cell populations become less susceptible to density-dependent inhibition of growth, reaching much higher population densities before cell division stops. When these different cell populations are tested for their ability to produce tumors in mice, tumor-forming ability is found to be directly related to the loss of density-dependent growth control; in other words, cells capable of growing to the highest population densities in culture are most effective at forming tumors in animals (Figure 2-2).

Cancer Cell Proliferation Is Anchorage-Independent

Another way in which the proliferation of cancer cells differs from that of normal cells involves the requirement for anchorage. Most normal cells will not proliferate if

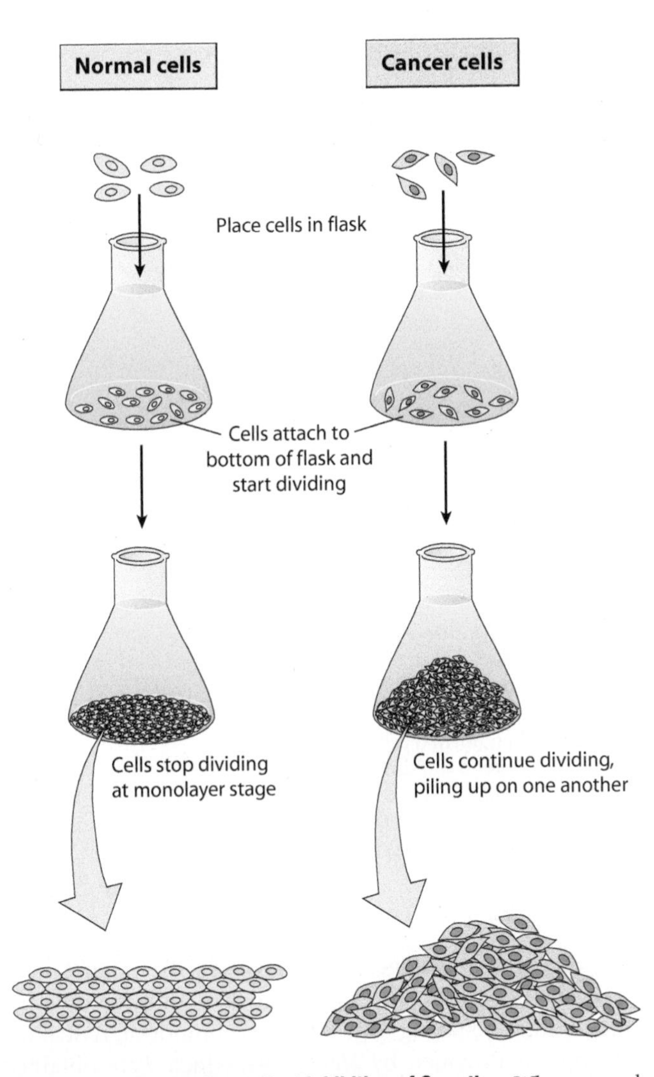

Figure 2-1 Density-Dependent Inhibition of Growth. When normal cells are placed in culture, they tend to divide until the surface of the container is covered by a single layer of cells (the monolayer stage). The inhibition of further cell division that occurs at the crowded, monolayer stage is called density-dependent inhibition of growth. In contrast to the behavior of normal cells, the proliferation of cancer cells tends to continue beyond the monolayer stage as dividing cells pile up on top of one another, forming multilayered aggregates.

they are put in a liquid growth medium and shaken or stirred to keep them in suspension, nor will they proliferate if they are placed in a semisolid medium such as soft agar. When they are provided with an appropriate solid surface to which they can adhere, however, the cells will attach to the surface, spread out, and begin to proliferate (Figure 2-3). The growth of normal cells is therefore said to be **anchorage-dependent**. In contrast, most cancer cells grow well not just when they are anchored to a solid surface, but also when they are suspended in a liquid or semisolid medium. The growth of cancer cells is therefore said to be **anchorage-independent**.

In intact organisms, the requirement that cells be anchored before they can reproduce is met by binding cells to the **extracellular matrix**, an insoluble meshwork of protein and polysaccharide fibers that fills the spaces between neighboring cells. Cells attach themselves to the extracellular matrix through cell surface proteins called *integrins*, which bind to molecules present in the matrix. If this attachment is prevented experimentally using chemicals that block the binding of integrins to the matrix, normal cells are prevented from dividing and may even commit suicide by *apoptosis*, an orchestrated program for cell death described later in the chapter. Apoptotic cell death triggered by lack of contact with the extracellular matrix is called **anoikis** (from the Greek word for "homelessness"). Anoikis is an important safeguard for maintaining tissue integrity because it prevents normal cells from floating away and setting up housekeeping in another tissue. The lack of anchorage simply causes cells to commit suicide along the way. Cancer cells are not subject to this normal safeguard because they are anchorage-independent and so can spread to distant sites without self-destructing.

Considerable evidence suggests that anchorage-independent growth exhibited by cells grown in culture is related to their ability to form tumors. One set of studies involved cells with many of the traits of cancer cells, including decreased density-dependent inhibition of growth, low requirements for external growth factors (p. 22), and anchorage-independent growth. Single cells were isolated from the original population and allowed to proliferate separately, thereby creating a series of **clones**, which are individual cell populations each derived from the proliferation of a single cell. Careful analysis of the clones revealed that some of them had lost one or more of the initial properties. When the ability of these clones to produce tumors in animals was compared, anchorage-independent growth was the only

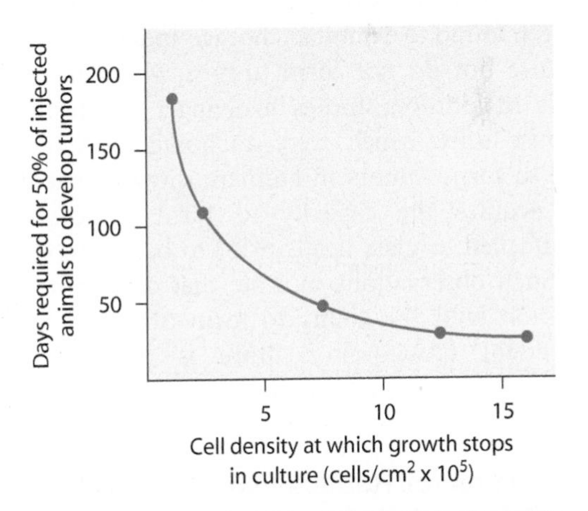

Figure 2-2 Tumor-Forming Ability of Mouse Cells That Differ in Their Susceptibility to Density-Dependent Inhibition of Growth. Each point represents a different cell population. The data show that cell lines that grow to higher population densities in culture—that is, cells that are less susceptible to density-dependent inhibition of growth—tend to produce tumors more rapidly when injected into mice. Note that the y-axis is measuring how long it takes for tumors to develop, so higher values indicate cells that are less efficient at forming tumors. [Adapted from S. A. Aaronson and G. J. Todaro, *Science* 162 (1968): 1024.]

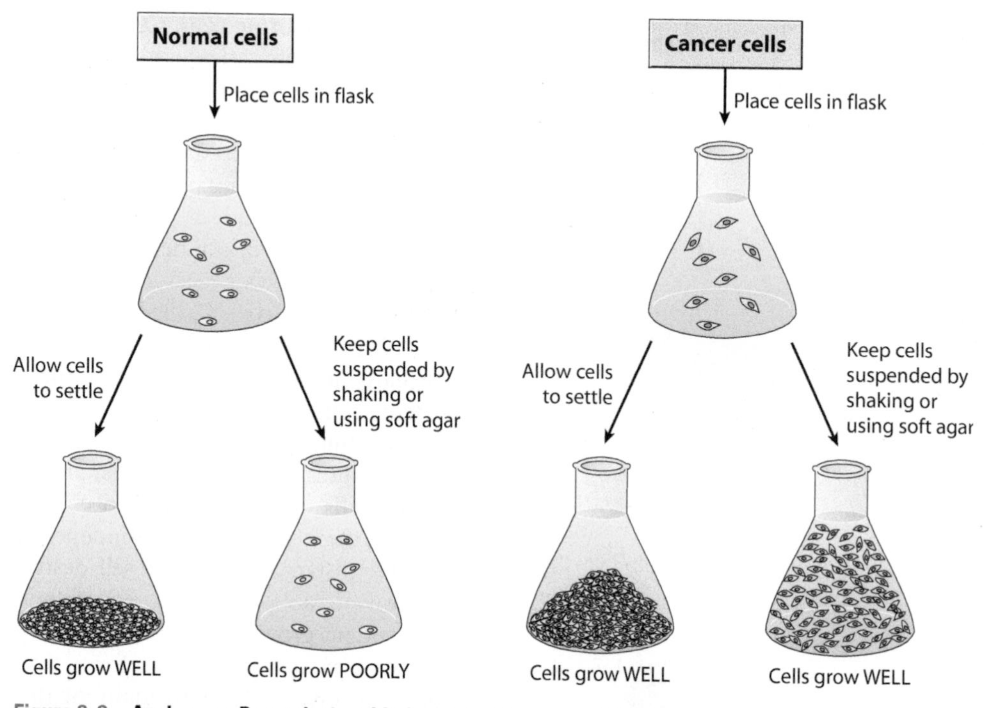

Figure 2-3 Anchorage-Dependent and Independent Growth. When normal cells are placed in culture, they only grow well when they are allowed to adhere to the surface of the culture flask; their growth is therefore said to be anchorage-dependent. In contrast, cancer cells grow well not just when they are anchored to a solid surface, but also when they are freely suspended in a liquid or a semisolid medium. The growth of cancer cells is therefore said to be anchorage-independent.

property consistently retained by all the clones that could produce tumors. In other words, the ability to form tumors appeared to require cells whose growth in culture is anchorage-independent.

This connection between anchorage-independent growth and tumor formation is not without its exceptions, however. Some cells exposed to cancer-causing chemicals have been found to exhibit anchorage-independent growth in culture but do not form tumors when injected into animals. In addition, studies involving a long-term culture of mouse cells, which were anchorage-dependent and unable to form tumors in animals, showed that the cells could acquire the capacity to form tumors if they were attached to glass beads prior to being implanted in mice. Such observations indicate that despite its general association with the ability to form tumors, anchorage-independent growth in culture is not an absolute prerequisite for tumor formation.

Mechanisms for Replenishing Telomeres Make Cancer Cells Immortal

One of the most striking differences between normal cells and cancer cells involves their reproductive lifespans. When normal cells are grown in culture, they usually divide for only a limited number of times. For example, human fibroblasts—a cell type whose behavior has been extensively studied—divide about 50 to 60 times when placed in culture and then stop dividing, undergo a variety of degenerative changes, and may even die (Figure 2-4). Cancer cells exhibit no such limit and continue dividing indefinitely, behaving as if they were immortal. A striking example is provided by *HeLa* cells, which were obtained from a malignant tumor of the uterus arising in a woman

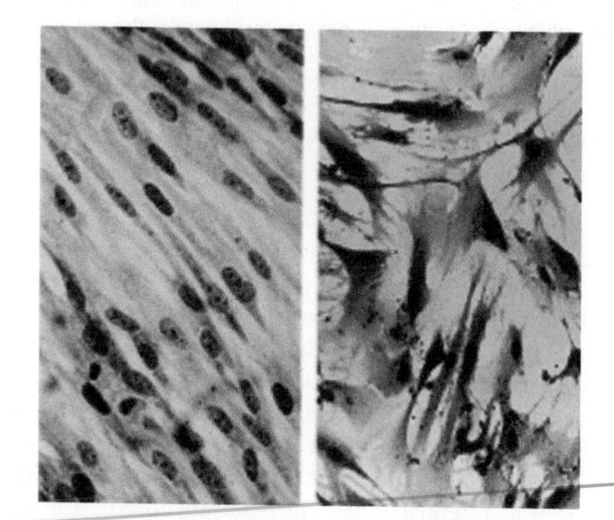

Figure 2-4 Microscopic Appearance of Young and Old Human Fibroblasts Growing in Culture. (*Left*) Young fibroblasts that have divided a relatively small number of times in culture exhibit a thin, elongated shape. (*Right*) After dividing about 50 times in culture, the cells stop dividing and undergo a variety of degenerative changes. Note the striking difference in appearance between the young (dividing) and older (nondividing) cells. [Courtesy of L. Hayflick.]

named Henrietta Lacks (hence the name "HeLa" cells). After removing the tumor in a cancer operation performed in 1951, doctors placed some of its cells in culture. The cultured cells began to grow and divide and have continued to do so for more than 50 years, dividing more than 18,000 times with no signs of stopping.

Why are cancer cells capable of reproducing indefinitely in culture, whereas most normal human cells divide no more than 50 or 60 times? The answer is related to the mechanism by which cells replicate their DNA. Each time a cell divides, its chromosomal DNA molecules must be duplicated so that a complete set of genetic instructions can be distributed to each of the two cells produced by cell division. However, the biochemical mechanism responsible for DNA replication has an inherent limitation: The enzymes that replicate DNA are unable to copy the very end of a linear DNA molecule, perhaps the final 50 to 100 nucleotides or so. As a result, each time a DNA molecule is replicated, it is in danger of losing a small amount of DNA at each of its two ends. If this trend were to continue indefinitely, DNA molecules would become shorter and shorter until there was nothing left, and we would not be here today!

To solve this so-called end-replication problem, cells place a special type of DNA sequence at the two ends, or **telomeres**, of each chromosomal DNA molecule. The special DNA consists of multiple copies of a short base sequence repeated over and over again. For example, in humans the six-base sequence TTAGGG is repeated about 2500 times in a row at the ends of each chromosomal DNA molecule at birth. These telomere sequences are protected by *telomere capping proteins*, and the DNA also loops back upon itself to protect the end of the chromosome even further (Figure 2-5). Unlike genes, whose DNA base sequences code for useful products, telomeric DNA does not code for anything but simply consists of the same six-base sequence repeated again and again. Placing such noncoding telomere DNA at the ends of each chromosome ensures that a cell will not lose any important genetic information when DNA molecules are shortened slightly during replication.

Since telomeres get shorter with each cell division, they provide a counting device for tracking how many times a cell has divided. If a cell divides too many times, the telomeres become extremely short and are in danger of disappearing entirely. When this happens, the telomeric DNA becomes too short to bind telomeric capping proteins or generate a loop, exposing a bare end of double-stranded DNA. Such unprotected DNA ends are very unstable and often fuse with each other, creating joined chromosomes that tend to become fragmented and separate improperly at the time of cell division. In normal cells, such a hazardous outcome is prevented by a mechanism in which the unprotected DNA at the end of a chromosome triggers a pathway that halts cell division or triggers cell death. This pathway helps protect organisms from any inappropriate, excessive proliferation of adult cells.

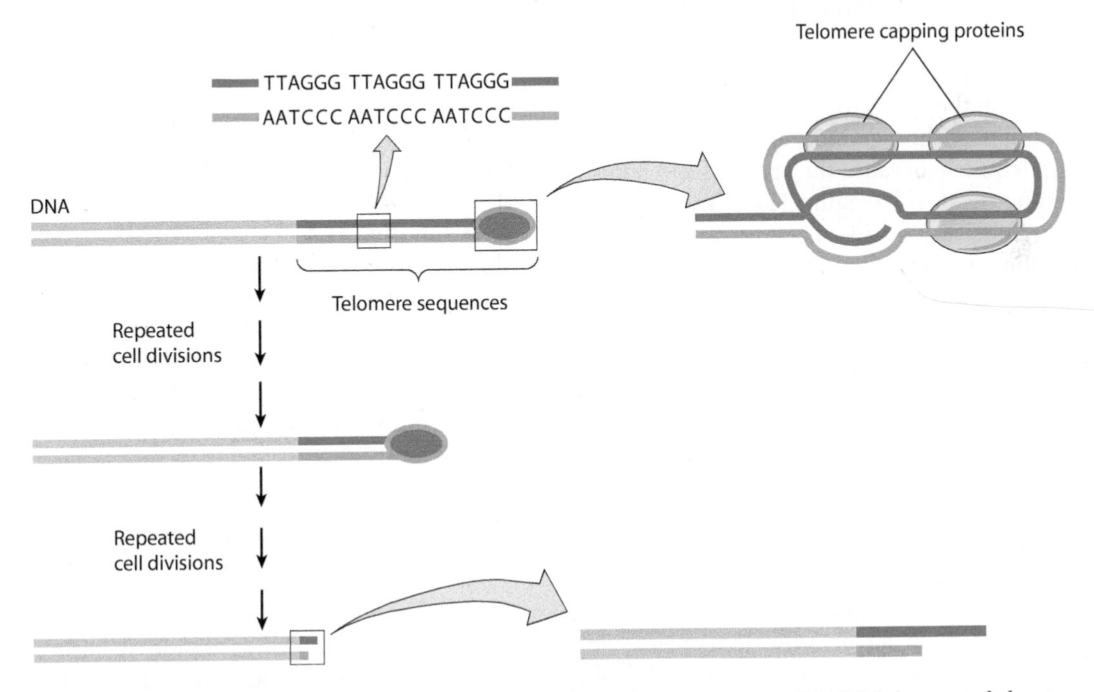

Figure 2-5 Telomere Structure in Human DNA. The six-base telomere sequence, TTAGGG, is repeated about 2500 times in a row at the ends of each chromosomal DNA molecule at birth. Because telomeric DNA contains the same sequence repeated over and over, the end of one DNA strand can loop back and form base pairs with an earlier repeat of the same complementary sequence in the opposite strand. Telomeres get shorter with each cell division, thereby providing a counting device for tracking how many times a cell has divided. If a cell divides too many times, the telomeres become too short to generate a loop or bind telomeric capping proteins. Such unprotected DNA ends are unstable and trigger pathways that halt cell division or trigger cell death.

But what happens with cells that must divide for prolonged periods of time, such as the *germ cells* that give rise to sperm and eggs or the *bone marrow cells* that continually produce new blood cells? Such cells prevent excessive telomere shortening by producing an enzyme called **telomerase**, which adds new copies of the telomeric repeat sequence to the ends of existing DNA molecules. The telomerase-catalyzed addition of new telomere repeat sequences prevents the gradual decline in telomere length that would otherwise occur at both ends of a chromosome during DNA replication. The presence of telomerase therefore allows cells to divide indefinitely without telomere shortening.

How do the preceding considerations apply to cancer cells? If cancer cells behaved like most normal cells, which do not produce telomerase, repeated cell divisions would cause the telomeres to become unusually short and the cells would eventually be destroyed. Most cancer cells circumvent this problem by activating the gene that produces telomerase, thereby causing new copies of the telomeric repeat sequence to be continually added to the ends of their DNA molecules. A few cancer cells activate an alternative mechanism for maintaining telomere sequences that involves the exchange of sequence information between chromosomes. By one mechanism or the other, cancer cells maintain telomere length above a critical threshold and can therefore divide indefinitely.

GROWTH FACTORS AND THE CELL CYCLE

We have now seen that cancer cells differ from most normal cells in that they grow to high population densities in culture, exhibit anchorage-independent proliferation, and divide indefinitely because they possess mechanisms for maintaining telomere length. These traits play an important permissive role in allowing cancer cells to continue dividing, but they do not actually cause cells to divide. The driving force for ongoing proliferation can be traced to abnormalities in the signaling systems that control cell division, the topic to which we now turn.

Cancer Cells Exhibit a Decreased Dependence on External Growth Factors

The cells of multicellular animals do not normally divide unless they are stimulated to do so by an appropriate signaling protein known as a **growth factor**. For example, if cells are isolated from an organism and placed in a culture medium containing nutrients and vitamins, they will not proliferate unless an appropriate growth factor is also provided. Growth media are therefore commonly supplemented with blood serum, which contains several growth factors that stimulate cell proliferation. One is **platelet-derived growth factor** (**PDGF**), a protein produced by blood platelets that stimulates the proliferation of connective tissue cells and smooth muscle cells. Another growth factor in blood serum, called **epidermal growth factor** (**EGF**), is also widely distributed in tissues. Some growth factors, such as EGF, stimulate the growth of a wide variety of cell types, whereas others act more selectively on particular target cells. Growth factors play important roles in stimulating tissue growth during embryonic and early childhood development, and during wound repair and cell replacement in adults. For example, release of the growth factor PDGF from blood platelets at wound sites is instrumental in stimulating the growth of tissue required for wound healing.

Growth factors exert their effects by binding to **receptor** proteins located in the **plasma membrane** that forms the outer boundary of all cells. Different cell types have different plasma membrane receptors and hence differ in the growth factors to which they respond (Figure 2-6). The binding of a growth factor to its corresponding receptor triggers a multistep cascade in which a series of *signal transduction proteins* relay the signal throughout the cell, triggering molecular changes that stimulate (or occasionally inhibit) cell growth and division. Cells will not normally divide unless they are stimulated by an appropriate growth factor, but this restraint is circumvented in cancer cells by various mechanisms that create a constant signal to divide, even in the absence of growth factors.

Some cancer cells achieve this autonomy by producing their own growth factors, thereby causing cell proliferation to be stimulated without the need for growth factors produced by other cells. Similarly, other cancer cells possess abnormal receptors that are permanently activated, causing cell division to occur whether growth factors are present or not. Cancer cells may also produce excessive quantities or hyperactive versions of other proteins involved in relaying signals from cell surface receptors to the cell division machinery in the cell's interior. The net effect of the preceding types of alterations is to cause the pathways that signal cell proliferation to become hyperactive or even autonomous, functioning in the absence of growth factors. In Chapters 9 and 10, the molecules involved in these pathways and the ways in which they malfunction in cancer cells will be described in detail.

The Cell Cycle Is Composed of G1, S, G2, and M Phases

To understand how pathways activated by growth factors ultimately cause a cell to divide, it is first necessary to review the events associated with cell division. In cells that are dividing, the nuclear DNA molecules must be duplicated and then distributed in a way that ensures that the two new cells each receive a complete set of genetic instructions. In preparing for and accomplishing these tasks, cells pass through a series of discrete stages called **G1 phase**, **S phase**, **G2 phase**, and **M phase**.

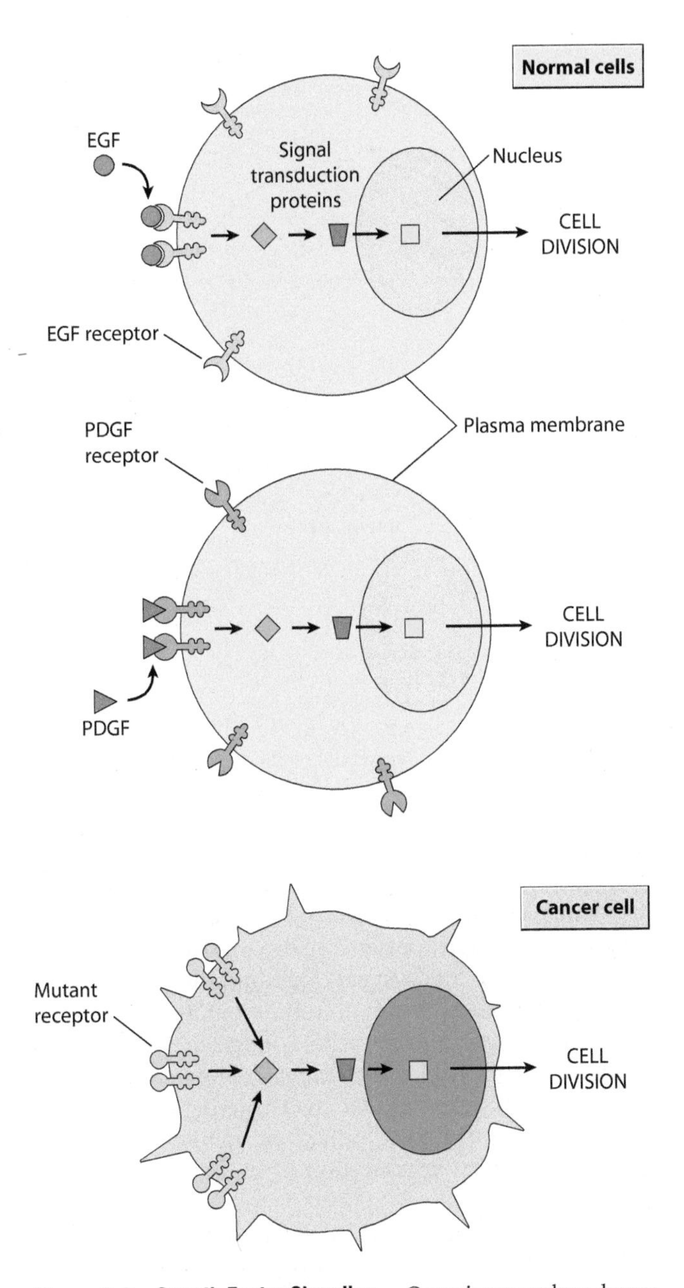

Figure 2-6 Growth Factor Signaling. Organisms produce dozens of different growth factors, each stimulating the proliferation of target cells that contain a receptor to which the growth factor can bind. The top two cells contain receptors for epidermal growth factor (EGF) and platelet derived growth factor (PDGF), respectively. The binding of a growth factor to its corresponding receptor triggers changes in signal transduction proteins that cause cells containing that particular receptor to divide. The cancer cell shown at the bottom contains a mutant receptor that is locked in an active configuration, causing the cell to continually divide (even in the absence of growth factor). Alterations in other proteins involved in growth factor signaling pathways can likewise cause cancer cells to divide in an uncontrolled fashion. The signal transduction proteins involved in such pathways will be described in Chapters 9 and 10 (e.g., see Figure 9-8).

The four phases are collectively referred to as the **cell cycle** (Figure 2-7).

G1—the first phase to occur after a cell has just divided—varies the most in length. A typical G1 phase lasts about 8 to 10 hours in human cells, but rapidly dividing cells may spend only a few minutes or hours in G1. Conversely, cells that divide very slowly may become arrested in G1 and spend weeks, months, or even years in the offshoot of G1 called the **G0 phase** (G zero). After completing G1, the cell enters S phase, a period of roughly 6 to 8 hours when the chromosomal DNA molecules are replicated. Next comes G2 phase, where 3 to 4 hours are spent making final preparations for cell division. The cell then enters M phase, which takes about an hour to physically divide the original cell into two new cells. The main events of M phase include division of the nucleus, or **mitosis**, followed by division of the cytoplasm, or **cytokinesis**. The two newly formed cells then enter again into G1 phase and begin preparations for another round of cell division.

Taken together, the G1, S, and G2 phases are collectively referred to as **interphase**. Besides providing the time needed for a cell to make copies of its DNA molecules, interphase is also a period of cell growth. Interphase occupies about 95% of a typical cell cycle, whereas the actual process of cell division (M phase) only takes about 5%. Overall, the time occupied by the various stages of the cycle allows a typical human cell to divide as often as once every 18 to 24 hours. However, the various cell types that make up the body differ greatly in cycle time, ranging from cells that divide very rapidly and continuously to differentiated cells that do not divide at all.

The variability observed in rates of cell division means that mechanisms must exist for regulating progression through the cell cycle. A key control point has been identified during late G1, where the cell cycle is usually halted in cells that stop dividing. For example, the division of cultured cells can be slowed down or stopped by allowing the cells to run out of either nutrients or growth factors, or by adding inhibitors of vital processes such as protein synthesis. In such cases, the cell cycle is halted in late G1 at a point referred to as the **restriction point**. Under normal conditions, the ability to pass through the restriction point is governed mainly by the presence of growth factors. Cells that successfully move through the restriction point are committed to S phase and the remainder of the cell cycle, whereas those that do not pass the restriction point enter into G0 and reside there for variable periods of time, awaiting a signal that will allow them to re-enter G1 and pass through the restriction point.

Progression Through the Cell Cycle Is Driven by Cyclin-Dependent Kinases

At the molecular level, passage through the restriction point and other key points in the cell cycle is controlled by proteins known as **cyclin-dependent kinases (Cdks)**. Cdks are **protein kinases**, a term referring to a class of enzymes that regulate the activity of targeted protein molecules by catalyzing their **phosphorylation** (attachment of phosphate groups to the targeted proteins). During protein phosphorylation reactions, the phosphate group is

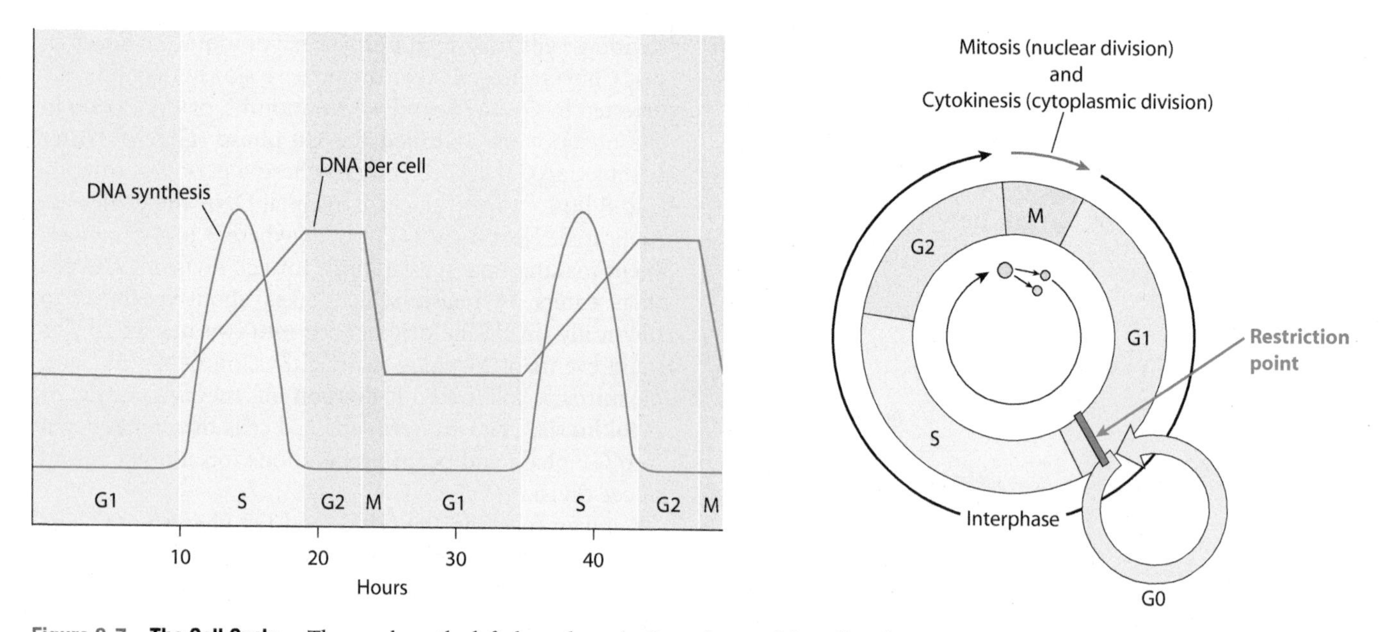

Figure 2-7 The Cell Cycle. The graph on the left shows how the four phases of the cell cycle (G1, S, G2, and M) are defined by two variables: DNA synthesis and the amount of DNA per cell. S phase is defined as the time during the cell cycle when DNA synthesis is taking place, leading to a doubling of the amount of DNA per cell. M phase is the time when the amount of DNA per cell drops in half as cells divide. G1 is defined as the interval between M phase and S phase, and G2 is defined as the interval between S phase and M phase. The cell cycle is commonly represented with a circular diagram like the one shown on the right. The restriction point is a control point near the end of G1 where the cell cycle can be halted until conditions are suitable for progression into S phase. Under normal conditions, the ability to pass through the restriction point is governed mainly by the presence of growth factors.

donated to the targeted protein by the high-energy compound *ATP* (adenosine triphosphate), which is converted to *ADP* (adenosine diphosphate) during the reaction. Cells contain dozens of different protein kinases, each designed to regulate the activity of a specific group of proteins by catalyzing their phosphorylation.

As the name implies, a cyclin-dependent kinase (Cdk) only exhibits protein kinase activity when it is bound to another type of protein called a **cyclin**. Progression through the cell cycle is controlled by several Cdks that bind to different cyclins, thereby creating a variety of Cdk-cyclin complexes. Cyclins involved in regulating the progression from G1 to S phase are called *G1 cyclins*, and the Cdk molecules to which they bind are known as *G1 Cdks*. Likewise, cyclins involved in regulating passage from G2 into M phase are called *mitotic cyclins*, and the Cdk molecules to which they bind are known as *mitotic Cdks*. Cdk-cyclin complexes act by phosphorylating specific target proteins whose actions are required for various stages of the cell cycle.

How do Cdk-cyclin complexes ensure that passage through key points in the cell cycle only occurs at the appropriate time? In addressing this question, we will briefly consider the behavior of the *mitotic Cdk-cyclin complex* (mitotic Cdk bound to mitotic cyclin), which regulates passage from G2 to M phase. Mitotic cyclin is continuously synthesized throughout interphase and gradually increases in concentration during G1, S, and G2, eventually reaching a concentration that is high enough to bind to mitotic Cdk (Figure 2-8). The resulting mitotic Cdk-cyclin triggers passage from G2 into M phase by phosphorylating key proteins involved in the early stages of mitosis. For example, proteins phosphorylated by mitotic Cdk-cyclin trigger nuclear envelope breakdown, chromosome condensation, and mitotic spindle formation. Shortly thereafter, mitotic cyclin is targeted for degradation by an enzyme called the *anaphase-promoting complex* and mitotic Cdk becomes inactive, triggering the exit from mitosis. During the next cell cycle, mitosis cannot be triggered until the concentration of mitotic cyclin builds up again.

Besides being regulated by the availability of cyclins, the activity of the various Cdk-cyclin complexes is controlled by reactions in which Cdk molecules are altered by *phosphorylation* (addition of phosphate groups) and *dephosphorylation* (removal of phosphate groups). Figure 2-9 illustrates how the mitotic Cdk-cyclin complex is regulated in this way. In step ①, the binding of mitotic cyclin to mitotic Cdk creates a complex that is initially inactive. Before it can trigger passage from G2 into M phase, the complex requires the addition of an activating phosphate group to a particular amino acid of the Cdk molecule. Prior to adding this phosphate, however, inhibitory phosphate groups are first attached to the Cdk molecule at two other locations, preventing the Cdk from functioning (step ②). The activating phosphate group, highlighted with yellow in step ③, is then added. The last step in the activation sequence is the removal of the inhibiting phosphates by a specific enzyme called a **protein phosphatase** (step ④). Once the phosphatase begins removing the inhibiting phosphates, a positive feedback loop is set up: The activated Cdk-cyclin complex

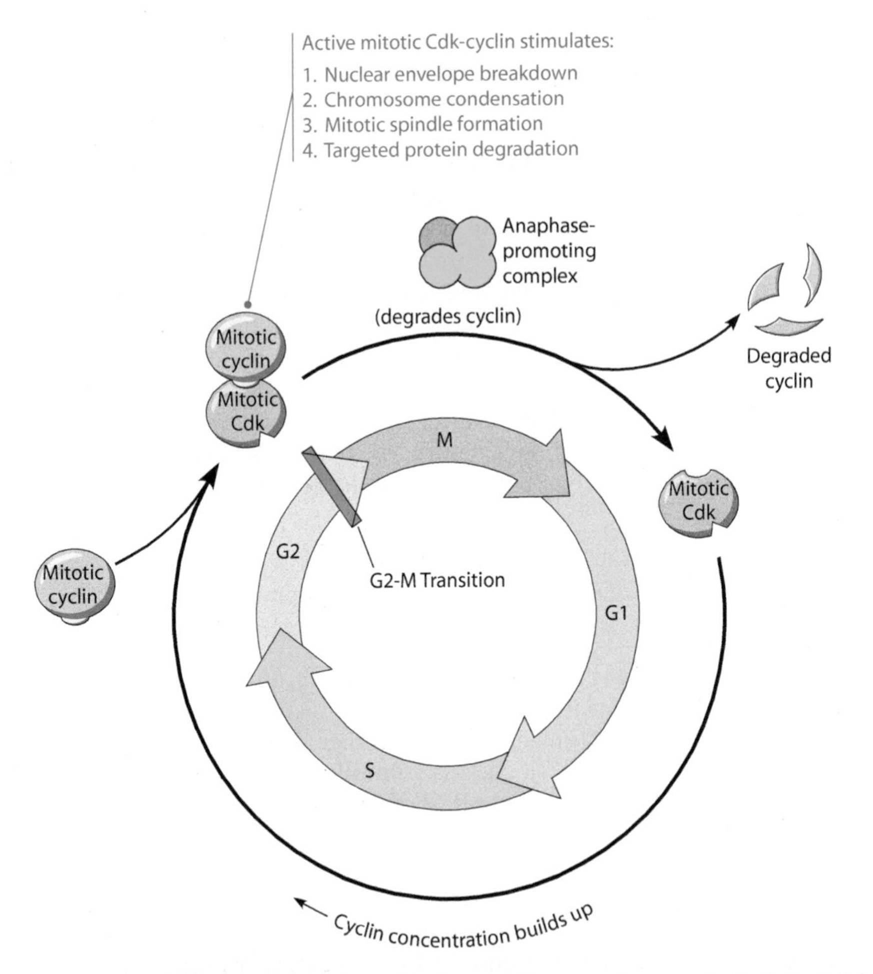

Figure 2-8 The Mitotic Cdk Cycle. This diagram illustrates the control of mitotic Cdk by mitotic cyclin during the cell cycle. During G1, S, and G2, the mitotic cyclin concentration gradually increases. Near the end of G2, mitotic Cdk and cyclin form an active complex that triggers passage from G2 into M phase by phosphorylating proteins involved in the mitotic events listed. By contributing to activation of the anaphase-promoting complex, which degrades cyclin, the mitotic Cdk-cyclin complex also brings about its own demise, allowing the completion of mitosis and entry into G1 of the next cell cycle.

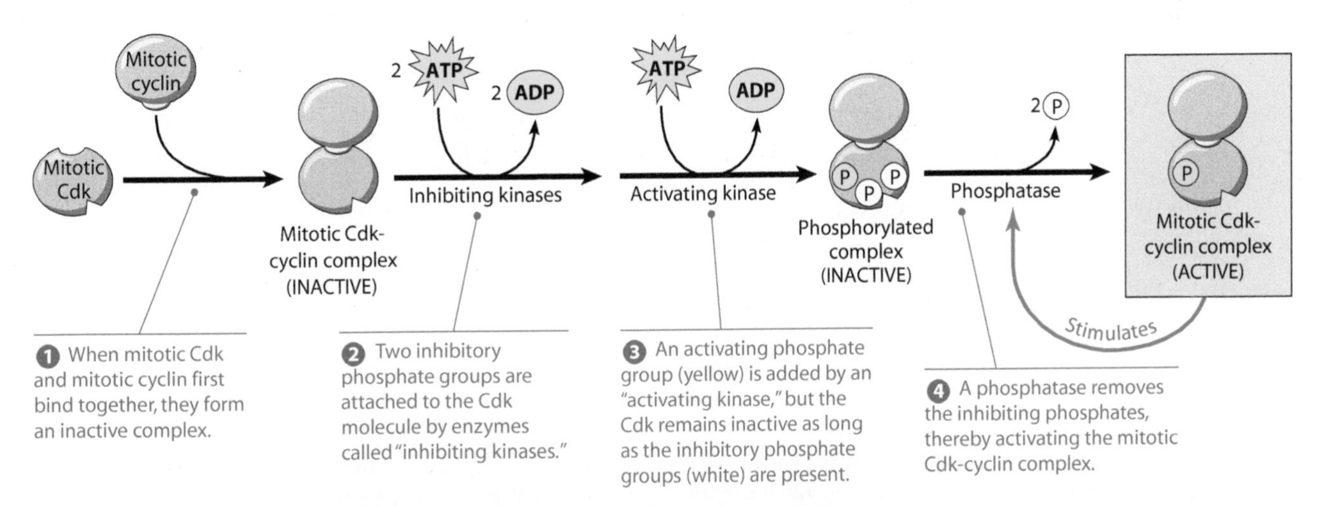

Figure 2-9 Regulation of Mitotic Cdk-Cyclin by Phosphorylation and Dephosphorylation. Activation of mitotic Cdk-cyclin involves the addition of inhibiting and activating phosphate groups, followed by removal of the inhibiting phosphate groups by a phosphatase. Once removal of the inhibiting phosphate groups has begun in step ④, a positive feedback loop is set up: The activated Cdk-cyclin complex generated by this reaction stimulates the phosphatase, thereby causing the activation process to proceed more rapidly.

generated by this reaction stimulates the phosphatase, thereby causing the activation process to proceed more rapidly. After being activated, the mitotic Cdk-cyclin complex triggers passage from G2 into M phase by catalyzing the phosphorylation of proteins required for the onset of mitosis.

Growth Factor Signaling Pathways Act on the Restriction Point by Stimulating Phosphorylation of the Rb Protein

Now that Cdk-cyclins have been introduced, we can explain how growth factors exert their control over cell proliferation. If normal cells are placed in a culture medium containing nutrients and vitamins but no growth factors, the cells become arrested at the restriction point. Subsequent addition of growth factors is sufficient to cause the cells to start dividing again.

How do growth factors cause G1-arrested cells to resume progression through the cell cycle? The binding of a growth factor to its corresponding cell surface receptor causes the receptor to become activated, and the activated receptor then triggers a complex pathway of reactions involving dozens of different cytoplasmic and nuclear molecules that relay the signal throughout the cell. A detailed description of the molecules involved in growth factor signaling pathways will be provided in Chapters 9 and 10, when we discuss how cancer-causing mutations affect components of these pathways. For now we are concerned only with the question: How do these pathways impinge on the cell cycle and cause cells to pass through the restriction point and into S phase?

The answer to this question is that growth factor signaling pathways trigger the production of Cdk-cyclins that in turn catalyze the phosphorylation of target proteins required for the transition into S phase. A key target is the **Rb protein**, a molecule that normally restrains cell proliferation by preventing passage through the restriction point (Figure 2-10). After Cdk-cyclin phosphorylates Rb, it can no longer exert this inhibitory influence and cells are free to pass through the restriction point and into S phase. The mechanism by which the Rb protein exerts its inhibitory control over the restriction point will be described in Chapter 10, where we examine the role played by Rb mutations in the development of cancer.

Checkpoint Pathways Monitor for DNA Replication, Chromosome-to-Spindle Attachments, and DNA Damage

The ability of growth factors to promote passage through the restriction point by stimulating the production of Cdk-cyclins that phosphorylate Rb is just one example of how the cell cycle is controlled by external and internal factors that determine whether or not a cell should divide. Another type of cell cycle control involves a series of **checkpoint** pathways that prevent cells from proceeding from one phase to the next before the preceding phase has been properly completed. These checkpoint pathways monitor conditions within the cell and transiently halt the

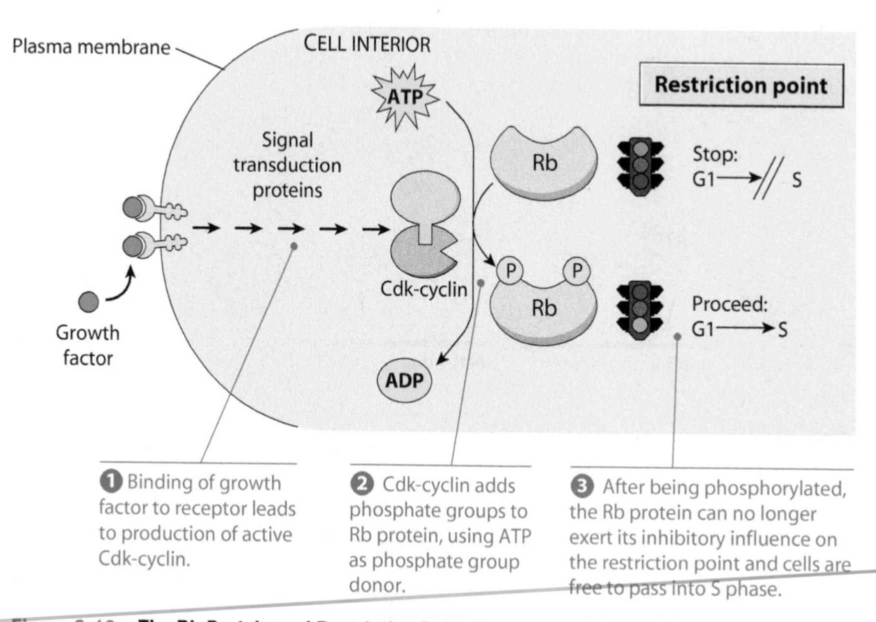

Figure 2-10 The Rb Protein and Restriction Point Control. The Rb protein normally prevents cells from passing through the restriction point in the absence an appropriate growth factor. Growth factors trigger the production of active Cdk-cyclin complexes that phosphorylate the Rb protein. After being phosphorylated, the Rb protein can no longer exert its inhibitory influence on the restriction point and cells are free to pass into S phase. (As will be described in Chapter 10, the Rb protein exerts its effects by binding to a protein called E2F, which regulates gene expression [see Figure 10-3].)

cell cycle at various points if conditions are not suitable for continuing (Figure 2-11).

One such mechanism, called the **DNA replication checkpoint**, monitors the state of DNA replication to ensure that DNA synthesis has been completed prior to proceeding with cell division. If DNA replication is not complete, the cell cycle is halted to allow DNA replication to be finished prior to entering M phase. The existence of the DNA replication checkpoint has been demonstrated by treating cells with inhibitors of DNA synthesis. Under such conditions the final dephosphorylation step involved in activating mitotic Cdk-cyclin (step ④ in Figure 2-9) is blocked through a series of events triggered by proteins associated with replicating DNA. The resulting lack of active mitotic Cdk-cyclin halts the cell cycle at the end of G2 until DNA replication is completed.

A second checkpoint mechanism, called the **spindle checkpoint**, acts between the *metaphase* and *anaphase* stages of mitosis, the point where the two duplicate sets of chromosomes are about to be parceled out to the two new cells being formed by the process of cell division. At the end of metaphase, the two sets of chromosomes are normally lined up at the center of the *mitotic spindle*, a structure composed of microtubules that attach to the chromosomes and eventually pull them into the two newly forming cells. Before chromosome movement begins (the event that marks the beginning of anaphase), the spindle checkpoint mechanism is invoked to make certain that the chromosomes are all properly attached to the spindle. If the chromosomes are not completely attached, the cell cycle is temporarily halted at this point to allow the process to be completed. In the absence of such a control mechanism for monitoring chromosome-to-spindle attachments, there would be no guarantee that each of the newly forming cells would receive a complete set of chromosomes. A detailed description of the spindle checkpoint and its role in the development of cancer will be provided in Chapter 10 (see Figure 10-16).

A third type of checkpoint is used to prevent cells with damaged DNA from proceeding through the cell cycle. In this case, a series of **DNA damage checkpoints** monitor for DNA damage and halt the cell cycle at various points—including late G1, S, and late G2—by inhibiting different Cdk-cyclin complexes. A molecule called the **p53 protein** plays a central role in these checkpoint pathways. In the presence of damaged DNA, the p53 protein accumulates and triggers cell cycle arrest to provide time for the DNA damage to be repaired. If the damage cannot be repaired, p53 may also trigger cell death by *apoptosis* (a process to be described shortly). The ability of p53 to trigger cell cycle arrest or cell death prevents cells with damaged DNA from proliferating and passing the damage on to succeeding generations of cells. The mechanisms involved in the p53 pathway, including the way in which DNA damage is detected and the way in which p53 triggers cell cycle arrest and cell death, will be described in Chapter 10 (see Figure 10-5).

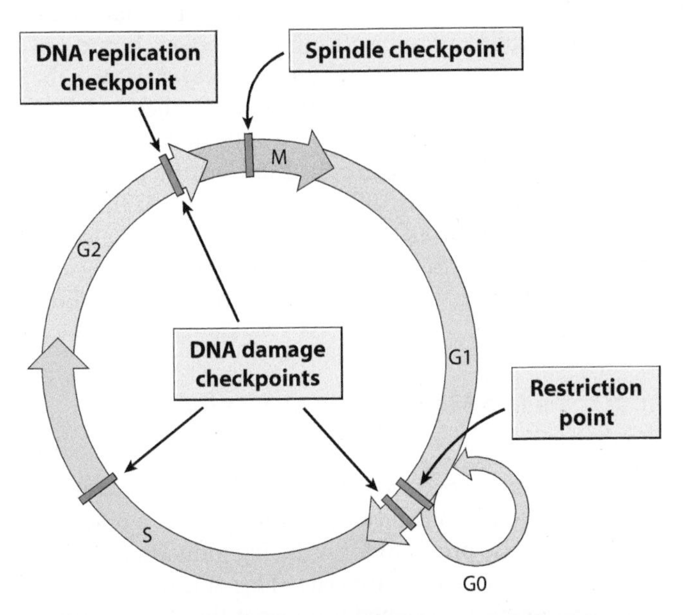

Figure 2-11 Control Points in the Cell Cycle. The ability of growth factors to promote passage through the restriction point illustrates how the cell cycle is regulated by external conditions. Another type of control involves checkpoint pathways that monitor conditions inside the cell and transiently halt the cell cycle at various points if conditions are not suitable for continuing. Checkpoint pathways monitor for DNA damage, DNA replication, and chromosome attachment to the spindle, transiently halting progression through the cell cycle at various points if conditions are not suitable for continuing.

Cell Cycle Control Mechanisms Are Defective in Cancer Cells

The preceding discussion of cell cycle control mechanisms has focused largely on the behavior of normal cells. How do these principles apply to the behavior of cancer cells, which grow and divide in an uncontrolled fashion? We have already seen that cancer cells often produce excessive amounts (or hyperactive versions) of growth factors, receptors, or other components of growth factor signaling pathways. Such alterations cause an excessive production of the Cdk-cyclins that phosphorylate the Rb protein, thereby providing an ongoing stimulus for cells to pass through the restriction point and divide.

The situation is made even worse by the fact that the restriction point often fails to function properly in cancer cells. When cancer cells are grown under suboptimal conditions—for example, insufficient growth factors, high cell density, lack of anchorage, or inadequate nutrients—that would cause normal cells to become arrested at the restriction point, cancer cells continue to grow and divide without halting at the restriction point. In other words, cancer cells exhibit a *loss of restriction point control*. Under extremely adverse conditions, such as severe nutritional deprivation, cancer cells die at random points in the cell cycle rather than arresting at the restriction point.

In addition to the loss of restriction point control, cancer cells frequently exhibit defects in the checkpoint

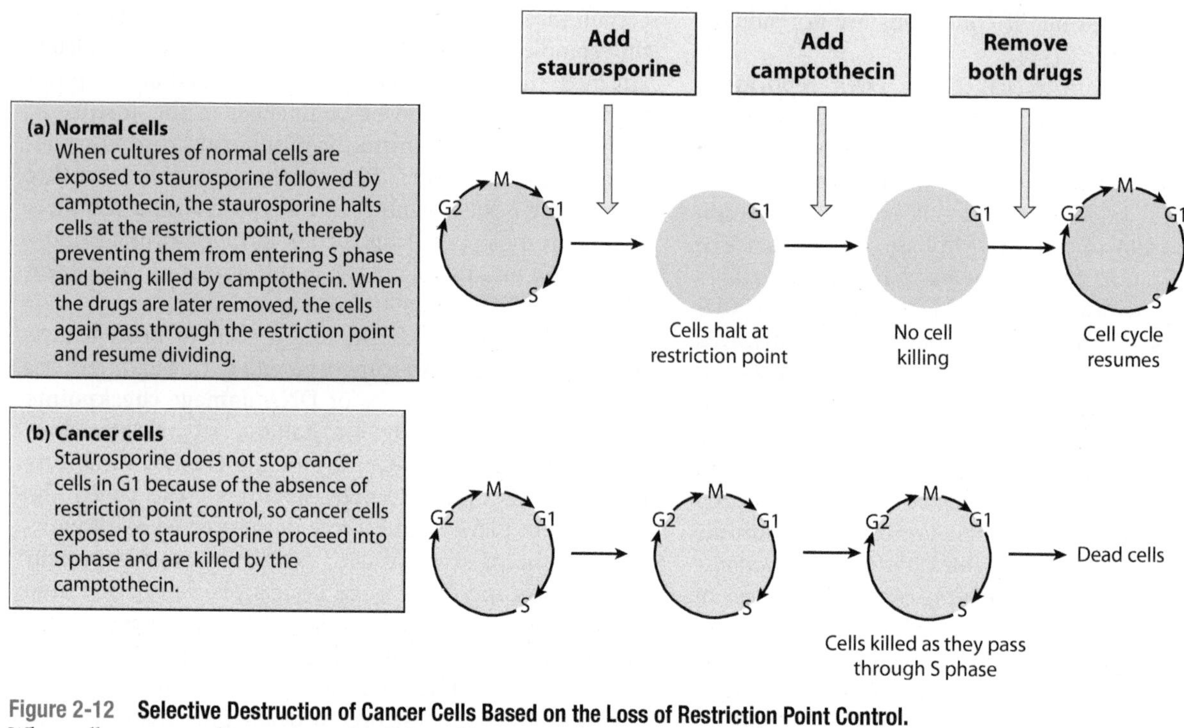

Figure 2-12 Selective Destruction of Cancer Cells Based on the Loss of Restriction Point Control.
When cells are exposed to staurosporine followed by camptothecin, cancer cells are killed but normal cells are not.
[Based on data from X. Chen et al., *J. Natl. Cancer Inst.* 92 (2000): 1999.]

pathways that would otherwise respond to internal problems, such as DNA damage, by halting the cell cycle. Failures in checkpoint pathways, along with the loss of restriction point control, allow cancer cells to continue proliferating under conditions in which the cell cycle of normal cells would stop. The molecular abnormalities responsible for the restriction point and checkpoint failures exhibited by cancer cells will be described in Chapters 9 and 10, which cover the genetic mutations that underlie the development of cancer.

The difference between cell cycle regulation in cancer cells and normal cells can be exploited experimentally using drugs that act at different points in the cycle. For example, *staurosporine* is a drug that halts the cell cycle at the restriction point, and *camptothecin* is a drug that kills cells in S phase by disrupting DNA synthesis. As shown in Figure 2-12, when cultures of normal cells are exposed to staurosporine followed by camptothecin, the staurosporine halts cells at the restriction point and thus prevents them from entering S phase and being killed by camptothecin. If the two drugs are later removed, the cells again pass through the restriction point and resume dividing. With cancer cells, the results are quite different. Staurosporine does not stop cancer cells in G1 because of the loss of restriction point control, so the cells proceed into S phase and are killed by camptothecin. This discovery that cancer cells can be killed by drug combinations that do not harm normal cells raises the possibility that similar strategies might eventually be devised for treating cancer patients.

APOPTOSIS AND CELL SURVIVAL

Thus far, this chapter has focused on traits that affect cancer cell proliferation by affecting the cell cycle and cell division. An unrestrained cell cycle, however, is not the only factor that contributes to the uncontrolled production of tumor cells. The number of cells that accumulate in a growing tumor is determined not just by the rate at which cells divide, but also by the rate at which they die. As in the case of the cell cycle, cell death is controlled by pathways that fail to function properly in cancer cells, thereby permitting the survival of cells that would otherwise be destroyed.

Apoptosis Is a Mechanism for Eliminating Unneeded or Defective Cells

Cell death seems like it would be a random, uncontrolled, undesirable event. In reality, organisms possess a precisely regulated genetic program for inducing individual cells to kill themselves when appropriate. This suicide program, called **apoptosis**, is designed to prevent the accumulation of unneeded or defective cells that arise during embryonic development as well as later in life. For example, you might think that embryos would produce only the exact number of cells they need, but that is hardly the case. Embryos produce many extra cells that will not form part of the final organ or tissue in which they arise. A case in point is the human hand, which starts off as a solid mass of tissue. The fingers are then carved out of the tissue by a process in which apoptosis is invoked to destroy the cells that would otherwise form a webbing between the fingers. Apoptosis is also important in the newly forming brain, where extra nerve cells created during embryonic development are destroyed by apoptosis during early infancy as the final network of nerve connections is established.

Another function of apoptosis is to rid the body of defective cells. For example, cells infected with viruses often invoke apoptosis to trigger their own destruction, thereby limiting reproduction and spread of the virus. Cells with damaged DNA may also trigger apoptosis, especially if the damage cannot be repaired. This ability to destroy genetically damaged cells is especially useful in helping avert the development of cancer.

Apoptosis Is Carried Out by a Caspase Cascade

Apoptosis is a unique type of cell death, quite different from what happens when cells are destroyed by physical injury or exposure to certain poisons. In response to such nonspecific damage, cells undergo *necrosis*, a slow type of death in which cells swell and eventually burst, spewing their contents into the surrounding tissue. Necrosis often results in an inflammatory reaction that can cause further cell destruction, which makes it potentially dangerous.

In contrast, apoptosis kills cells in a quick and neat fashion without causing damage to surrounding tissue. The process involves a carefully orchestrated sequence of intracellular events that systematically dismantle the cell (Figure 2-13). The first observable change in a cell undergoing apoptosis is cell shrinkage. Next, small bubble-like protrusions of cytoplasm ("blebs") start forming at the cell surface as the nucleus and other cellular structures begin to disintegrate. The chromosomal DNA is then degraded into small pieces and the entire cell breaks apart, forming small fragments known as **apoptotic bodies.** Finally, the apoptotic bodies are swallowed up by neighboring cells called *phagocytes*, which are specialized for ingesting foreign matter and breaking it down into molecules that can be recycled for other purposes.

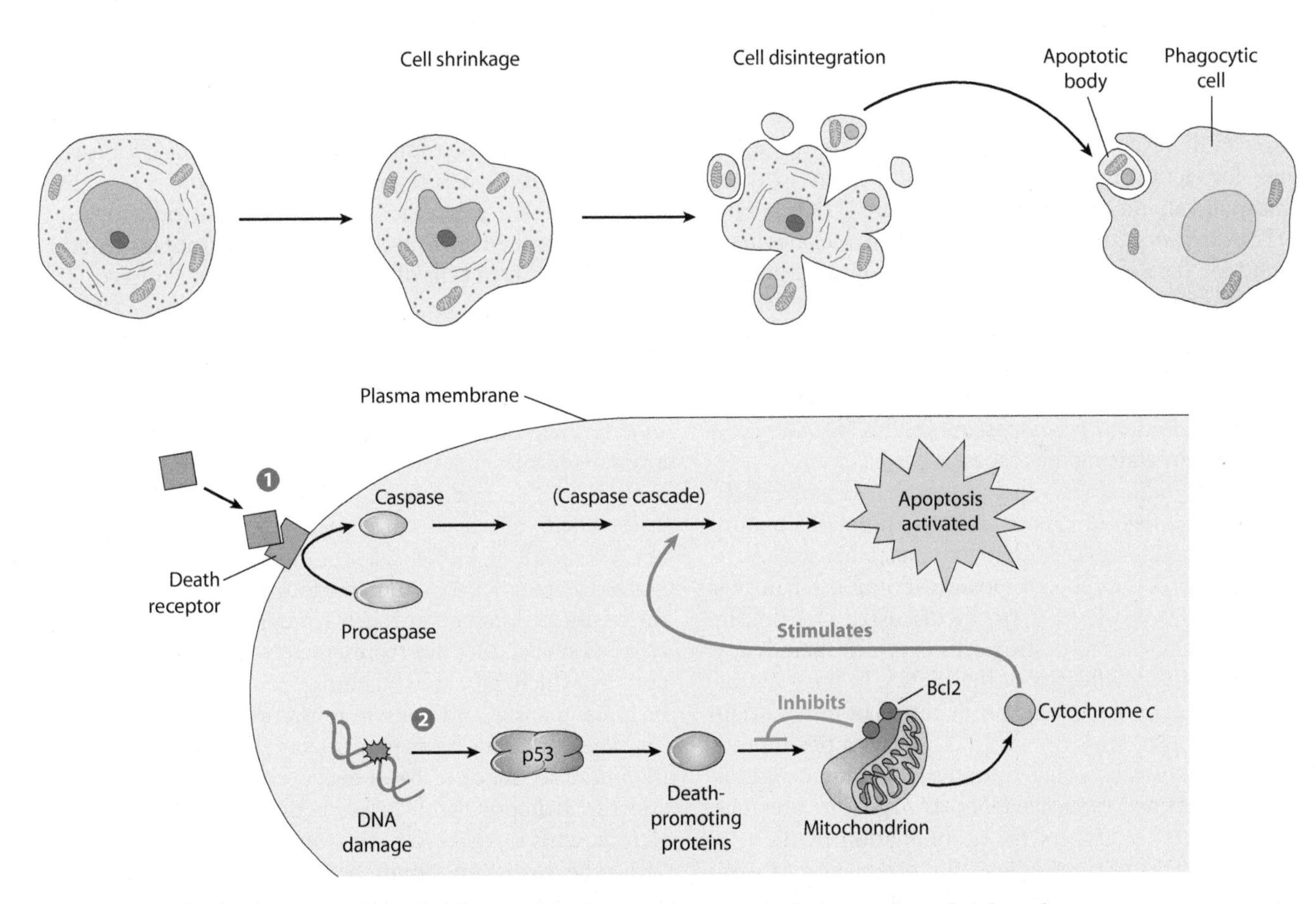

Figure 2-13 Main Steps in Apoptosis. (*Top*) As a cell begins to undergo apoptosis, its cytoplasm shrinks and bubble-like protrusions of cytoplasm form at the cell surface. The nucleus and other cellular structures then disintegrate and the entire cell breaks apart, forming small apoptotic bodies that are engulfed by neighboring phagocytic cells. (*Bottom*) The two main routes for triggering apoptosis are the external pathway and the internal pathway. ① In the external pathway, external molecules bind to death receptors on the outer surface of the targeted cell. The activated death receptors then trigger the caspase cascade. ② In the internal pathway, damaged DNA triggers accumulation of the p53 protein, which simulates the production of death-promoting proteins that alter the permeability of mitochondrial membranes. This event leads to the release of a group of mitochondrial proteins, including cytochrome *c*, that activate the caspase cascade.

Apoptosis is carried out by a series of protein-degrading enzymes known as **caspases**. Normally, caspases reside in cells in the form of inactive precursors called *procaspases*. When a cell receives a signal to commit suicide, an initiating member of the procaspase family is converted into an active caspase. The activated caspase catalyzes the conversion of another procaspase into an active caspase, which activates yet another procaspase, and so forth. Some members of this *caspase cascade* destroy key cellular proteins. For example, one caspase degrades a protein involved in maintaining the structural integrity of the nucleus, and another caspase degrades a protein whose destruction releases an enzyme that causes fragmentation of chromosomal DNA. Hence, the net effect of the caspase cascade is the activation of a series of enzymes that degrade the cell's main components, thereby leading to an orderly disassembly of the dying cell.

Cancer Cells Are Able to Evade Apoptosis

The presence of procaspases within a cell means that the cell is programmed with the seeds of its own destruction, ready to commit suicide quickly if so required. It is therefore crucial that the mechanisms employed to control caspase activation are precisely and carefully regulated and are called into play only when there is a legitimate need to destroy an unneeded or defective cell. There are two main routes for activating the caspase cascade, an external pathway and an internal pathway (see Figure 2-13, *bottom*).

The *external pathway* is employed when a cell has been targeted for destruction by other cells in the surrounding tissue. In such cases, neighboring cells produce molecules that transmit a "death signal" by binding to *death receptors* present on the outer surface of the targeted cell. The activated death receptors then interact with, and trigger activation of, initiator procaspase molecules located inside the cell, thereby starting the caspase cascade.

The *internal pathway*—a pathway that is particularly relevant to the field of cancer biology—functions mainly in the destruction of cells that have sustained extensive DNA damage. Although cells possess several mechanisms for repairing DNA damage (to be discussed shortly), in many cases it is safer to destroy cells in which there is any question about the integrity of their DNA. In this way, the potential danger posed by the proliferation of mutant cells is minimized. The p53 protein (p. 27) plays a pivotal role in the mechanism by which apoptosis is induced in cells that have sustained extensive DNA damage. The presence of damaged DNA triggers the accumulation of the p53 protein, which in turn stimulates the production of proteins that alter the permeability of mitochondrial membranes. The altered mitochondria then release a group of proteins, especially *cytochrome c*, that activate the caspase cascade and thereby cause the cell to be destroyed by apoptosis.

Given that killing defective cells is one of the main functions of apoptosis, why aren't cancer cells destroyed? After all, cancer cells fit the definition of defective cells: They grow in an uncontrolled fashion and, as you will learn shortly, possess DNA mutations and other chromosomal abnormalities. The reason cancer cells are still able to survive is that they have developed ways of avoiding apoptosis. One common mechanism is that many cancer cells have mutations that disable the gene coding for p53, thereby disrupting the main internal pathway for triggering apoptosis. As you will learn in Chapter 10, which provides a detailed description of the p53 pathway, mutations in the *p53* gene are the most common genetic defect observed in human cancers. Other genes involved in apoptosis may also be altered in cancer cells. For example, the gene coding for the *Bcl2 protein*, a naturally occurring inhibitor of apoptosis, is altered in some cancers in a way that causes too much Bcl2 to be produced, thereby blocking apoptosis.

DNA DAMAGE AND REPAIR

As an alternative to committing suicide by apoptosis, cells with damaged DNA may try to repair the damage. DNA repair is a topic of great importance to the field of cancer biology because DNA **mutations**—defined as any change in DNA base sequence—play a central role in causing the uncontrolled proliferation of cancer cells. Many of the specific genes whose mutation can foster the development of cancer will be described in Chapters 9 and 10. In this chapter, we are concerned mainly with introducing the kinds of mechanisms that generate DNA mutations and seeing why cancer cells often fail to repair the damage.

DNA Mutations Arise Spontaneously and in Response to Mutagens

In describing how mutations arise, we first need to review a few basic principles regarding DNA structure and replication. DNA molecules are constructed from two intertwined, helical strands, each consisting of a linear chain of building blocks called *nucleotides*. As was pointed out in Chapter 1, DNA contains four types of nucleotides, each with a different nitrogen-containing base. The four types of nucleotides are represented by the letters A, T, G, and C, which are abbreviations for the bases *adenine*, *thymine*, *guanine*, and *cytosine*, respectively (see Figure 1-8).

The nucleotide bases play two crucial roles in DNA: (1) the linear sequence of bases encodes genetic information, and (2) *hydrogen bonds* between the bases hold the two DNA strands together to form a **double helix** (Figure 2-14, *left*). The hydrogen bonds that hold the double helix together only fit when they are formed between the base A in one DNA strand and T in the other, or between the base G in one DNA strand and C in the other. The interaction between A and T, or between G and C, is known as **complementary base pairing**. During DNA replication, the two strands of the double helix become separated and each strand functions as a *template* that dictates the synthesis of a new complementary DNA strand using the base-pairing

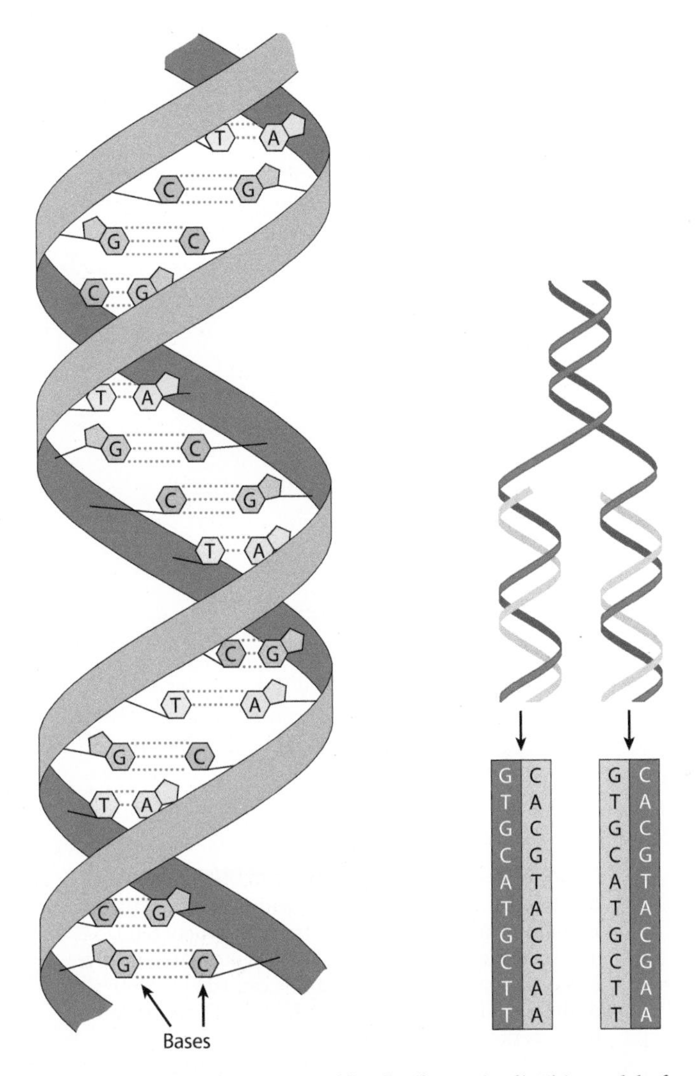

Figure 2-14 DNA Structure and Replication. (*Left*) This model of the DNA double helix shows that the two intertwined strands of the double helix are held together by hydrogen bonding between the bases adenine (A) and thymine (T), and between the bases guanine (G) and cytosine (C). (*Right*) During DNA replication, each strand acts as a template that dictates the synthesis of a new complementary DNA strand using the base-pairing rules. In other words, the base A in the template strand specifies insertion of the base T in the newly forming strand, G specifies insertion of C, T specifies insertion of A, and C specifies insertion of G.

rules. In other words, the base A in the template strand specifies insertion of the base T in the newly forming strand, G specifies insertion of C, T specifies insertion of A, and C specifies insertion of G (Figure 2-14, *right*).

Mutations are changes in DNA base sequence that arise either spontaneously or as a result of exposure to mutation-causing agents in the environment. The most common types of spontaneous mutations are caused by random interactions between DNA and the water molecules around it. For example, *depurination* is the loss of the base adenine or guanine caused by spontaneous *hydrolysis* (cleavage by water) of the bond linking the base to the DNA chain (Figure 2-15a). This bond is so susceptible to being broken that the DNA of a human cell may lose thousands of adenines and guanines every day.

Another spontaneous reaction, *deamination*, involves removal of a base's amino group (see Figure 2-15b). Deamination, which affects the bases C, A, and G, alters the base-pairing properties of the affected bases. Like depurination, deamination is usually caused by random collision of a water molecule with the bond that attaches the amino group to the base. In a typical human cell, the rate of DNA damage by this means is approximately 100 deaminations per day.

If a DNA strand with missing adenines or guanines, or with deaminated bases, is not repaired, an erroneous base sequence may be propagated when the strand serves as a template during DNA replication. For example, if the base C loses its amino group by deamination, the result is an altered structure that will pair with the base A—rather than with the correct base G—during the next round of DNA replication. The net effect is a mutation in which the base G has been replaced by the base A in the newly formed DNA strand.

In addition to spontaneous mutations, DNA errors can also be created by mutation-causing agents, or **mutagens**, in the environment. Environmental mutagens fall into two major categories: *chemicals* and *radiation*. Mutagenic chemicals alter DNA structure by a variety of mechanisms. For example, *base analogs* are chemicals resembling normal DNA bases in structure that are incorporated into DNA in their place; substances known as *base-modifying agents* react chemically with DNA bases to alter their structures; and molecules called *intercalating agents* insert themselves between adjacent bases of the double helix, thereby distorting DNA structure and increasing the chance that a nucleotide will be deleted or inserted during DNA replication.

DNA mutations can also be caused by several types of radiation. Sunlight is a strong source of *ultraviolet radiation*, which alters DNA at regions containing two adjacent *pyrimidines*—a class of single-ring bases that includes the bases C and T. In locations where two pyrimidines lie next to each other, absorption of ultraviolet radiation can trigger the formation of a covalent bond between the adjacent bases, creating a unique mutation called a **pyrimidine dimer** (see Figure 2-15c). Pyrimidine dimers distort DNA structure, thereby causing incorrect nucleotides to be inserted during DNA replication. Mutations can also be caused by X-rays and forms of radiation emitted by radioactive substances. These types of radiation are referred to as *ionizing radiation* because they remove electrons from biological molecules, thereby generating highly reactive intermediates that cause various types of DNA damage.

Mutations Involving Abnormal Bases Are Corrected by Translesion Synthesis or Excision Repair

As is perhaps not surprising for a molecule so important to an organism's health and survival, a variety of mechanisms have evolved for repairing damaged DNA. In the case of

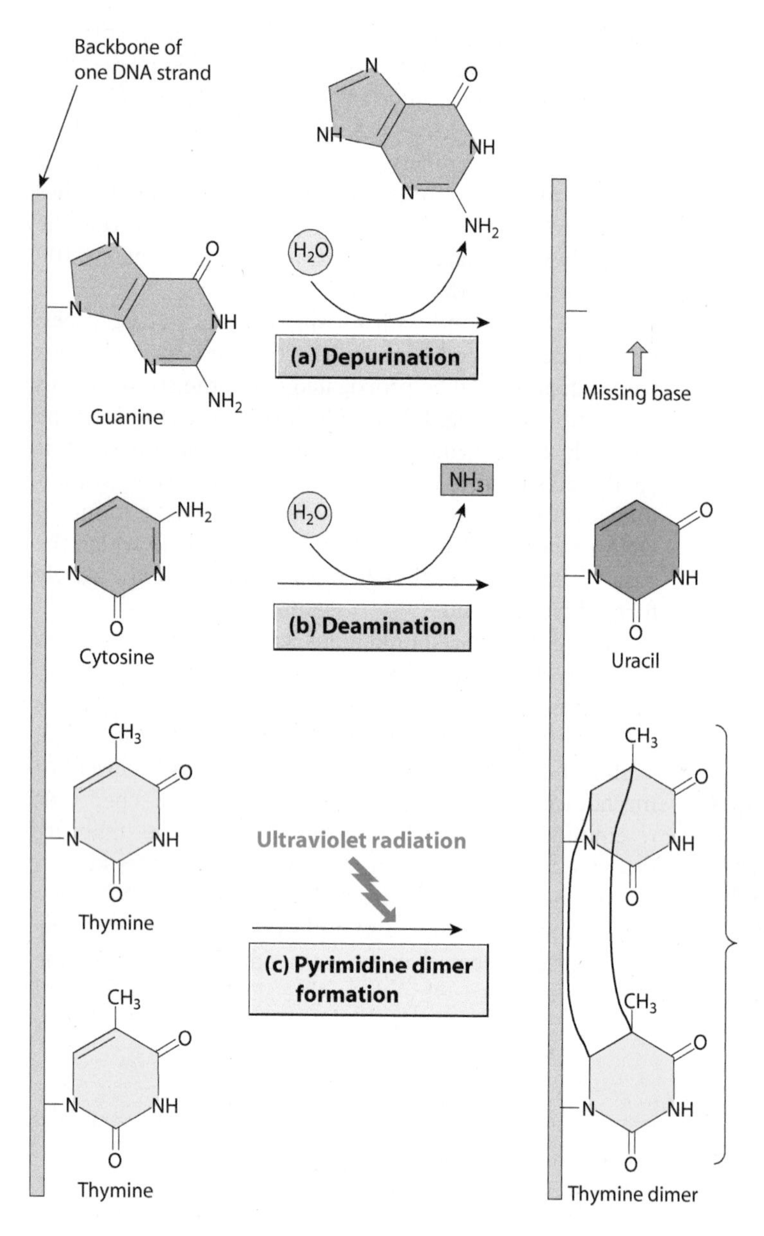

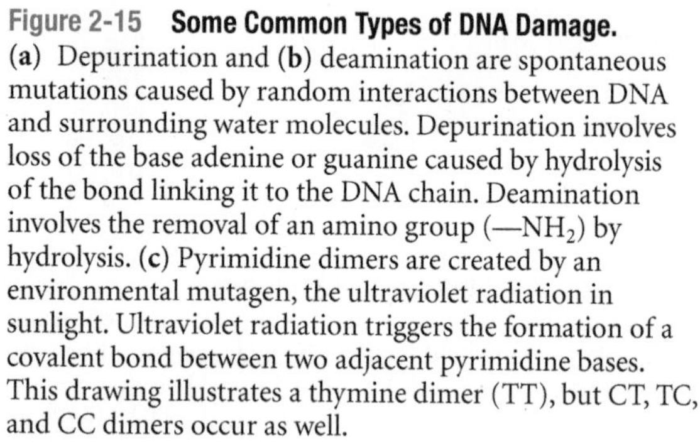

Figure 2-15 Some Common Types of DNA Damage. (**a**) Depurination and (**b**) deamination are spontaneous mutations caused by random interactions between DNA and surrounding water molecules. Depurination involves loss of the base adenine or guanine caused by hydrolysis of the bond linking it to the DNA chain. Deamination involves the removal of an amino group (—NH_2) by hydrolysis. (**c**) Pyrimidine dimers are created by an environmental mutagen, the ultraviolet radiation in sunlight. Ultraviolet radiation triggers the formation of a covalent bond between two adjacent pyrimidine bases. This drawing illustrates a thymine dimer (TT), but CT, TC, and CC dimers occur as well.

abnormal bases, the damage is sometimes handled during DNA replication using special forms of **DNA polymerase**, the enzyme that catalyzes DNA synthesis. Whereas the main types of DNA polymerase often halt when they encounter regions of damaged DNA, certain forms of the enzyme carry out **translesion synthesis**, the synthesis of new DNA across regions in which the DNA template strand is damaged. For example, a form of the enzyme called DNA polymerase η (eta) can catalyze DNA synthesis across a region containing a pyrimidine dimer composed of two covalently linked T's, correctly inserting two new A's opposite the dimer. Because the mutation is eliminated from the new strand but not from the template strand, translesion synthesis is a damage-tolerance mechanism that prevents an initial mutation from being transmitted to newly forming DNA strands.

Errors left behind after DNA replication, as well as abnormalities that subsequently arise, become the province of a group of more than a hundred different enzymes and proteins that are involved in removing and replacing abnormal nucleotides. These proteins are components of **excision repair** pathways, which correct defects using a three-step process (Figure 2-16). In step ①, the defective DNA region is removed from one strand of the double helix by enzymes called *repair endonucleases*, which are recruited by proteins that recognize sites of DNA damage. Repair endonucleases cleave the DNA backbone adjacent to the damage site, and other enzymes then facilitate removal of the defective DNA. During step ②, the excised nucleotides are replaced with the correct ones in a reaction catalyzed by DNA polymerase. The base sequence of the

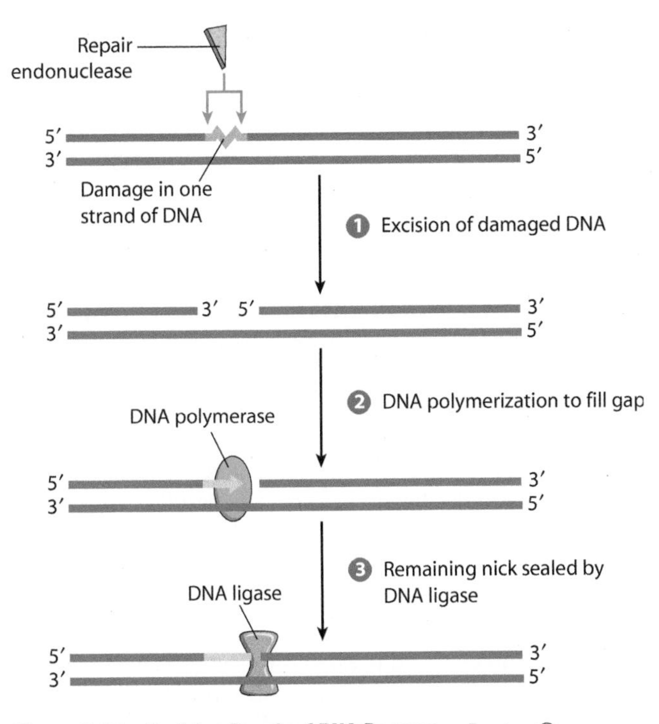

Figure 2-16 Excision Repair of DNA Damage. In step ①, damaged nucleotides are removed from one strand of a DNA double helix. During step ②, the missing nucleotides are replaced with the correct ones in a reaction catalyzed by DNA polymerase. The base sequence of the undamaged strand serves as a template to ensure insertion of the correct bases by complementary base pairing. Finally, in step ③ the enzyme DNA ligase closes the break in the repaired strand.

complementary strand serves as a template to ensure correct base insertion, just as it does during DNA replication. Finally, in step ③, an enzyme called *DNA ligase* closes the break in the repaired strand.

Excision repair pathways are of two main types. The first type, called *base excision repair*, corrects single damaged bases in DNA. For example, deaminated bases are detected by enzymes called *DNA glycosylases*, which recognize a specific deaminated base and remove it from the DNA molecule by cleaving the bond between the base and the DNA backbone. Repair endonucleases then cleave the DNA backbone and repair is completed as in steps ② and ③ above.

The other type of excision repair, known as *nucleotide excision repair* (*NER*), is used for removing pyrimidine dimers and other bulkier lesions in DNA. The NER system utilizes proteins that recognize major distortions in the DNA double helix and recruit an enzyme called an *NER endonuclease*, which makes two cuts in the DNA backbone, one on either side of the distortion. The double helix is then unwound, and the gap is filled in by DNA polymerase and sealed by DNA ligase. The NER system is the most versatile of a cell's DNA repair systems, recognizing and correcting many types of damage that cannot otherwise be repaired. The importance of NER is underscored by the plight of people who have mutations affecting this pathway. For example, a disease called *xeroderma pigmentosum* is caused by an inherited mutation in any of seven genes coding for components of the NER system. As we will see in Chapter 8, individuals with xeroderma pigmentosum exhibit an extremely high risk of developing skin cancer because they cannot repair the DNA damage that is caused by exposure to the ultraviolet radiation in sunlight.

Mutations Involving Noncomplementary Base Pairs Are Corrected by Mismatch Repair

Another type of DNA damage needing repair involves bases that, although structurally normal, are inappropriately paired with each other. Such errors frequently arise during DNA replication. For example, let us consider the fate of a single AT base pair. Suppose that during DNA replication the base A becomes incorrectly paired with the base C in one of the newly forming strands (rather than pairing with the correct base, T). The result is an abnormal AC base pair. If the DNA molecule is allowed to replicate again without first repairing the mismatched AC base pair, the resulting mutation becomes very difficult to correct. Figure 2-17 illustrates why: When the DNA molecule containing the mismatched AC base pair is allowed to replicate (step ② in Figure 2-17), the incorrect base (C) will serve as a template for its complement (G), yielding a new DNA double helix with a CG base pair. Hence, the net result is a CG base pair where an AT base pair had originally been located. Because a CG base pair is a properly matched base pair, DNA repair mechanisms would not recognize anything abnormal and the mutation would now be replicated indefinitely. Thus, mismatched base pairs (such as AC) need to be repaired quickly before subsequent rounds of DNA replication yield mutations that are difficult to detect and repair.

Fortunately, the presence of abnormally paired bases in DNA can be detected and corrected by a process known as **mismatch repair**. To operate properly, mismatch repair must solve a problem that puzzled biologists for many years: How is the incorrect member of an abnormal base pair distinguished from the correct member? Unlike the situation in excision repair, neither of the bases in a mismatched pair exhibits any structural alteration that would allow it to be recognized as an abnormal base. The pair is simply composed of two normal bases that are inappropriately paired with each other, such as A paired with C (step ① in Figure 2-17). If the incorrect member of an AC base pair were the base C and the repair system instead removed and "corrected" the base A, the repair system would create a permanent mutation instead of correcting a mismatched base pair! One way of solving this problem is to carry out mismatch repair in close conjunction with DNA replication, when the new DNA strand can be

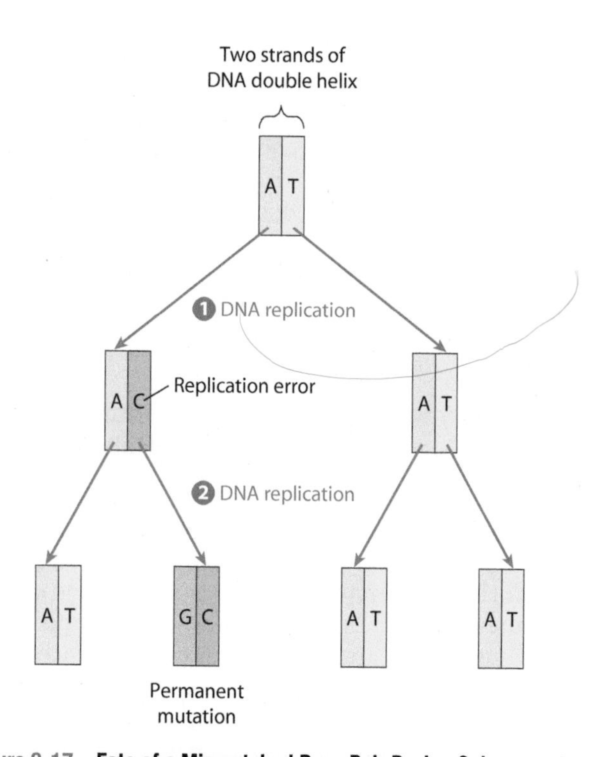

Figure 2-17 Fate of a Mismatched Base Pair During Subsequent Rounds of DNA Replication. This example focuses on the fate of a mismatched AC base pair created during the first round of DNA replication (step ①). If this error is not corrected by mismatch repair, the incorrect base in this pair (the base C) will serve as a template for its complement (the base G) during the second round of DNA replication (step ②), thereby creating a DNA molecule with a CG base pair where an AT base pair had originally been located. Since DNA repair mechanisms would not recognize anything abnormal about a CG base pair, the mutation will now be replicated indefinitely.

recognized because it is still in the process of being synthesized. The mismatch repair system excises the new strand in any region containing a mismatch and then replaces the missing segment, using the base sequence of the original strand as template (just as in steps ② and ③ of excision repair, shown in Figure 2-16).

A striking illustration of the importance of mismatch repair is provided by *hereditary nonpolyposis colon cancer* (*HNPCC*), an inherited disease in which people exhibit an abnormally high risk of developing colon cancer. The reason for the increased cancer risk in HNPCC is an inherited mutation in one of the genes involved in mismatch repair. Since people normally inherit two copies of every gene (one from each parent), inheriting a single defective gene is usually not enough to disrupt mismatch repair because the second copy of the gene is functional. However, if the remaining, functional copy of the mismatch repair gene subsequently becomes mutated in just a single proliferating cell, that cell will become deficient in mismatch repair and will give rise to cells that accumulate mutations at a higher rate than normal. The ensuing buildup of mutations makes cells more susceptible to becoming cancerous because, as you will learn in Chapter 10, most forms of cancer are associated with the accumulation of multiple mutations.

Double-Strand DNA Breaks Are Repaired by Nonhomologous End-Joining or Homologous Recombination

The repair mechanisms described thus far—excision and mismatch repair—are effective in correcting DNA damage involving chemically altered or incorrect bases. In such cases a "cut-and-patch" pathway removes the damaged or incorrect nucleotides from one DNA strand, and the resulting gap is filled using the intact strand as template. Certain types of damage, such as double-strand breaks in the DNA double helix, cannot be handled this way, however. Repair is more difficult for double-strand breaks because with other types of DNA damage, one strand of the double helix remains undamaged and can serve as a template for aligning and repairing the defective strand. In contrast, double-strand breaks completely cleave the DNA double helix into two separate fragments; the repair machinery is therefore confronted with the problem of identifying the correct two fragments and rejoining their broken ends without losing any nucleotides.

Two main pathways are employed in such cases. One, called **nonhomologous end-joining**, uses a set of proteins that bind to the ends of the two broken DNA fragments and join them together (Figure 2-18, *left*). Unfortunately, this mechanism is error-prone because it cannot prevent the loss of nucleotides from the broken ends and has no way of ensuring that the correct two DNA fragments are being joined to each other. A more precise method for fixing double-strand breaks, called **homologous recombination**, exploits the fact that cells generally possess two copies of each chromosome; if the DNA molecule in one chromosome incurs a double-strand break, another intact copy of the chromosomal DNA is still available to serve as a template for guiding the repair of the broken chromosome (see Figure 2-18, *right*).

As in the case of excision and mismatch repair, inherited defects that disrupt the repair of double-strand breaks can increase a person's risk of developing cancer. For example, women who inherit mutations in genes called *BRCA1* or *BRCA2* incur a high risk for breast and ovarian cancer. As we will see in Chapter 10, the *BRCA1* and *BRCA2* genes produce proteins involved in the pathway that repairs double-strand breaks by homologous recombination.

Cancer Cells Are Genetically Unstable and Often Exhibit Gross Chromosomal Abnormalities

The number of mutations accumulated by cancer cells is generally greater than would be expected in comparable populations of normal cells. This tendency to accumulate an excessive number of mutations and other kinds of

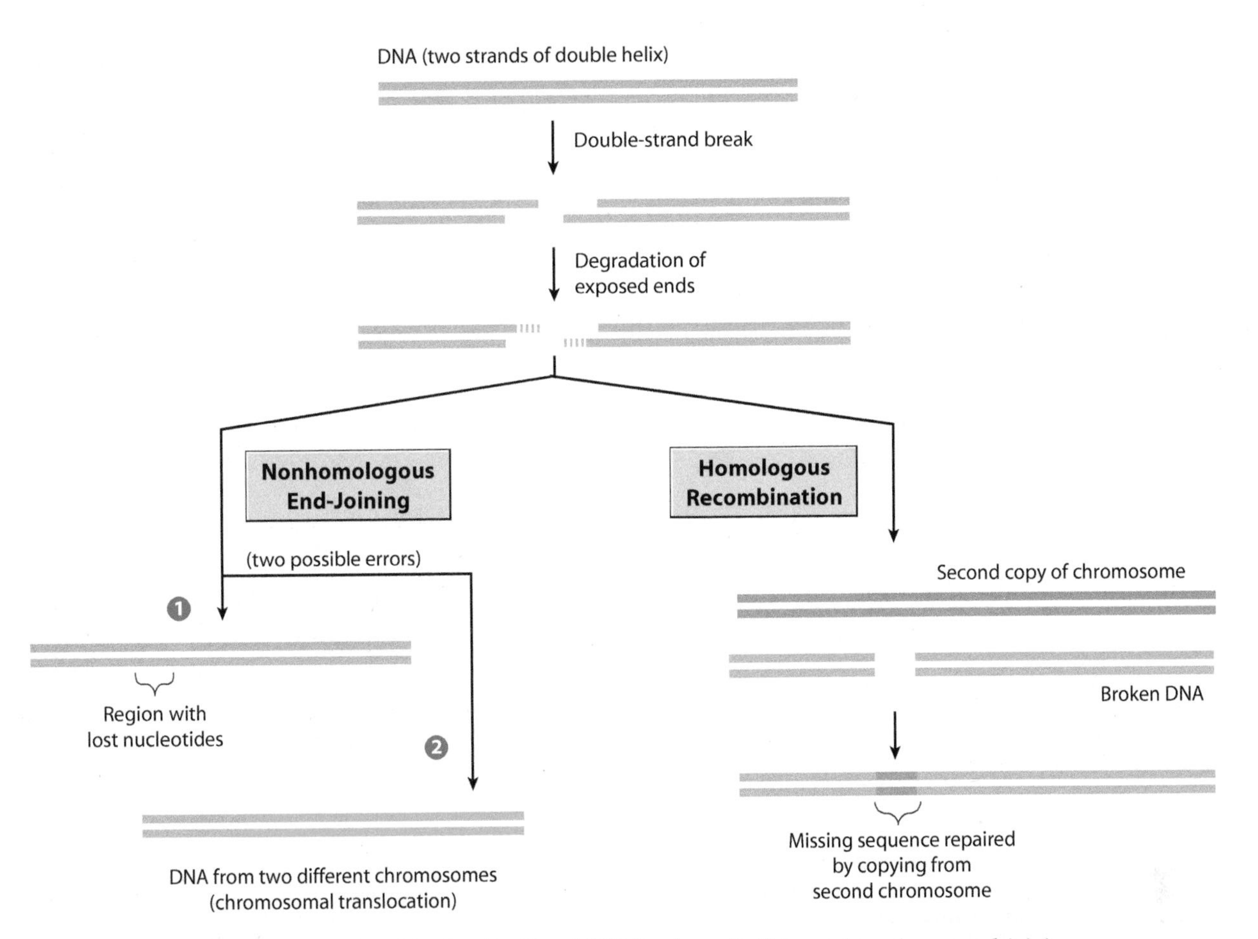

Figure 2-18 Two Approaches for Repairing Double-Strand DNA Breaks. (*Left*) In nonhomologous end-joining, double-strand breaks are repaired by directly joining the broken ends together. This approach is error-prone, however, because it cannot prevent ① loss of nucleotides from the broken ends or ② joining together two fragments from different chromosomes. (*Right*) Homologous recombination is a more precise mechanism of repair that takes advantage of the fact that cells generally possess two copies of each chromosome. If a DNA molecule in one chromosome incurs a double-strand break, another intact copy of the DNA molecule is available to serve as a template for guiding repair of the broken DNA.

DNA damage is called **genetic instability**. As we have just seen, one reason for genetic instability is that cancer cells may exhibit defects in DNA repair that diminish their ability to correct DNA mutations. The net result is elevated mutation rates that can be hundreds or even thousands of times higher than normal. Another factor is that cancer cells often exhibit defects in the pathways that trigger apoptosis in response to DNA damage. As a result, cells that have incurred extensive DNA damage that is beyond repair do not self-destruct by apoptosis as would normally be expected.

Mistakes in the handling of chromosomes during cell division also contribute to genetic instability by creating gross abnormalities in chromosome structure and number. Normally, human cells other than sperm and eggs possess 23 pairs of chromosomes, or a total of 46 chromosomes per cell. Such cells are said to be **diploid** (from the Greek word *diplous*, meaning "double") because two copies of each type of chromosome are present, one derived from each parent. In contrast, cancer cells are often **aneuploid**, which means that they possess an abnormal number of chromosomes. Aneuploidy usually involves both the loss of some chromosomes and extra copies of others (see Figure 10-14).

In addition to an abnormal *number* of chromosomes, cancer cells often possess chromosomes whose structure has been altered by *deletions* (loss of long stretches of DNA) and *translocations* (exchange of long stretches of DNA between different chromosomes). One of the first chromosomal abnormalities to be consistently observed in any type of cancer was the **Philadelphia chromosome**, an oddly shaped chromosome present in the cancer cells of nearly 90% of all individuals with chronic myelogenous leukemia (Figure 2-19). The Philadelphia chromosome is produced by DNA breakage near the ends of chromosomes 9 and 22, followed by reciprocal exchange of DNA between the two chromosomes. A similar phenomenon is observed in Burkitt's lymphoma, a cancer of human lymphocytes, in which segments derived from chromosomes 8 and 14 are exchanged. We will see in Chapter 9 that in both of these situations, scientists have identified the specific gene whose alteration by chromosomal translocation leads to cancer.

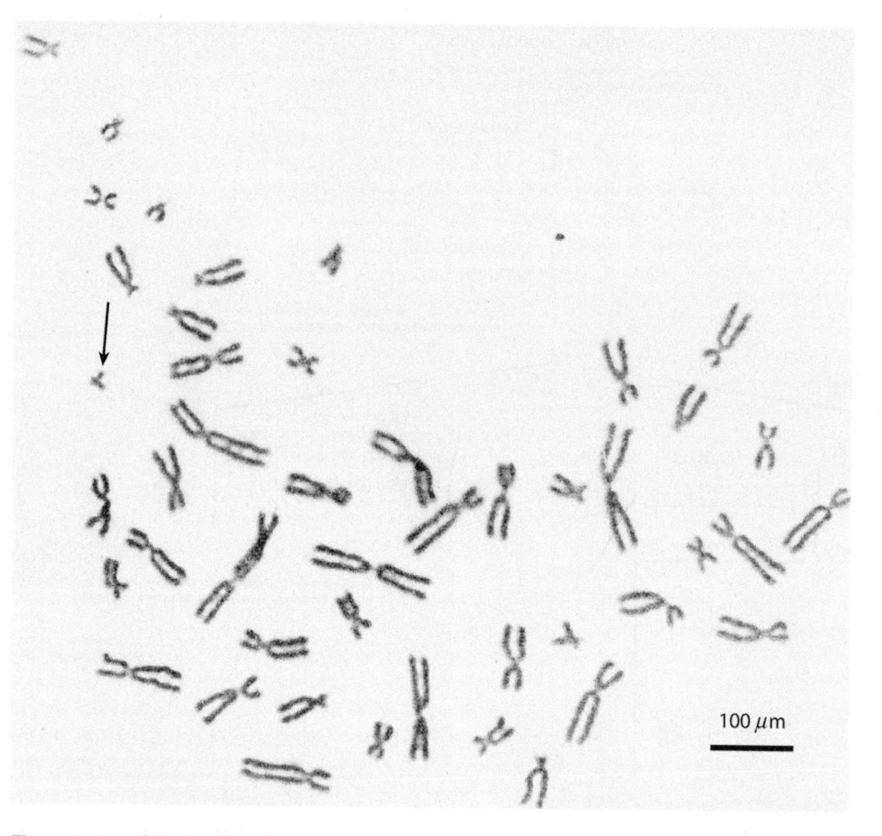

Figure 2-19 Philadelphia Chromosome. This set of chromosomes was present in a cancer cell obtained from a patient with chronic myelogenous leukemia. The arrow points to the Philadelphia chromosome, a tiny remnant of chromosome 22 that remains after exchange of genetic material with chromosome 9. [Courtesy of P. C. Nowell.]

Abnormalities in chromosome number and structure contribute to cancer development in various ways: Some crucial genes may be lost, other genes may become overactive, and yet other genes may become structurally altered and produce abnormal products. In contrast to mutations involving short stretches of DNA, gross chromosomal defects are extremely difficult, if not impossible, to repair. The best solution for cells exhibiting such chromosomal problems is suicide by apoptosis. However, if cells have genetic defects that disrupt the pathways responsible for carrying out apoptosis, self-inflicted suicide is not an option and the genetically damaged cells will continue to proliferate.

DNA Mutations Can Lead to Cancer

The presence of gross chromosomal defects in cancer cells was first reported almost a hundred years ago, and the smaller changes in base sequence that occur in the DNA of cancer cells have also been recognized for many years. For a long time, it was difficult to distinguish whether such DNA abnormalities are responsible for causing cancer or whether they simply represent secondary changes that arise because cancer cells are dividing in a rapid, uncontrolled fashion. In other words, the question was much like the classic problem of "which came first, the chicken or the egg?" Do DNA abnormalities cause cancer to arise, or does cancer cause DNA abnormalities to arise?

This issue was finally resolved in the early 1980s, when DNAs isolated from several different human cancers were shown to cause cancer under laboratory conditions. In the first studies of this type (Figure 2-20), DNA was extracted from human bladder cancer tissue and applied to a culture of normal mouse cells under experimental conditions that favor incorporation of the foreign DNA into the cells' chromosomes. The uptake of foreign DNA by cells under such artificial laboratory conditions is called **transfection**. In response, some of the cultured mouse cells began to proliferate excessively. When these proliferating cells were injected back into mice, the animals developed cancer. Similar experiments using DNA extracted from normal human tissues did not produce mouse cells capable of forming tumors. It was therefore concluded that DNA obtained from human bladder cancer contains gene sequences, not present in normal DNA, that are capable of causing cancer.

Similar results were subsequently obtained using DNA isolated from other human cancers. In Chapter 9, we will see that such studies have facilitated the identification of a number of specific genes that contribute to cancer development.

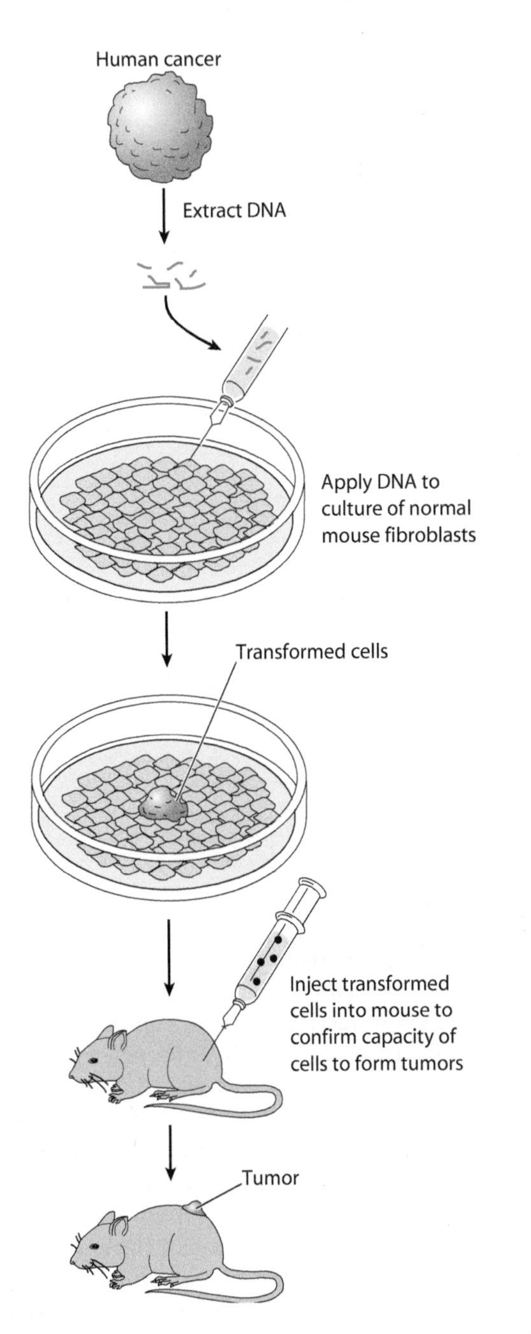

Figure 2-20 Inducing Cancer by DNA Transfection. When DNA that has been extracted from certain kinds of human cancer cells is introduced into normal mouse cells growing in culture, some of the mouse cells begin to proliferate excessively. When these cells are injected back into mice, they develop into malignant tumors.

TUMOR IMMUNOLOGY

The ability of cancer cells to proliferate in an uncontrolled fashion, combined with their capacity to spread through the body, makes them a potentially lethal hazard. Are any of the body's normal defense mechanisms capable of protecting against such a threat? The immune system is designed to defend against infection by potentially harmful agents, such as bacteria, viruses, fungi, parasites, and it also attacks foreign tissues and cells. Is the immune system also capable of recognizing cancer cells, and if so, how do cancer cells frequently manage to thrive despite the immune system? In addressing these questions, let us start by reviewing the basic mechanisms involved in an immune response.

Immune Responses Are Carried Out by B Lymphocytes, T Lymphocytes, and NK Cells

Molecules capable of provoking an immune response are referred to as **antigens**. To function as an antigen, a substance must be recognized as being "foreign"—that is, different from molecules normally found in a person's body. The more a molecule differs in structure from normal tissue constituents, the greater the intensity of the immune response mounted against that substance.

Antigens also need to be susceptible to degradation and processing by cells. This requirement explains why nondegradable foreign materials, such as stainless steel pins and plastic valves, can be surgically implanted into humans without eliciting an immune response. To trigger an efficient immune response, an antigen must be degraded and processed by specialized **antigen-presenting cells** that "present" antigens to cells of the immune system in a way designed to activate an immune response.

Macrophages and *dendritic cells* are among the most commonly encountered antigen-presenting cells. As shown in Figure 2-21, antigens engulfed by these cells are degraded into small fragments that eventually become bound to cell surface proteins called **major histocompatibility complex (MHC)** molecules. When an antigen fragment bound to an MHC molecule is present at the surface of an antigen-presenting cell, the MHC-antigen complex stimulates cells called **lymphocytes** to mount an attack against that particular antigen. The stimulated lymphocytes attack foreign antigens in two different ways. One group of lymphocytes, called **B lymphocytes**, produce proteins called **antibodies**, which circulate in the bloodstream and penetrate into extracellular fluids, where they bind to the foreign antigen that induced the immune response. Another group of lymphocytes, called **cytotoxic T lymphocytes (CTLs)**, bind to cells exhibiting foreign antigens on their surface and kill the targeted cells by causing them to burst.

In addition to B and T lymphocytes, a small fraction of the total lymphocyte population consists of *natural killer (NK) cells* that possess the intrinsic ability to recognize and kill certain kinds of tumor cells (as well as virus-infected cells). In contrast to a typical immune response involving antibody formation or cytotoxic T lymphocytes, NK cells do not need to recognize a specific antigen before attacking a target cell. Instead, they are programmed to attack a broad spectrum of abnormal cells in a relatively indiscriminate fashion, while leaving normal cells unharmed.

Some Cancer Cells Possess Antigens That Trigger an Immune Response

Because so many people develop cancer, it is clear that the ability of NK cells to attack tumors is often overwhelmed. A more powerful and efficient immune response requires

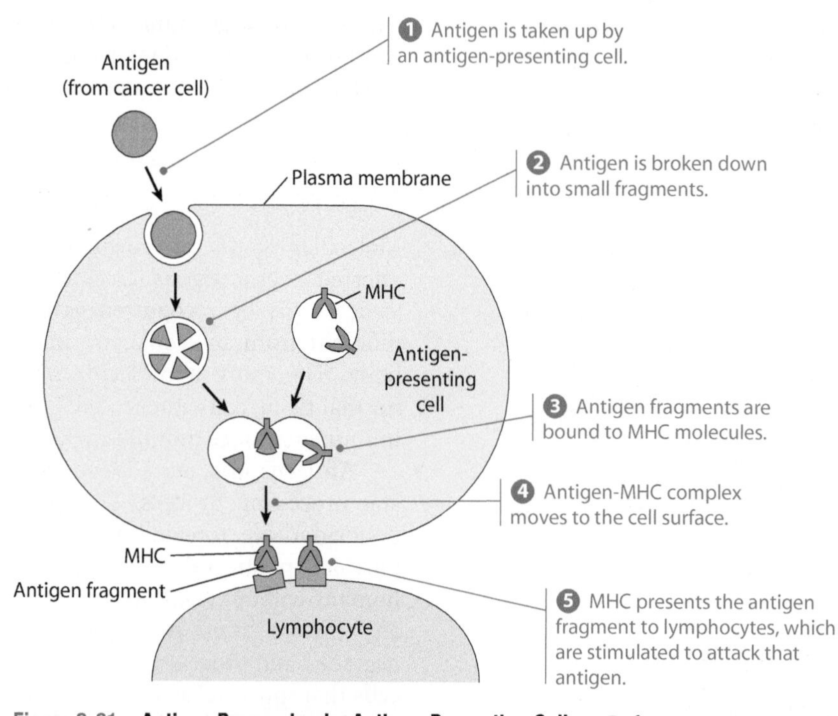

Figure 2-21 Antigen Processing by Antigen-Presenting Cells. Before an antigen can elicit an efficient immune response, it must be processed and presented to lymphocytes of the immune system by antigen-presenting cells such as macrophages and dendritic cells. If the antigen is derived from a cancer cell, the cytotoxic T lymphocytes activated in step ⑤ will selectively attack cancer cells containing that antigen.

the participation of cytotoxic T lymphocytes, whose killing activity is selectively directed at cells containing a specific antigen. For cytotoxic T lymphocytes to become involved, however, they need to recognize an antigen as being foreign or abnormal.

The question of whether cancer cells exhibit unique antigens that can elicit such an immune response has had a long and controversial history. Part of the difficulty in reaching a consensus arises because cancers exhibit a variety of antigenic changes. In human melanomas, for example, at least three classes of antigens have been detected. Antigens of the first type are specific both for melanomas and for the person from whom a particular melanoma is obtained. Antigens of the second type are specific for melanomas but not for the particular person who has the melanoma. Antigens of the two preceding types cannot be detected in normal cells and are therefore examples of *tumor-specific antigens*. Antigens of the third type are present in both normal and melanoma cells, although their concentration in melanoma cells is greater. Such antigens, which are present in higher concentration in a tumor but are not unique to tumors, are more accurately referred to as *tumor-associated antigens*.

The distinction between tumor-specific and tumor-associated antigens is sometimes difficult to make. For example, a group of molecules called *MAGE antigens* are expressed in melanomas and several other cancers but not in most normal tissues. The MAGE antigens are therefore close to being "tumor-specific," and an immune response directed against them would be expected to be reasonably selective, causing minimum damage to normal tissues. Other tumor-associated antigens, such as the *prostate-specific antigen* (*PSA*) produced by prostate cancer cells, are unique to a specific tissue. Although antigens of this type are produced by normal cells as well, an immune response directed against them would be relatively selective in that it is only directed against a single tissue.

Antigens that are genuinely tumor-specific occur in cancers that produce structurally abnormal proteins. Such proteins do not appear in normal cells and thus can be recognized by the immune system as being "foreign." In Chapters 9 and 10, we will encounter several examples of mutant cancer cell genes that produce abnormal proteins. These molecules can act as antigens that elicit a highly selective, cytotoxic T cell response against the tumor cells in which they are found, provided that the mutant proteins are processed and presented to the immune system in the appropriate fashion.

The Immune Surveillance Theory Postulates That the Immune System Is Able to Protect Against Cancer

The existence of tumor-specific antigens raises an interesting question: Why don't people with cancer reject their own tumors? The **immune surveillance theory** postulates that immune destruction of newly forming cancer cells is in fact a routine event in healthy individuals, and cancer simply reflects the occasional failure of an adequate immune response to be mounted against aberrant cells.

The validity of this theory has been debated for many years, with various kinds of evidence being cited both for and against it.

Some of the evidence involves organ transplant patients who take *immunosuppressive* drugs, which depress immune function and thereby decrease the risk of immune rejection of the transplanted organ. As would be predicted by the immune surveillance theory, individuals treated with immunosuppressive drugs develop many cancers at higher rates than normal (Figure 2-22). Although this finding appears to support the idea that the immune system normally helps prevent cancers from developing, it is also possible that the immunosuppressive drugs are acting directly to trigger the development of cancer. For example, *cyclosporin*—one of the most effective and commonly used immunosuppressive drugs—has been shown to stimulate the proliferation and motility of isolated cancer cells growing in culture. These results indicate that direct effects of immunosuppressive drugs on newly forming cancer cells may contribute to the increased tumor growth observed in individuals taking such drugs.

A more direct approach for evaluating the immune surveillance theory involves the use of animals that have been genetically altered to introduce specific defects in the immune system. One study of this type employed mutant mice containing disruptions in *Rag2*, a gene expressed only in lymphocytes. The mutant mice, which produce no functional lymphocytes, were found to develop cancer more frequently than do normal mice. An increased cancer risk was observed both for cancers that arise spontaneously and for cancers that were induced by injecting animals with a cancer-causing chemical. Such results indicate that a normally functioning immune system helps protect mice against the development of cancer.

Nonetheless, the question still exists as to the relevance of these findings to human cancers. If the immune system plays a significant role in protecting humans from common cancers, you would expect to see a dramatic increase in overall cancer rates in AIDS patients with severely depressed immune function. While people with AIDS do exhibit higher rates for a few types of cancer, especially Kaposi's sarcoma and lymphomas, increased rates for the more common forms of cancer have not been observed. Most of the cancers that do occur in higher rates in AIDS patients are known to be caused by viruses (see Figure 7-5). Such observations suggest that immune surveillance may play an important role in protecting humans from virally induced cancers but that it is less effective in preventing the more common forms of cancer.

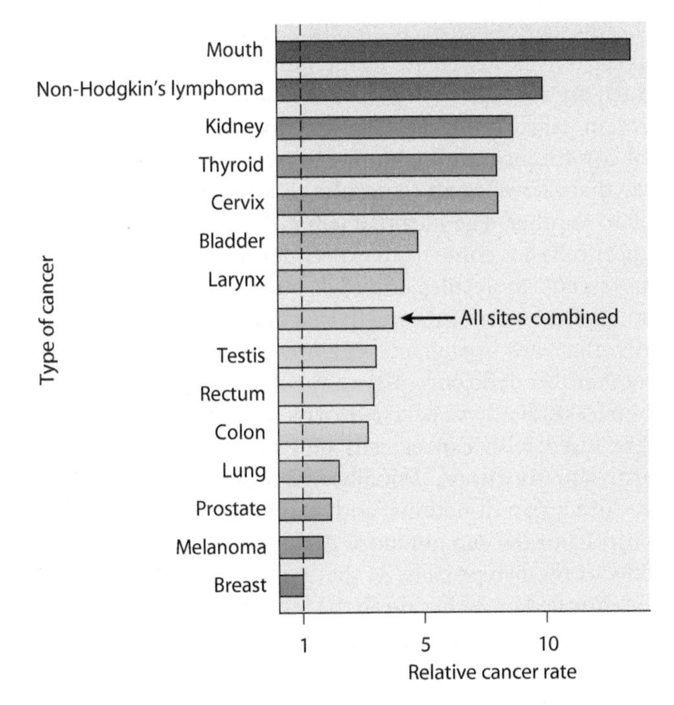

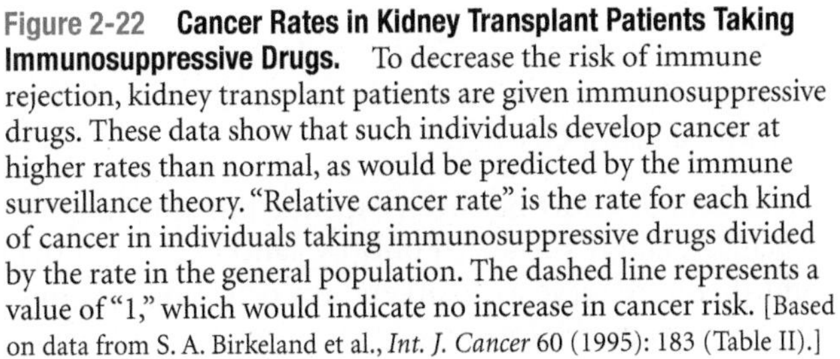
Figure 2-22 Cancer Rates in Kidney Transplant Patients Taking Immunosuppressive Drugs. To decrease the risk of immune rejection, kidney transplant patients are given immunosuppressive drugs. These data show that such individuals develop cancer at higher rates than normal, as would be predicted by the immune surveillance theory. "Relative cancer rate" is the rate for each kind of cancer in individuals taking immunosuppressive drugs divided by the rate in the general population. The dashed line represents a value of "1," which would indicate no increase in cancer risk. [Based on data from S. A. Birkeland et al., *Int. J. Cancer* 60 (1995): 183 (Table II).]

Cancer Cells Have Various Ways of Evading the Immune System

Based on the large number of people who develop cancer each year, it is clear that tumors routinely find ways of evading destruction by the immune system. One mechanism is based on *tumor progression*, which refers to the gradual changes in the makeup of cancer cell populations that occur over time as natural selection favors the survival of cells that are more aggressive and aberrant. During tumor progression, cells containing antigens that elicit a strong immune response are most likely to be attacked and destroyed. Conversely, cells that either lack or produce smaller quantities of antigens marking them for destruction are more likely to survive and proliferate. So as tumor progression proceeds, there is a continual selection for cells that invoke less of an immune response.

Cancer cells have also devised ways of actively confronting and overcoming the immune system. For example, some cancer cells produce molecules that kill T lymphocytes or disrupt their ability to function. Tumors may also surround themselves with a dense layer of supporting tissue that shields them from immune attack. And some cancer cells simply divide so quickly that the immune system cannot destroy them fast enough to keep tumor growth in check. Consequently, the larger a tumor grows, the easier it becomes to overwhelm the immune system.

Although tumors are often successful at evading immune attack, immune rejection is not necessarily an unattainable objective. Experiments in mice have shown that immunizing an animal with tumor antigens can trigger an effective immune response under conditions in which the tumor growing in the animal had not elicited any response on its own. Observations like this one suggest that it may be possible to stop or even prevent the development of cancer by stimulating a person's immune system to attack cancer cells. Current approaches for pursuing such goals will be discussed more fully in Chapter 11.

OTHER MOLECULAR CHANGES

This chapter has now introduced most of the distinctive features of cancer cells that are related to cell proliferation and survival. In addition to these behavioral traits, cancers exhibit numerous alterations in the concentrations of individual proteins. Some of these alterations involve unique or abnormal proteins that are not ordinarily encountered in adult tissues, but most of the molecular changes seen in cancer cells are increases or decreases in the quantity or activity of normal proteins. In concluding this chapter, we will examine a few such changes that are especially noteworthy.

Cancer Cells Exhibit Cell Surface Alterations That Affect Adhesiveness and Cell-Cell Communication

Alterations in the makeup of the outer cell membrane (*plasma membrane*) are almost universally observed in cancer cells, yet it is often difficult to assess the significance of such changes because similar alterations occur in normal cells when they begin to proliferate. For example, membrane transport proteins responsible for the uptake of sugars, amino acids, and other nutrients are frequently activated in tumor cells, but similar increases are observed in normal cells that have been stimulated to divide by adding nutrients or growth factors. It is therefore important to distinguish between membrane changes that play unique roles in cancer cells and membrane changes that are characteristic of dividing cells in general.

Among the cell surface changes that are particularly distinctive and important for the behavior of cancer cells are those that influence adhesiveness. In normal tissues, cell-cell adhesion helps keep cells in their place; in cancers, this adhesiveness is diminished or missing entirely. The reduced adhesiveness of cancer cells can often be traced to defects in *E-cadherin*, a cell-cell adhesion protein that is located at the cell surface. In Chapter 3, we will see how defects in cell-cell adhesion facilitate invasion and metastasis by allowing cancer cells to wander away from one another and invade into surrounding tissues.

A second cell surface property altered in cancer cells is their enhanced tendency to clump together when exposed in the laboratory to proteins called **lectins**. A lectin is a carbohydrate-binding protein possessing two or more carbohydrate-binding sites, which means that a single lectin molecule can link two cells together by binding to carbohydrate groups exposed on the surface of each cell. As a result, when lectins are added to a suspension of isolated cells, the lectin molecules link the cells together to form large clumps (Figure 2-23). It was initially thought

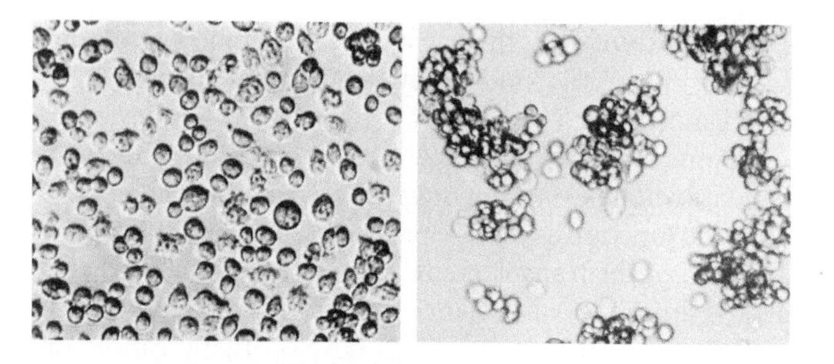

Figure 2-23 Behavior of Normal and Cancer Cells Exposed to Lectin Molecules. (*Left*) Normal cells exposed to a lectin protein called concanavalin A tend to remain separate from one another. (*Right*) Cancer cells exposed to concanavalin A exhibit an enhanced tendency to adhere to each other and form clumps. [Courtesy of L. Sachs.]

that the increased susceptibility to lectin-induced clumping meant that cancer cells possess more cell surface carbohydrate groups to which lectins can bind. Careful measurements, however, have led to the conclusion that the total number of these carbohydrate groups tends to be similar in normal and cancer cells. What does appear to differ is the mobility of carbohydrate groups within the plasma membrane, which tends to be greater in cancer cells than in normal cells. This ability of cell surface carbohydrate groups to move more readily in cancer cells apparently increases the rate at which they can bind to added lectins.

Another cell surface alteration commonly observed in cancer cells is a decrease in the number of **gap junctions**, which are specialized cell surface structures composed of a protein called *connexin*. Gap junctions play a role in cell-cell communication by joining adjacent cells together in a way that allows small molecules to pass directly from one cell to another. The idea that gap junctions are deficient in cancer cells has come from experiments showing that small fluorescent molecules injected into normal cells move rapidly into surrounding cells that are normal but not into cells that are malignant (Figure 2-24). To investigate whether this deficiency plays any role in the loss of growth control, studies have been performed in which normal cells were fused with cancer cells that had lost the ability to form gap junctions. Initially, the resulting hybrid cells produced gap junctions and exhibited normal growth control, but the gap junctions eventually disappeared from some of the hybrid cells. At that point the cells reverted to uncontrolled growth, raising the interesting possibility that normal growth control is influenced by the ability of cells to communicate through gap junctions.

Cancer Cells Produce Embryonic Proteins, Proteases, and Stimulators of Blood Vessel Growth

A great deal of effort has been expended in searching for molecules that are produced only by cancer cells and that might therefore serve as "markers" for detecting the presence of cancer. Unfortunately, few of the molecules

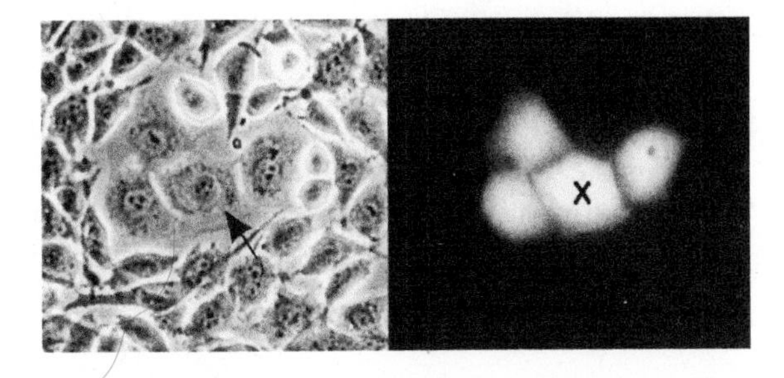

Figure 2-24 Evidence for Decreased Cell-Cell Communication in Cancer Cells. (*Left*) This photograph shows a group of cultured cells in which four normal liver cells are surrounded by numerous cancer cells. The arrow points to where a fluorescent dye was injected into one of the four normal cells. (*Right*) When the same cells are subsequently viewed with ultraviolet light to visualize the fluorescent dye, it can be seen that the dye molecules have diffused from the injected cell (marked with an "X") into the three adjacent normal cells but not into the surrounding cancer cells. This decrease in cell-cell communication is caused by a loss of gap junctions in the cancer cells. [Courtesy of W. R. Loewenstein.]

identified thus far are broadly useful as unique identifiers of cancer cells, although a number of them can be employed for detecting the presence of specific kinds of cancer. For example, some cancer cells manufacture and secrete proteins that are usually found only in embryos. One such protein, *alpha-fetoprotein*, is produced by embryonic liver cells but is detectable in only trace amounts in normal adults. In people with liver cancer, the concentration of alpha-fetoprotein in the blood increases dramatically. *Carcinoembryonic antigen (CEA)*, a protein produced in the embryonic digestive tract, and fetal hormones such as *chorionic gonadotropin* and *placental lactogen*, are also secreted by some cancers. Blood tests for embryonic markers such as alpha-fetoprotein and carcinoembryonic antigen can therefore be used to monitor the presence of certain kinds of cancer, but the fact that these substances are made by only a few tumor types limits the applicability of this approach.

Other proteins produced by cancer cells are not unique to such cells but have provided some important insights into the behavior of malignant tumors. For example, cancer cells tend to produce *proteases* (protein-degrading enzymes) that facilitate the breakdown of structures that would otherwise represent barriers to cancer cell movement and invasion. Although proteases are also secreted by certain kinds of normal cells, their enhanced production by cancer cells facilitates the ability of malignant tumors to invade surrounding tissues and enter the circulatory system. Cancer cells also produce proteins that stimulate the growth of blood vessels, thereby helping ensure that tumors have a sufficient blood supply. In the next chapter, we will see that proteases and stimulators of blood vessel growth play crucial roles in allowing tumors to grow to a large size and spread through the body.

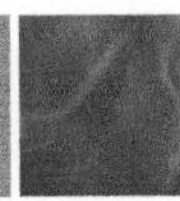

Summary of Main Concepts

- Cancer cells exhibit a distinctive collection of abnormal traits, many related to uncontrolled cell proliferation and survival, that together constitute the "profile" of a typical cancer cell.
- Isolated cancer cells can be identified by their ability to form tumors when injected into immunologically deficient mice. Cancer cells grown in culture are anchorage-independent, exhibit a decreased susceptibility to density-dependent inhibition of growth, and proliferate indefinitely because they possess mechanisms for replenishing telomere sequences.
- Cell proliferation is normally controlled by growth factors, which bind to cell surface receptors that trigger intracellular signaling pathways leading to cell growth and division. These signaling pathways act in part by fostering the production and activation of Cdk-cyclin complexes. One type of Cdk-cyclin phosphorylates the Rb protein, which controls passage through the restriction point and into S phase. Cells will not normally divide unless they are stimulated by an appropriate growth factor, but this restraint is circumvented in cancer cells by various mechanisms that create a constant activation of growth signaling pathways, even in the absence of growth factors.
- Checkpoint pathways monitor the cell for DNA damage, DNA replication, and chromosome attachment to the spindle, and transiently halt the cell cycle if conditions are not suitable for continuing. Cancer cells exhibit defects in checkpoint pathways as well as a loss of restriction point control. As a result, cancer cells continue proliferating under conditions in which the cell cycle of normal cells would stop.
- Apoptosis is a cell death pathway that is invoked for destroying unneeded or defective cells. During apoptosis, cells are destroyed by protein-degrading enzymes called caspases. The p53 protein plays a key role in triggering apoptosis in cells that have incurred DNA damage. In cancer cells, defects in p53 or other components of the apoptotic pathway allow the cells to evade apoptosis.
- Mutations play a central role in triggering the uncontrolled proliferation of cancer cells. Mutations are changes in the base sequence of DNA that arise

spontaneously or in response to mutagenic chemicals or radiation. Translesion synthesis of DNA and excision repair are used to correct mutations involving abnormal bases, mismatch repair is employed to repair mutations involving mismatched bases, and nonhomologous end-joining and homologous recombination are used to repair double-strand DNA breaks. Cancer cells often exhibit defects in these DNA repair mechanisms, allowing cells to accumulate mutations at rates that are significantly higher than normal.

- Cancer cells frequently exhibit gross chromosomal abnormalities, including deleted chromosome segments, chromosomal translocations, and the loss of some chromosomes and extra copies of others. DNA isolated from the chromosomes of some human cancer cells has been shown to cause cancer when transferred into normal cells under laboratory conditions.

- Certain cancers exhibit tumor-specific antigens that are capable of eliciting an immune response. The immune surveillance theory postulates that such an immune response can protect people from developing cancer. Many cancer cells, however, have devised ways of evading immune attack.

- Other molecular changes detected in cancer cells include a decreased production of proteins involved in cell-cell adhesion and gap junction formation, and an increased production of embryonic proteins, proteases, and proteins that stimulate blood vessel growth.

Key Terms for Self-Testing

Features of Cancer Cell Proliferation

density-dependent inhibition of growth (p. 18)
anchorage-dependent (p. 19)
anchorage-independent (p. 19)
extracellular matrix (p. 19)
anoikis (p. 19)
clone (p. 19)
telomere (p. 21)
telomerase (p. 22)

Growth Factors and the Cell Cycle

growth factor (p. 22)
platelet-derived growth factor (PDGF) (p. 22)
epidermal growth factor (EGF) (p. 22)
receptor (p. 22)
plasma membrane (p. 22)
G1 phase (p. 22)
S phase (p. 22)
G2 phase (p. 22)
M phase (p. 22)
cell cycle (p. 23)
G0 phase (G zero) (p. 23)
mitosis (p. 23)
cytokinesis (p. 23)
interphase (p. 23)
restriction point (p. 23)
cyclin-dependent kinase (Cdk) (p. 23)
protein kinase (p. 23)
phosphorylation (p. 23)
cyclin (p. 24)
protein phosphatase (p. 24)
Rb protein (p. 26)
checkpoint (p. 26)
DNA replication checkpoint (p. 27)
spindle checkpoint (p. 27)
DNA damage checkpoint (p. 27)
p53 protein (p. 27)

Apoptosis and Cell Survival

apoptosis (p. 28)
apoptotic bodies (p. 29)
caspase (p. 30)

DNA Damage and Repair

mutation (p. 30)
double helix (p. 30)
complementary base pairing (p. 30)
mutagen (p. 31)
pyrimidine dimer (p. 31)
DNA polymerase (p. 32)
translesion synthesis (p. 32)
excision repair (p. 32)
mismatch repair (p. 33)
nonhomologous end-joining (p. 34)
homologous recombination (p. 34)
genetic instability (p. 35)
diploid (p. 35)
aneuploid (p. 35)
Philadelphia chromosome (p. 35)
transfection (p. 36)

Tumor Immunology

antigen (p. 37)
antigen-presenting cell (p. 37)
major histocompatibility complex (MHC) (p. 37)
lymphocyte (p. 37)
B lymphocyte (p. 37)
antibody (p. 37)
cytotoxic T lymphocyte (CTL) (p. 37)
immune surveillance theory (p. 38)

Other Molecular Changes

lectin (p. 40)
gap junction (p. 40)

Suggested Reading

Cancer Cell Proliferation, Growth Factors, and the Cell Cycle

Bedi, A., and M. B. Kastan. The Cell Cycle and Cancer. In *Clinical Oncology* (M. D. Abeloff, ed.). New York: Churchill Livingstone, 2000, Chapter 2.

Blagosklonny, M. V., and A. B. Pardee. Exploiting cancer cell cycling for selective protection of normal cells. *Cancer Research* 61 (2001): 4301.

Blasco, M. A. Telomerase beyond telomere. *Nature Reviews Cancer* 2 (2002): 627.

Frisch, S. M., and R. A. Screaton. Anoikis mechanisms. *Curr. Opinion Cell Biol.* 13 (2001): 555.

Masters, J. R. HeLa cells 50 years on: the good, the bad and the ugly. *Nature Reviews Cancer* 2 (2002): 315.

Neumann, A. A., and R. R. Reddel. Telomere maintenance and cancer—look, no telomerase. *Nature Reviews Cancer* 2 (2002): 879.

Ruddon, R. W. Biochemistry of Cancer. In *Holland-Frei Cancer Medicine*, 5th ed. (R. C. Bast et al., eds.). Lewiston, NY: Decker, 2000, Chapter 7.

Ruoslahti, E., and J. C. Reed. Anchorage dependence, integrins, and apoptosis. *Cell* 77 (1994): 477.

Apoptosis and Cell Survival

Igney, F. H., and P. H. Krammer. Death and anti-death: tumour resistance to apoptosis. *Nature Reviews Cancer* 2 (2002): 277.

Lockshin, R. A., and Z. Zakeri. Programmed cell death and apoptosis: origins of the theory. *Nature Reviews Molecular Cell Biology* 2 (2001): 545.

DNA Damage and Repair

Friedberg, E. C. DNA damage and repair. *Nature* 421 (2003): 436.

Lindahl, T., and R. D. Wood. Quality control by DNA repair. *Science* 286 (1999): 1897.

Rowley, J. D. Chromosome translocations: dangerous liaisons revisited. *Nature Reviews Cancer* 1 (2001): 245.

Weinberg, R. A. A molecular basis of cancer. *Sci. Amer.* 249 (May 1983): 126.

Tumor Immunology

Shankaran, V., et al. IFNγ and lymphocytes prevent primary tumour development and shape tumour immunogenicity. *Nature* 410 (2001): 1107.

Smyth, M. J., D. I. Godfrey, and J. A. Trapani. A fresh look at tumor immunosurveillance and immunotherapy. *Nature Immunol.* 2 (2001): 293.

Trapani, J. A. Tumor-mediated apoptosis of cancer-specific T lymphocytes—reversing the "kiss of death"? *Cancer Cell* 2 (2002): 169.

Van den Eynde, B. J., and P. van der Bruggen. T cell defined tumor antigens. *Current Opinion Immunol.* 9 (1997): 684.

Other Molecular Changes

Sidransky, D. Emerging molecular markers of cancer. *Nature Reviews Cancer* 2 (2002): 210.

3 How Cancers Spread

When a doctor tells a person that he or she has cancer, one of the most important questions that needs to be addressed is whether the cancer cells are still confined to their initial location. Once a tumor has invaded neighboring tissues and begins to spread to other regions of the body, it becomes more difficult to treat. The term *metastasis* refers to the spread of cancer cells via the bloodstream or lymphatic system to distant sites, where they form secondary tumors—called *metastases*—that are not physically connected to the primary tumor. Because they can arise in almost any vital organ, metastases rather than primary tumors are responsible for most cancer deaths.

Scientists have expended much effort in studying the cellular properties responsible for metastasis with the hope that a better understanding of the mechanisms involved will eventually lead to better treatments. Such studies have revealed that metastasis is a complex process that can be subdivided into several distinct stages, each involving a different set of cellular traits and interactions. Among the earliest steps in converting a tiny localized mass of cancer cells into an invasive, metastasizing tumor is the growth of blood vessels that penetrate into the tumor, supplying nutrients and oxygen to the cancer cells and removing waste products. In the absence of such a network of blood vessels, tumors cannot grow beyond a few millimeters in size and would not be a major health hazard. Since the development of a sustaining network of blood vessels is therefore a crucial step in converting a tiny group of cancer cells into a larger tumor capable of spreading to distant sites, this chapter will begin with a discussion of blood vessel growth and will then cover the topics of tumor invasion and metastasis. ■

TUMOR ANGIOGENESIS

To survive and grow, all body tissues require a continual supply of oxygen and nutrients accompanied by the removal of carbon dioxide and other waste products. These needs are met by a system of blood vessels comprised of *arteries* that carry blood from the heart to the rest of the body, *veins* that carry blood from the body back toward the heart, and tiny *capillaries* that connect the smallest arteries and veins. The wall of a capillary is only a single cell layer thick, so oxygen and nutrients carried in the bloodstream can easily diffuse through capillary walls and nourish the surrounding tissues, and carbon dioxide and other waste products produced by tissues diffuse back into the capillaries for removal from the body.

Like the cells of any other tissue, tumor cells require a network of blood vessels to perform these same tasks. The vessels that feed and sustain tumors are produced by **angiogenesis**, a term that refers to the process by which new blood vessels sprout and grow from pre-existing vessels in the surrounding normal tissues. To understand how a tumor causes surrounding tissues to provide it with such a growing network of blood vessels, we first need to describe the process of normal angiogenesis and the factors that control it.

Angiogenesis Is Prominent in Embryos but Relatively Infrequent in Adults

Angiogenesis is a normal biological event that occurs at specific times for specific purposes. For example, a developing embryo in a mother's womb must create the vast network of arteries, veins, and capillaries that are needed for a mature circulatory system. To initiate blood vessel formation, the embryo first creates a primary population of cells, called **endothelial cells**, that form the inner lining of blood vessels. As part of this process of *vasculogenesis*, the newly created endothelial cells are organized into a primitive network of channels representing the major blood vessels of the circulatory system. Once the primordial network of vessels has been created, angiogenesis takes over. Angiogenesis involves an extensive phase of growth and proliferation of pre-existing endothelial cells, which form buds that sprout from existing vessels and develop into an interconnected network of new vessels (Figure 3-1).

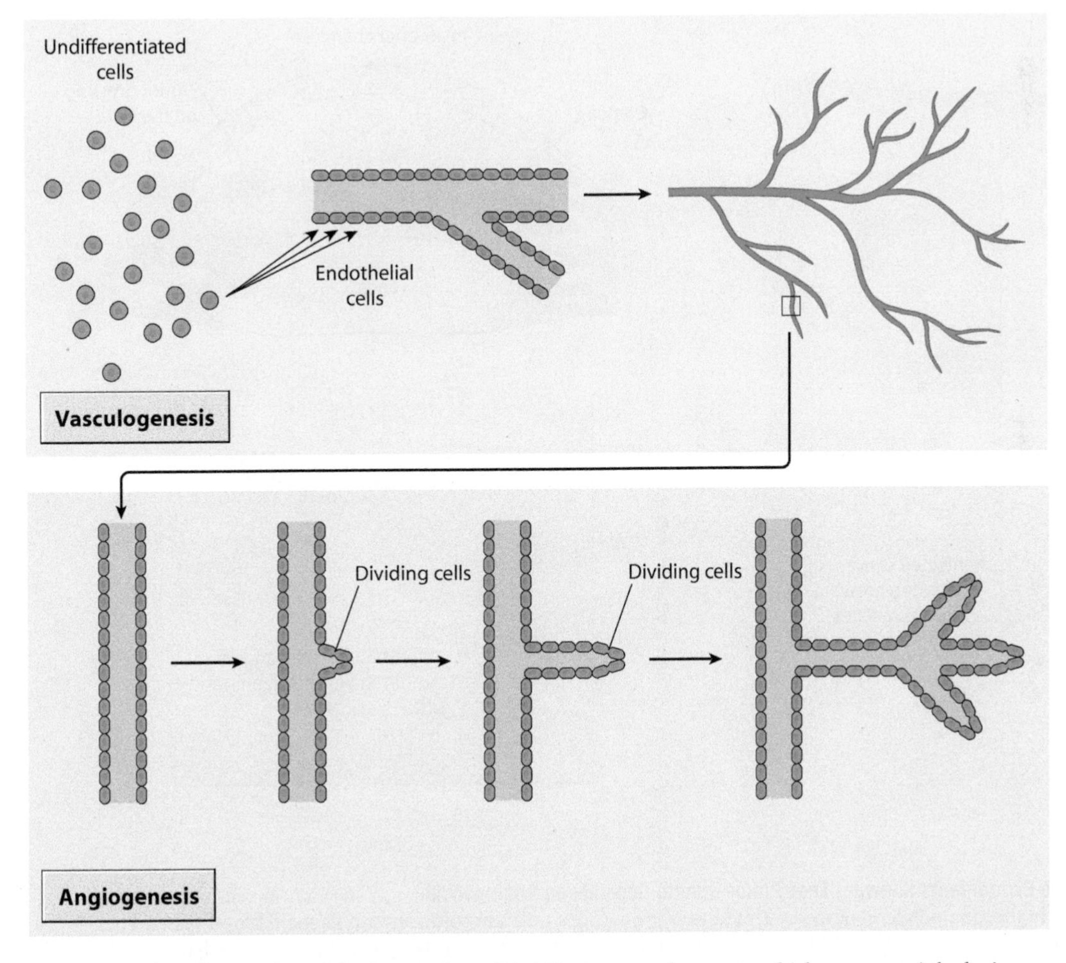

Figure 3-1 Vasculogenesis and Angiogenesis. (*Top*) During vasculogenesis, which occurs mainly during embryonic development, undifferentiated cells are converted into endothelial cells that organize themselves into a network of channels representing the major blood vessels. (*Bottom*) Angiogenesis refers to the growth and proliferation of the endothelial cells that line the inner surface of existing blood vessels, forming buds that sprout from the vessel wall and develop into new vessels.

Although vasculogenesis is restricted to early embryonic development, angiogenesis continues to occur after birth when additional blood vessels are required. In adults, who have a fully formed circulatory system, this need for new vessels is limited to a few special situations and endothelial cells rarely divide, doing so about once every three years on average. When new vessels are required, however, endothelial cell division can be stimulated and angiogenesis will take place. For example, blood vessel growth is needed each month in the inner lining of the uterus as part of the normal menstrual cycle. Angiogenesis is therefore activated a few days each month in the uterine lining of women of reproductive age. In both males and females, angiogenesis is also called upon anytime an injury requires new blood vessels for wound healing and tissue repair.

Angiogenesis needs to be precisely regulated in such cases, turning on for a short time and then stopping. Regulation is accomplished through the use of both activator and inhibitor molecules. Normally the inhibitors predominate, blocking angiogenesis. When a need for new blood vessels arises, angiogenesis activators increase in concentration and inhibitors decrease, triggering the proliferation of endothelial cells and the formation of new vessels. As we will see shortly, many of these regulatory molecules were first identified through the study of angiogenesis triggered by cancer cells.

Angiogenesis Is Required for Tumors to Grow Beyond a Few Millimeters in Diameter

For more than 100 years, scientists have known that tumors are supplied with a dense network of blood vessels. Some investigators initially believed that these blood vessels were pre-existing vessels that had expanded in response either to the increased metabolic activity of tumors or to toxic products that tumors release. Others thought that the vessels were new structures formed as part of an inflammatory response designed to defend the host against the tumor. Then in 1971, Judah Folkman

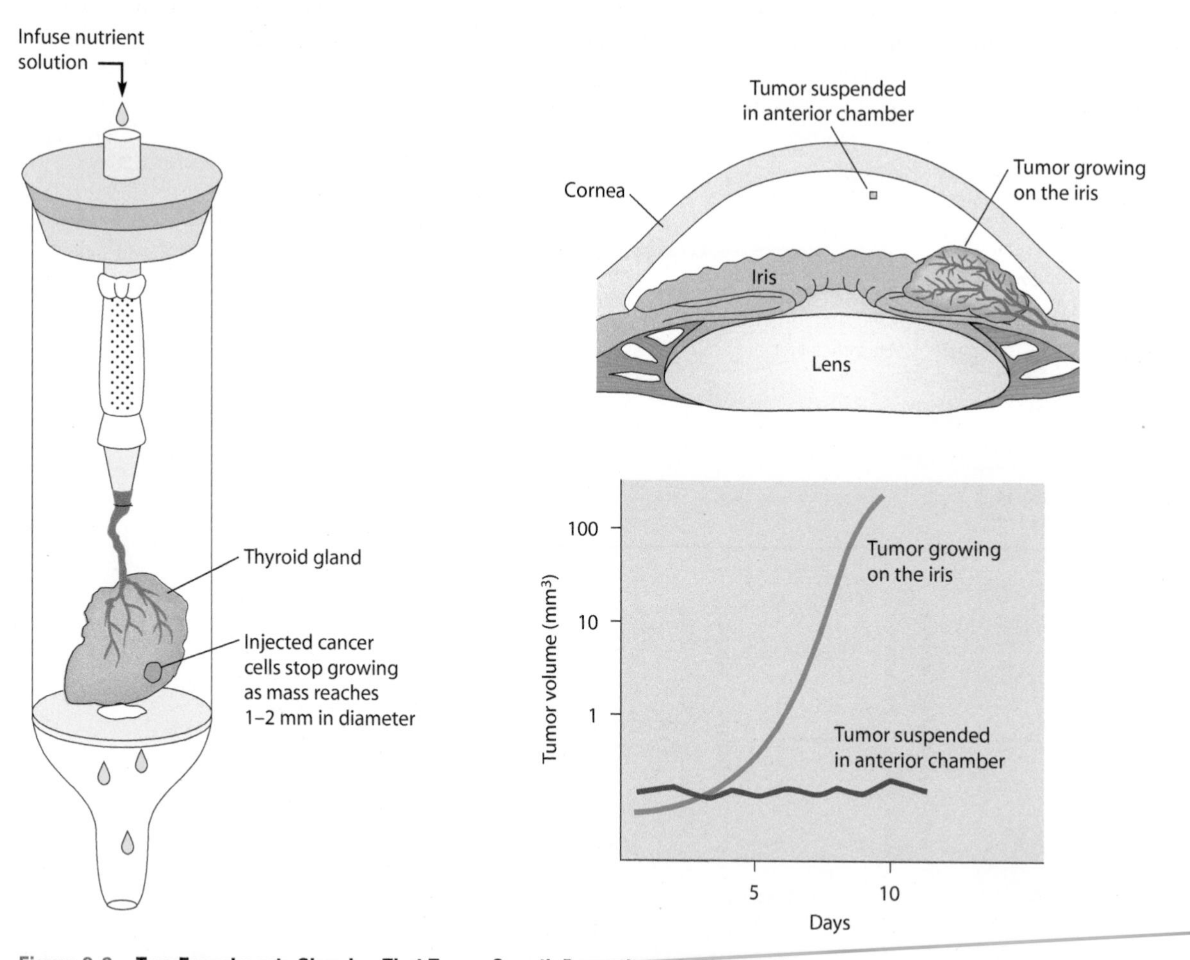

Figure 3-2 **Two Experiments Showing That Tumor Growth Depends on Angiogenesis.** (*Left*) Cancer cells were injected into an isolated rabbit thyroid gland that was kept alive by pumping a nutrient solution into its main blood vessel. The tumor cells fail to link up to the organ's blood vessels and the tumor mass stops growing when it reaches a diameter of roughly 1 to 2 millimeters. (*Right*) Cancer cells were either injected into the liquid-filled anterior chamber of a rabbit's eye, where there are no blood vessels, or were placed directly on the iris. Tumor cells in the anterior chamber, nourished solely by diffusion, remain alive but stop growing before the tumor mass reaches 1 millimeter in diameter. In contrast, blood vessels quickly infiltrate the cancer cells implanted on the iris, allowing the tumors to grow to thousands of times their original mass. [Adapted from J. Folkman, *Sci. Amer.* 234 (May 1976): 58.]

proposed a radical new idea regarding the significance of blood vessels in tumor development. He suggested that tumors release signaling molecules that trigger the growth of new blood vessels in the surrounding host tissues and that these new vessels are required to sustain tumor growth.

This concept was initially based on experiments in which cancer cells were grown in isolated organs under artificial laboratory conditions. In one such experiment, illustrated in Figure 3-2 (*left*), a normal thyroid gland was removed from a rabbit and placed in a glass chamber. A small number of cancer cells were then injected into the gland and a nutrient solution was pumped into the organ to keep it alive. The cancer cells divided for a few days but suddenly stopped when the tumor mass reached a diameter of 1 to 2 millimeters. Virtually every tumor stopped growing at exactly the same size, suggesting that some kind of limitation allowed them to grow only so far.

When tumor cells were removed from the thyroid gland and injected back into animals, cell proliferation resumed and massive tumors developed. Why did the tumors stop growing at a tiny size in the isolated thyroid gland and yet grow in an unrestrained fashion in live animals? On closer examination, a possible explanation became apparent. The tiny tumors, alive but dormant in the isolated thyroid gland, had failed to link up to the organ's blood vessels; as a result, the tumors stopped growing when they reached a diameter of 1 to 2 mm. When injected into live animals, these same tumors became infiltrated with blood vessels and grew to an enormous size.

To test the theory that blood vessels are needed to sustain tumor growth, Folkman implanted tumor cells in the anterior chamber of a rabbit's eye, where there is no blood supply. As shown in Figure 3-2 (*right*), cancer cells placed in this location survived and formed tiny tumors, but blood vessels from the nearby iris could not reach the cells and the tumors quickly stopped growing. When the same cells were implanted directly on the iris tissue, blood vessels from the iris quickly infiltrated the tumor cells and the mass of each tumor grew to thousands of times its original size. Once again, it appeared that tumors must trigger development of a blood supply before they can grow beyond a tiny mass. This process by which cancer cells stimulate the development of a blood supply is called **tumor angiogenesis** (Figure 3-3).

Angiogenesis Is Controlled by the Balance Between Angiogenesis Activators and Inhibitors

If tumors require blood vessels to sustain tumor growth, how do they ensure that this need is met? The first hint came from studies in which cancer cells were placed inside a chamber surrounded by a filter possessing tiny pores through which cells cannot pass (Figure 3-4). When such chambers are implanted into animals, new capillaries begin to proliferate in the surrounding host tissue. In contrast, normal cells placed in the same type of chamber do not stimulate blood vessel growth. The most straightforward interpretation is that cancer cells produce molecules that diffuse through the tiny pores in the filter and activate angiogenesis in the surrounding host tissue.

The next job was to identify the molecules responsible for stimulating angiogenesis, a task that occupied many investigators over a span of several decades. This intensive effort eventually led to the identification of more than a

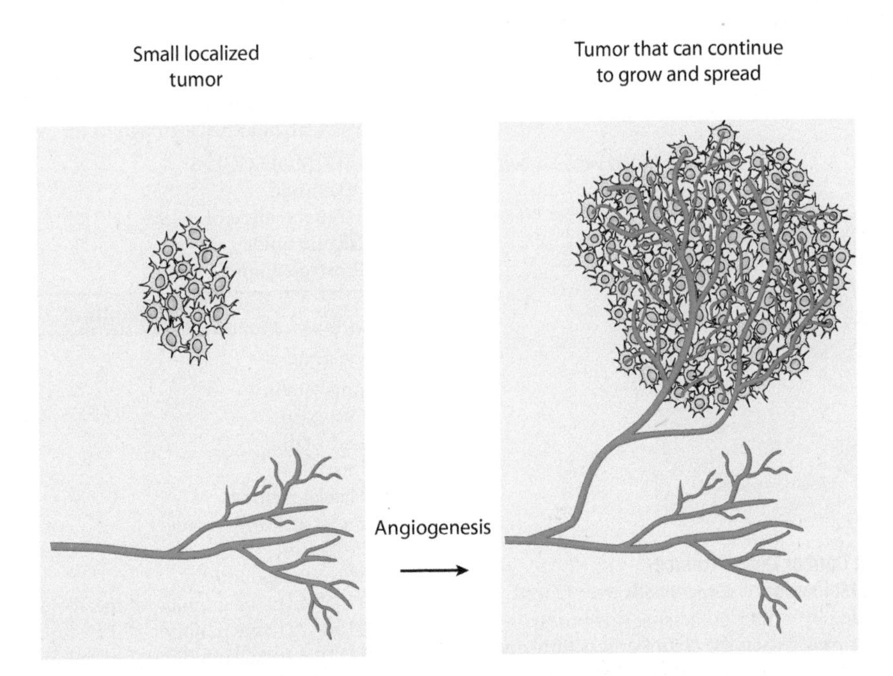

Figure 3-3 Tumor Angiogenesis. In the absence of their own blood supply, tumors remain as small, localized masses that do not generally grow beyond 1 to 2 millimeters in diameter.

dozen proteins, as well as several smaller molecules, that can activate angiogenesis (Table 3-1). Two of the proteins, **vascular endothelial growth factor** (**VEGF**) and **fibroblast growth factor** (**FGF**), appear to be especially important for sustaining tumor growth. VEGF and FGF are produced by many kinds of cancer cells (and certain types of normal cells), and they trigger angiogenesis by binding to specific receptor proteins located on the surface of endothelial cells.

To see how this process works, let us briefly focus on VEGF (Figure 3-5). VEGF is produced by the majority of tumors and is secreted into the surrounding tissues. When VEGF molecules encounter an endothelial cell, they bind to and activate **VEGF receptors** located on the endothelial cell surface. The signal is then relayed from the activated receptors to a sequential pathway of signal transduction proteins that trigger changes in cell behavior and gene expression. As a result, endothelial cells begin to proliferate and to produce **matrix metalloproteinases** (**MMPs**), a group of protein-degrading enzymes that are released into the surrounding tissue. The MMPs break down components of the *extracellular matrix* that fills the spaces between neighboring cells (p. 19), thereby allowing the endothelial cells to migrate into the surrounding tissues. As they migrate, the proliferating endothelial cells become organized into hollow tubes that evolve into new networks of blood vessels.

Although many tumors produce VEGF or FGF, they are not the sole explanation for the activation of angiogenesis. For angiogenesis to proceed, these molecules must overcome the effects of angiogenesis *inhibitors* that normally restrain the growth of blood vessels. More than a dozen naturally occurring inhibitors of angiogenesis have been identified (see Table 3-1), including the proteins

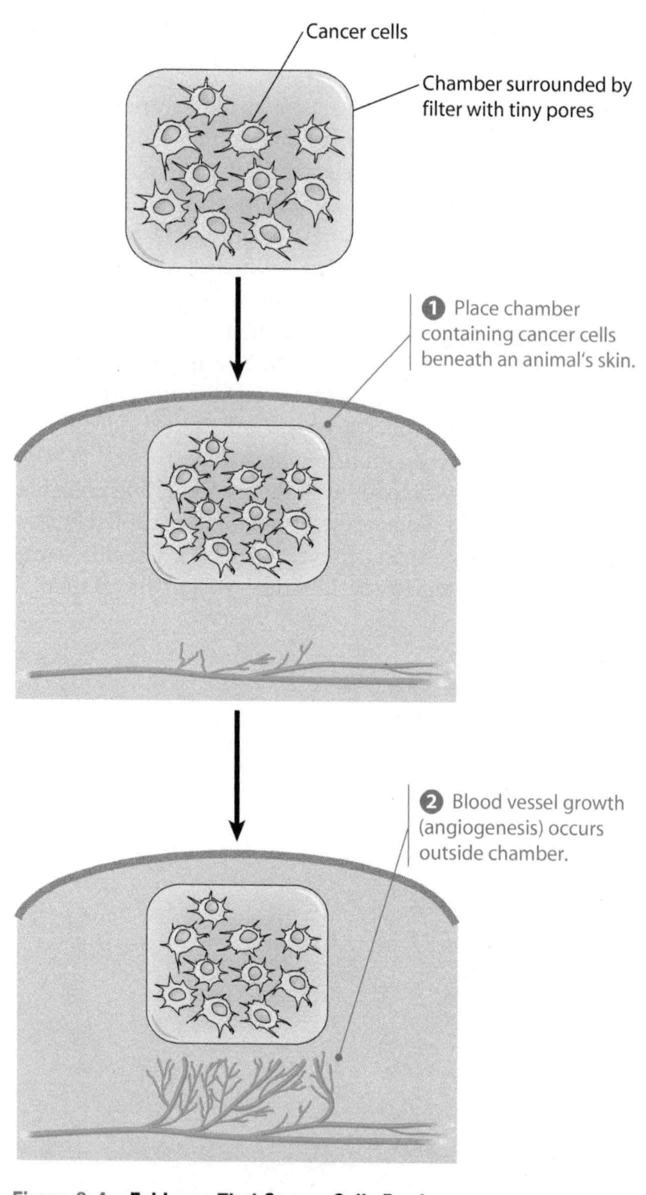

Figure 3-4 **Evidence That Cancer Cells Produce Angiogenesis-Stimulating Molecules.** Cancer cells were placed inside a chamber surrounded by a filter possessing tiny pores through which cells cannot pass. When the chamber was implanted into animals, new capillaries began to proliferate in the surrounding host tissue. This observation suggests that cancer cells produce molecules that pass through the filter and stimulate angiogenesis in the surrounding tissue.

Table 3-1 Some Natural Stimulators and Inhibitors of Angiogenesis

Stimulators
PROTEINS
Angiogenin
Epidermal growth factor
Fibroblast growth factor (FGF)
Granulocyte colony-stimulating factor
Hepatocyte growth factor
Interleukin 8
Placental growth factor
Platelet-derived endothelial growth factor
Transforming growth factor alpha
Tumor necrosis factor alpha
Vascular endothelial growth factor (VEGF)
SMALL MOLECULES
Adenosine
1-Butyryl glycerol
Nicotinamide
Prostaglandins E1 and E2
Inhibitors*
PROTEINS
Angiostatin
Canstatin
Endostatin
Interferons
Platelet factor 4
Prolactin 16Kd fragment
Protamine
Thrombospondin
TIMP-1 (tissue inhibitor of metalloproteinase-1)
TIMP-2 (tissue inhibitor of metalloproteinase-2)
TIMP-3 (tissue inhibitor of metalloproteinase-3)
Tumstatin

*Some synthetic molecules that have been developed for use as angiogenesis-inhibiting drugs will be described in Chapter 11.

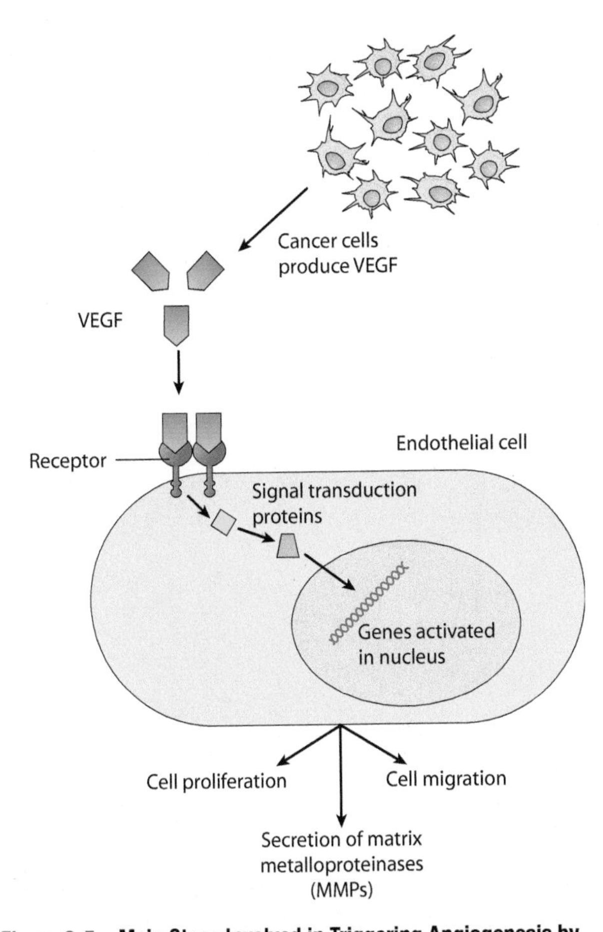

Figure 3-5 Main Steps Involved in Triggering Angiogenesis by VEGF. Cancer cells secrete VEGF molecules that bind to receptor proteins located on the surface of endothelial cells. The binding of VEGF to its receptor leads to the activation of a series of signal transduction proteins that trigger changes in gene expression and cell behavior. The net result is the stimulation of endothelial cell proliferation and migration as well as the secretion of MMPs that degrade components of the extracellular matrix. The proliferating endothelial cells are gradually organized into new networks of blood vessels. (The drawing of the endothelial cell is enlarged relative to the size of the cancer cells to illustrate the signaling pathway.)

angiostatin, **endostatin**, and **thrombospondin**. A finely tuned balance between the concentration of these angiogenesis inhibitors and the concentration of activators (such as VEGF and FGF) determines whether a tumor will induce the growth of new blood vessels. When tumors trigger angiogenesis, it is usually accomplished by increasing the production of angiogenesis activators and, at the same time, decreasing the production of angiogenesis inhibitors.

Inhibitors of Angiogenesis Can Restrain Tumor Growth and Spread

The discovery of angiogenesis inhibitors has caused scientists to speculate about the potential usefulness of such molecules. If sustained tumor development requires the proliferation of new blood vessels, using angiogenesis inhibitors to block vessel formation might be useful for slowing tumor growth. This approach has been quite effective when tested in mice. In one striking study, mice with several types of cancer were injected with the angiogenesis inhibitor endostatin. After a few cycles of treatment, the primary tumors virtually disappeared (Figure 3-6).

Studies involving mice with mutations that hinder angiogenesis have provided additional support for the idea that tumor growth can be restrained by inhibiting angiogenesis. As shown in Figure 3-7 (*left*), injecting breast cancer cells into such angiogenesis-deficient mutant mice leads to the formation of tumors that grow for a short time and then completely regress. In contrast, normal mice injected with the same breast cancer cells die of cancer within a few weeks. When lung cancer cells are injected into the same mutant mice, the results are slightly different. Unlike breast cancer cells, lung cancer cells do develop into tumors in the angiogenesis-deficient mice, but the tumors grow more slowly than in normal mice and fail to metastasize to other organs.

The failure of these lung cancer cells to metastasize in angiogenesis-deficient mice raises the possibility that angiogenesis-inhibiting drugs might be useful in preventing metastasis. In an experiment designed to address this issue, shown in Figure 3-7 (*right*), cancer cells were injected beneath the skin of laboratory mice and allowed to grow for two weeks. The primary tumors were then removed and the animals were monitored for several weeks to see whether visible metastases would appear in other organs. Within a few weeks the average mouse developed almost 50 lung tumors, which arose from cancer cells that had spread to the lungs prior to removal of the primary tumor. In contrast, mice treated with angiostatin developed an average of only two or three

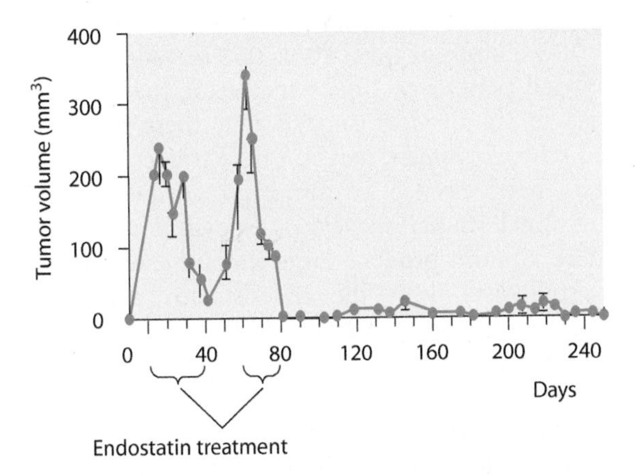

Figure 3-6 Treating Cancer by Inhibiting Angiogenesis. In this experiment, cancer cells were allowed to grow for about ten days to form a large tumor in mice. The mice were then injected with an angiogenesis inhibitor, endostatin, until the tumor regressed. After allowing the tumor to grow again in the absence of endostatin, a second treatment cycle was given. After the second treatment was stopped, the tumor no longer grew again. [Data from T. Boehm et al., *Nature* 390 (1997): 404.]

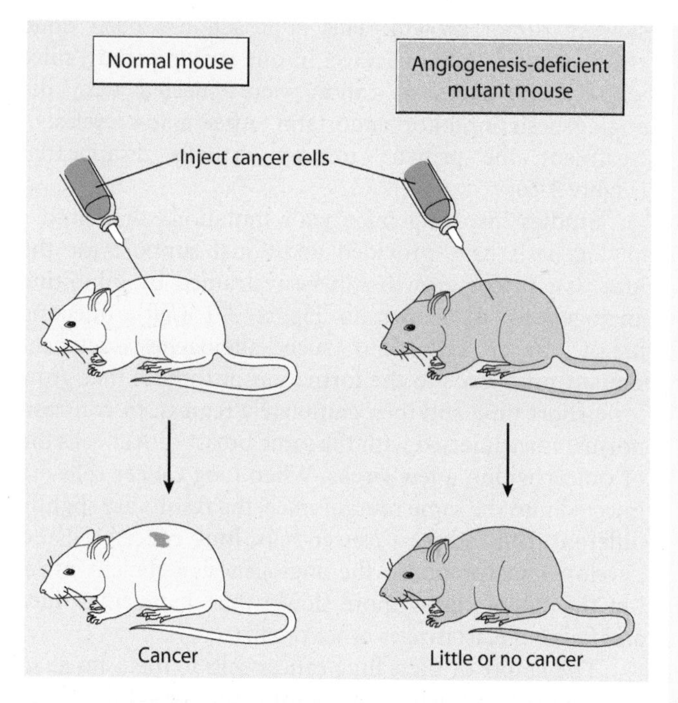

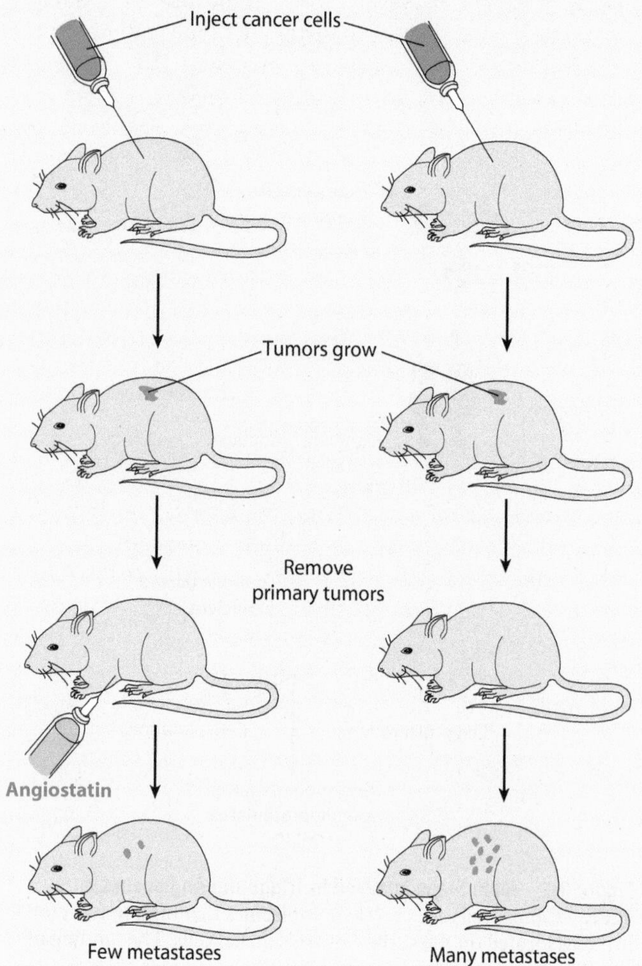

Figure 3-7 Effect of Angiogenesis Inhibition on Tumor Growth and Spread. (*Left*) When cancer cells are injected into mutant mice that are deficient in their ability to carry out angiogenesis, there is little or no growth of tumors. (*Right*) In a set of experiments designed to investigate the effects of angiogenesis inhibitors on metastasis, cancer cells were injected into mice and allowed to grow into primary tumors that were subsequently removed. The development of subsequent metastases from cancer cells that had already spread through the body can be inhibited by treating mice during this ensuing period with the angiogenesis inhibitor, angiostatin.

tumors in their lungs, indicating that the angiogenesis inhibitor had reduced the rate of metastasis about 18-fold.

Such observations provide a possible explanation for the phenomenon of **tumor dormancy**, in which cancer cells spread from a primary tumor to another organ and form tiny clumps of cancer cells that remain dormant for prolonged periods of time. One factor that might contribute to tumor dormancy is that these tiny tumors, called *micrometastases*, may not have triggered the angiogenesis that is needed for tumor growth beyond a small size. Animal studies suggest one possible reason: Some primary tumors produce large amounts of angiostatin that spill over into the bloodstream and circulate throughout the body. It has been hypothesized that this circulating angiostatin can inhibit angiogenesis at other sites, thereby preventing micrometastases from growing into visible tumors.

Why wouldn't the angiostatin also prevent the angiogenesis that is needed for growth of the primary tumor? Most likely, the inhibitory effects of angiostatin on the primary tumor are overcome by the stimulatory effects of VEGF, which is also produced in large amounts by many primary tumors. Unlike angiostatin, however, VEGF is quickly destroyed when it enters the bloodstream. So angiostatin, but not VEGF, circulates in the bloodstream and might block angiogenesis at sites of micrometastases, which do not yet produce enough VEGF of their own to overcome the inhibitory effects of the circulating angiostatin.

Although the proposed role of angiogenesis inhibitors in tumor dormancy remains to be firmly established, the overall body of evidence strongly supports the idea that angiogenesis inhibitors can restrain the growth and spread of cancer cells in animals. Are such findings relevant to humans? To address this question, numerous angiogenesis-inhibiting drugs are being tested in cancer patients and one such drug, called *Avastin*, has already been approved for use in treating colon cancer. The future prospects for this approach, called *anti-angiogenic therapy*, will be discussed in more detail when the topic of cancer treatment is covered in Chapter 11.

INVASION AND METASTASIS

Once angiogenesis has occurred at the site of a primary tumor, the stage is set for tumor cells to invade neighboring tissues and spread to distant sites. A few kinds of cancer, such as nonmelanoma skin cancers, rarely invade and metastasize. About half the people who develop other forms of cancer, however, have tumors that have already begun to spread beyond the site of origin by the time the cancer is diagnosed. Because cancer is much harder to treat after it has spread, this alarming statistic points to the importance of better procedures for early cancer detection. It might also be possible to develop improved treatments for cancer if we better understood the mechanisms that allow cancer cells to spread, the topic to which we now turn.

Spreading of Cancer Cells by Invasion and Metastasis Is a Complex, Multistep Process

Cancers spread through the body via two distinct mechanisms: invasion and metastasis. **Invasion** refers to the direct migration and penetration of cancer cells into neighboring tissues, whereas **metastasis** involves the ability of cancer cells to enter the bloodstream (or other body fluids) and travel to distant sites, where they form new tumors that are not physically contiguous with the primary tumor.

Metastasis involves a complex cascade of events, beginning with the process of angiogenesis that has already been described. The events following angiogenesis can be grouped into three main steps. First, cancer cells invade surrounding tissues and penetrate through the walls of lymphatic and blood vessels, thereby gaining access to the bloodstream. Second, these cancer cells are then transported by the circulatory system throughout the body. And third, the cancer cells leave the bloodstream and enter particular organs, where they establish new metastatic tumors (Figure 3-8). If cells from the initial tumor fail to complete any of these steps, or if any of the steps can be prevented, metastasis will not occur. It is therefore crucial to understand how the properties of cancer cells make these three steps possible.

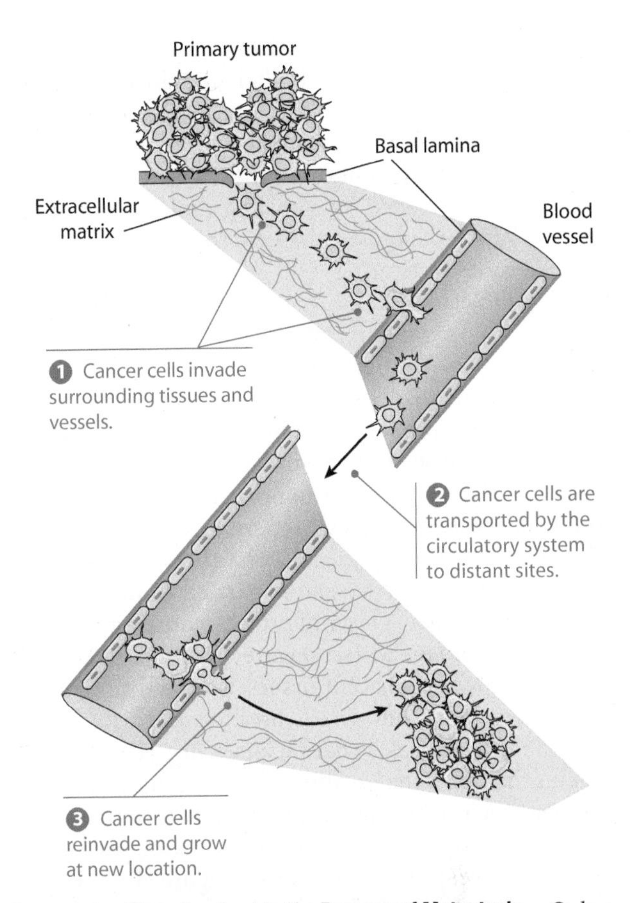

Figure 3-8 Steps Involved in the Process of Metastasis. Only a small fraction of the cells in a primary malignant tumor can successfully carry out all the steps required for metastasis. During step ①, cancer cells adhere to and invade through the basal lamina, degrade the extracellular matrix of the underlying tissue to open up a path through which the cells can move, and finally adhere to and invade through the basal lamina of a tiny vessel, thereby entering the circulatory system. During step ②, the cancer cells are transported via the bloodstream to distant sites. In step ③, the cancer cells become arrested in a capillary located in another organ, adhere to and invade through the basal lamina of the capillary wall, enter the surrounding tissue, and start to grow again.

Changes in Cell Adhesion, Motility, and Protease Production Allow Cancer Cells to Invade Surrounding Tissues and Vessels

The initial step leading to metastasis is the invasion of surrounding tissues and vessels by cancer cells (see Figure 3-8, step ①). Thus, unlike the cells of benign tumors or most normal cells, which remain together in the location where they are formed, cancer cells are capable of leaving their original site and penetrating through surrounding tissue barriers, eventually entering the circulatory system.

Several mechanisms make this invasive behavior possible. The first involves changes in the adhesive forces between cells. In most tissues, adjoining cells are held together by binding interactions between **cell-cell adhesion proteins** found on the outer surface of each cell. These adhesion molecules, which normally function to keep cells in place, are often missing or deficient in cancer cells, thereby allowing cells to separate from the main tumor mass more readily. The diminished adhesiveness of cancer cells can be readily demonstrated by a simple experiment: If you take a sample of cancer tissue, suspend it in a fluid-filled flask, and shake the flask vigorously, the cells will separate from one another more readily than cells obtained from a comparable sample of normal tissue.

In many cases, the reduced adhesiveness of cancer cells can be traced to the loss of **E-cadherin**, a cell-cell adhesion protein that normally binds epithelial cells to one another. Highly invasive cancers usually have less E-cadherin than normal cells, suggesting a relationship between the loss of cadherins and the ability to invade. Support for this idea has come from studies in which non-invasive populations of cancer cells were treated with antibodies that block the function of E-cadherin. Such treatment gives cancer cells the ability to invade. Conversely, restoring E-cadherin to cancer cells lacking this molecule inhibits their ability to form invasive tumors when the cells are injected back into animals.

A second property involved in tumor invasion is cell motility, which is activated after the loss of cell-cell adhesion permits cancer cells to detach from one another. Cancer cells possess all the normal cytoplasmic machinery required for cell locomotion, but their actual movement needs to be stimulated by signaling molecules produced either by surrounding host tissues or by cancer cells themselves. Besides activating cell motility, some of these signaling molecules act as *chemoattractants* that guide cell movement by serving as attracting signals toward which the cancer cells will migrate.

In addition to decreased adhesiveness and activated motility, a third property involved in invasion is the production of **proteases** (protein-degrading enzymes). The purpose of these enzymes is to break down structures that would otherwise represent barriers to cancer cell movement. For example, epithelial cells, the source of about 90% of all human cancers, are separated from underlying tissues by a thin, dense layer of protein-containing material called the *basal lamina* (see Figure 1-4). Before epithelial cancers can invade adjacent tissues, the basal lamina must first be breached. Cancer cells break through this barrier by producing proteases that facilitate degradation of the proteins that form the backbone of the basal lamina.

One such protease is **plasminogen activator**, an enzyme that converts the inactive precursor *plasminogen* into the active protease **plasmin** (Figure 3-9). Because high concentrations of plasminogen are present in almost all tissues, small amounts of plasminogen activator released by cancer cells can quickly catalyze the formation of large quantities of plasmin. The plasmin in turn performs two tasks: (1) it degrades components of the basal lamina and the extracellular matrix, thereby facilitating tumor invasion; and (2) it cleaves inactive precursors of *matrix metalloproteinases (MMPs)*, produced mainly by surrounding host cells, into active enzymes that also degrade components of the basal lamina and extracellular matrix. (These MMPs are the same enzymes that were mentioned earlier in the chapter in the description of how endothelial cells degrade the extracellular matrix during angiogenesis.)

After proteases allow the basal lamina to be penetrated, they degrade the matrix of the underlying tissues to open up paths through which the cancer cells can move. The cancer cells migrate until they reach tiny blood or lymphatic vessels, which are also surrounded by a basal lamina. The proteases then digest holes in this second basal lamina, allowing cancer cells to pass through it and through the layer of endothelial cells that form the vessel's inner lining, at which point the cancer cells have finally gained entry into the circulatory system.

Relatively Few Cancer Cells Survive the Voyage Through the Bloodstream

Cancer cells that penetrate through the walls of tiny blood vessels gain direct entry to the bloodstream, which then transports the cells to distant parts of the body (Figure 3-8, step ②). When cancer cells initially penetrate the walls of lymphatic vessels rather than blood capillaries, the cells are first carried to regional lymph nodes, where they may become lodged and grow. For this reason, regional lymph nodes are a common site for the initial spread of cancer. Nonetheless, lymphatic vessels have numerous interconnections with blood vessels, so cancer cells that initially enter into the lymphatic system eventually arrive in the bloodstream and circulate throughout the body.

Regardless of their initial entry route, the net result is often large numbers of cancer cells in the bloodstream. Even a tiny malignant tumor weighing only a few grams

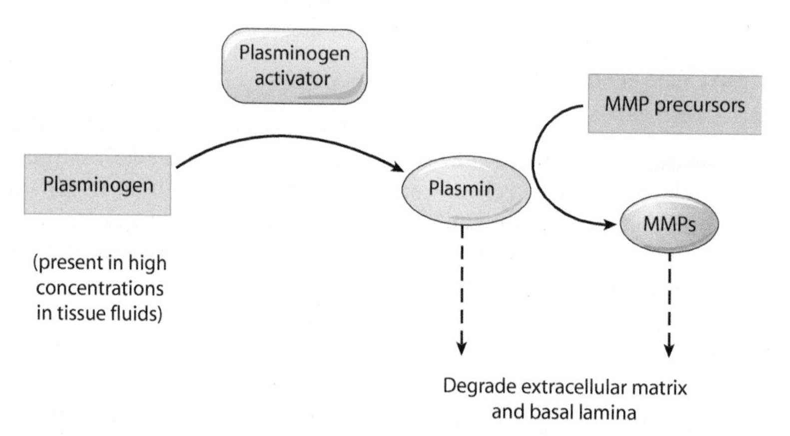

Figure 3-9 Proteases Involved in Degrading the Basal Lamina and Extracellular Matrix. Most cancer cells produce plasminogen activator, an enzyme that catalyzes the conversion of the inactive precursor, plasminogen, into the active protease, plasmin. Because plasminogen is present in high concentrations in almost all tissues, a small amount of plasminogen activator can produce large quantities of plasmin. The plasmin degrades components of the basal lamina and the extracellular matrix, thereby promoting invasion, and also converts inactive precursors of matrix metalloproteinases (MMPs) into active forms, thereby creating additional proteases that can degrade the basal lamina and extracellular matrix. The MMP precursors are produced to a large extent by surrounding host cells rather than by tumor cells.

can release several million cancer cells into the circulatory system each day. To produce metastases, however, the cells must survive the trip through the circulatory system, and most cancer cells do not thrive in such an environment. Evidence for this conclusion has come from experiments in which cancer cells were radioactively labeled (to allow them to be identified) and then injected into the bloodstream of laboratory animals. After a few weeks, less than one in a thousand radioactive cells was found to be alive. Apparently the bloodstream is an inhospitable place for most cancer cells and only a tiny number survive the trip to potential sites of metastasis.

The Ability to Metastasize Differs Among Cancer Cells and Tumors

Since only a tiny fraction of the cancer cells that enter the bloodstream survive and establish metastases, the question arises as to whether these cells are random members of the original tumor population or specialized cells better suited for metastasis. Figure 3-10 illustrates an experiment designed by Isaiah Fidler to address this question. In these studies, mouse melanoma cells were injected into the bloodstream of healthy mice to study the ability of the cells to metastasize. A few weeks after the injections, metastases were detected in a variety of locations, but mainly the lungs. Cells from the lung metastases were removed and injected into another mouse, leading to the production of more lung metastases. By repeating the same procedure many times in succession, Fidler eventually obtained a population of cancer cells that formed greater numbers of lung metastases than did the original tumor cell population.

The most straightforward interpretation was that the initial melanoma consisted of a heterogeneous population of cells with differing metastatic capabilities and that the successive experiments had gradually selected for those cells that are especially well suited for metastasizing. To test this hypothesis, studies were subsequently performed in which single cells were isolated from a primary melanoma and each isolated cell was allowed to grow in culture to form a separate population of cells. Such cell populations, each derived from the proliferation of a single initial cell, are referred to as **clones**. When the various cloned populations of cells were injected into animals, some of the clones produced few metastases, some produced numerous metastases, and some fell in between. Since each clone was derived from a different cell in the original tumor, the results support the idea that the cells in a primary tumor differ in their ability to metastasize.

It has been known for many years that human cancers of the same type can differ significantly in their ability to metastasize. In one set of experiments, investigators analyzed gene activity in lung cancers and discovered a pattern involving the expression of 17 genes that predicted whether a primary lung cancer was likely to metastasize. This same gene-expression "signature" also predicts the metastatic behavior of other types of cancer. For example, prostate or breast cancer patients whose primary tumors exhibit the 17-gene signature are more likely to develop metastases than individuals whose tumors do not exhibit the 17-gene signature. Thus, the likelihood that metastasis will occur appears to be genetically programmed into the cells of the primary tumor.

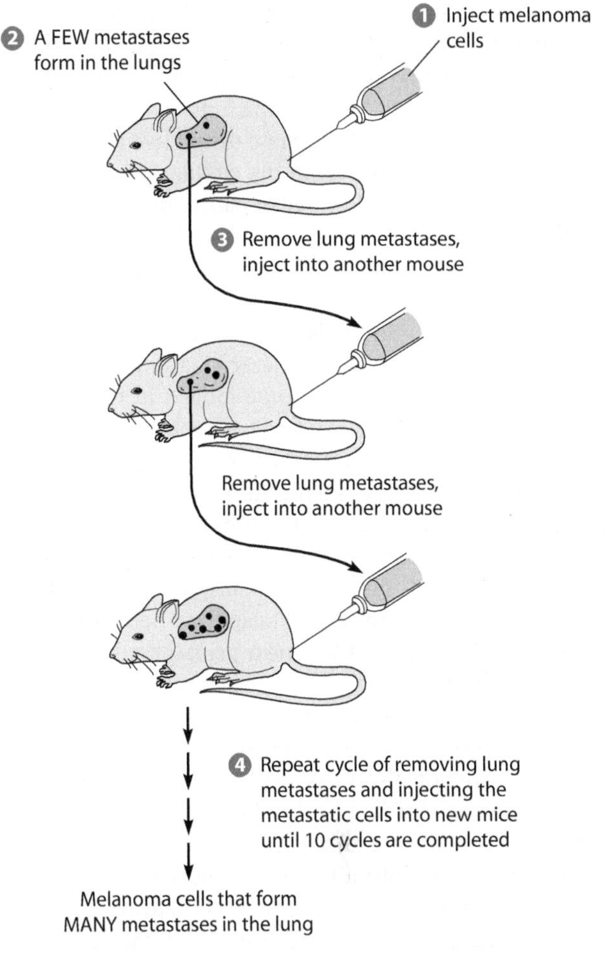

Figure 3-10 Selection of Melanoma Cells Exhibiting an Enhanced Ability to Metastasize. In this experiment, mouse melanoma cells were injected into the tail vein of a mouse. After a small number of metastatic tumors developed in the lungs, cells from some of the lung metastases were removed and injected into another mouse. When this cycle was repeated ten times, the final population of melanoma cells was found to produce many more lung metastases than the original cell population produced.

Blood-Flow Patterns Often Dictate Where Cancer Cells Will Metastasize

After traveling through the circulatory system to distant parts of the body, cancer cells exit from the bloodstream and invade organs that may be located far from the primary tumor (Figure 3-8, step ③). Although the bloodstream carries cancer cells everywhere in the body, the final distribution of metastases is not random, nor is it the same for every type of cancer. Instead, cancers arising in each organ preferentially metastasize to particular locations. For example, stomach and colon cancers frequently metastasize to the liver, prostate and breast cancers often metastasize to bone, and many forms of cancer tend to metastasize to the lungs.

One factor underlying these distinctive relationships is the pattern of blood flow in the circulatory system, which determines where cancer cells floating in the bloodstream are likely to become lodged. Based solely on size considerations, the most probable site for circulating cancer cells to become stuck is in capillaries—the tiny vessels whose diameter is generally no larger than that of a single blood cell. Because they are usually larger than blood cells, circulating cancer cells often become lodged in tiny capillaries as they move through the body. The arrested cancer cells may then adhere to and penetrate through the capillary walls and invade the surrounding organ, beginning the formation of a new metastatic tumor.

The preceding scenario suggests that after a cancer cell has entered the bloodstream, it is susceptible to becoming arrested in the first capillary bed it encounters. Figure 3-11 shows that for most primary tumors, the first such capillary bed will be in the lungs, which may help explain why the lungs are a relatively frequent site of metastasis for many kinds of cancer. However, blood-flow patterns do not always point to the lungs. For cancers arising in the stomach and colon, cancer cells that enter the bloodstream are first carried to the liver, where the vessels break up into a bed of capillaries. As a result, the liver is a common site of metastasis for stomach and colon cancers. Finally, when cancer arises in the lung, cancer cells entering the bloodstream will flow first to the left side of the heart and from there to capillary beds located throughout the body. It is therefore not surprising that lung cancers metastasize to many different organs, including the liver, bones, brain, kidney, adrenal gland, thyroid, and spleen.

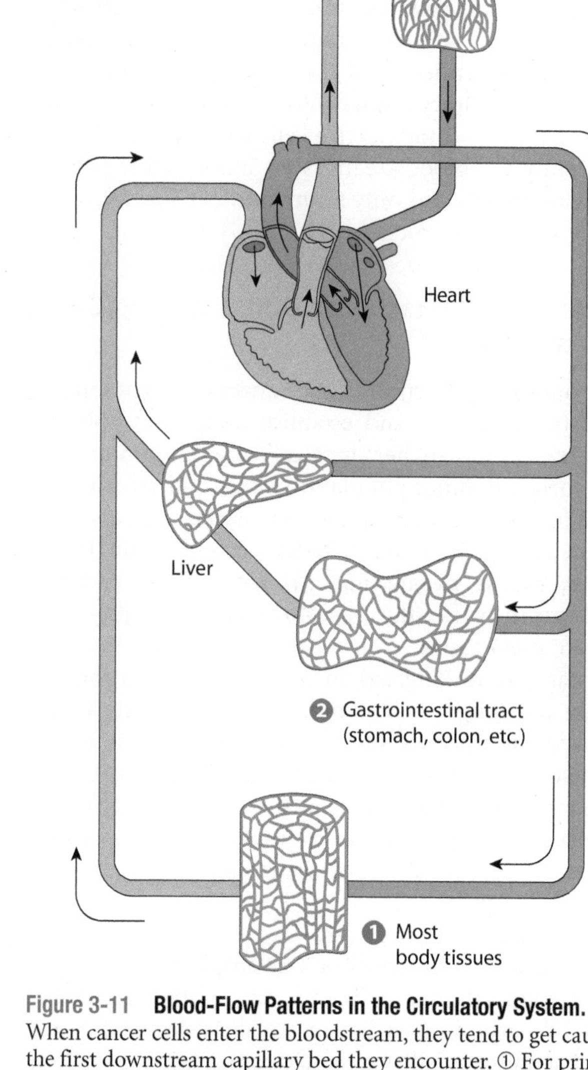

Figure 3-11 Blood-Flow Patterns in the Circulatory System. When cancer cells enter the bloodstream, they tend to get caught in the first downstream capillary bed they encounter. ① For primary tumors arising in most locations, the bloodstream first carries the cells to the heart and from there to the lungs, where the first capillary bed is encountered. ② For cancers arising in the gastrointestinal tract, the bloodstream first carries the cells to the liver, where the vessels break up into capillaries. This anatomy explains why the liver is a common site of metastases for stomach and colon cancers. ③ For cancers arising in the lung, the bloodstream carries the cells to the left side of the heart and from there to capillary beds that can be located anywhere in the body.

Organ-Specific Factors Play a Role in Determining Where Cancer Cells Will Metastasize

Although blood-flow patterns are clearly important, they do not always explain the observed distribution of metastases. As early as 1889, Stephen Paget proposed that the nonrandom distribution of metastases arises in part because individual cancer cells have a special affinity for the environment provided by particular organs. Paget's idea is often referred to as the "seed and soil" hypothesis, based on his analogy that when a plant produces seeds, they are carried by the wind in all directions but they only grow if they fall on congenial soil. According to this view, cancer cells are carried to a variety of organs by the bloodstream, but only a few sites provide an optimal environment for the growth of a particular type of cell. In other words, metastasis only takes place where the seed (a cancer cell) and the soil (a particular organ) are compatible.

Support for the "seed and soil" concept has come from a systematic analysis of the sites to which various human cancers tend to metastasize. For roughly two-thirds of the cancer types examined, the rates of metastasis to each organ can be explained solely on the basis of blood-flow patterns. Of the remaining cases, some kinds of cancers metastasize to particular organs less frequently than would be expected and others metastasize to particular organs more frequently than would be expected.

Animal experiments have revealed that this nonrandom behavior can be traced to the properties of individual cancer cells. In one set of experiments, similar to those shown in Figure 3-10, mouse melanoma cells were injected into normal mice and metastases were isolated from the brain instead of the lung. The metastatic cells were then injected into another healthy mouse, and the same cycle was repeated again and again. Even though the melanoma cells employed in this study initially metastasized more often to the lung than they did to the brain, the repeated selection of metastatic cells from the brain eventually led to the isolation of cells that preferentially metastasize to the brain rather than to the lung. Similar experiments involving the selection of cells derived from ovarian metastases yielded cells that preferentially metastasize to the ovary rather than to the lung or brain. Hence the initial tumor must have consisted of a heterogeneous population of cells that differ in the sites to which they tend to metastasize.

Why do individual cancer cells grow best at particular sites? The general answer is that the ability to grow in different locations is affected by interactions between cancer cells and molecules present in the organs to which they are delivered. An example of this principle is provided by prostate cancer cells, which preferentially metastasize to bone (a pattern that would not be predicted based on blood flow). To investigate the reason for this preference, experiments have been performed in which prostate cancer cells were mixed together with cells from various locations—including bone, lung, and kidney—and the cell mixtures were then injected into animals. It was found that the ability of prostate cancer cells to develop into tumors was stimulated by the presence of cells derived from bone, but not from lung or kidney. Subsequent studies uncovered a possible explanation: Bone cells produce growth factors that stimulate the proliferation of prostate cancer cells. This example illustrates just one of several kinds of molecular interactions that influence the ability of cancer cells to grow in particular organs.

Some of the Cellular Properties Involved in Invasion and Metastasis Arise During Tumor Progression

After tumor cells have left the bloodstream and invaded an appropriate organ where conditions for their growth are favorable, the cells begin proliferating at the new site. Before a metastatic tumor can grow beyond a few millimeters in diameter, however, it must first trigger angiogenesis (as was the case for the primary tumor). Some metastatic tumors are slow to elicit angiogenesis and remain dormant for significant periods of time, whereas others are quick to trigger angiogenesis and may grow so rapidly that they soon exceed the size of the primary tumor from which they were derived. In the latter case, the first sign that a person has cancer may arise from metastases rather than from the primary tumor. For example, the earliest symptom noticed by a person with lung cancer could be back pain triggered by cancer cells that have metastasized to the bones of the spinal column. Or a person might develop liver failure caused by cancer cells that metastasized to the liver from a smaller primary tumor located in the esophagus, stomach, or colon.

It may seem surprising that tumor metastases can be larger than the primary tumor from which they originated. Remember, however, that primary tumors consist of a heterogeneous population of cells that differ in their capacity to metastasize and in their ability to grow at particular sites. Those cells that successfully manage to metastasize to a particular organ may represent a subpopulation of the original cancer cells that are especially well suited for creating metastases. Such cells may be faster growing, more invasive, better at triggering angiogenesis, and generally more aggressive than the average cell of the initial tumor population. Metastatic tumors may therefore end up growing and invading more efficiently than the original primary tumor, shedding more cells into the bloodstream and generating further metastases that are even more aggressive.

Cancer is thus a disease whose properties change with time. In the beginning stages of tumor formation, the properties required for malignant growth are not fully developed. Instead, as a tumor begins to grow, individual cells often acquire gene mutations and alter the genes they express, turning on some genes and turning off others. Such alterations create a population of cells whose properties, including the ability to invade and metastasize, gradually change over time. Cells acquiring traits that confer a selective advantage—such as increased growth rate, increased invasiveness, ability to survive in the bloodstream, resistance to immune attack, ability to grow in other organs, resistance to drugs used in cancer treatment, and evasion of apoptosis—will be more successful than cells lacking these traits and so will gradually tend to predominate. This gradual change in the properties of a tumor cell population, as cells acquire more and more aberrant traits and become increasingly aggressive, is known as **tumor progression**. We will return to the topic of tumor progression in later chapters when we discuss the identity of the genes whose alterations are associated with cancer development and progression.

The Immune System Can Inhibit the Process of Metastasis

Given the life-threatening nature of metastasis, the question arises as to whether the body has any defenses against it. One possibility is the immune system, which has the ability to attack and destroy foreign cells. When cancer cells circulate in the bloodstream, where cells of the immune system travel in large numbers, they are especially vulnerable to attack. Of course, cancer cells are not literally of "foreign" origin, but they often exhibit molecular changes that allow the immune system to recognize the cells as being abnormal.

Animal experiments suggest that in some cases, attack by the immune system does limit the process of metastasis. One such study involved two strains of mouse lung cancer cells: *D122* cells that metastasize with high frequency and *A9* cells that rarely metastasize. In general, the ability of the immune system to recognize cells as being foreign or abnormal requires the involvement of cell surface proteins called **major histocompatibility complex** (**MHC**) molecules (see Figure 2-21). The MHC molecules carried by the two lines of lung cancer cells exhibit a prominent difference: A9 cells carry two types of MHC (called H-2K and H-2D), whereas the D122 cells express only one form (H-2D).

The discovery that D122 and A9 cells carry different cell surface MHC molecules raises an important question: Is the differing metastatic behavior of the two cells related to the immune system's ability to recognize and attack the two cell types? This issue was investigated by injecting A9 and D122 cells into separate groups of animals and monitoring the production of **cytotoxic T lymphocytes** (**CTLs**), a class of immune cells specialized for attacking foreign and abnormal cells. The animals were found to produce numerous CTLs targeted against A9 cancer cells, but few CTLs targeted against D122 cells.

Why do CTLs attack A9 cells more readily than D122 cells? The most obvious possibility is that the immune system recognizes the H-2K MHC molecules, which are carried by A9 cells but not by D122 cells. This hypothesis was tested by introducing purified DNA containing the H-2K gene into D122 cells, thereby causing the D122 cells to produce the H-2K form of MHC (Figure 3-12). As predicted, the altered D122 cells expressing H-2K exhibited a reduced capacity to metastasize when injected into mice, suggesting that the presence of H-2K made the cells more susceptible to immune attack. However, the primary tumor at the site of injection grew normally, implying that tumor cells are more susceptible to immune attack when they are circulating in the bloodstream, where large numbers of immune cells reside.

Invasion and Metastasis Involve a Variety of Tumor-Host Interactions

The ability of the immune system to inhibit the process of metastasis illustrates an important principle: The behavior of malignant tumors depends not just on the traits of tumor cells, but also on interactions between tumor cells and normal cells of the surrounding host tissues. As is summarized in Figure 3-13, this chapter has already covered several examples of such **tumor-host interactions**. For example, angiogenesis is triggered by growth factors released by tumor cells that act on normal endothelial cells of the surrounding host tissue, thereby stimulating the proliferation of new blood vessels. Invasion is facilitated by both tumor- and host-derived proteases that degrade normal extracellular structures such as the basal lamina and the extracellular matrix. The motility of cancer cells and the direction in which they migrate is influenced by signaling molecules made by normal cells of the surrounding tissues. Penetration through capillaries involves adhesion of cancer cells to molecules present in the basal lamina. And finally, the

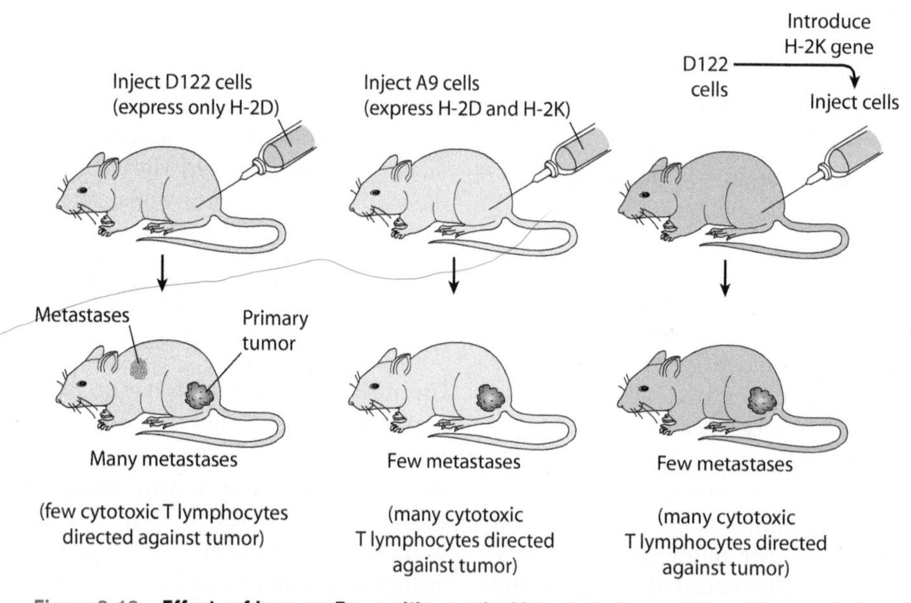

Figure 3-12 **Effects of Immune Recognition on the Metastatic Competence of Cancer Cells.** (*Left*) D122 lung cancer cells, which express cell surface H-2D molecules, produce numerous metastases. (*Middle*) A9 lung cancer cells, which express both H-2D and H-2K molecules, metastasize poorly. (*Right*) If the H-2K gene is introduced into D122 cells, the cells lose their ability to metastasize. Because H-2K is a cell surface MHC molecule recognized by the immune system, such studies suggest that immune recognition influences the ability of cancer cells to metastasize.

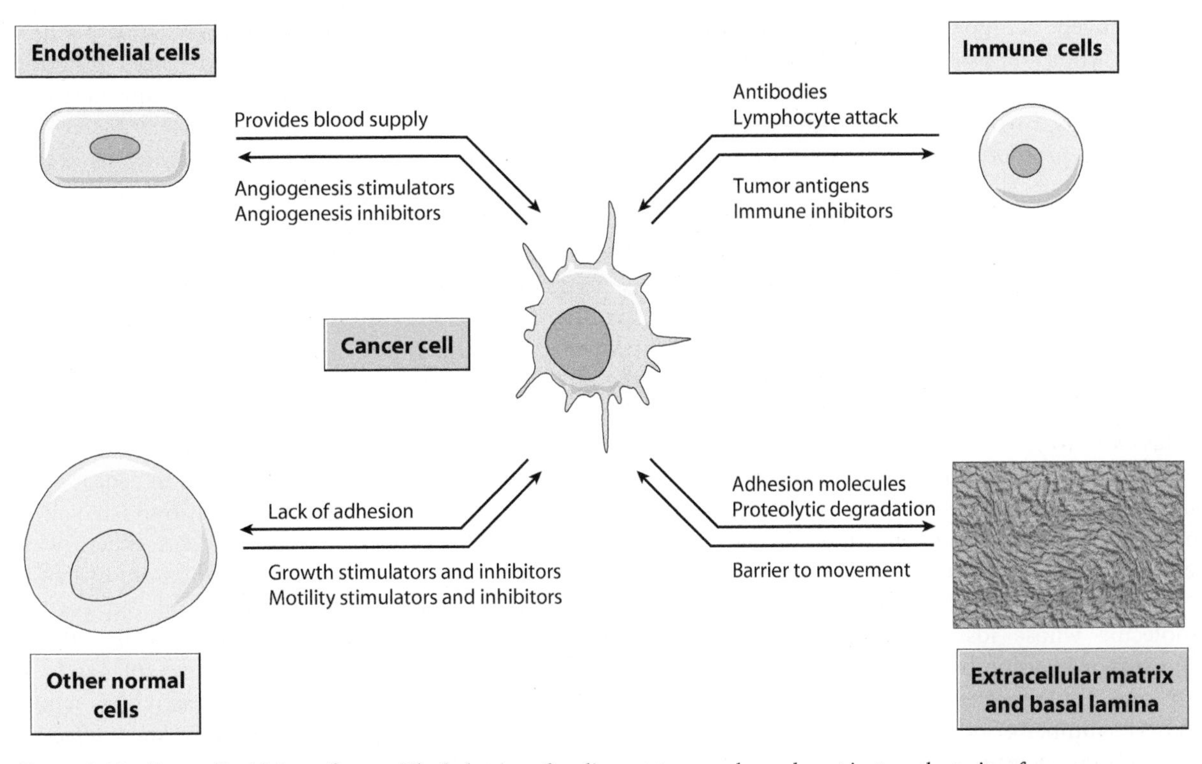

Figure 3-13 Tumor-Host Interactions. The behavior of malignant tumors depends not just on the traits of cancer cells, but also on interactions between cancer cells and normal cells of the host. This diagram summarizes a few of the many such interactions that affect the development of cancer.

growth of metastases at distant sites is simulated by growth factors and other molecules produced by cells residing in the organs being invaded.

Normal tissues also contain cells and molecules that are capable of hindering invasion and metastasis. For example, we have already seen that immune lymphocytes are capable of attacking and destroying cancer cells, thereby limiting their ability to metastasize. In addition, normal tissues produce *protease inhibitors* that reduce the activity of the proteases that cancer cells require for degrading the basal lamina and extracellular matrix. The invasiveness of cancer cells therefore reflects a competition between proteases produced by tumor cells and protease inhibitors produced by surrounding normal cells. These are just a few of the numerous examples in which tumor-host interactions influence the ability of cancer cells to invade neighboring tissues and metastasize to distant sites.

Specific Genes Promote or Suppress the Ability of Cancer Cells to Metastasize

Because metastasis is the property that makes cancer so dangerous, scientists have been trying to identify some of the key molecules involved in metastasis that might serve as useful targets for new anticancer drugs. The matrix metalloproteinases (MMPs), which facilitate destruction of the basal lamina and extracellular matrix, are one possibility. MMPs are involved in angiogenesis as well as invasion and metastasis, so drugs that inhibit MMP activity could conceivably interfere with cancer progression in multiple ways. MMP inhibitors are therefore being tested in human cancer patients to see if they can slow down or stop the spread of cancer. Another group of attractive targets are those molecules found on the cancer cell surface that help cancer cells adhere to capillary walls. If inhibitors could be found that block the activity of these cell surface molecules, it might be possible to hinder the interactions between cancer cells and capillaries that allow cancer cells to pass into and out of the bloodstream. Attempts are also being made to target the growth factors that stimulate cancer cell proliferation in specific target organs, such as the growth factor produced in bone tissue that stimulates the proliferation of metastatic prostate cancer cells.

Genetic analysis is beginning to facilitate the identification of molecules that might be good targets for drugs designed to halt cancer spread. Through this approach, researchers have identified dozens of genes that influence the ability of cancer cells to metastasize, some exerting positive effects and others exerting negative effects. The positively acting genes, called **metastasis promoting genes**, code for proteins that stimulate events associated with invasion and metastasis. While it might seem surprising that cells possess genes that promote invasion and metastasis, several traits exhibited by invasive cancer cells—for example, decreased adhesiveness and enhanced motility—are also important for certain kinds of normal cells, such as embryonic cells and cells of the immune system. The negatively acting genes, called **metastasis suppressor genes**, code for proteins that inhibit events associated with invasion and metastasis. Acquiring the ability to metastasize is usually associated with enhanced activity of metastasis promoting genes as well as diminished activity of metastasis suppressor genes.

Of the two classes of genes, metastasis suppressors are easier to identify because it often takes the action of only one of these genes to block metastasis. To determine whether a gene is a metastasis suppressor, a normal copy of the gene is simply introduced into a population of cancer cells that already possess the ability to metastasize. If activation of the newly introduced gene blocks the ability to metastasize without inhibiting the ability of the cells to form tumors, it is classified as a metastasis suppressor gene. Several metastasis suppressors have been identified using such approaches. One of the best understood is the *CAD1* gene, which codes for E-cadherin. We saw earlier in the chapter that E-cadherin is a cell-cell adhesion molecule whose loss from the cell surface contributes to tumor invasion by allowing cancer cells to detach from one another and move away from the primary tumor. A number of other metastasis suppressor genes have also been identified, including genes called *NM23*, *KiSS1*, *KAI1*, *BRMS1*, and *MKK4*. Some of these genes code for other proteins involved in cell adhesion and motility, but additional mechanisms appear to be involved as well.

Progress has also been made in identifying genes that promote rather than suppress metastasis. An especially interesting example is the gene coding for a protein called *Twist*, which regulates the activity of a specific group of genes during embryonic development. Genes activated by the Twist protein cause cells to lose their adhesive properties, become motile, and migrate from one part of the embryo to another. After embryonic development is complete, the Twist protein is no longer needed and its production is shut down in most tissues. However, production of the Twist protein is reactivated in cancer cells, allowing them to reacquire the embryonic traits that allow cells to move throughout the body. Recent experiments have shown that introducing an inhibitor of Twist production into mouse breast cancer cells reduces the ability of the cells to metastasize. It is therefore hoped that further study of metastasis promoting genes such as *Twist* will help identify those events and activities whose disruption by appropriate drugs would be most effective at preventing metastasis.

Summary of Main Concepts

- Angiogenesis is required before tumors can grow beyond a few millimeters in diameter. Control over angiogenesis is exerted by a finely tuned balance between activator proteins (such as VEGF and FGF) and inhibitor proteins (such as endostatin, angiostatin, and thrombospondin).
- VEGF and FGF produced by tumor cells activate angiogenesis by binding to receptors present on the surface of endothelial cells, thereby stimulating the cells to proliferate, produce matrix metalloproteinases, migrate into surrounding tissues, and assemble into new vessels. Treating animals with angiogenesis inhibitors can inhibit the growth of primary tumors as well as the formation of metastases.
- After angiogenesis has occurred, cancer cells invade surrounding tissues, enter the circulatory system, and metastasize to distant sites. Each of these steps involves numerous interactions between cancer cells and normal cells of the host.
- The ability to invade neighboring tissues and penetrate vessel walls is facilitated by three properties exhibited by cancer cells: decreased cell-cell adhesion (often traced to a loss of E-cadherin), increased motility, and secretion of proteases that facilitate degradation of the extracellular matrix and basal lamina.
- Tumors consist of heterogeneous populations of cells that vary in their ability to metastasize, their susceptibility to immune attack, and the sites to which they preferentially metastasize. Only a tiny fraction of the cancer cells that enter the circulatory system survive the trip and establish successful metastases.
- Cancer cells tend to preferentially metastasize to the organ containing the first capillary bed encountered after entering the bloodstream. For this reason, cancers arising in the gastrointestinal tract often metastasize to the liver, and most other cancers frequently metastasize to the lungs. Many cancers also exhibit an organ-specific pattern of spread, preferentially metastasizing to particular sites that provide a congenial environment for their growth.
- Cancer is a disease whose properties change with time as tumor progression yields cells with increased growth rate, invasiveness, ability to survive in the bloodstream and evade immune attack, ability to grow in other organs, and resistance to drugs used in cancer treatment.
- The study of metastasis promoting and metastasis suppressor genes is helping scientists identify potential targets for the development of new drugs designed to halt cancer spread.

Key Terms for Self-Testing

Tumor Angiogenesis

angiogenesis (p. 45)
endothelial cell (p. 45)
tumor angiogenesis (p. 47)
vascular endothelial growth factor (VEGF) (p. 48)
fibroblast growth factor (FGF) (p. 48)
VEGF receptor (p. 48)
matrix metalloproteinases (MMPs) (p. 48)
angiostatin (p. 49)
endostatin (p. 49)
thrombospondin (p. 49)
tumor dormancy (p. 50)

Invasion and Metastasis

invasion (p. 51)
metastasis (p. 51)
cell-cell adhesion protein (p. 51)
E-cadherin (p. 51)
protease (p. 52)
plasminogen activator (p. 52)
plasmin (p. 52)
clone (p. 53)
tumor progression (p. 55)
major histocompatibility complex (MHC) (p. 56)
cytotoxic T lymphocyte (CTL) (p. 56)
tumor-host interaction (p. 56)
metastasis promoting gene (p. 57)
metastasis suppressor gene (p. 58)

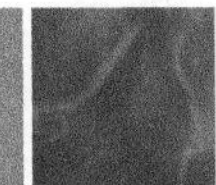

Suggested Reading

Tumor Angiogenesis

Bergers, G., and L. E. Benjamin. Tumorigenesis and the angiogenic switch. *Nature Reviews Cancer* 3 (2003): 410.

Carmeliet, P. Mechanisms of angiogenesis and arteriogenesis. *Nature Medicine* 6 (2000): 389.

Carmeliet, P., and R. K. Jain. Angiogenesis in cancer and other diseases. *Nature* 407 (2000): 249.

Ferrara, N. VEGF and the quest for tumour angiogenesis factors. *Nature Reviews Cancer* 2 (2002): 795.

Folkman, J. The vascularization of tumors. *Sci. Amer.* 234 (May 1976): 58.

Folkman, J. Tumor Angiogenesis. In *Holland-Frei Cancer Medicine*, 5th ed. (R. C. Bast et al., eds.). Lewiston, NY: Decker, 2000, Chapter 9.

Lyden, D., et al. Id1 and Id3 are required for neurogenesis, angiogenesis and vascularization of tumour xenografts. *Nature* 401 (1999): 670.

Yancopoulos, G. D., et al. Vascular-specific growth factors and blood vessel formation. *Nature* 407 (2000): 242.

Invasion and Metastasis

Chambers, A. F., A. C. Groom, and C. MacDonald. Dissemination and growth of cancer cells in metastatic sites. *Nature Reviews Cancer* 2 (2002): 563.

Feldman, M., and L. Eisenbach. What makes a tumor cell metastatic? *Sci. Amer.* 259 (November 1988): 60.

Fidler, I. J. Cancer biology: Invasion and Metastasis. In *Clinical Oncology* (M. D. Abeloff, ed.). New York: Churchill Livingstone, 2000, Chapter 3.

Fidler, I. J. The pathogenesis of cancer metastasis: the 'seed and soil' hypothesis revisited. *Nature Reviews Cancer* 3 (2003): 453.

Friedl, P., and K. Wolf. Tumour-cell invasion and migration: Diversity and escape mechanisms. *Nature Reviews Cancer* 3 (2003): 362.

Gleave, M., et al. Acceleration of human prostate cancer growth in vivo by factors produced by prostate and bone fibroblasts. *Cancer Res.* 51 (1991): 3753.

Liotta, L. A. Cancer cell invasion and metastasis. *Sci. Amer.* 266 (February 1992): 54.

Liotta, L. A. The microenvironment of the tumour-host interface. *Nature* 411 (2001): 375.

Logothetis, C. J., and S.-H. Lin. Osteoblasts in prostate cancer metastasis to bone. *Nature Reviews Cancer* 5 (2005): 21.

Mueller, M. M., and N. E. Fusenig. Friends or foes—Bipolar effects of the tumour stroma in cancer. *Nature Reviews Cancer* 4 (2004): 839.

Mundy, G. R. Metastasis to bone: Causes, consequences and therapeutic opportunities. *Nature Reviews Cancer* 2 (2002): 584.

Nicolson, G. L. Cancer metastasis. *Sci. Amer.* 240 (March 1979): 66.

Paget, S. The distribution of secondary growths in cancer of the breast. *Lancet* 1 (1889): 571.

Pantel, K., and R. H. Brakenhoff. Dissecting the metastatic cascade. *Nature Reviews Cancer* 4 (2004): 448.

Ramaswamy, S., et al. A molecular signature of metastasis in primary solid tumors. *Nature Genetics* 33 (2003): 49.

Ruoslahti, E. How cancer spreads. *Sci. Amer.* 275 (September 1996): 72.

Steeg, P. S. Metastasis suppressors alter the signal transduction of cancer cells. *Nature Reviews Cancer* 3 (2003): 55.

Yang, J., et al. Twist, a master regulator of morphogenesis, plays an essential role in tumor metastasis. *Cell* 117 (2004): 927.

4 Identifying the Causes of Cancer

The uncontrolled proliferation of cancer cells, combined with their ability to spread to distant sites, makes cancer a potentially life-threatening disease. What causes the emergence of such cells that have the ability to destroy the organism in which they reside? It seems almost paradoxical that cancer cells acquire traits that can lead to the death of the host organism, thereby simultaneously ensuring destruction of the cancer cells as well.

The pathway by which normal cells are converted into cancer cells is a complex, multistep process that typically unfolds over a period of many years. Yet despite this complexity, many of the underlying causes that trigger the development of cancer are well understood. Of course, the general public often views cancer as a mysterious disease that strikes randomly and without known cause, but such misconceptions fail to consider the results of thousands of scientific investigations, some dating back more than two hundred years. The inescapable conclusion to emerge from these studies is that most human cancers are caused by an identifiable group of environmental and lifestyle factors.

In the next several chapters, we will examine the causes of cancer and describe how they work. We will begin by considering the various ways in which scientists go about establishing cause-and-effect relationships, followed by an introduction to the major types of cancer-causing agents that have been identified.

THE EPIDEMIOLOGICAL APPROACH

Hardly a week seems to go by without another new claim appearing in the media about the latest cause of cancer. Unfortunately, the news is often contradictory. In recent years, conflicting reports have appeared as to whether radon, pesticides, electromagnetic fields, hair dyes, cell phones, coffee, birth control pills, and numerous other agents are significant causes of cancer. In some cases an apparently definitive news report claiming that a particular agent causes cancer is published one year, only to be contradicted by another study reported the following year. In other cases, one need wait only a few weeks for the contradictory report to appear!

Given the often-conflicting nature of such reports, how can you assess the validity of every new claim and make reasoned decisions as to whether or not a proposed cause of cancer actually represents a significant danger? To answer this question, it is first necessary to understand how scientists go about establishing cause-and-effect relationships that underlie human disease.

Epidemiologists Study the Patterns of Disease Distribution in Human Populations

The first indication that a particular agent may cause cancer is usually provided by an approach called **epidemiology**, the branch of medical science that investigates the frequency and distribution of diseases in human populations. By tracking and analyzing who develops what kind of cancer under which conditions, epidemiologists gather clues about the possible origins of the disease. To illustrate how this process works, let us briefly consider a hypothetical example involving two adjoining cities of comparable size named "Chemtown" and "Cleantown" (Figure 4-1). Suppose you read in a newspaper that researchers have discovered that bladder cancer is twice as frequent in Chemtown as it is in Cleantown. Moreover, the report also notes a possible reason for the elevated cancer rate: A chemical factory in Chemtown pollutes the air so badly that people can smell a strong chemical odor just by driving into the city. In contrast, Cleantown has no chemical plant and the air smells fresh and clean.

Epidemiological evidence like this can be used to support the *hypothesis* that chemical pollutants in the air cause people to develop bladder cancer. Many news reports about suspected causes of cancer involve similar kinds of epidemiological observations. Can such data actually prove that something causes cancer, and if so, what criteria must first be met?

The *p* Value Is a Statistical Tool for Assessing Whether Differences Observed in Epidemiological Studies Are Likely to Be Genuine

The first question to be addressed when evaluating epidemiological data is whether an observed trend represents a genuine phenomenon or a random fluctuation encountered by chance. For example, the hypothetical news report cited above stated that cancer rates were twice as frequent in Chemtown than in Cleantown. Expressing data using *relative numbers* such as ratios (e.g., "twice as frequent") or percentages (e.g., "a 100% increase") can be misleading because such statements hide the underlying *absolute numbers* upon which the ratios or percentages were calculated. A cancer incidence that is "twice as frequent" might simply reflect 2 cases of cancer measured in one population versus 1 case measured in another population of comparable size. If so, the apparent doubling of the cancer rate would have been caused by 1 extra case of cancer, an event of questionable significance that could easily arise by random chance!

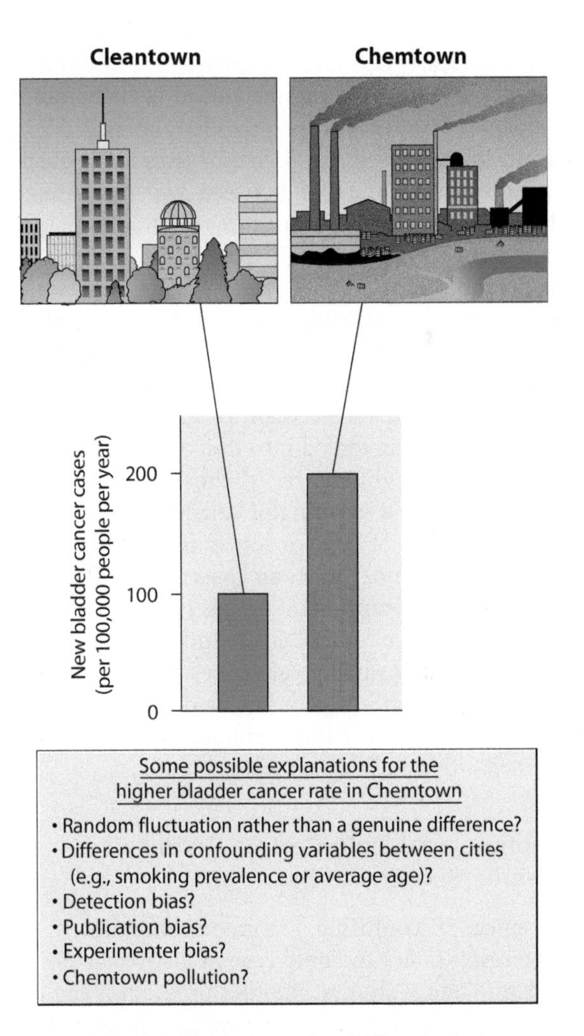

Figure 4-1 **Some Problems Encountered When Analyzing Epidemiological Data.** This hypothetical scenario, involving two cities with different cancer rates, illustrates some of the difficulties encountered when interpreting epidemiological data. If air pollution is more prevalent in the city exhibiting the higher cancer rate, it is tempting to attribute the increased cancer rate to the effects of pollution. However, care must be taken to rule out the possibility that random fluctuations in cancer rates, various sources of bias, or differences in confounding variables are responsible for the differing cancer rates. Methods for addressing such issues are covered in the first part of this chapter.

The larger the sample size, the less likely it is that differences will arise solely through random fluctuations. This principle can be illustrated using the familiar example of flipping a coin. A coin toss has an equal probability of coming up heads or tails, and yet flipping a coin 6 times could easily yield 4 heads and 2 tails—that is, twice as many heads as tails. To find the true probability of heads versus tails, a larger sample is needed. After one thousand coin tosses, for example, the result will probably be much closer to the expected 1:1 ratio of heads and tails. So the larger the sample, the more likely it is that the true value will be revealed.

How large is large enough? Returning to our hypothetical example involving the apparent doubling of bladder cancer rates in Chemtown, the data would obviously be questionable if scientists had measured only 2 cases of cancer in Chemtown and 1 case in Cleantown. Measuring one hundred, or even one thousand, cancer cases would clearly be more impressive, but would that be sufficient? How large do numbers need to be before they are reliable? The answer is provided by calculating a statistic called the ***p* value**, which estimates the probability that an observed difference between two measurements would appear by chance, even if there were, in reality, no such difference at all.

For example, if statistical analysis of the raw data from Chemtown and Cleantown yielded a p value of 0.2, it would indicate a 0.2 (or 20%) likelihood that the difference in cancer rates observed between the two cities was a random fluctuation rather than a genuine difference. A 20% chance of being fooled into thinking a random fluctuation is a real difference is considered to be too unreliable to draw a meaningful conclusion. In common practice, a p value of 0.05 or less is required before scientists will conclude that an observed difference is **statistically significant**—that is, likely to be a genuine numerical difference rather than a random fluctuation. Of course, even such a small p value does not guarantee that an observed difference is genuine. By definition, a p value of 0.05 only means that the likelihood of being fooled is quite small—that is, 5%.

Epidemiological Data Can Be Distorted by Various Sources of Bias

The chance of confusing a random fluctuation for a genuine result is not the only type of error that can arise in epidemiological studies. Another problem is the possibility of **bias**, which refers to the systematic distortion of results in a certain direction caused by the failure to account for some influencing factor. There are many types of bias, but they all share one feature in common: the ability to cause an observed experimental result to deviate from the true value.

One easily understood type of bias is **experimenter bias**, which arises when the expectations of the scientist carrying out the investigation interfere with objective evaluation of the data. For example, if an investigator believes that a particular chemical causes cancer, this belief may affect his or her assessment of how many people exposed to the chemical have actually developed cancer. Evaluation of medical records in such studies should therefore be carried out under "blinded" conditions in which the scientist analyzing the data does not know how much exposure each person has had to the agent being investigated.

Epidemiological studies can also be distorted by **detection bias**, a phenomenon encountered when scientists fail to ensure that equivalent procedures are being used to measure cancer rates in the populations being compared. For example, in the hypothetical situation involving the apparent doubling of bladder cancer rates in Chemtown, let us suppose that the owners of the chemical plant in Chemtown require workers to have frequent medical examinations because of their workplace exposure to toxic chemicals. These frequent exams might allow more cases of bladder cancer to be detected than would have been found in the absence of such rigorous medical screening. The underlying bladder cancer rates in Chemtown and Cleantown could therefore be exactly the same, but more cases would be reported in Chemtown because of greater access to medical detection procedures. In such a situation, the apparent "doubling" of the cancer rate in Chemtown would be a false impression generated by the phenomenon of detection bias.

Another type of distortion, called **recall bias**, is a common problem when scientists rely on individuals to self-report about their past behaviors and exposures to suspected cancer-causing agents. For example, people who have already developed cancer are usually more motivated than other individuals to recall past situations that exposed them to specific chemicals. As a result, cancer may appear to be associated with exposure to a particular chemical, when in fact people who develop cancer may be simply more likely to remember and report being exposed to that agent.

Selection bias is a type of error that arises when people volunteer nonrandomly for research studies. Consider, for example, investigations into the relationship between diet and cancer risk. When individuals with cancer are asked to participate in such studies, 90% or more will typically agree; in contrast, only 50% of the people recruited for the control group (people without cancer) may participate. The potential pitfall is that the 50% who agree to serve in the control group may not be random. If these self-selected participants without cancer happen to volunteer because they have a healthier life style and are interested in dietary studies, it may introduce a bias in which people without cancer appear to have a different diet than those with cancer. Thus, selection bias causes a group of participants to appear to have a certain characteristic simply because people with that trait volunteer in disproportionate numbers.

A final type of distortion, known as **publication bias**, arises from a common practice of scientific journals: They rarely publish studies in which investigators have failed to detect some kind of relationship. Thus, 20 different studies might have been carried out on bladder cancer rates in Chemtown and Cleantown, with 19 of them not being published because they revealed no difference in cancer rates between the two cities. If the 20th study did find bladder cancer rates to be higher in Chemtown, with a statistically significant p value of 0.05, this sole positive study might be the only one to be published. Recall, though, that a p value of 0.05 means that we will be misled 5% of the time—that is, 1 time in 20—into accepting a random fluctuation as being a real difference. The problem created by publication bias is that the 1 study out of 20 containing the misleading result may be the only study to be published!

Confounders Can Lead to the Incorrect Identification of Agents That Supposedly Cause Cancer

A phenomenon called **confounding** can also lead to the incorrect identification of agents that supposedly cause cancer. A confounding variable is a factor that (1) affects the risk of developing cancer and (2) is linked in some way to the factor being investigated. Let us illustrate this concept with a simple example: People who carry matches have elevated rates of lung cancer, but this does not mean that matches cause cancer. People who carry matches are simply more likely to smoke cigarettes, which is the real cause of their elevated cancer rates.

In reality, smoking is a confounding factor in many epidemiological studies because tobacco is a potent cancer-causing agent and a variety of other behaviors are commonly associated with smoking. For example, cigarette smokers are more likely to drink alcohol than are nonsmokers. Therefore, studies designed to investigate the connection between alcohol and cancer must control for the confounding variable, cigarette smoking. Otherwise, data showing an association between alcohol consumption and a particular kind of cancer might be pointing to an effect that is caused by cigarettes rather than alcohol.

Another important confounding variable is age. Cancer rates increase dramatically as people get older, which complicates the study of any agent whose exposure is not randomly distributed among different age groups. For example, suppose that you are interested in determining whether a particular type of prescription drug causes cancer. The most straightforward approach is to simply compare cancer rates in people who take the drug to cancer rates in people not taking the drug. However, the data might be confounded by differences in the average age of the two populations, especially because older people tend to use more prescription drugs than younger people do. If such a study found cancer rates to be higher in people taking a particular drug, the higher cancer rates might simply reflect the older average age of the people taking the drug. Care must therefore be taken to use *age-adjusted incidence rates* (p. 2) that correct for age differences in the populations being compared. Otherwise, the drug rather than age could have been incorrectly identified as the factor responsible for the increased cancer rate.

Retrospective Studies Evaluate Past Events and Prospective Studies Monitor People into the Future

Epidemiological investigations into the causes of cancer can be divided into two general categories: **retrospective studies** that assess the past exposures of people who have already developed cancer and **prospective studies** that follow people into the future to see who will develop cancer. The retrospective approach is a natural extension of a practice that is familiar to doctors, namely the taking of medical histories. In fact, the first suggestion that smoking might cause lung cancer came from doctors who asked lung cancer patients about their past history and discovered that they were virtually all cigarette smokers. Such retrospective "evidence" is just anecdotal, however, unless care is taken to compare data collected from people who have developed cancer with data obtained from a comparable group of people who have not developed cancer. This approach is often called the *case-control* method because individuals with cancer (the "cases") are being compared with an equivalent group of people without cancer (the "controls").

In contrast, prospective studies (also called *cohort studies*) involve a group of people who are selected before anyone has developed cancer and are then followed into the future to see who develops the disease. The initial group may be a random population or may consist of groups that have been selected based on differing exposures to a suspected cancer-causing agent. One advantage of prospective studies is that they are less susceptible to recall bias than are retrospective studies, where the existence of a serious disease like cancer can affect the reliability of recollection and reporting of information about past activities. The main obstacle encountered with prospective studies is that they require a very large sample size. Otherwise, it is unlikely that a sufficient number of cancer cases will arise during the limited amount of time typically available for following a group of individuals into the future.

Several Criteria Must Be Met Before Inferring Cause-and-Effect Relationships from Epidemiological Data

Epidemiological studies have played a central role in the discovery of some of the most important causes of human cancer. A notable example is the identification of cigarette smoking as the primary cause of lung cancer, a topic to be discussed shortly. The main advantage of the epidemiological approach is that it is based on real-life human experiences and is therefore likely to identify agents that are relevant to human disease. However, associations detected using the epidemiological approach may or may not represent cause-and-effect relationships. Recall the connection between carrying matches and developing lung cancer. Although the two events are strongly linked, this linkage does not mean that carrying matches causes a person to develop lung cancer. When a person incorrectly concludes that two events are connected by a cause-and-effect relationship just because they are associated with each other, this error in reasoning is referred to as the **post hoc fallacy**.

So how do scientists determine whether epidemiological studies have uncovered something that actually causes cancer? Rather than providing absolute proof, epidemiology should be viewed as providing evidence that lies on a continuum, with the likelihood of a causal relationship moving forward or backward as evidence accumulates either for or against the hypothesis being investigated. In analyzing the evidence, a number of important criteria are used.

First, statistical significance needs to be demonstrated and the common sources of confounding and bias need to be identified and eliminated. Next, the evidence can be strengthened by demonstrating the existence of a **dose-response relationship**—that is, by showing that the risk of developing cancer increases in direct relationship to the amount of exposure to the agent being investigated (Figure 4-2, *top*). A dose-response relationship is considered to provide strong (but not absolute) evidence for a cause-and-effect relationship because it is unlikely that common sources of confounding or bias would generate data in which cancer rates progressively increase as exposure to a particular agent increases.

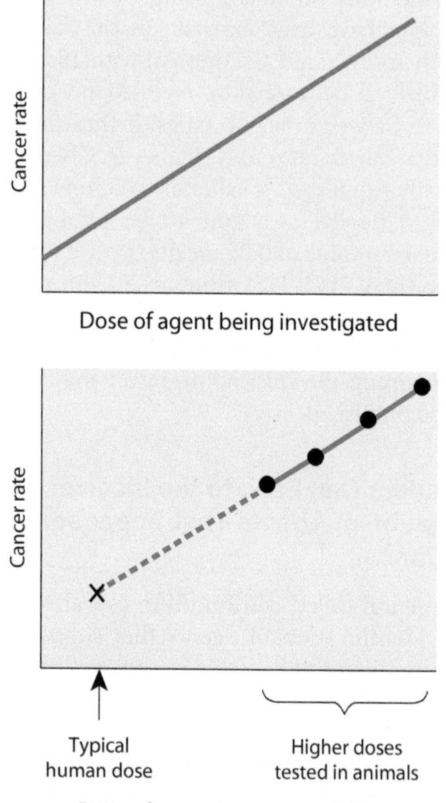

Figure 4-2 The Dose-Response Concept. (*Top*) If cancer rates go up as the exposure dose to a particular agent is increased, it supports the conclusion that the agent in question is causing cancer. (*Bottom*) Suspected carcinogens must sometimes be administered to animals in high doses to observe enough cancer cases to be reliably measured. If the dose-response curve is linear, the results obtained at higher doses in animals (dark circles) can be extrapolated back to the typical human dose (represented by the "X"). Such extrapolations assume that humans and animals are equally susceptible to the carcinogen being evaluated, an assumption that is not always valid. Assuming the existence of a linear relationship is not always valid either (see Figures 5-5 and 5-6).

In addition to the requirement for a dose-response relationship, a number of other criteria must be met before concluding that the evidence for a cause-and-effect relationship is strong. One important criterion is that exposure to the agent in question needs to occur *before*, rather than after, a person develops cancer. Although the reason for such a requirement is obvious, it has not always been met. For example, early investigations into the connection between psychological factors and cancer revealed that cancer patients are more likely than other individuals to be depressed and anxious. These observations were initially interpreted to mean that psychological stress can cause cancer, but a more straightforward interpretation of the data would be exactly the opposite: Depression and anxiety are likely to be triggered by the realization that a person has cancer, occurring *after* the disease arises rather than being the underlying cause!

Another factor affecting the reliability of epidemiological studies is the *magnitude* of the effect being observed. The greater the increase in cancer rates detected in a population exposed to a particular agent, the less likely that the observed increase is caused by confounding or bias. In the case of cigarette smokers, for example, increases of 2500% (25-fold) or more in lung cancer rates have been reported in epidemiological studies, providing strong evidence for a causal relationship. Although effects of this magnitude are relatively rare, suspected cancer-causing agents identified in epidemiological studies generally need to cause *at least a doubling of the cancer rate* before the evidence is likely to be reliable. Cancer-causing agents responsible for smaller increases in cancer rates might still have a significant impact on human health, but these smaller effects are difficult to detect in a reliable fashion using the epidemiological approach.

Finally, evidence for a causal relationship is strengthened when an association between a particular factor and increased cancer rates has been observed repeatedly by different investigators using different study designs and with different populations. And when possible, the argument can be further reinforced if a known biological mechanism exists to explain how a particular agent might cause cancer. If the various criteria discussed in this section have all been met, a strong case can be made for inferring a cause-and-effect relationship based on epidemiological data. In common practice, however, the best arguments for cause-and-effect relationships involve the use of epidemiological data combined with additional evidence obtained from laboratory experiments, our next topic of discussion.

THE EXPERIMENTAL APPROACH

The biggest shortcoming of the epidemiological approach is that researchers are functioning simply as passive observers who monitor and analyze normally occurring human events. To obtain definitive evidence that something causes cancer in humans, an active experiment would need to be performed in which an investigator randomly assigns people to receive varying doses (or no dose at all) of a suspected cancer-causing agent and then waits to see who develops cancer. Needless to say, such an approach—called a **randomized trial**—is not possible in humans because it would be unethical to deliberately expose people to an agent suspected of causing cancer. (We will see in Chapters 11 and 12 that randomized trials do have a role in investigating agents that might cure or prevent cancer, since such experiments involve treatments thought to be helpful rather than harmful.)

Because suspected cancer-causing agents cannot be directly evaluated in human subjects, scientists have turned to animal testing as an alternative. Animal testing and epidemiology are complementary approaches. When epidemiological data indicate that an agent might be causing cancer in humans, animal testing is carried out to confirm that the agent in question can actually cause cancer. Conversely, agents that cause cancer may be initially identified by animal testing, but epidemiological data are then required to determine whether exposure to the agent in question is actually associated with increased cancer rates in humans.

Animal Testing Is Useful for the Preliminary Identification of Possible Human Carcinogens

Animal testing has been used with hundreds of different chemicals to determine which ones are **carcinogens** (cancer-causing agents). By law, any new chemical introduced for human consumption in the United States must first be tested in animals to check for possible carcinogenic effects. Such tests are usually carried out in rats or mice, whose life expectancy is short enough to permit the effects of chemical exposures to be monitored throughout their lifetimes.

The main advantage of animal testing is that experiments can be carried out under precisely controlled conditions in which different groups of animals are exposed to different doses of the chemical being examined. If cancer rates are found to increase as the chemical dose is increased, it provides strong evidence that the chemical causes cancer in animals. In general, a high level of agreement has been detected between the ability of substances to cause cancer in humans and in animals (Table 4-1).

Nonetheless, extrapolating test results from animals to humans is not a straightforward task. First of all, the existence of numerous physiological differences between humans and laboratory animals means that there is no guarantee that a substance found to cause cancer in animals will

Table 4-1 Some Examples of Proven Human Carcinogens That Also Cause Cancer in Animals

Substance*	Main Cancers in Humans	Main Cancers in Animals
Aflatoxins	Liver	Liver
4-Aminobiphenyl	Bladder	Bladder, liver
Arsenic compounds	Skin, lung	Lung
Asbestos	Lung, mesothelioma	Lung, mesothelioma
Benzene	Leukemia	Lymphoma, skin, lung, breast
Benzidine	Bladder	Bladder, liver, breast
Chlorambucil	Leukemia	Leukemia, lung
Coal tar	Skin, lung	Skin, lung
Cyclophosphamide	Bladder, leukemia	Bladder
Diethylstilbestrol	Vagina and cervix**	Breast, uterus, others
Mustard gas	Lung	Lung
2-Naphthylamine	Bladder	Bladder, liver
Tobacco smoke	Lung, bladder, pancreas, others	Lung, lymphomas
Vinyl chloride	Liver, lung, brain, leukemia	Breast, lung, liver

*Sources of human exposure to many of these substances are described in Chapter 5. Data for this table were obtained from *11th Report on Carcinogens, 2005* (U.S. Department of Health and Human Services, Public Health Service, National Toxicology Program).

**Cancer occurs in daughters who were exposed during embryonic development to diethylstilbestrol taken by their mothers while pregnant (see Chapter 5).

always do so in humans (or that a substance that fails to cause cancer in animals will always fail to do so in humans). An even bigger problem concerns the issue of dosage. To illustrate, let us consider a hypothetical example in which 1 case of human cancer per 10,000 people is caused by exposure to a particular dose of a carcinogenic chemical. If everyone in the United States (current population roughly 300 million) were exposed to this same dose, it would cause a total of $1/10{,}000 \times 300{,}000{,}000 = 30{,}000$ cases of cancer. Identifying an agent that causes so many cases of cancer would obviously be important. But how many animals would need to be tested using the typical human dose (adjusted for the smaller body weight of rats or mice)? If animals react the same as humans, 10,000 animals given the carcinogenic chemical would yield just 1 animal with cancer. To obtain statistically reliable results, many more cases of cancer would need to be detected. Such a study would therefore require hundreds of thousands of animals, a requirement that is clearly impractical.

To overcome this problem, animal studies often employ high doses of chemicals to obtain a sufficient number of cancer cases. If the dose-response curve is linear and if animals and humans respond the same way to the chemical in question, cancer rates observed in animals exposed to these higher doses can be used to calculate the number of cancer cases expected at lower doses in humans (Figure 4-2, *bottom*). Unfortunately, we will see in Chapter 5 that such assumptions are not always valid, making it difficult to extrapolate accurately from high-dose animal experiments to low-dose human effects.

Despite these drawbacks, animal testing has one powerful advantage over the epidemiological approach: It allows suspected carcinogens to be tentatively identified and studied in animals before they have caused large numbers of cancers in humans. For this reason, U.S. law requires that new chemicals be tested for cancer-causing ability in animals prior to being introduced for mass human use.

The Ames Test Permits Rapid Screening for Chemical Carcinogens That Are Mutagens

It typically costs about $1 million and several years to thoroughly assess the cancer-causing ability of a single chemical compound using laboratory animals. Given the thousands of chemicals present in the environment, animal testing by itself is therefore an impractical approach for assessing the potential dangers of all the substances to which we are exposed. In trying to devise a test that is faster and less expensive than animal testing, scientists have looked for properties of cancer-causing agents that can serve as a simple substitute for the complex task of monitoring tumor formation. The most successful test is based on the idea, initially proposed around 1950, that cancer-causing agents often act as **mutagens**—that is, they cause cancer by triggering DNA mutations (changes in DNA base sequence). This hypothesis arose from observations dating from the early 1900s, when it was discovered that X-rays cause mutations as well as cancer in animals. By the late 1940s, the ability of certain chemicals to cause both mutations and cancer in animals had also been noted.

The suggestion that carcinogens act by causing mutations was quite speculative when it was initially proposed in the 1950s because nobody had systematically compared the mutagenic potency of different chemicals with their ability to cause cancer. The need for such information inspired Bruce Ames to develop a simple, rapid laboratory test for measuring a chemical's mutagenic activity. The procedure he developed, called the **Ames test**, utilizes bacteria as a test organism because they can be quickly grown in enormous numbers in culture. The bacteria used for the Ames test are a special strain that lacks the ability to synthesize the amino acid *histidine*. As shown in Figure 4-3 (*top*), the bacteria are placed in a culture dish containing a growth medium without histidine, along with the chemical being tested for mutagenic activity. Normally, the bacteria would not grow in the absence of histidine. If the chemical being tested is mutagenic, however, it will trigger random mutations, some of which might restore the ability to synthesize histidine. Each bacterium acquiring such a mutation will grow into a visible colony, so the total number of colonies observed in the presence of an added chemical provides a measure of the mutagenic potency of the substance being investigated.

In Chapter 5, we will see that many chemicals that cause cancer only become carcinogenic after they have been modified by metabolic reactions occurring in the liver; the Ames test therefore includes a step in which the chemical being tested is first incubated with an extract of liver cells to mimic the reactions that normally occur in that organ. The resulting chemical mixture is then tested for its ability to cause bacterial mutations. When the Ames test is performed in this way, a strong correlation is observed between a chemical's ability to cause mutations and its ability to cause cancer (see Figure 4-3, *bottom*). In other words, chemicals that are highly effective mutagens in the Ames test usually turn out to be potent carcinogens.

The Ames test is one of the least expensive and fastest screening tests available, taking only a few days to detect a property—the ability to cause mutations—common to many carcinogens. Unfortunately, the Ames test also gives a significant number of false positive and false negative results, so it cannot be relied on as a sole indicator of whether or not a substance causes cancer. Additional laboratory tests have therefore been devised to measure the ability of suspected carcinogens to trigger mutations, chromosome breakage, and changes in the growth properties of mammalian cells growing in culture. To make a strong argument for (or against) the ability of a substance to cause cancer, it is necessary to have supporting data from several of these additional tests, as well as from animal testing and human epidemiological studies.

The fact that potent mutagens in the Ames test usually turn out to be potent carcinogens does not mean that every chemical that increases cancer risk is a mutagen. Another class of chemical agents, called *promoters*, increase cancer risk but do not cause mutations. A detailed description of the roles played by mutagenic as well as nonmutagenic chemicals in causing cancer will be provided in Chapter 5, "Chemicals and Cancer."

THE MAIN CAUSES OF HUMAN CANCER

Now that the approaches used by scientists to identify the causes of cancer have been described, we are ready to discuss what these studies have revealed. In other words, we can tackle the big question: What causes cancer? The remainder of the chapter will introduce the main categories of cancer-causing agents that have been uncovered to date. Succeeding chapters will then describe how these agents work.

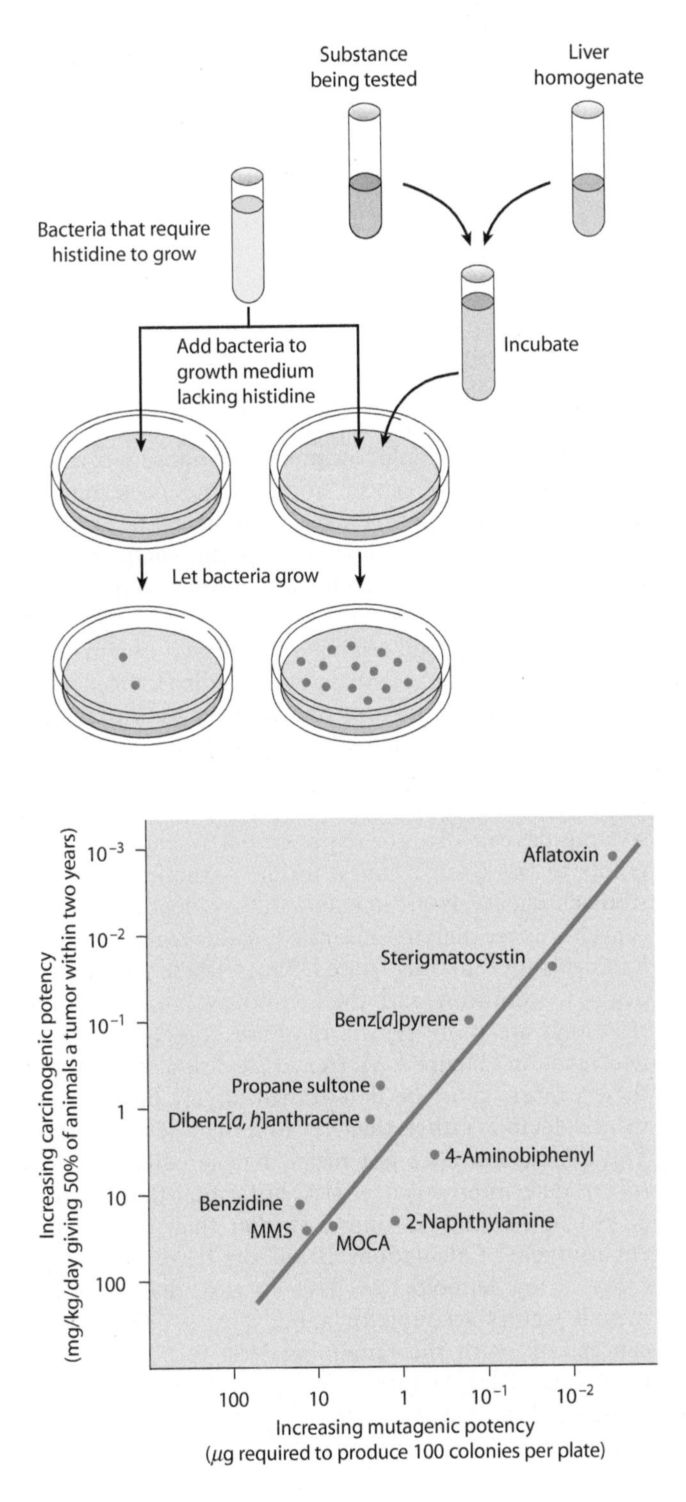

Cancer Rates Increase Dramatically with Age

Many factors affect the likelihood that a person will develop cancer, so we will begin our discussion by trying to identify the factors that have the greatest impact. In terms of its overall effects, the risk factor with the greatest influence in determining whether a person will develop cancer is age. Figure 4-4 illustrates the extent to which the risk of developing several types of cancer increases as a person gets older. For some forms of the disease, the risk of developing cancer after age 60 is more than 1000 times greater than for people under age 40.

In contrast, cancer is relatively rare in children, accounting for less than 1% of the total number of cancer cases diagnosed each year. Still, cancer is second only to accidental death as a cause of childhood fatalities. Leukemias and lymphomas are the most frequent cancers in children, accounting for about half of all childhood malignancies. The remaining childhood cancers arise mainly in the nervous system, kidney, bone, and other supporting tissues, which are forms of cancer that tend to be infrequent in adults. Many of the cancers arising in children respond well to treatment, and cures are fairly common.

When the relationship between aging and cancer was first revealed by epidemiological data, some scientists suggested that cancer might be a direct consequence of aging, somehow built into our genes. Although you might think that evolution would select against any such genetic

Figure 4-3 Ames Test for Identifying Potential Carcinogens. (*Top*) In the Ames test, the ability of chemicals to induce mutations is measured in bacteria that lack the ability to synthesize the amino acid histidine. (There is no special significance to the focus on histidine. Mutations in the gene required for histidine synthesis are simply being measured as a general indicator of mutation rates. The mutation rate of any other gene could have been measured instead.) When they are placed in a medium that lacks histidine, the only bacteria that will grow are those that acquire a mutation that allows them to make histidine. A few such mutations may occur spontaneously (left side of diagram), but more mutations will arise in the presence of a mutagenic chemical (right side of diagram). The number of bacterial colonies that grow in the presence of an added chemical is therefore a measure of the mutagenic potency of the substance being tested. Chemicals being investigated with the Ames test are first incubated with a liver homogenate because many of the chemicals to which humans are exposed only become carcinogenic after they have undergone biochemical modification in the liver. (*Bottom*) The data in the graph reveal that substances that exhibit strong mutagenic activity in the Ames test also tend to be strong carcinogens. Among this particular group of substances, aflatoxin is the most potent mutagen and the most potent carcinogen. *Abbreviations:* MOCA = 4-4′-methylenebis(2-chloroaniline), MMS = methyl methanesulfonate. [Data from S. Meselson and L. Russell in *Origins of Human Cancers* (H. H. Hiatt et al., eds., Cold Spring Harbor, NY: Cold Spring Harbor Laboratory, 1977), pp. 1473–1482.]

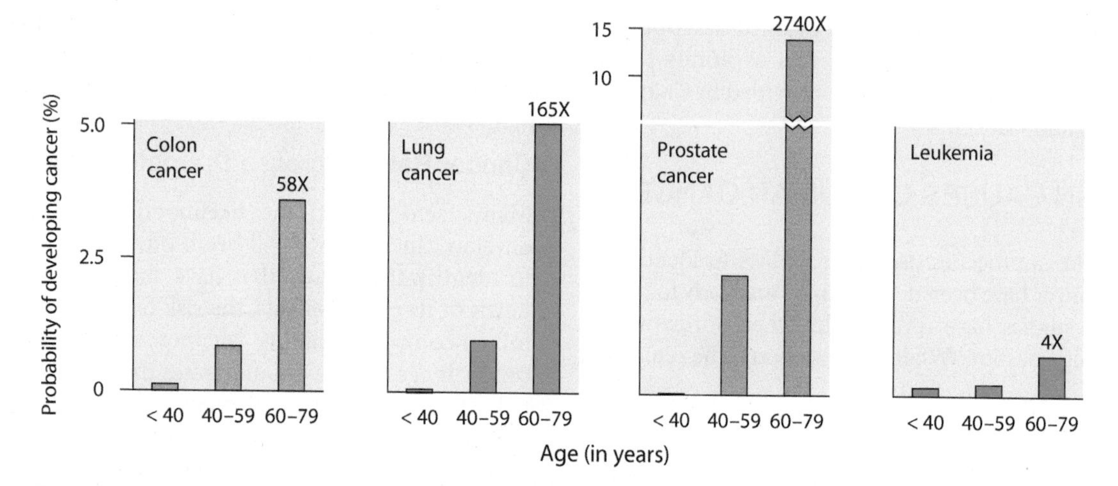

Figure 4-4 Relationship Between Age and Cancer Risk. In terms of overall magnitude, age is the variable with the greatest impact on a person's risk of developing cancer. The relationship between age and cancer risk is shown for several different cancers. Note that the incidence of colon, lung, and prostate cancers goes up dramatically after age 60, whereas the increase in leukemia rates is much smaller. The reason for this difference is that leukemia is one of the few cancers to be commonly observed in young children. [Based on data from *Cancer Facts & Figures 2003* (Atlanta, GA: American Cancer Society, 2003), p. 14.]

program for cancer, remember that most people who develop cancer do so after they have passed through their reproductive years. As a result, any genes that impart a susceptibility to cancer would already have been passed on to a person's children before the negative consequences of cancer appear. Some scientists have even suggested that cancer might serve a useful evolutionary purpose by limiting the number of older people who might otherwise drain resources from succeeding generations of offspring, who are just entering the reproductive phase of their life.

Obviously, such speculations cannot resolve the question of whether or not aging plays a direct role in causing cancer. Yet as we will now see, a fairly clear answer is suggested by epidemiological observations on the geographical distribution of various forms of the disease.

Environmental and Lifestyle Factors Play a Prominent Role in Triggering the Development of Cancer

One of the more striking findings to emerge from epidemiological studies of cancer incidence is the extent to which particular kinds of cancer arise with differing frequencies in different parts of the world (Figure 4-5). For example, we already saw in Chapter 1 that liver cancer is especially prominent in Africa and Southeast Asia, stomach cancer is frequent in Japan, and prostate, colon, breast, and lung cancers are common in the United States. These differences are based on age-adjusted rates, which means that the likelihood of developing a particular kind of cancer for a person of a given age, say 50 years old, is significantly affected by the country in which that person lives.

If a 50-year-old person's risk of developing various types of cancer depends on where he or she lives, it suggests that the causes of cancer are to be found in factors that differ between countries rather than in the process of aging per se. So we must look to variables that differ between geographic locations, and not simply aging, to find the causes of cancer. In theory, these determining variables might involve either environmental factors that vary between countries, differences in the genetic makeup of people living in those countries, or some combination of the two.

To investigate the relative roles played by environment and heredity, scientists have studied cancer rates in people who move from one country to another. In such cases, a person's heredity stays the same but his or her environment changes, so the relative contributions of the two factors can be evaluated. Some of the most striking data involve the experience of Japanese immigrants to the United States. In Japan, the incidence of stomach cancer is greater and the incidence of colon cancer is lower than in the United States. When Japanese families move to the United States, these differences begin to diminish and their children acquire cancer risks that are more typical of those observed in their new location (Figure 4-6). Hence the risk of developing these cancers must be determined largely by environmental factors rather than by inherited susceptibility. This conclusion does not mean that heredity plays no role in determining cancer risk, but rather that the role played by heredity is simply smaller than that of the environment. Calculations based on these and other types of epidemiological data suggest that environmental factors account for about 80% to 90% of our cancer risk, with the remaining 10% to 20% coming from hereditary factors. (The term *environmental* is being used here in its broadest sense and encompasses not just the physical environment, but also the social

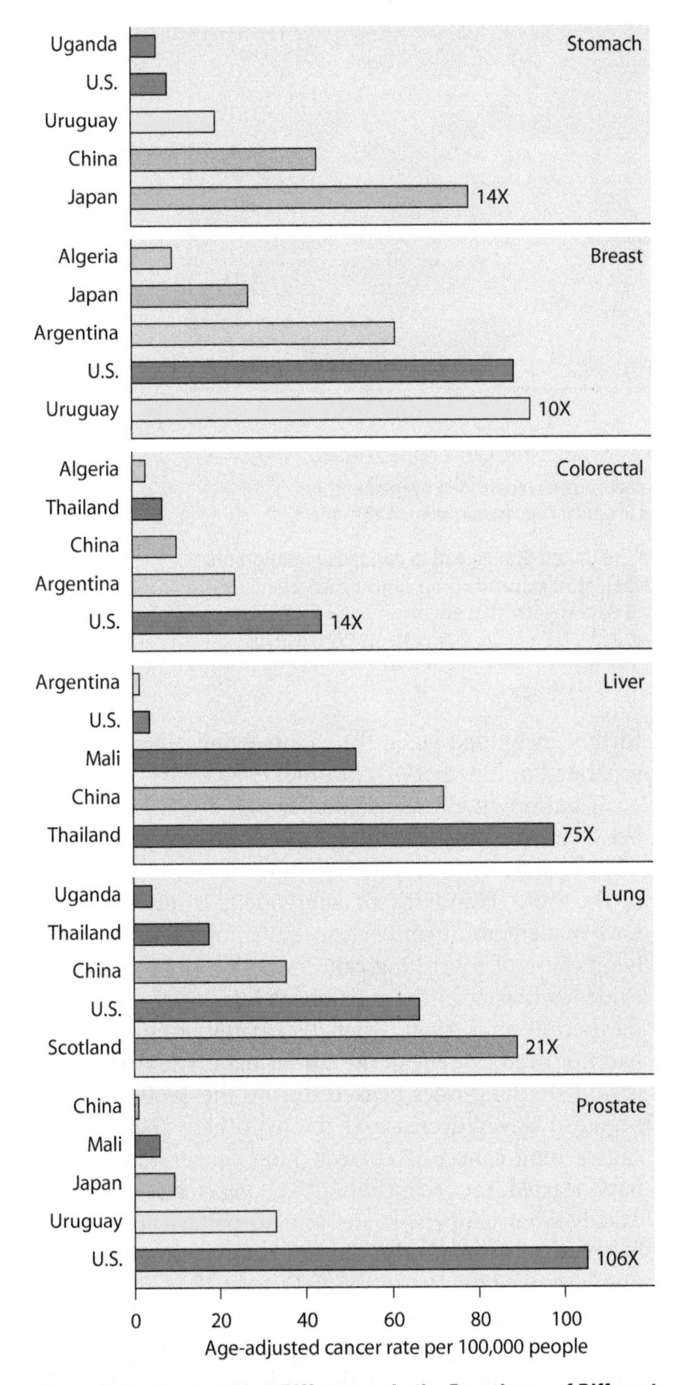

Figure 4-5 Geographical Differences in the Prevalence of Different Kinds of Cancer. The six graphs show the countries with the highest and lowest rates for each of six common kinds of cancer. Several countries with intermediate values are included for each cancer. The number to the right of the bottom bar in each graph indicates the ratio of the cancer rate in the highest-rate country to that in the lowest-rate country. In other words, stomach cancer is 14 times more frequent in Japan than in Uganda and breast cancer is 10 times more frequent in Uruguay than in Algeria. [Based on data from D. M. Parkin et al., *Cancer Incidence in Five Continents*, Vol. VII (Lyon, France: IARC Press, 1997).]

and cultural environment and its associated effect on lifestyle choices.)

If the risk of developing cancer is determined to a large extent by environmental factors, how do we explain the dramatic increase in cancer rates seen in older people? The answer to this question involves at least two components. First, the longer a person lives, the greater the accumulated lifetime exposure to those environmental agents that cause cancer and thus the greater the chance that cancer will eventually arise. Second, the development of cancer is usually a prolonged, multistep process that evolves over a period of several decades, making the disease less likely to appear in younger people. The nature of the steps involved in this multistep process will be described in the next chapter (see Figure 5-17).

Tobacco Smoke Is Responsible for Approximately One-Third of All Cancer Deaths

Once the importance of environmental factors in triggering the development of cancer had been established, the question arose as to the identity of the agents involved. At the top of the list is tobacco smoke, now known to be responsible for roughly one of every three cancer deaths. The first indication of the dangers posed by tobacco smoke came from the emergence of lung cancer as a major disease. Lung cancer was one of the rarest forms of cancer prior to the twentieth century, with only 140 cases reported in the entire world medical literature up until 1898. As late as the 1920s, doctors were still being called in to observe any case that did arise because of the belief that they might never have an opportunity to see a patient with lung cancer again! Today lung cancer is among the most frequently encountered cancers, causing more deaths than any other type.

So what happened during the twentieth century to convert a rare cancer into the number one cancer killer? When doctors started asking questions of the increasing number of lung cancer patients that came to their offices in the early 1900s, it was discovered that virtually all individuals with lung cancer shared one thing in common: They smoked cigarettes. Moreover, cigarette smoking was a relatively new habit. During the 1800s tobacco had been consumed in relatively small amounts, mainly in the form of pipe tobacco, chewing tobacco, and cigars. Habits began to change with the invention of the cigarette rolling machine in 1881, followed by the introduction of safety matches shortly thereafter. Both developments encouraged the smoking of tobacco, and the number of cigarettes consumed per year went from a few dozen per person in 1900 to an average of more than 4000 per person in 1963, the year that smoking rates peaked in the United States.

Figure 4-7 illustrates the relationship between this explosive growth in cigarette smoking and the ensuing epidemic of lung cancer. Examination of this graph reveals that a time lag of about 25 years transpired between the increase in smoking rates and the subsequent increase in lung cancer rates. We now know that such a long delay is typical of the behavior of human cancers, which often require many years after carcinogen exposure to complete the steps involved in creating a malignant tumor.

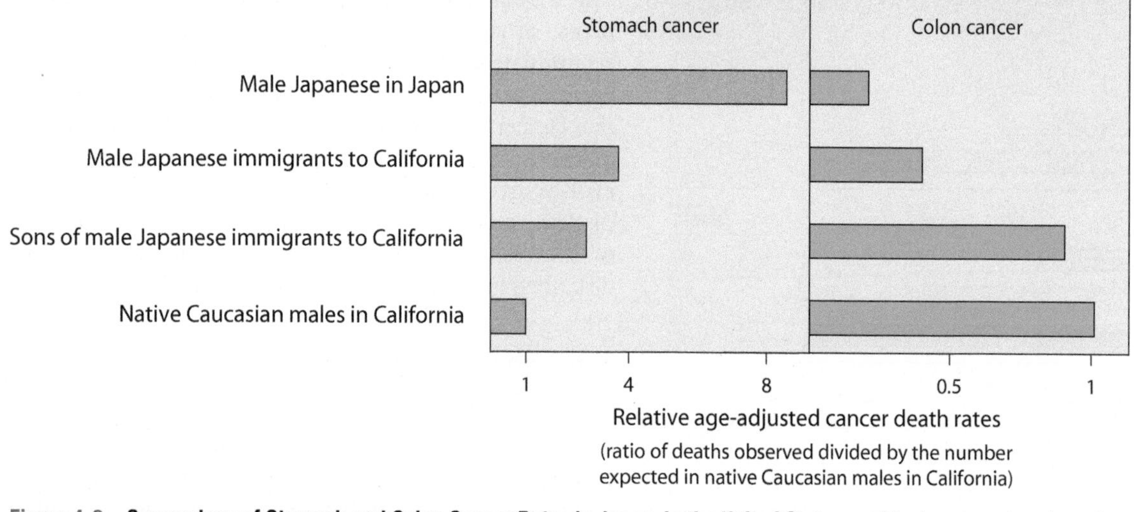

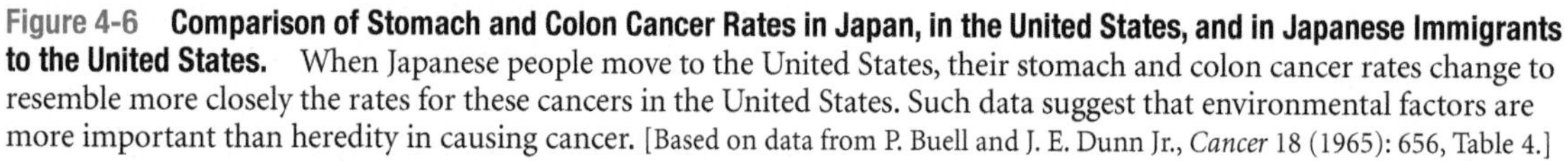

Figure 4-6 Comparison of Stomach and Colon Cancer Rates in Japan, in the United States, and in Japanese Immigrants to the United States. When Japanese people move to the United States, their stomach and colon cancer rates change to resemble more closely the rates for these cancers in the United States. Such data suggest that environmental factors are more important than heredity in causing cancer. [Based on data from P. Buell and J. E. Dunn Jr., *Cancer* 18 (1965): 656, Table 4.]

When the link between cigarette smoking and lung cancer was first widely publicized in the 1960s, some scientists (and virtually all representatives of the tobacco industry) questioned whether epidemiological data pointing to events separated by 25 years could demonstrate that smoking and cancer risk were connected in any way. After all, it could be easily argued that the epidemic of lung cancer that began in the 1940s was caused by some environmental factor that appeared at the same time, such as air pollution, rather than being triggered by cigarette smoking 25 years earlier. As we saw earlier in this chapter, it is inherently difficult to infer cause-and-effect relationships from epidemiological data, and the problem is further magnified when the events being investigated are separated by long periods of time.

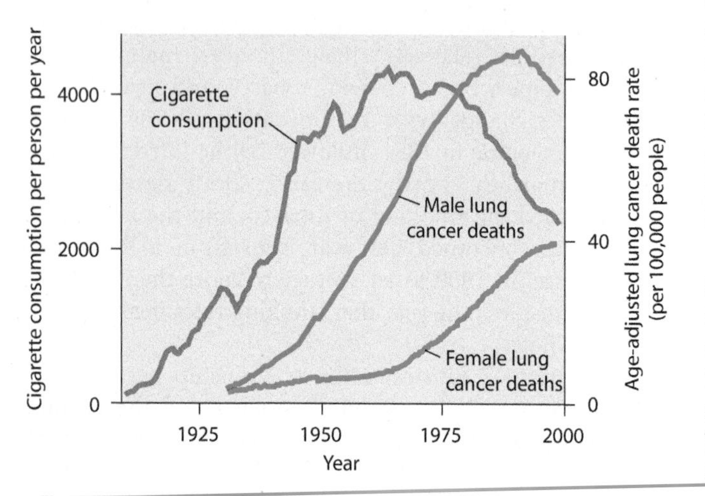

Figure 4-7 Relationship Between Cigarette Smoking and Lung Cancer in the United States. Lung cancer rates in the United States began to increase about 25 years after the increase in smoking rates. Women began to smoke in large numbers several decades after men had started, so lung cancer rates increased later in women. [Adapted from *Cancer Prevention & Early Detection Facts & Figures 2003* (Atlanta, GA: American Cancer Society, 2003), p. 3]

Caution in drawing conclusions about the linkage between smoking and lung cancer was appropriate in 1960 because of the limited amount of evidence available at the time. Hundreds of subsequent studies, however, have made it abundantly clear that smoking is the underlying cause of most lung cancers. Some of the additional evidence is seen in the portion of Figure 4-7 that covers the period after 1960, when two patterns emerged that had not been evident in the earlier data. The first pattern is that smoking rates peaked during the 1960s and then began to slowly decrease. If the hypothesis that smoking causes lung cancer is correct, lung cancer rates should have started to decline about 25 years later, which is exactly what happened. The second pattern involves the behavior of women, who did not begin to smoke in large numbers until the 1940s, about 20 years after smoking had become popular among men. If smoking causes lung cancer, we would expect lung cancer rates to start increasing in women later than it had in men. Again, this is exactly what the data in Figure 4-7 show.

The difference in timing between the onset of smoking as a common habit among men and women fostered a myth, often heard in the 1960s, which claimed that cigarettes cause lung cancer in men but not in women. The origin of such a myth is easy to understand. Many men and women were smoking cigarettes on a regular basis by the early 1960s, but lung cancer was seen mainly in men because they had started smoking decades earlier and it takes about 25 years for lung cancer to develop. The difference in lung cancer rates in men and women was relatively short lived. As expected, female lung cancer rates began to increase in the 1970s, and by 1987, lung cancer had surpassed breast cancer as the number one cancer killer in women (Figure 4-8).

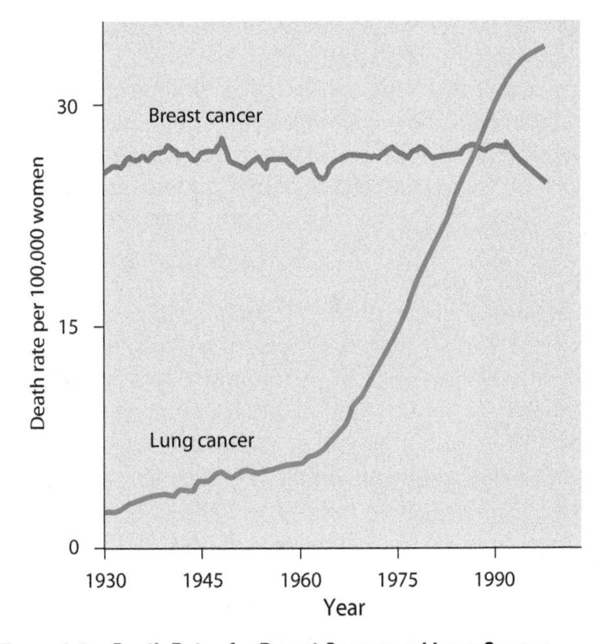

Figure 4-8 Death Rates for Breast Cancer and Lung Cancer in Women. Breast cancer was once responsible for the most cancer deaths among women in the United States. In the late 1980s, however, lung cancer surpassed breast cancer as the number one cancer killer in women. [Adapted from *Women and Smoking, A Report of the Surgeon General–2001* (Rockville, MD: U.S. Department of Health and Human Services, 2001), p. 194.]

A Dose-Response Relationship Exists Between Exposure to Tobacco Smoke and the Risk of Developing Cancer

Additional support for the conclusion that smoking causes lung cancer has come from dose-response data relating cancer risk to how much people smoke. Such data have shown that lung cancer rates are directly proportional to the number of cigarettes smoked per day (Figure 4-9, *left*). A similar relationship has been demonstrated by analyzing the age at which people began smoking. In this case, the data show that long-term smokers develop lung cancer more frequently than do short-term smokers (Figure 4-9, *middle*). Finally, people who inhale deeply develop lung cancer more frequently than smokers who do not inhale deeply, again revealing a dose-response relationship between tobacco smoke exposure and lung cancer (Figure 4-9, *right*). Overall, a typical heavy smoker incurs roughly a 2500% increase in lung cancer risk, which is enormous compared with the risks posed by most of the other causes of cancer.

The carcinogenic properties of tobacco smoke can be traced to a specific group of chemical compounds. Tobacco smoke contains more than 4000 different chemicals, more than 40 of which are carcinogenic when administered to animals (Table 4-2). Laboratory studies have shown that many of these chemicals cause DNA damage and trigger gene mutations, which are early steps in the development of cancer. The carcinogens benzo[*a*]pyrene and 2-naphthylamine, which will be discussed in Chapter 5, are examples of the numerous potent carcinogens in tobacco smoke that cause mutations by forming chemical linkages to DNA (see Figure 5-10). Ironically, strict safety regulations control the use of benzo[*a*]pyrene, 2-naphthylamine, and many of the other substances found in tobacco smoke when these chemicals are handled in the workplace, but the exact same chemicals can be freely inhaled in large concentration by anyone who chooses to smoke cigarettes.

Current evidence suggests that in addition to being responsible for about 85% of all lung cancers, cigarette smoking can also cause cancers of the mouth, pharynx, larynx, esophagus, stomach, pancreas, uterine cervix, kidney, bladder, and colon, as well as leukemias. Depending on the particular cancer involved, smoking may be a major risk

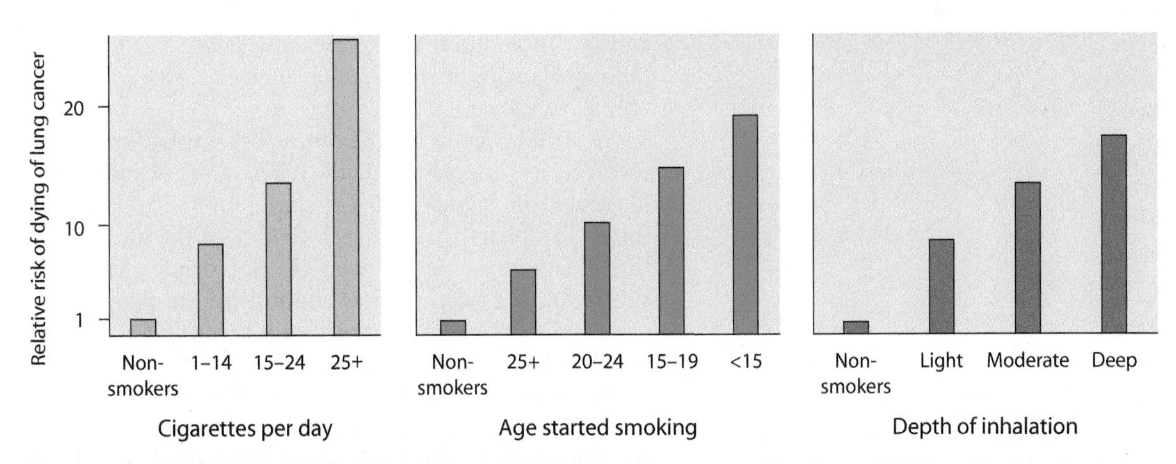

Figure 4-9 Dose-Response Relationships Between Cigarette Smoking and Lung Cancer. (*Left*) The risk of dying of lung cancer increases as a function of the number of cigarettes smoked per day. (*Middle*) The risk of dying of lung cancer increases in direct relation to how many years a person has been a cigarette smoker. (*Right*) The risk of dying of lung cancer increases as a function of how deeply a person inhales cigarette smoke. "Relative risk" is the ratio obtained by dividing the cancer death rate in each group by the corresponding rate in nonsmokers. [Based on data from *The Health Consequences of Smoking—Cancer: A Report of the Surgeon General—1982* (Rockville, MD: U.S. Department of Health and Human Services, 1982), Tables 6, 7, and 8.]

Table 4-2 Some Chemical Carcinogens Found in Tobacco Smoke

Carcinogen*
Acetaldehyde
Acrylonitrile
Aldehydes
4-Aminobiphenyl
Arsenic
Benz[*a*]anthracene
Benzene
Benzo[*a*]pyrene
Benzo[*b*]fluoranthene
Benzo[*f*]fluoranthene
Benzo[*k*]fluoranthene
Cadmium
Chromium
Chrysene
Crotonaldehyde
Dibenz[*a,h*]acridine
Dibenz[*a,h*]anthracene
Dibenz[*a,j*]acridine
Dibenzo[*a,l*]pyrene
7H-Dibenzo[*c,g*]carbazole
1,1-Dimethylhydrazine
Ethylcarbamate
Formaldehyde
Hydrazine
Indeno[1,2,3][*c,d*]pyrene
Lead
5-Methylchrysene
4-(Methylnitrosamino)-1-(3-pyridyl)-1-butanone
2-Naphthylamine
Nickel
2-Nitropropane
N-Nitrosoethylmethylamine
N′-Nitrosoanabasine
N-Nitrosodiethanolamine
N-Nitrosodiethylamine
N-Nitrosodimethylamine
N′-Nitrosonomicotine
N-Nitrosomorpholine
N-Nitrosopyrrolidine
Polonium-210
Quinoline
2-Toluidine
Vinyl chloride

*From *1989 Surgeon General Report: Reducing the Health Consequences of Smoking* (Rockville, MD: U.S. Department of Health and Human Services, Public Health Service, Office on Smoking and Health, 1989), Table 7, pp. 86–87.

factor (e.g., cancer of the lung, mouth, pharynx, and larynx) or a smaller risk factor (e.g., colon cancer and leukemia). When the cancer risks associated with smoking are combined with other tobacco-related illnesses, such as stroke and heart disease, smoking accounts for at least 400,000 of the roughly two million deaths occurring in the United States each year. This statistic makes tobacco the number one cause of preventable death, killing close to half of all regular smokers.

Because tobacco smoke is carcinogenic, exposure of nonsmokers to *secondhand smoke* (tobacco smoke generated by others) is also a potential cancer hazard. Fortunately, the amount of secondhand smoke inhaled by nonsmokers is usually rather small. Current estimates suggest that secondhand smoke typically causes no more than a 30% increase in lung cancer risk, compared with the 2500% increase experienced by the average heavy smoker. Nonetheless, secondhand smoke is thought to contribute to the 15% of lung cancer cases that occur in people who have never smoked. In addition, animal studies have shown that exposure to secondhand smoke increases the concentration of the angiogenesis stimulator *VEGF* in the bloodstream and speeds tumor growth in animals that had been previously injected with cancer cells. Such findings suggest that secondhand smoke may pose cancer-related risks that go beyond the carcinogenic chemicals it contains.

Alcohol Acts Synergistically with Tobacco to Increase Cancer Risk

The use of alcohol is associated with an increased risk of developing several types of cancer, especially in tissues where alcohol makes direct contact with an epithelium. The most common problems are cancers of the mouth, throat, larynx, esophagus, and stomach. Excessive drinking is also the main cause of *cirrhosis*, a degenerative disease in which liver cells are destroyed and replaced by scar tissue and fat. In addition to experiencing poor liver function, individuals with cirrhosis are at increased risk for developing liver cancer.

A striking feature of the cancer risk created by alcohol involves its interaction with tobacco, a phenomenon illustrated in Figure 4-10 for cancer of the mouth and throat (oropharyngeal cancer). Cancer of the mouth and throat is rare in people who do not drink alcohol or smoke, but its rate increases about 6-fold in people who are regular drinkers (but not smokers) and 7-fold in people who are regular smokers (but not drinkers). You might therefore expect that for people who drink *and* smoke, the increase in risk would be no more than 13-fold (i.e., the alcohol-related risk added to the smoking-related risk). In reality, people who are both drinkers and smokers exhibit a 38-fold increase in cancer incidence. This means that the interaction between alcohol and tobacco is **synergistic**; that is, the two agents act together in a

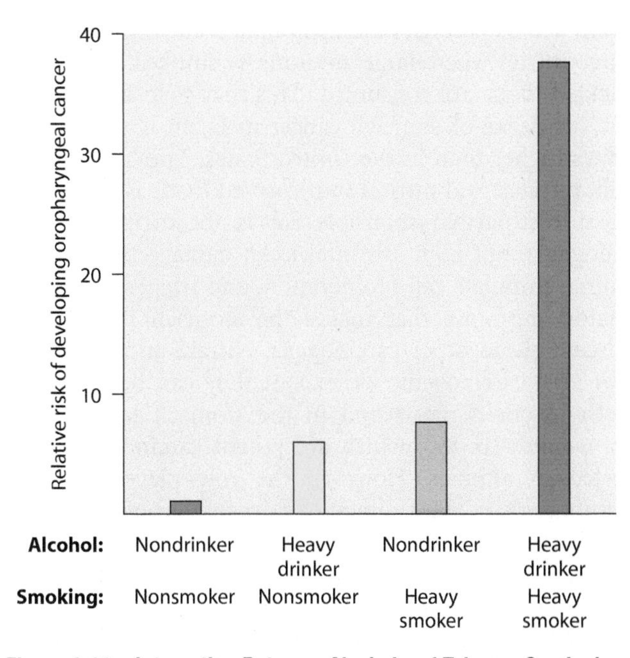

Figure 4-10 Interaction Between Alcohol and Tobacco Smoke in Causing Cancer of the Mouth and Throat. Cancer of the mouth and throat (oropharyngeal cancer) is rare in people who do not drink alcohol or smoke cigarettes, but increases significantly in individuals who do drink, smoke, or both. In these data, heavy drinking is defined as 30 drinks per week and heavy smoking is defined as 2 packs of cigarettes per day for 20 years. Note that the combination of drinking and smoking is synergistic, creating a cancer risk that is greater than the sum of the effects produced by each acting alone. "Relative risk" is the ratio obtained by dividing the cancer rate in each group by the cancer rate in individuals who neither smoke nor drink. [Based on data from W. J. Blot et al., *Cancer Research* 48 (1988): 3285 (Table 4).]

cooperative way to produce an effect that is greater than the sum of the effects produced by each acting alone.

The existence of synergism between alcohol and tobacco can be explained by the fact that the two substances trigger different events in the pathway leading to cancer. We have already seen that tobacco smoke contains dozens of chemicals that are potent carcinogens and mutagens. Alcohol, on the other hand, is only a weak carcinogen when tested in animals and does not seem to cause mutations. Repeated exposure to alcohol, however, causes tissue damage and cell death in regions of the body that are in direct contact with the alcohol. In response to such damage, the surviving cells grow and divide to replace the cells that have been destroyed. It is this ability to trigger cell proliferation (rather than mutation) that appears to explain the role played by alcohol in the development of cancer. Alcohol and tobacco are therefore a deadly combination because tobacco smoke contains potent carcinogens that create DNA mutations and alcohol then stimulates the proliferation of the genetically damaged cells. This involvement of DNA mutation coupled with cell proliferation in the pathway leading to cancer will be discussed in greater depth in Chapter 5 (see Figure 5-17).

Food Contains Carcinogens as Well as Anticarcinogens

Besides tobacco and alcohol, where else do people encounter chemicals that may be carcinogenic? Of the thousands of chemicals we encounter each day, most of the chemicals that enter our bodies do so because we deliberately ingest them in the form of food. Although it is not often viewed that way, food is nothing more than a complex mixture of thousands of different chemicals. So it is obvious to ask whether any of these chemicals can cause cancer.

The narrow question of whether food contains any carcinogens is fairly easy to answer. Laboratory studies employing the Ames test to screen for chemicals that cause mutations, combined with animal testing to see if particular chemicals trigger tumor formation, have led to the identification of dozens of chemicals in food that can cause cancer. Particular attention has been paid to artificial additives and pesticides, although the potential hazards from such substances are probably overemphasized. It has been calculated, for example, that our daily diet is contaminated by about 0.05 to 0.10 milligrams of potentially carcinogenic *synthetic pesticides.* To put this number in proper perspective, the same diet contains about 1500 milligrams of *natural pesticides*—that is, chemicals produced by plants (fruits and vegetables) to protect themselves against insect and animal predators. Thus, 99.99% of the pesticides we consume on a daily basis are natural components of fruits and vegetables.

When these natural pesticides are tested in animals, many turn out to be carcinogenic (Table 4-3). Such natural carcinogens are found in a wide variety of fruits and vegetables, including apples, apricots, bananas, broccoli, Brussels sprouts, cabbage, cantaloupe, carrots,

Table 4-3 Examples of Some Natural Chemicals in Fruits and Vegetables That Are Carcinogenic When Tested in Animals

Substance	Sources
Aflatoxin	Nuts
Allyl isothiocyanate	Mustard, collard greens, brussels sprouts
Benzyl acetate	Jasmine tea, basil, honey
Caffeic acid	Apple, carrot, celery, cherry, grape, lettuce, pear, plum, potato
Coumarin	Cinnamon
Estragole	Basil, fennel
Ethyl acrylate	Pineapple
Hydrazines	Mushrooms
D-Limonene	Black pepper, mango, orange juice
Psoralen	Parsnip, celery, parsley
Safrole	Nutmeg, black pepper

Data from B. N. Ames et al., *Proc. Natl. Acad. Sci. USA* 87 (1990): 7777; and L. S. Gold et al., *Drug Metab. Reviews* 30 (1998): 359.

cauliflower, celery, cherries, grapefruit, grapes, honeydew, lentils, lettuce, mushrooms, mustard, oranges, parsley, peaches, pears, peas, pineapples, plums, potatoes, radishes, raspberries, and tomatoes, to name but a few! In fact, probably every fruit and vegetable sold at the grocery store contains natural plant chemicals that can cause cancer in animals, and the overall concentration of these molecules is thousands of times higher than the concentration of the synthetic pesticides that are present as contaminants.

Despite the large number of natural carcinogens present in fruits and vegetables, dozens of epidemiological studies have shown that people who eat lots of fruits and vegetables actually have a *decreased* risk of developing cancer—especially stomach and colon cancers—rather than an increased risk! One possible explanation for the apparent paradox is that fruits and vegetables contain thousands of different chemicals, and while some have the potential to act as carcinogens, others may function as **anticarcinogens** that protect against the development of cancer. So eating fruits and vegetables involves the consumption of a vast array of chemicals, and the potentially carcinogenic effects of some may be blocked by the protective effects of others. *Lycopene* in tomatoes, *epigallocatechin gallate* in green tea, *resveratrol* in the skin of red grapes, and *sulfides* in garlic are among the numerous natural substances currently being investigated for possible anticarcinogenic activity (see Figure 12-10). These molecules exhibit a broad range of properties with potential cancer-fighting relevance, but the exact role they play, if any, in protecting against cancer is yet to be determined. Current progress in identifying anticarcinogens and elucidating their mechanisms of action will be covered in Chapter 12, "Preventing Cancer."

Given that eating fruits and vegetables tends to decrease rather than increase cancer risk, it seems unlikely that natural pesticides, or tiny residues of synthetic pesticides, represent much of a cancer hazard. However, at least one natural substance found in some foods—a molecule called **aflatoxin**—does pose a significant cancer risk. Aflatoxin is a toxic chemical produced by the mold *Aspergillus*, which grows on grains and nuts stored under humid conditions. Contamination of food with aflatoxin is prevalent in certain areas of Africa and Asia, where its presence correlates with high rates of liver cancer. Laboratory tests have shown that aflatoxin is a strong mutagen and one of the most potent carcinogens ever studied. A total dose of only 0.0001 gram of aflatoxin given to rats spread out over 16 months is sufficient to cause every single animal to develop liver cancer! Fortunately, aflatoxin contamination is only a minor problem in the United States. At least part of the reason can be traced to the widespread use of pesticides that kill the mold responsible for producing aflatoxin. In this particular case, using pesticides may actually be decreasing a carcinogenic hazard by preventing aflatoxin contamination of grains and nuts.

Food preservation techniques are another potential source of carcinogenic chemicals. It has been known for many years that high rates of stomach cancer are common in countries where large amounts of smoked, cured, and pickled foods are consumed. This may help explain why the incidence of stomach cancer in Japan is roughly ten times higher than in the United States. The heavy use of salt, nitrates, and nitrites in preserved foods is thought to be at least partly responsible. Salt is not intrinsically carcinogenic, but high salt intake can damage the stomach lining, stimulate cell proliferation, and trigger an inflammatory response that makes the stomach lining more susceptible to other carcinogens. Nitrate and nitrite are not very carcinogenic either, but they can be converted both in cured meats and in the stomach to *N-nitroso compounds* (p. 89), which are potent carcinogens when tested in animals. However, the role played by such compounds in the etiology of human cancer is far from clear, especially since the data linking cancer risk to preserved foods may be confounded by other variables. For example, countries with a higher consumption of preserved foods also tend to have a lower consumption of fresh fruits and vegetables, which might contribute to the high rates of stomach cancer.

Red Meat, Saturated Fat, Excess Calories, and Obesity May All Contribute to Cancer Risk

One of the better-documented relationships between eating habits and cancer risk involves the connection between meat consumption and colon cancer. As shown in Figure 4-11, countries with high levels of meat

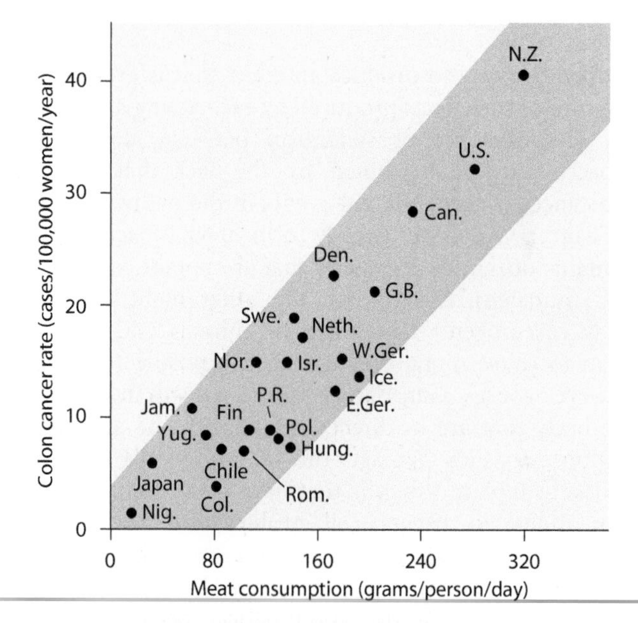

Figure 4-11 Correlation Between Meat Consumption and Colon Cancer. In this graph, where each point represents a different country, colon cancer rates are seen to increase in direct relation to the amount of meat consumed in each country (blue shading is used to highlight the trend). Although such epidemiological data cannot prove a cause-and-effect relationship, experimental studies support the idea that the high fat content of meat increases the risk of developing colon cancer. [Adapted from B. Armstrong and R. Doll, *Int. J. Cancer* 15 (1975): 617.]

consumption typically have higher colon cancer rates than countries where less meat is eaten. However, confounding variables might be responsible for the observed trend, so caution is needed in drawing any cause-and-effect conclusions. For example, people who consume large amounts of meat might also eat fewer fruits and vegetables, in which case the high rate of colon cancer could stem from eating too few fruits and vegetables rather than, or in addition to, eating too much meat. The main usefulness of epidemiological data like those presented in Figure 4-11 is not in establishing cause and effect but in generating hypotheses about possible dietary risk factors that can be tested through additional epidemiological and experimental studies.

In the case of meat consumption and colon cancer, additional evidence has tended to support the idea that eating red meat (but not poultry or fish) contributes to colon cancer, perhaps through the large amount of saturated fat it contains. (The term *saturated* refers to a fat molecule in which all carbon atoms are bound to the maximum number of hydrogen atoms.) Diets rich in saturated fat have frequently (although not always) been associated with increased rates of colon cancer in human epidemiological studies. Animal studies have confirmed that dietary fat is capable of increasing colon cancer rates, but the way in which fat increases cancer risk is not well understood and may involve indirect mechanisms. For example, high-fat diets are known to cause the liver to secrete large amounts of bile acids into the intestines. One of these bile acids, *lithocholic acid*, can produce DNA damage and has been found to induce colon cancer when injected into animals.

Another possible indirect mechanism is related to the fact that fats contain more calories per gram than do proteins or carbohydrates. Animal studies have revealed that cancer rates go up as the number of calories in the diet is increased, so the elevated cancer rates seen in people who eat high-fat diets may be caused by the consumption of excess calories. Support for such an interpretation has come from epidemiological data showing that people who are overweight exhibit an increased cancer risk (p. 249). In fact, some studies suggest that up to one-third of all cancers of the colon, breast, kidney, and digestive tract are linked to obesity and lack of exercise.

In addition to the possible roles played by fat and calories, cooking techniques may also influence the cancer risks associated with meat consumption. Grilling meat at high temperatures over an open flame causes fat droppings on the hot fire to create smoke containing *polycyclic aromatic hydrocarbons* (p. 89), which are carcinogenic chemicals that adhere to the surface of the meat. Carcinogenic *aromatic amines* (p. 89) are also produced under such conditions or even by regular cooking methods when meats are cooked at high temperatures or for a long time. The idea that cooking techniques can influence the production of carcinogens is supported by studies reporting elevated rates of colon cancer in people who prefer eating well-done meat with a darkly browned surface.

Diet Appears to Be an Important Factor in Determining Cancer Risk but Is Difficult to Study

From this brief overview of dietary risk factors, it should be apparent that the human diet consists of a complex mixture of foods that are difficult to study in a systematic way. Consider how much easier it is to study smoking: Scientists simply collect data on people who smoke varying numbers of cigarettes per day and look for a dose-response relationship between smoking and lung cancer. In contrast, it is difficult to find people whose diets differ in the quantity of only a single type of food and to study this difference systemically over many years. Moreover, the effect of any single food on cancer risk tends to be much smaller than the impact of cigarette smoke, making it more difficult to measure its influence reliably.

Dietary studies are also hampered by a lack of reliable methods for determining the kinds of foods that people eat. Consider, for example, the question of whether saturated fat in the diet increases the risk of breast cancer. A number of early retrospective studies suggested a link between fat consumption and breast cancer, but later prospective studies have generally failed to detect such a relationship. One possible explanation for the conflicting data is related to the way in which fat intake is measured. The most commonly used approach is a *food-frequency questionnaire* that asks people to recall the kinds of foods they have been eating. A less widely used, but more accurate approach, is to ask participants to keep a *food diary* in which they record what they eat each day. When the two approaches were compared in 25,000 participants involved in a long term study called the *European Prospective Investigation into Cancer and Nutrition*, the data showed that breast cancer risk was associated with saturated fat intake when measured using a food diary, but this connection was not detected in the same individuals when fat intake was measured using the food-frequency questionnaire. Such observations suggest that imprecise methods for assessing eating habits may be hampering our ability to identify dietary risk factors.

Given such obstacles, it is easy to understand why the history of dietary research has included many conflicting claims about the cancer hazards associated with particular foods. Of course, the difficulty in identifying the risks or benefits associated with individual foods does not mean that diet is unimportant. To the contrary, the differences in cancer rates exhibited by populations with differing diets have led epidemiologists to estimate that diet plays a role in about 30% of all fatal cancers. This value, if correct, would make diet almost as important as tobacco smoke in determining a person's overall risk of dying of cancer. However, assessments of dietary impact involve data that are much less precise than data related to tobacco use, making the 30% estimate no more than a rough approximation. The limited nature of our current understanding concerning the role played by diet, combined with its potential importance if the 30% estimate turns out to be correct, makes dietary studies a vital area for future research. We will return to this topic again in Chapter 12, which deals with the topic of cancer prevention.

Several Types of Radiation Cause Cancer

Like carcinogenic chemicals, radiation is another source of cancer risk routinely encountered in the environment. **Radiation** is defined simply as energy traveling through space. There are many different types of radiation, each defined by its wavelength and energy content. Natural sources of radiation include ultraviolet radiation from the sun, cosmic rays from outer space, and emissions from naturally occurring radioactive elements. Medical, industrial, and military activities have created additional sources of radiation, mainly in the form of X-rays and radioactivity. Among the various types of radiation, two main classes have been clearly identified as causes of cancer: ultraviolet radiation and ionizing radiation.

The ability of *ultraviolet radiation* to cause cancer was first deduced from the observation that skin cancer is most prevalent in people who spend long hours in the sun and is more frequent in geographical areas where the sunlight is especially intense. Because ultraviolet radiation is absorbed by normal skin pigments, dark-skinned individuals have lower rates of skin cancer than do fair-skinned individuals. Exposure to sunlight rarely causes any type of malignancy other than skin cancer because ultraviolet radiation is too weak to pass through the skin and into the interior of the body. Fortunately, the most common types of skin cancer rarely metastasize, and their superficial location makes these cancers relatively easy to remove surgically.

Ionizing radiation poses a more serious cancer hazard because it is strong enough to penetrate through the skin and reach internal organs. The first type of ionizing radiation found to be a cancer hazard was X-rays, which were discovered in 1895 by Wilhelm Roentgen. Shortly thereafter, people working with X-rays began to develop cancer in unexpectedly high numbers. The suspicion that radiation had caused these cancers was soon confirmed by studies showing that animals exposed to X-rays develop cancer at rates that are directly related to the dose of radiation received. Another form of ionizing radiation, called *nuclear radiation*, is emitted by radioactive elements. Among the earliest scientists to work with nuclear radiation was Marie Curie who, along with her husband Pierre Curie, discovered the radioactive elements polonium and radium in 1898. She later died of a form of leukemia that we now know can be caused by high-dose exposure to ionizing radiation. Some human populations, such as the atomic bomb survivors in Japan, have experienced very high doses of ionizing radiation, but the main exposure for most people comes from low-dose sources of natural background radiation, such as the radioactive radon gas that seeps from the earth's crust.

Like most carcinogenic chemicals, cancer-causing forms of radiation create DNA damage and mutations. A description of how this property allows radiation to cause cancer, as well as a detailed account of the types of radiation involved, will be provided in Chapter 6 ("Radiation and Cancer").

Viruses and Other Infectious Agents Can Cause Cancer

The ability of chemicals and radiation to cause cancer was widely recognized by the early 1900s, but the possibility that infectious agents might also be involved was not seriously considered at that time because cancer does not generally behave as if it were contagious. However, in 1911 Peyton Rous carried out a series of studies on sick chickens brought to him by local farmers that showed for the first time that cancer can behave like an infectious disease (Figure 4-12). The chickens brought to Rous turned out to have malignant tumors of connective tissue

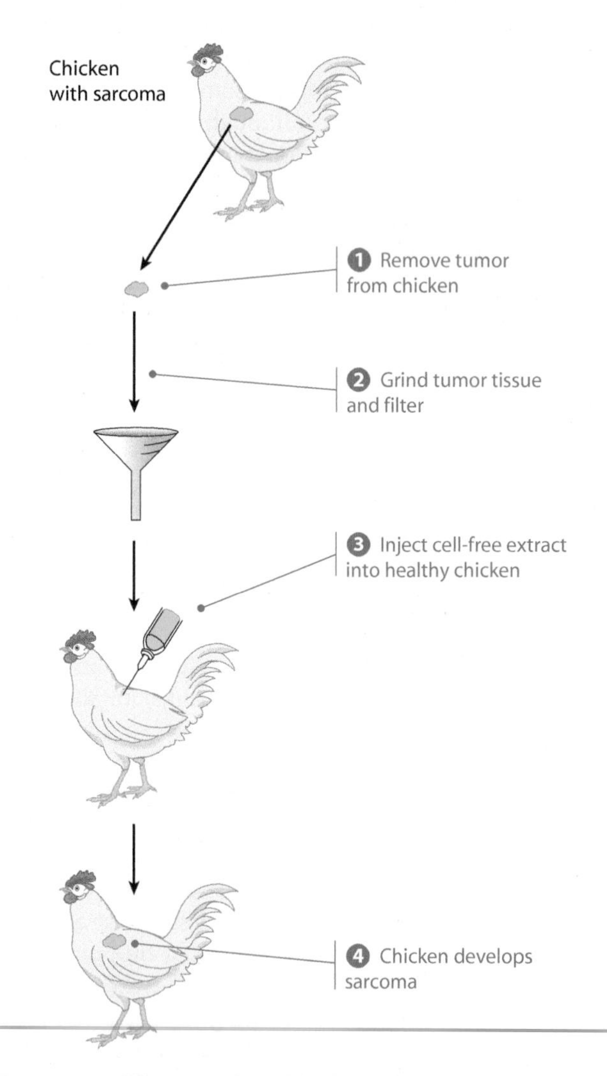

Figure 4-12 Discovery of the Rous Sarcoma Virus. In this experiment carried out by Peyton Rous, sarcoma tissue was ① removed from a sick chicken and the tumor was ② ground up and passed through a filter with tiny pores too small for any cells to pass through. The resulting cell-free extract was then ③ injected into healthy chickens, which ④ subsequently developed sarcomas. Rous concluded that sarcomas are transmitted by an infectious agent that is smaller than a bacterial cell—in other words, a virus.

origin, or *sarcomas*. To investigate the origin of the tumors, Rous ground up samples of tumor tissue obtained from the sick chickens and passed the material through a filter whose pores were so small that not even bacterial cells could pass through it. When he injected this clear, cell-free extract into healthy chickens, Rous found that these chickens also developed sarcomas. Since no cancer cells had been injected into the healthy chickens, Rous concluded that the sarcomas were transmitted by an infectious agent that was smaller than a bacterial cell. This study represented the first demonstration of the existence of an **oncogenic virus**—that is, a virus that causes cancer.

Despite the clarity of the Rous experiments, his conclusions were initially greeted with skepticism. At one point he was told that "this can't be cancer because you know its cause," and it took many years before the existence of cancer-causing viruses came to be widely accepted. Convincing proof required the isolation and characterization of a large number of viruses that trigger different kinds of cancer. Scientists began to discover such viruses in the 1930s, and the list of oncogenic viruses today has grown to include dozens of examples that cause cancer in animals. Several viruses have also been implicated in the development of human cancers. Included in this group are the *Epstein-Barr virus* associated with Burkitt's lymphoma, *human papillomaviruses* associated with cervical cancer, and the *hepatitis B* and *hepatitis C* viruses associated with liver cancer. As a group, these viruses are responsible for about 10% of all cancers worldwide.

Unlike chemicals and radiation, whose carcinogenic effects are based on their ability to cause random DNA damage, some cancer-inducing viruses act by introducing specific genes into target cells. This property has made cancer viruses important to the field of cancer biology not just for the few kinds of human cancer they cause but also as research tools for identifying the types of genes that underlie the development of cancer. In recognition of the numerous contributions to cancer biology that have emerged from the study of cancer viruses, Rous was finally awarded a Nobel Prize in Medicine in 1966. Although Nobel Prizes were originally intended to recognize recent scientific achievements, an 87-year-old Rous received the prize more than 50 years after his pioneering discovery of the virus, now called the **Rous sarcoma virus**, that causes cancer in chickens. Sometimes it takes science a long time to recognize the achievements of the pioneers who open up new fields of investigation. In Chapter 7, we will examine our current understanding of the roles played by viruses, as well as other infectious agents, in the development of cancer.

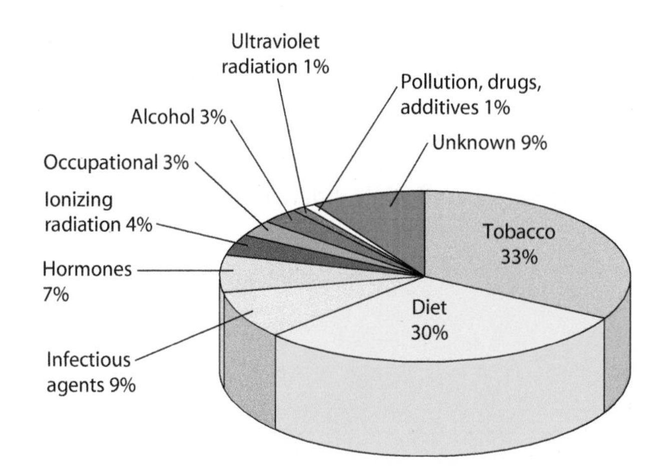

Figure 4-13 Estimate of Cancer Deaths Attributable to Different Environmental Causes. This pie chart is an estimate of the number of cancer fatalities thought to arise from different underlying causes. Ultraviolet radiation from sunlight actually causes more cases of cancer than any other agent, but its value is low in this chart because the skin cancers caused by ultraviolet radiation are rarely fatal. The value for diet is based on data that are less precise than the data for tobacco, making the 30% estimate for dietary impact a rough approximation. [Based on data from R. Doll and R. Peto, in *Oxford Textbook of Medicine* (D. J. Weatherall et al., eds., New York: Oxford University Press, 1996), p. 221 (Table 7).]

Summing Up: The Main Causes of Cancer Involve Chemicals, Radiation, Infectious Agents, and Heredity

The purpose of this introductory chapter on the causes of cancer has been to describe the methods that scientists use to establish cause-and-effect relationships and to introduce the main groups of cancer-causing agents that have been identified using such methods. An estimate of the relative importance of the main causes of cancer, based on the number of fatal cancers produced, is provided in Figure 4-13. In the next four chapters, each of these causes of cancer will be discussed in detail.

Chapter 5 begins with a description of the role played by chemicals in causing cancer. We have already encountered several examples of cancer-causing chemicals in this chapter, including those found in tobacco smoke, alcohol, and the diet. In Chapter 5 the list will be expanded to include the various kinds of carcinogenic chemicals encountered in our homes, workplaces, air, and water. We will assess the risks posed by these carcinogenic chemicals and the mechanisms by which they are thought to act. Chapter 6 examines a similar set of questions for the types of radiation that cause cancer, and Chapter 7 deals in a comparable fashion with the role of oncogenic viruses and other infectious agents.

Finally, Chapter 8 addresses the relationship between heredity and cancer. Although the immigration studies discussed earlier in this chapter (p. 68) indicate that the causes of cancer can be traced largely to environmental and lifestyle factors, heredity plays a significant role in affecting susceptibility to certain forms of cancer. In fact, the inheritance of a single gene mutation may be sufficient to raise a person's risk of developing cancer within his or her lifetime to virtually 100%. Besides describing such hereditary forms of cancer, Chapter 8 will also describe how the study of hereditary risk has contributed to our understanding of nonhereditary cancers.

Even though the causes of cancer are covered in four separate chapters devoted to chemicals, radiation, infectious agents, and heredity, it is important to understand that cancers typically have *multifactorial origins*; that is, several of these factors can contribute to the development of a single cancer. And while different cancers may be triggered by different causes, the ultimate result is always a population of cancer cells whose aberrant behavior can be traced to gene mutations and abnormal gene expression. The nature of the affected genes will be described in Chapters 9 and 10, which are devoted to oncogenes and tumor suppressor genes, respectively.

Summary of Main Concepts

- Many of the factors responsible for causing cancer in humans were initially identified through epidemiological studies in which people who had been exposed to particular agents were later found to exhibit elevated cancer rates.
- Good epidemiological studies require that several criteria be met, including statistical reliability and the elimination of sources of bias such as experimenter bias, detection bias, recall bias, selection bias, and publication bias. Care must also be taken to consider confounding variables, such as smoking and age, that might otherwise distort the data.
- Retrospective (case-control) studies assess the past experiences of people who have already developed cancer, whereas prospective (cohort) studies follow people into the future to see who will develop cancer.
- Epidemiological data cannot provide unequivocal proof for the existence of a cause-and-effect relationship, but the argument for cause and effect is strengthened when the data include a dose-response effect, reveal at least a doubling of cancer rates, show that cancer developed *after* exposure to the agent being investigated, and involve multiple independent studies carried out by different investigators.
- The best proof for the existence of cause-and-effect relationships requires the experimental approach, in which varying doses of a suspected carcinogen are administered to randomly chosen populations. Because such experiments would be unethical in humans, they are performed using laboratory animals. Animal studies can prove that something causes cancer in animals, but extrapolating the conclusion to humans is only valid if the dose-response curve is linear and if susceptibility to the agent in question is similar in animals and humans.
- The Ames test is a rapid screening technique for the identification of potential carcinogens that is based on the rationale that most carcinogens are mutagens.
- Age is the variable with the greatest effect on cancer risk. The longer a person lives, the greater his or her lifetime exposure to carcinogens and the more time available for the multistep process of carcinogenesis to unfold.
- The study of cancer rates in people who move from one country to another has led to the conclusion that environmental and lifestyle factors are responsible for 80% to 90% of a population's overall cancer risk.
- In terms of the total number of cancer fatalities involved, the most important carcinogens are the chemicals found in tobacco smoke. These chemicals, many of which cause DNA mutations, are responsible for roughly one of every three fatal cancers, including most cases of lung cancer and many cancers of the mouth, pharynx, larynx, esophagus, stomach, pancreas, uterine cervix, kidney, bladder, and colon, as well as some leukemias. Tobacco also acts synergistically with alcohol to dramatically increase the risk for certain cancers.
- Diet is thought to be an important contributor to cancer risk, but few definitive conclusions are possible because controlled studies of the human diet are so difficult to carry out. Most of the carcinogenic chemicals that we consume on a daily basis are natural components of fruits and vegetables, and yet paradoxically, people who eat large amounts of fruits and vegetables exhibit a lower rather than an elevated cancer risk, especially for colon and stomach cancers. A possible explanation is that fruits and vegetables may also contain anticarcinogens that protect against the development of cancer.
- Dietary intake of red meat and saturated fats has been correlated with increased risk for certain cancers, although the risk may come from the high calorie content of such foods rather than from fat itself. High-calorie diets and lack of exercise both contribute to obesity, which is a strong risk factor for several forms of cancer.
- Two types of radiation are clearly established causes of cancer: ultraviolet radiation (from sunlight) and ionizing radiation (from X-rays and radioactive elements). Like many cancer-causing chemicals, these forms of radiation trigger DNA damage and mutation.

- Numerous viruses cause cancer in animals, and a smaller number have been implicated in the development of human cancers, including the Epstein-Barr virus associated with Burkitt's lymphoma, human papillomaviruses associated with uterine cervical cancer, and hepatitis B and hepatitis C viruses associated with liver cancer.

Key Terms for Self-Testing

The Epidemiological Approach

epidemiology (p. 61)
p value (p. 62)
statistically significant (p. 62)
bias (p. 62)
experimenter bias (p. 62)
detection bias (p. 62)
recall bias (p. 62)
selection bias (p. 62)
publication bias (p. 62)
confounding (p. 63)
retrospective study (p. 63)
prospective study (p. 63)
post hoc fallacy (p. 64)
dose-response relationship (p. 64)

The Experimental Approach

randomized trial (p. 65)
carcinogen (p. 65)
mutagen (p. 66)
Ames test (p. 66)

The Main Causes of Human Cancer

synergistic (p. 72)
anticarcinogen (p. 74)
aflatoxin (p. 74)
radiation (p. 76)
oncogenic virus (p. 77)
Rous sarcoma virus (p. 77)

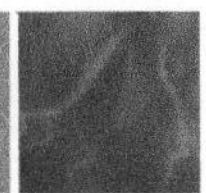

Suggested Reading

The Epidemiological and Experimental Approaches

Adami, H.-O., P. Lagiou, and D. Trichopoulos. Identifying Cancer Causes Through Epidemiology. In *The Cancer Handbook* (M. Alison, ed.). London: Nature Publishing Group, 2002, Chapter 19.

Ames, B. N., W. E. Durston, E. Yamasaki, and F. D. Lee. Carcinogens as mutagens: A simple test system combining liver homogenates for activation and bacteria for detection. *Proc. Natl. Acad. Sci. USA* 70 (1973): 2281.

Benson, B. A., and A. J. Hartz. A comparison of observational studies and randomized, controlled trials. *New England J. Med.* 342 (2000): 1878.

Feinstein, A. R. Scientific standards in epidemiologic studies of the menace of daily life. *Science* 242 (1988): 1257.

Huff, D., and I. Geis (illustrator). *How to Lie with Statistics.* 1954. Reprint, New York: Norton, 1993.

Peto, J. Cancer epidemiology in the last century and the next decade. *Nature* 411 (2001): 390.

Taubes, G. Epidemiology faces its limits. *Science* 269 (1995): 164.

The Main Causes of Human Cancer

Ames, B. N., M. Profet, and L. S. Gold. Dietary pesticides (99.99% all natural). *Proc. Natl. Acad. Sci. USA* 87 (1990): 7777.

Bartecchi, C. E., T. D. MacKenzie, and R. W. Schrier. The global tobacco epidemic. *Sci. Amer.* 272 (May 1995): 44.

Bingham, S. A., et al. Are imprecise methods obscuring a relation between fat and breast cancer? *Lancet* 362 (2003): 212.

Bingham, S., and E. Riboli. Diet and cancer—The European Prospective Investigation into Cancer and Nutrition. *Nature Reviews Cancer* 4 (2004): 206.

Block, G., B. Patterson, and A. Subar. Fruit, vegetables, and cancer prevention: A review of the epidemiological evidence. *Nutrition and Cancer* 18 (1992): 1.

Blot, W. J., et al. Smoking and drinking in relation to oral and pharyngeal cancer. *Cancer Research* 48 (1988): 3282.

Doll, R., and R. Peto. Epidemiology of Cancer. In *Oxford Textbook of Medicine* (D. J. Weatherall, J. G. G. Ledingham, and D. A. Warrell, eds.). New York: Oxford University Press, pp. 197–222.

Hecht, S. S. Tobacco carcinogens, their biomarkers and tobacco-induced cancer. *Nature Reviews Cancer* 3 (2003): 733.

Key, T. J., et al. The effect of diet on risk of cancer. *Lancet* 360 (2002): 861.

Proctor, R. N. Tobacco and the global lung cancer epidemic. *Nature Reviews Cancer* 1 (2001): 82.

Willett, W. C. Diet and cancer: An evolving picture. *JAMA* 293 (2005): 233.

World Cancer Research Fund. *Food, Nutrition and the Prevention of Cancer: A Global Perspective.* Washington, DC: American Institute for Cancer Research, 1997.

5 Chemicals and Cancer

The idea that chemicals can cause cancer was first proposed more than two hundred years ago based on the discovery that people exposed to tobacco snuff or chimney soot develop cancer at rates that are higher than normal. Thousands of subsequent studies, involving both epidemiological and experimental approaches, have confirmed that exposure to certain chemicals can lead to cancer. As a consequence, the U.S. government now routinely publishes a list of several hundred chemicals that are known to be, or reasonably anticipated to be, human carcinogens (see Appendix B).

This list represents only a small fraction of the more than 75,000 chemicals currently in widespread use, and most have never been tested for carcinogenic properties. While the ability to cause cancer is not very common among those chemicals that have been tested, it is a virtual certainty that additional carcinogens remain to be discovered from among the massive list of chemicals that we routinely encounter. Individually analyzing each of these substances to determine which ones are carcinogenic is clearly an overwhelming and impractical task. Nonetheless, considerable progress has been made in determining the types of molecules that are likely to be carcinogenic and the mechanisms by which they act.

IDENTIFYING CHEMICALS THAT CAUSE CANCER

We seem to be surrounded by a sea of chemical carcinogens: They are found in the air we breathe, the food we eat, the water and beverages we drink, the medications we take, the places where we work, and the homes in which we live. However, this assessment—while technically correct—conveys the misimpression that we are faced with severe hazards everywhere we look and that these dangers cannot be avoided. In fact, many of the carcinogens we normally encounter are only weakly carcinogenic, and most of the more potent ones can be easily avoided by the general public. So rather than lumping all chemical carcinogens together, we need to consider them as individual molecules and make informed judgments about the dangers posed by each one.

Chemical Carcinogens Were First Discovered More Than Two Hundred Years Ago

The first indication that chemicals might cause cancer came from the observations of doctors who, in their struggle to understand the nature of the disease, asked cancer patients a variety of questions about their backgrounds, experiences, and habits. This allowed physicians to gain some impressions, if not firm evidence, about the possible causes of cancer. Such an approach led a London doctor, John Hill, to point to chemicals as a probable cause of cancer more than two hundred years ago. In 1761, Hill reported that people who routinely use snuff—a powdered form of tobacco that is inhaled—suffered an abnormally high incidence of nasal cancer, suggesting the existence of one or more cancer-causing chemicals in tobacco.

Several years later Percival Pott, another British physician, reported an unusual prevalence of oozing sores on the scrotums of men coming to his medical practice in London. While a less astute observer might have thought it was just one of the venereal diseases that were widespread at the time, Pott's close examination of the sores revealed that they were actually a form of skin cancer. Careful questioning revealed that the men with this condition shared something in common: They had all served as chimney sweepers in their youth. It was common practice at the time to employ young boys to clean chimneys because they fit into narrow spaces more readily than adults. Pott therefore speculated that chimney soot chemicals had become dissolved in the natural oils of the scrotum, irritating the skin and eventually triggering the development of cancer. These ideas led to the first successful public health campaign for preventing a particular type of cancer: Scrotal cancer was virtually eliminated by promoting the use of protective clothing and regular bathing practices among chimney sweeps.

In the years since these pioneering observations, it has become increasingly evident that certain kinds of chemicals can cause cancer. Unfortunately, the ability of a particular chemical to cause cancer has often become apparent only after large numbers of cancers arise in people exposed to that chemical on a regular basis. For example, in the early 1900s elevated rates of skin cancer were noted among workers in the coal tar industry, and an increased incidence of bladder cancer was seen in factories that produced aniline dyes. The experience in the aniline dye industry was especially dramatic and led to the discovery of several basic principles of chemical carcinogenesis, as will now be described.

Workers in the Aniline Dye Industry Developed the First Cancers Known to Be Caused by a Specific Chemical

The late 1800s witnessed the birth of a series of new chemical industries that for the first time exposed large numbers of workers to high concentrations of toxic substances. A prominent example involved the industrial production of dyes used to color clothing and other fabrics. Prior to the mid-1800s, most dyes were natural substances extracted from vegetable or animal sources. An accidental discovery made in 1856 by William Perkin, however, led to the birth of the synthetic dye industry and the first mass exposure of workers to potent carcinogenic chemicals.

Perkin was attempting to synthesize quinine, a drug for treating malaria, by carrying out chemical reactions on substances present in coal tar (a thick, black liquid formed during the distillation of coal). In one experiment, he extracted *aniline* from coal tar and oxidized it with potassium dichromate. The result was a dark brown precipitate. Most nineteenth-century chemists would have discarded any such dark masses of material because scientists were generally interested in clear, crystalline products. But Perkin was instinctively curious and decided to investigate further. To his pleasant surprise, dissolving the dark sludge in alcohol yielded an intense purple solution that exhibited strong dying properties. Perkin had discovered *aniline purple*, the first synthetic dye.

Within a few years, coal tar had yielded several other dyes and the aniline dye industry was born. Chemists quickly discovered that a compound related to aniline called *2-naphthylamine* is an ideal starting material for the synthesis of many dyes, and large-scale production began in Germany around 1890. Unfortunately, factory employees working with 2-naphthylamine soon began developing bladder cancer in alarming numbers (Figure 5-1). In one small factory, all 15 workers developed the disease. A vigorous and protracted debate ensued as to whether 2-naphthylamine was actually responsible because bladder cancer also occurs among the general public and there was little precedence for using epidemiological data to infer cause and effect. Eventually the cancer-causing ability of 2-naphthylamine was demonstrated in convincing fashion, using both epidemiological and animal data, but it took almost 50 years before the large-scale production of this highly potent carcinogen was stopped.

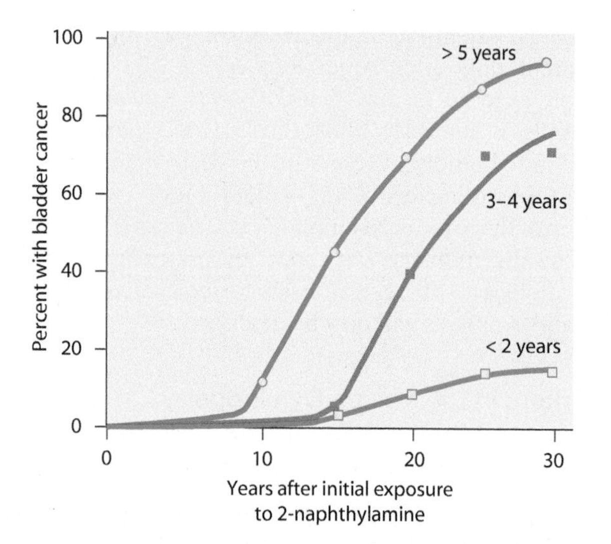

Figure 5-1 Relationship Between 2-Naphthylamine Exposure and Bladder Cancer in Factory Workers. These data show bladder cancer rates in three groups of men exposed for varying amounts of time to 2-naphthylamine in the workplace. Note that in the group of men with the longest exposure (more than 5 years), almost every person eventually developed bladder cancer. [Data from J. Cairns, *Cancer: Science and Society* (San Francisco: Freeman, 1978), p. 56.]

Bladder cancer triggered by occupational exposure to 2-naphthylamine was the first example of a human cancer known to be caused by a specific chemical compound. The introduction of 2-naphthylamine into the workplace was also a key event because it marked the beginning of a massive increase in industrial chemical production and illustrated some important principles that are widely relevant to chemical carcinogenesis. One of these principles involves the long delay that is typically observed between exposure to a chemical carcinogen and the onset of cancer. Few cases of bladder cancer were seen in factory workers until 10 years after initial exposure to 2-naphthylamine, and most cases took between 15 and 30 years to appear. Such a long delay is typical of chemical carcinogenesis and reflects the multiple events that take place on the road to developing cancer. Another principle illustrated by the experience with 2-naphthylamine is dose dependence: Workers who were exposed to the chemical over a longer period of time, and hence had a larger total exposure, exhibited higher bladder cancer rates than workers with shorter exposures. Figure 5-1 reveals that almost every worker in the group with the longest exposure to 2-naphthylamine eventually developed bladder cancer. If virtually everyone develops cancer under such conditions, it means that hereditary differences did not play a significant role in determining risk.

Finally, the experience with 2-naphthylamine illustrated the organ specificity that is a common feature of chemical carcinogenesis. Instead of causing cancer in general, chemical carcinogens tend to preferentially cause a few particular types of cancer. In the case of 2-naphthylamine, the bladder is the prime target. We have already encountered other examples of organ specificity—such as aflatoxin and liver cancer mentioned in Chapter 4—and will encounter additional examples later. Organ specificity is generally caused by the selective ways in which chemicals make contact with, or accumulate in, certain body tissues. For example, chemicals that become concentrated in the urine are likely to produce bladder cancer, and carcinogens that are inhaled tend to cause lung cancer. The ability of cigarette smoke to cause many kinds of cancer in addition to lung cancer may seem to violate this principle; tobacco smoke, however, contains more than 40 different carcinogens, and some of these accumulate in tissues other than the lung.

Asbestos Is the Second Most Lethal Commercial Product (After Tobacco) in Causing Cancer Deaths

The natural mineral **asbestos** is a particularly striking example of an organ-specific carcinogen. Commercial use of asbestos began in the late 1800s, when large deposits of asbestos rock were discovered in Canada and shipped to the United States and to the newly industrializing countries in Europe. Crushing the rock yields a mixture of fine fibers that can be woven into materials that exhibit excellent insulating and fire-retarding properties. The most commonly used form of asbestos has the formula $(Mg,Fe)_3Si_2O_5(OH)_4$, but many chemical variations exist. Numerous fireproof products, ranging from oven mittens and fireproof clothing to various kinds of construction materials, have been manufactured with asbestos.

Unfortunately, the widespread use of asbestos has had severe health consequences. Asbestos readily breaks down into a fine dust containing numerous sharp, needle-like fibers that are so tiny that they can only be seen with an electron microscope. These "needles of death" are easily inhaled and become lodged in the lung, where they cause scarring that kills people through suffocation. Shortly after this disease, called *asbestosis*, was first recognized among asbestos workers in the 1920s, the same workers began to develop lung cancer. Because cigarette smoking was not yet popular, lung cancer was still rare, and the connection between the cancer outbreak and exposure to asbestos was therefore easy to detect. Scientists eventually found that asbestos and cigarette smoke interact synergistically in causing lung cancer. As a result, smokers who have been heavily exposed to asbestos exhibit lung cancer rates that are 50 times higher than is observed in people who do not smoke or have significant exposure to asbestos.

An unusual property of asbestos is its ability to cause **mesothelioma**, a rare form of cancer derived from the *mesothelial cells* that cover the interior surfaces of the chest and abdominal cavities. This type of cancer was uncommon prior to the 1950s, when the first mesothelioma epidemic was reported in and around a group of asbestos mines in South Africa. Mesotheliomas were subsequently detected in many locations around the world, virtually everywhere that asbestos is used. An increased risk for mesothelioma is exhibited mainly by asbestos workers and by individuals who experience significant exposure to asbestos either by living in neighborhoods surrounding asbestos factories or by working or living in asbestos-insulated buildings. At present, asbestos is the only clearly established cause of mesothelioma.

Microscopic examination of lung tissue obtained from asbestos workers has revealed that mesothelioma is caused by a rather unusual mechanism. Tiny, microscopic fibers of inhaled asbestos become embedded in the lung and gradually penetrate completely through the lung tissue, emerging into the chest cavity. Here the asbestos fibers trigger a chronic irritation and inflammation that promotes the development of cancer in the mesothelial cells that cover the lungs and line the interior chest wall. In a similar fashion, asbestos fibers that have been inadvertently ingested can penetrate through the walls of the stomach and intestines, emerging into the abdominal cavity and triggering the development of abdominal mesotheliomas. When the fatalities caused by asbestos-induced mesotheliomas and lung cancers are combined, asbestos ranks as the second most lethal commercial product (after tobacco) in terms of the number of cancer deaths caused.

Governmental actions to regulate the production and use of asbestos began in earnest in the 1960s and have become progressively more restrictive in many countries, so the incidence of asbestos-induced cancers should eventually begin to decline. Nonetheless, mesothelioma deaths are still rising (Figure 5-2) and will probably continue to do so for several decades because a lag period of 30 or more years can intervene between asbestos exposure and developing cancer. Moreover, countries that formerly used asbestos still contain vast reservoirs of the carcinogen in existing buildings.

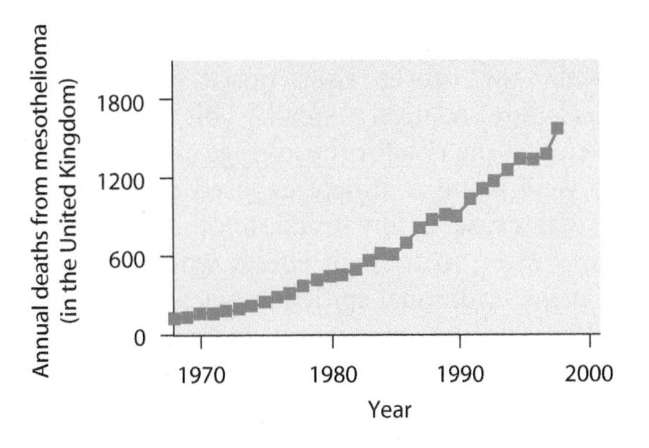

Figure 5-2 Death Rate Trends for Mesothelioma. Despite increasing restrictions on the production and use of asbestos, deaths from mesothelioma (a cancer induced almost solely by asbestos exposure) continue to rise. The most likely reasons are that a long delay can intervene between asbestos exposure and developing cancer and that vast reservoirs of asbestos remain behind in existing buildings. Data are for the United Kingdom. [Data from G. Tweedale, *Nature Reviews Cancer* 2 (2002): 311 (Figure 2).]

Workplace Exposure to Chemical Carcinogens Occurs in Numerous Industries

As in the case of 2-naphthylamine and asbestos, carcinogenic chemicals are often identified only after a particular type of cancer starts to appear in people exposed to a specific substance in high doses. Once such observations point to a particular chemical as a potential carcinogen, follow-up animal testing is carried out to determine whether the substance really causes cancer. Beginning around 1900 with 2-naphthylamine in the aniline dye industry, the list of known chemical carcinogens grew progressively as the Industrial Revolution proceeded throughout the twentieth century and unusual cancer patterns began to emerge among workers in various industries. Table 5-1 lists some of the main occupational carcinogens that were eventually discovered, including examples from the rubber, chemical, plastic, mining, fuel, and dye industries.

Workplace exposure to occupational carcinogens was substantial in the first half of the twentieth century before the cancer risks from such agents came to be fully appreciated. In a few extreme cases, all the workers exposed to the chemicals present in certain factories eventually developed cancer. However, most of the currently known occupational carcinogens were identified by the 1970s and relatively few new ones have been identified since then. In 1970, an act of the United States Congress created the *Occupational Safety and Health Administration* (*OSHA*) to formulate regulations designed to protect the safety and health of workers. OSHA has worked to eliminate the most dangerous chemicals from the workplace and to limit worker exposure to other chemicals. As a result, many occupationally induced cancers that were once prevalent in the United States have declined in frequency, and workplace exposure to carcinogens now accounts for less than 5% of all cancer deaths.

While a similar pattern is evident in many other industrialized nations, progress has been far from uniform. To illustrate some of the disparities, the use of asbestos in Nordic countries has decreased dramatically in recent decades, falling to a negligible value of 4 grams per person in 1996; in that same year, asbestos use in the former Soviet Union was 600 times higher at a value of 2400 grams per person. In general, exposure to industrial carcinogens is a greater problem in developing countries, where less progress has been made in regulating the workplace use of toxic chemicals.

Environmental Pollution Is Not a Major Source of Cancer Risk

Cancers arising in the workplace are usually triggered by sustained high-dose exposures to specific chemical carcinogens. Small amounts of the same chemicals are also released inadvertently into the environment, where they contaminate the air we breathe, the water we drink, and the food we eat. It has therefore become fashionable to blame industrial pollution for creating a growing cancer threat of epidemic proportions.

Table 5-1 Some Human Carcinogens for Which Exposure Is Mainly Occupational

Carcinogen	Industry	Type of Cancer
4-Aminobiphenyl	Rubber	Bladder
Arsenic	Glass, metals, pesticides	Lung, skin
Asbestos	Insulation, construction	Lung, pleura
Benzene	Solvent, fuel	Leukemia
Benzidine	Dye	Bladder
Bis(chloromethyl) ether*	Chemical	Lung
Chloromethyl methyl ether*	Chemical	Lung
Cadmium	Pigment, battery	Lung
Chromium	Metal plating, dye	Nasal cavity, lung
Coal-tar pitches	Construction, electrodes	Skin, lung, bladder
Coal-tars	Fuel	Skin, lung
Ethylene oxide	Chemical, sterilant	Leukemia
Mineral oils	Lubricant	Skin
Mustard gas*	Chemical weapon	Pharynx, lung
2-Naphthylamine*	Dye	Bladder
Silica	Construction, mining	Lung
Soots	Dye	Skin, lung
Sulfuric acid mist	Chemical	Larynx, lung
Talc	Paper, paint	Lung
Vinyl chloride	Plastic	Liver

*Mainly of historical interest.

However, there are reasons to question such a far-reaching conclusion. First, cancer risk is related to carcinogen dose, and the doses of industrial carcinogens to which the public is exposed are generally orders of magnitude lower than is encountered in the workplace. For example, consider the pesticide *ethylene dibromide (EDB)*, which is designated by the U.S. government as a probable human carcinogen based on its ability to cause cancer in animals. Workers who have experienced high-dose exposures to this suspected carcinogen encountered about 150 milligrams (mg) per day, whereas EDB residues in food (before EDB was banned in 1984) exposed the average person to a daily dose of 0.00042 mg, which is 300,000 times less than the workers' daily intake. A similar situation exists for most of the other chemical contaminants in our food, air, or water, which do not represent a major cancer threat because they are present in concentrations thousands of times lower than typical industrial exposures.

Another reason to question the assumption that pollution represents a major cancer hazard is based on historical trends. If industrial pollution were a major cancer threat to the general public, one would have expected a significant increase in overall cancer rates during the twentieth century in response to the explosive growth in the use of industrial chemicals. For example, the yearly production rates of plastics, pesticides, and synthetic rubber in the United States increased more than 100-fold between the 1940s and the 1970s. Any impact the chemicals used in these industries might have had in causing a cancer epidemic through environmental pollution should have been apparent by now. In fact, when the data are adjusted for the increasing average age of the population, it is clear that a significant growth in age-adjusted cancer rates has not occurred (Figure 5-3). Most of the cancers that are common today were also common one hundred years ago, and the main exception, lung cancer, is triggered by cigarette smoke and has little to do with industrial pollution.

Risks from Low-Dose Exposures to Chemical Carcinogens Are Difficult to Assess

The preceding arguments suggest that for most people other than those who receive high-dose exposures by working with or living near a concentrated source of toxic chemicals, the cancer risks posed by environmental chemicals are relatively small. This does not mean, however, that the risk for the average citizen is zero. If the public were being routinely exposed to low carcinogen doses that cause a tiny fraction of the population to develop cancer, such small effects would be difficult to detect using traditional epidemiological methods. In fact, a weak carcinogen present in the environment could theoretically cause hundreds or even thousands of cancer cases each year in a country of several hundred million people without being noticed.

Assessing the actual risk, if any, from low-dose exposures to known or suspected carcinogens is a difficult task. To illustrate, let us briefly consider the *dioxins*, a family of chlorinated chemicals produced as a by-product during the burning of municipal wastes, the bleaching of paper, and the production of herbicides. Several epidemiological studies have demonstrated increased rates of cancer in

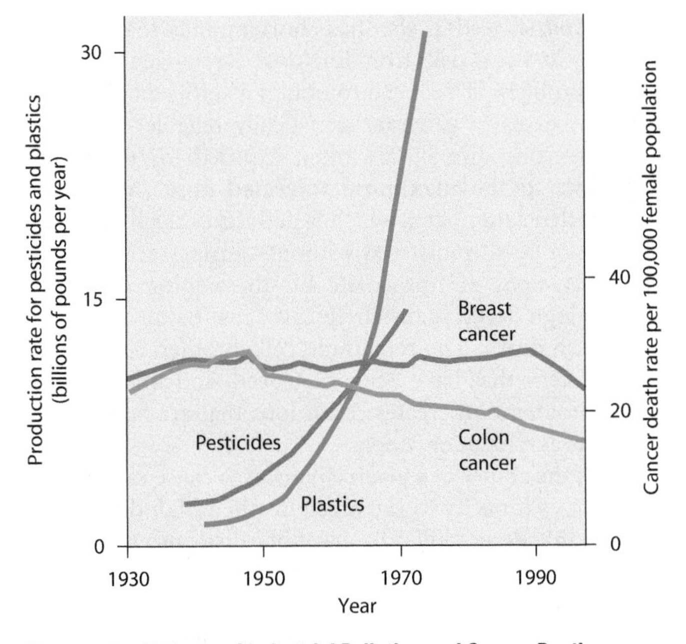

Figure 5-3 Patterns of Industrial Pollution and Cancer Death Rates in the United States. Industrial use of chemical carcinogens, as occurs in the plastics and pesticide industries, increased dramatically between the 1940s and 1970s. For most types of cancer, there was no corresponding increase in cancer incidence during the following decades. If industrial pollution had a major effect on cancer rates, it should have been evident by now. Mortality rates for breast and colon cancer in women are shown, but many cancers exhibit a similar pattern. The only cancer that has shown a major increase in incidence is lung cancer, which is caused mainly by tobacco smoke rather than industrial pollution. [Based on data from *Cancer Facts & Figures 2002* (Atlanta, GA: American Cancer Society, 2002), p. 3; and R. H. Harris et al. in *Origins of Human Cancers* (H. H. Hiatt et al., eds., Cold Spring Harbor, NY: CSHL Press, 1977), pp. 309–330 (Figure 2).]

workers exposed to large concentrations of dioxin, and high-dose animal studies have also shown increased cancer rates. Although the quantities released into the environment are quite small, dioxins are stable molecules that tend to persist for long periods of time, accumulating in the food chain. For humans the main source of exposure is through the food we eat, especially fatty meats. The ingested dioxin molecules are fat soluble and are metabolized slowly, so they tend to accumulate in our body fat. The net result is that even tiny exposures to dioxin can lead to significant concentrations inside the body.

The large unanswered question is whether this accumulated dioxin represents a significant cancer risk. The most conservative approach has been to assume that even one molecule of a carcinogen can cause cancer—in other words, that there is no safe dose of dioxin. However, testing in rats has revealed that while high doses of dioxin cause cancer, low doses can sometimes *decrease* cancer rates compared to those observed in control animals. Such data indicate that it is possible for low-dose carcinogen exposures to be safe, and perhaps even beneficial! Unfortunately, differences between humans and rodents in mode of exposure, metabolism, and genetics make it difficult to extrapolate such results to humans exposed to tiny amounts of dioxin. In other words, we don't really know whether the tiny amount of dioxin that we typically encounter is a small cancer risk, poses no cancer risk, or perhaps even decreases our risk of developing cancer.

Another family of environmental contaminants that might represent a cancer hazard are the *organochlorine* pesticides, a group that includes the now-banned insecticide *dichlorodiphenyltrichloroethane* (*DDT*). Compounds of this type can mimic the action of estrogen, which is known to promote the development of breast cancer. Animals exposed to organochlorines exhibit increased cancer rates, and several widely quoted epidemiological studies have detected a correlation between exposure to organochlorine pesticides and breast cancer rates in women. However, most epidemiological studies have failed to detect such a relationship. Some especially interesting data emerged from a study of several thousand women in Long Island, an area in which organochlorines were extensively used and in which breast cancer rates are higher than the national average. To precisely quantify exposure to organochlorine pesticides, blood samples were obtained from breast cancer patients and from a group of comparable women without the disease. Measurements of the concentration of organochlorines in the blood failed to reveal any relationship between exposure to organochlorine compounds and the development of breast cancer.

Pollution of Outdoor and Indoor Air Creates Small Cancer Risks

The general topic of air pollution provides yet another example of the difficulties encountered when trying to assess environmental cancer hazards involving low-dose exposures. In many cities, both large and small, the air is contaminated with fine particles of airborne soot emitted by cars and trucks, power plants, and factories. A recent epidemiological study of 500,000 adults living in dozens of cities across the United States revealed that people located in cities with the largest amounts of this fine-particle soot have lung cancer death rates roughly 10% higher than in cities with minimal pollution. Of course, a 10% increase is not very much; for comparison purposes, cigarette smokers increase their risk of developing lung cancer by 2500% or more, a value that is 250 times higher than the small increase in lung cancer risk that might be associated with air pollution.

Although discussions of air pollution usually focus on outdoor air, our main exposure to polluted air may be indoors. In studies involving more than a dozen different cities, researchers equipped people with air-quality monitoring devices that were small enough to carry around as people performed their normal daily activities. For the average citizen, the greatest exposure to toxic airborne chemicals turned out to occur inside their homes (Figure 5-4). The sources of this indoor air pollution included ordinary consumer products such as cleaning compounds, paints, carpeting, gasoline, air fresheners, dry cleaning, and disinfectants. Even in cities

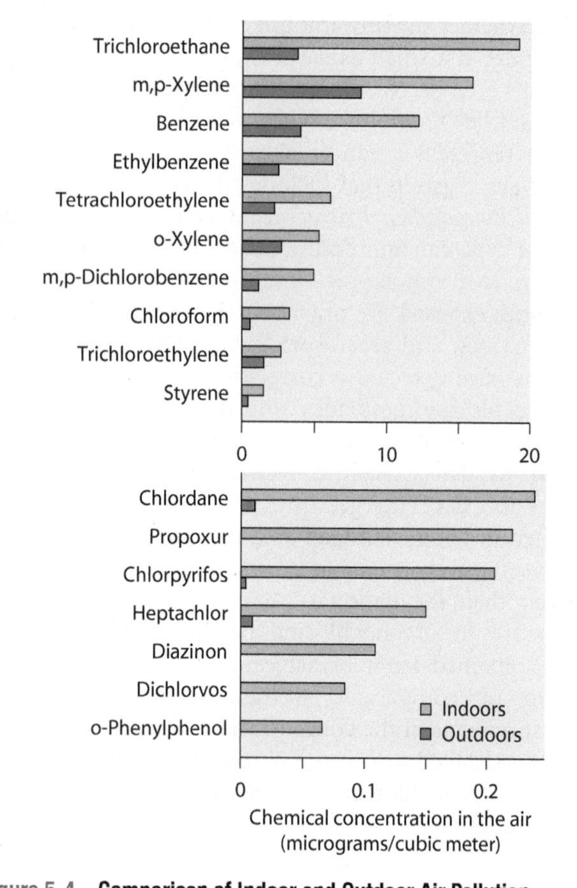

Figure 5-4 Comparison of Indoor and Outdoor Air Pollution. Pollution data were obtained from people equipped with portable air-quality monitoring devices designed to measure the concentration of toxic volatile organic chemicals (*top*) and pesticides (*bottom*) in indoor and outdoor air. Results from studies involving more than a dozen cities in the United States, including cities with chemical processing plants, have revealed that the air inside a person's home usually contains higher concentrations of potential carcinogens than are present in outdoor air. [Data from W. R. Ott and J. W. Roberts, *Sci. Amer.* 278 (February 1998): 88.]

where the outdoor air was polluted by emissions from chemical processing plants, the concentration of many airborne carcinogens was higher inside homes than outdoors. However, these indoor concentrations were still much lower than those typically encountered in industrial workplaces, and it is difficult to know whether such low-dose exposures pose any cancer risks.

Thresholds Can Cause Animal Studies to Overestimate Human Cancer Risks

The difficulty in assessing the hazards of low-dose chemical exposures arises from limitations that are inherent to epidemiological and animal testing. The main problem with the epidemiological approach is that it is not sensitive enough to reliably detect small increases (less than a doubling) in cancer incidence, which is the type of effect that might be expected from low-dose carcinogen exposures. As a consequence, scientists often turn to animal testing.

Animal testing also has shortcomings that limit the ability to assess risks from low-dose carcinogen exposures. One problem is the need to obtain a sufficient number of cancer cases to generate statistically reliable results. For this reason, animals are often exposed for their entire lifetime to the **maximum tolerated dose** (**MTD**) of a suspected carcinogen, which is defined as the highest dose that can be administered without causing serious weight loss or signs of immediate life-threatening toxicity. At these high doses, many chemicals cause tissue destruction and cell death. The remaining cells proliferate to replace those cells that have been destroyed, and this enhanced cell proliferation creates conditions that are favorable for the development of cancer.

If the ability of a given chemical to cause cancer stems from this capacity to cause cell death at high doses, lower doses that do not kill cells may not cause cancer. The dose-response curve for such a carcinogen would exhibit a **threshold**—that is, a dose that must be exceeded before cancer rates begin to rise. Doses below the threshold would be safe in terms of cancer risk. Figure 5-5 shows how the existence of a threshold can cause the cancer risk of low-dose exposures to be overestimated. The data in Figure 5-5 are for benzo[*a*]pyrene, a carcinogen present in gasoline exhaust fumes and in smoke generated by burning organic matter, including tobacco smoke. The graph on the left shows cancer rates for animals exposed to high doses of benzo[*a*]pyrene. If this data were the only information available, a straight line could be drawn through the data points and extrapolated to lower doses to estimate cancer risk for low-dose exposures to benzo[*a*]pyrene. The graph on the right, however, shows what happens when actual experiments are performed using lower doses of benzo[*a*]pyrene. The shape of the overall curve exhibits a threshold, and the actual cancer risk associated with low-dose benzo[*a*]pyrene exposure is much lower than predicted.

The preceding example highlights that cancer biologists have traditionally had two ways of viewing the relationship between high-dose and low-dose cancer risks: the linear model and the threshold model. As shown in Figure 5-6, the **linear model** assumes a linear dose-response relationship with no threshold, whereas the **threshold model** assumes no cancer risk at lower doses followed by a linear dose-response relationship at higher doses. A third possibility, called the **hormetic model**, has also begun to receive some attention. The hormetic model proposes that dose-response curves can also be U-shaped. The U-shape, known as *hormesis*, reflects a situation in which cancer rates actually decline at very low doses of carcinogen and then begin to go up as the dose is further increased. While it is not clear how widely this model might apply to cancer risk, several carcinogenic agents have been reported to reduce cancer rates when administered to animals at low doses (recall the example involving dioxin mentioned earlier).

The hormetic and threshold models both include the concept of a "threshold"—that is, a dose that must be

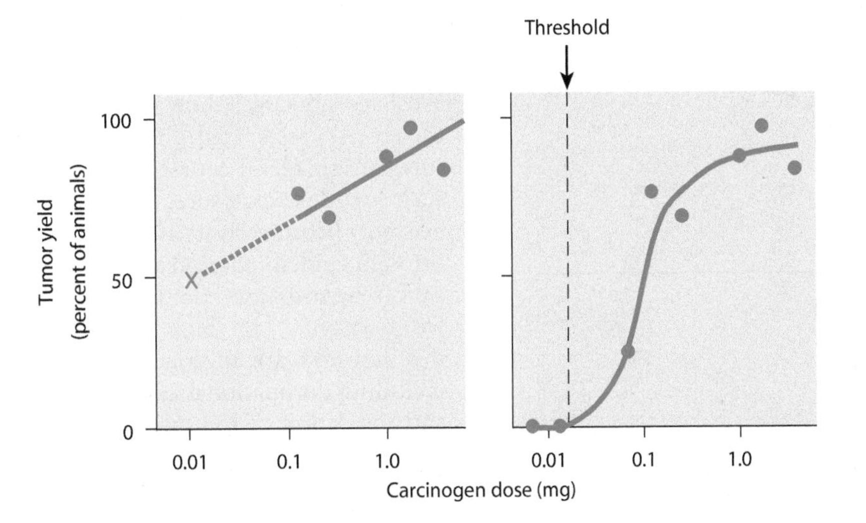

Figure 5-5 Possibility of Overestimating Cancer Risk When Extrapolating from High-Dose Data. Dose-response curves are illustrated for sarcomas arising in mice after a single injection of benzo[*a*]pyrene. (*Left*) The graph on the left is restricted to high doses of carcinogen (> 0.1 mg). The results appear to be roughly linear, and a straight line can be drawn through the data points to estimate the cancer risk for a lower dose (0.01 mg) of benzo[*a*]pyrene. This estimated cancer risk is indicated by the "X". (*Right*) When actual experiments are carried out that include lower doses of benzo[*a*]pyrene, the shape of the overall curve is seen to exhibit a threshold. Note that the actual cancer risk associated with a 0.01 mg dose of benzo[*a*]pyrene shown in the graph on the right is much lower than the risk estimated by the linear extrapolation derived from the high-dose data shown in the graph on the left. [Based on data from W. R. Bryan and M. B. Shimkin, *J. Natl. Cancer Inst.* 3 (1943): 503.]

exceeded before cancer rates begin to rise. One possible explanation for the existence of thresholds is the ability of high-dose exposures to cause tissue destruction and cell death, which creates a unique set of conditions that do not exist at lower doses. Another possible reason for thresholds is that carcinogens often act by causing DNA damage. As we saw in Chapter 2, the presence of damaged DNA triggers a group of repair mechanisms that attempt to correct the problem. Such repair mechanisms might be able to fix small amounts of DNA damage caused by low doses of carcinogens and may even help prevent cancer from arising in response to subsequent exposures to carcinogenic agents. According to this view, carcinogen-induced mutations only begin to accumulate and initiate the development of cancer after a threshold dose has been exceeded and the capacity of these DNA repair pathways is overcome by massive DNA damage.

Humans and Animals Differ in Their Susceptibilities to Some Carcinogens

Another problem that can complicate the extrapolation of animal data to humans is that animals often differ from one another, as well as from humans, in their susceptibility to different carcinogens. Consider the behavior of *2-acetylaminofluorene* (*AAF*), which is a potent carcinogen in rats but does not cause cancer in guinea pigs. Based on this information alone, it would be difficult to predict whether or not AAF is likely to be carcinogenic in humans. The reason for the differing behavior of AAF in rats and guinea pigs became apparent when it was discovered that AAF is actually a "precarcinogen" that needs to be metabolically activated before it can cause cancer. Rats, but not guinea pigs, contain the enzyme that catalyzes this metabolic activation. Biochemical analysis of human tissues has revealed that we also contain the activating enzyme, so it is likely that AAF is carcinogenic in humans just as it is in rats. Of course, if AAF had only been tested in guinea pigs, it never would have been suspected of being a carcinogen in the first place!

The artificial sweetener *saccharin* provides another illustration of the difficulties that can arise when extrapolating data from animal studies to humans. At the peak of its use in the 1970s, Americans consumed more than five million pounds of saccharin per year in artificially sweetened foods and beverages. In 1981, the U.S. government labeled saccharin as a suspected human carcinogen and attempted to ban its use as a food additive because saccharin causes bladder cancer when fed to rats. Subsequent investigations, however, have revealed that saccharin causes bladder cancer in rats for reasons that do not apply to humans. When rats ingest large amounts of any sodium salt, including sodium saccharin, a crystalline precipitate forms in the bladder that irritates the bladder lining, triggering cell proliferation and increasing the risk of developing cancer. But the precipitate only forms when there are large amounts of protein in the urine, and the urine protein concentration in rats is 100 to 1000 times greater than in humans. Subsequent studies have shown that other laboratory animals, such as hamsters, guinea

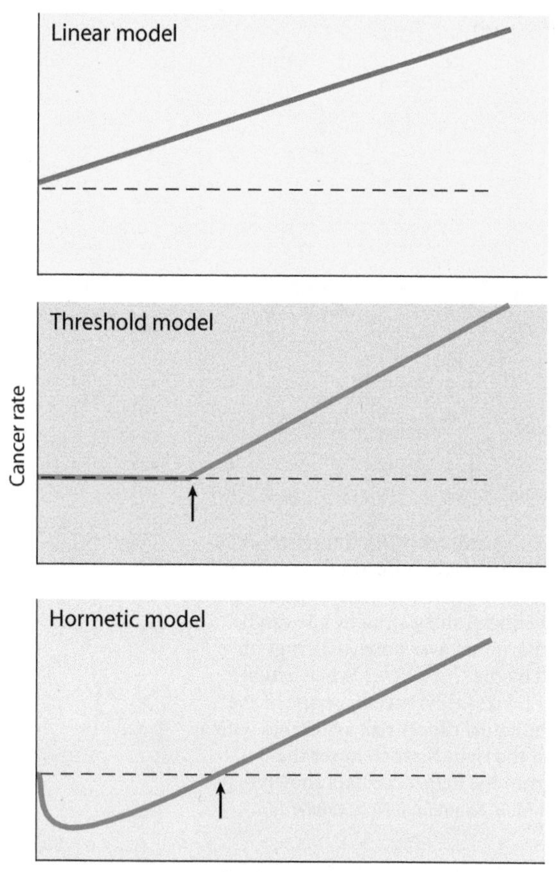

Figure 5-6 Three Models for the Relationship Between Carcinogen Dose and Cancer Risk. The dashed lines represent the background cancer rate in the absence of carcinogen. Note that the threshold and hormetic models both involve a threshold dose (*arrow*) that must be exceeded before cancer rates begin to rise. [Adapted from E. J. Calabrese, *Mutation Res.* 511 (2002): 181 (Figure 1).]

pigs, and mice, do not develop bladder cancer when fed saccharin. As a consequence, saccharin was recently taken off the government's list of suspected human carcinogens.

Because of the uncertainties involved in applying animal data to humans, caution is needed in labeling substances as human carcinogens when the information has been derived largely from animal studies. The U.S. government therefore publishes a list that subdivides carcinogens into two distinct categories: (1) *known to be human carcinogens* and (2) *reasonably anticipated to be human carcinogens* (see Appendix B). The list of "known" human carcinogens contains several dozen chemicals for which the data from animal studies have been supplemented with enough human data to clearly establish a cancer risk for humans. The list of "anticipated" human carcinogens includes more than 100 additional chemicals for which the potential cancer hazard has been extrapolated largely from animal studies. While many of these substances will almost certainly turn out to be human carcinogens, mistakes are possible because of the heavy reliance on animal data. For example, saccharin was listed as an "anticipated" human carcinogen for about 20 years before eventually being removed from the list.

Some Medications and Hormones Can Cause Cancer

We have now seen how difficult it can be to measure cancer risks associated with chemicals to which our exposures are small. Of course, this means that the hazards of such low-dose exposures must be rather small (or nonexistent) because larger risks would be readily detectable through epidemiological and animal testing. The greatest cancer hazards are posed by chemical carcinogens that we encounter in high concentrations. Included in this category are several situations already mentioned, including occupational exposure to industrial chemicals and inhalation of the carcinogenic chemicals present in tobacco smoke.

Another type of high-dose exposure to specific chemicals comes from the use of prescription drugs for treating certain illnesses. Prescription drugs are often taken for prolonged periods, so it is important to know whether the resulting high-dose chemical exposures can cause cancer. One tragic example involves *diethylstilbestrol (DES)*, a synthetic estrogen that was prescribed to pregnant women starting in the 1940s as a way of preventing miscarriages. Several decades later, women whose mothers had taken DES during pregnancy began developing vaginal cancer at alarmingly high rates. By that time, roughly five million women in the United States had already taken DES. This episode illustrates the difficulty in establishing the risks associated with ingesting any new chemical compound, even when it appears to be safe and is prescribed for a specific medical purpose. In the case of DES, the drug's ability to cause cancer did not become apparent until several decades after women had used DES, and the cancer did not affect the person who took the drug, but rather her children.

Although DES has now been banned as a prescription drug, a number of other medications can also cause cancer (Table 5-2). Most are prescribed for serious medical problems where the potential benefits of the drug in question are thought to outweigh the risk that cancer might arise. For example, some drugs used for slowing or stopping tumor growth in cancer patients can themselves trigger development of a new cancer as a side effect. In a person who already has cancer, the small risk of causing another cancer (often many years in the future) is far outweighed by the possible benefits to be gained from a drug that might cure an existing cancer.

A similar situation exists with **immunosuppressive drugs**, which inhibit immune function and are given to organ transplant patients to prevent rejection of a transplanted organ, such as a heart or kidney. Two of the most commonly used immunosuppressive drugs, *azathioprine* and *cyclosporin*, increase the risk of developing cancer (see Figure 2-22), but organ transplant patients depend on the transplanted organ for survival and the benefits of these drugs are thought to outweigh the risk of developing cancer. Nonetheless, cancer has turned out to be a major cause of death in organ transplant patients and a need

Table 5-2 Some Medications That Can Cause Cancer

Medication	Type of Cancer Caused
Analgesic:	
Phenacetin	Kidney
Cancer chemotherapy:	
Chlorambucil	Leukemia
Cyclophosphamide	Bladder, leukemia
Melphalan	Leukemia
Thiotepa	Leukemia
Hormones:	
Estrogens	Breast, uterus, vagina
Oxymetholone	Liver
Immunosuppressive drugs:	
Azathioprine	Lymphoma, skin, liver
Cyclosporin	Lymphoma, skin
Skin treatments:	
Arsenic compounds	Skin, liver, lung
Methoxypsoralen	Skin

therefore exists for better immunosuppressive drugs that do not increase cancer risk. One drug under current investigation is *rapamycin* (also called *sirolimus*), an antibiotic with immunosuppressive activity. Animal studies have shown that besides suppressing immune function, rapamycin inhibits tumor growth under conditions in which another immunosuppressive drug, cyclosporin, stimulates tumor growth. A possible explanation for the antitumor effect of rapamycin has come from the discovery that it inhibits angiogenesis, both by depressing the production of VEGF and by inhibiting the ability of endothelial cells to respond to VEGF (the normal role of VEGF in promoting angiogenesis is described on p. 48). Suppression of angiogenesis by rapamycin may therefore limit the ability of newly forming tumors to obtain the blood supply they require for growth beyond a tiny size.

MECHANISMS OF CHEMICAL CARCINOGENESIS

As the list of substances known to cause cancer has grown over the years, it has become increasingly apparent that carcinogens exhibit wide variations in structure and potency. At first this variability complicated our thinking about the origins of cancer because it was difficult to envision how such a diverse array of chemical substances could cause the same disease. Through an extensive series of studies, however, a common set of mechanisms and principles has begun to emerge that helps explain how the various kinds of carcinogens work.

Chemical Carcinogens Can Be Grouped into Several Distinct Categories

Despite their structural diversity, chemical carcinogens can be grouped into a relatively small number of categories (Figure 5-7). Most are natural or synthetic **organic chemicals**—that is, carbon-containing compounds. They range from small organic molecules containing only a few carbon atoms to large, complex molecules constructed from multiple carbon-containing rings. The vast majority fall into one of the following five categories.

1. Carcinogenic **polycyclic aromatic hydrocarbons** (or simply **polycyclic hydrocarbons**) are a diverse group of compounds constructed from multiple, fused benzene rings. Polycyclic hydrocarbons are natural components of coal tars, soots, and oils, and are also produced during the incomplete combustion of coal, oil, tobacco, meat, and just about any other organic material that can be burned. The carcinogenic potency of polycyclic hydrocarbons varies widely, from weak or noncarcinogenic molecules to very potent carcinogens. The polycyclic hydrocarbons *benzo[a]pyrene* and *dibenz[a,h]anthracene*, isolated from coal tar in the 1930s, were the first purified chemical carcinogens of any kind to be identified.

2. Carcinogenic **aromatic amines** are organic molecules that possess an *amino group* (—NH_2) attached to a carbon backbone containing one or more benzene rings. Some aromatic amines are *aminoazo compounds*, which means that they contain an *azo group* (N=N) as well as an amino group. Among the carcinogens in these categories are the aromatic amines *benzidine*, *2-naphthylamine*, *2-acetylaminofluorene*, and *4-aminobiphenyl*, and the aminoazo dyes *4-dimethylaminoazobenzene* and *o-aminoazotoluene*. Many of these compounds were once employed in the manufacturing of dyes, although most are no longer used in significant quantities because of their toxicity. Some aromatic amines, such as 2-naphthylamine and 4-aminobiphenyl, are components of tobacco smoke. As in the case of polycyclic hydrocarbons, the carcinogenic potency of aromatic amines and aminoazo dyes varies from substances that are strongly carcinogenic to substances that are not carcinogenic at all.

3. Carcinogenic **N-nitroso compounds** are organic chemicals that contain a *nitroso group* (N=O) joined to a nitrogen atom. Members of this group include the *nitrosamines* and *nitrosoureas*, which are potent carcinogens when tested in animals. Most of these compounds are industrial or research chemicals encountered mainly in the workplace, although a few are present in cigarette smoke. Nitrates and nitrites used in the curing of meats, which are not carcinogenic in themselves, can be converted in the stomach into nitrosamines, but no consistent relationship between these compounds and human cancer has been established.

4. Carcinogenic **alkylating agents** are molecules that readily undergo reactions in which they attach a carbon-containing chemical group to some other molecule. Unlike the three preceding groups of carcinogens, which are defined by their chemical structures (i.e., the presence of multiple benzene rings, amino groups, or nitroso groups), alkylating agents are defined not by their structural features but by their chemical reactivity—that is, their ability to join a chemical group to another molecule. The N-nitroso compounds, discussed in the

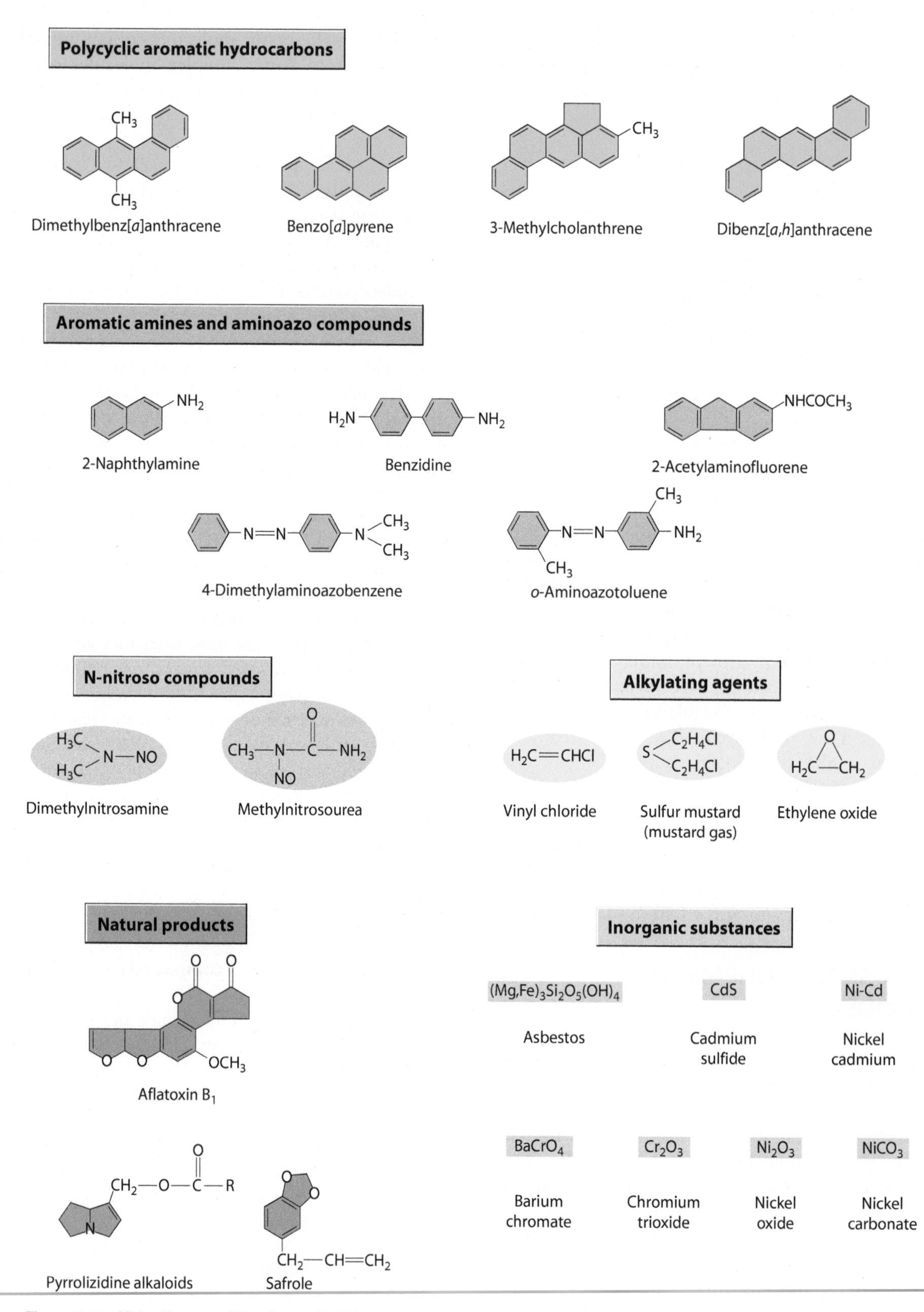

Figure 5-7 Main Classes of Carcinogenic Chemicals. Selected examples are illustrated for each of the main classes of cancer-causing chemicals. Some of these molecules are precarcinogens that need to be metabolically activated before they can cause cancer, whereas others are direct-acting carcinogens that do not require metabolic activation.

preceding paragraph, are examples of carcinogens that function as alkylating agents. Other examples include *vinyl chloride* (used in the production of plastics) and *ethylene oxide* (used in the production of antifreeze and other chemicals). Vinyl chloride and ethylene oxide are among the highest-volume chemicals produced in the United States. Other carcinogenic alkylating agents include *sulfur mustard* (a chemical warfare agent) and several drugs used in cancer chemotherapy (p. 213).

5. Carcinogenic **natural products** are a structurally diverse group of cancer-causing molecules produced by biological organisms, mainly microorganisms and plants. Included in this category is *aflatoxin*, a carcinogenic chemical made by the mold *Aspergillus*. One of the most potent carcinogens known, aflatoxin sometimes contaminates grains and nuts that have been stored under humid conditions. Other carcinogenic natural products include plant-derived molecules such as *safrole*, a major component of sassafras root bark, and *pyrrolizidine alkaloids*, produced by a variety of different plants.

In addition to the preceding five classes of organic molecules, a small number of **inorganic substances** (compounds without carbon and hydrogen) are carcinogenic. Included in this group are compounds containing the metals *cadmium*, *chromium*, and *nickel*. Some inorganic substances appear to be carcinogenic in the absence of chemical reactivity. For example, asbestos is a mineral composed of silicon, oxygen, magnesium, and iron, but its ability to cause cancer is related to the crystal structure and size of the microscopic fibers it forms rather than their precise chemical makeup.

Some Carcinogens Need to Be Activated by Metabolic Reactions Occurring in the Liver

The chemicals illustrated in Figure 5-7 are considered to be "carcinogens" because humans or animals develop cancer when exposed to them. This designation does not mean, however, that every carcinogen triggers cancer directly. For example, consider the behavior of 2-naphthylamine, whose ability to cause bladder cancer in industrial workers was described earlier in the chapter. As might be expected, feeding 2-naphthylamine to laboratory animals induces a high incidence of bladder cancer, but cancer rarely arises when 2-naphthylamine is directly inserted into an animal's bladder. The reason for this discrepancy is that when 2-naphthylamine is ingested (by animals) or inhaled (by humans), it first passes through the liver and is metabolically converted into chemical compounds that are the actual causes of cancer. Inserting 2-naphthylamine directly into the bladder bypasses the liver and the molecule is never activated, so cancer does not arise.

Many carcinogens share a similar need for metabolic activation before they can cause cancer. Carcinogens exhibiting such behavior are more accurately called **precarcinogens**, a term referring to any substance that is capable of causing cancer only after it has been metabolically activated. The activation of precarcinogens is generally carried out by liver proteins that are members of the **cytochrome P450** enzyme family. One function of these liver enzymes is to catalyze the oxidation of ingested foreign chemicals, such as drugs and pollutants, with the aim of making molecules less toxic and easier to excrete from the body. The hydroxylation reaction illustrated in Figure 5-8 is one of several ways in which cytochrome P450 oxidizes foreign chemicals to make them more water soluble, thereby facilitating their excretion in the urine. Occasionally, however, oxidation reactions catalyzed by cytochrome P450 accidentally convert substances into carcinogens, a phenomenon known as **carcinogen activation.**

Evidence that cytochrome P450 is involved in carcinogen activation has come from numerous animal studies. One set of experiments involved a mutant strain of mice that produce abnormally large amounts of *cytochrome P450 1A1*, a form of cytochrome P450 that oxidizes polycyclic hydrocarbons. As would be expected if

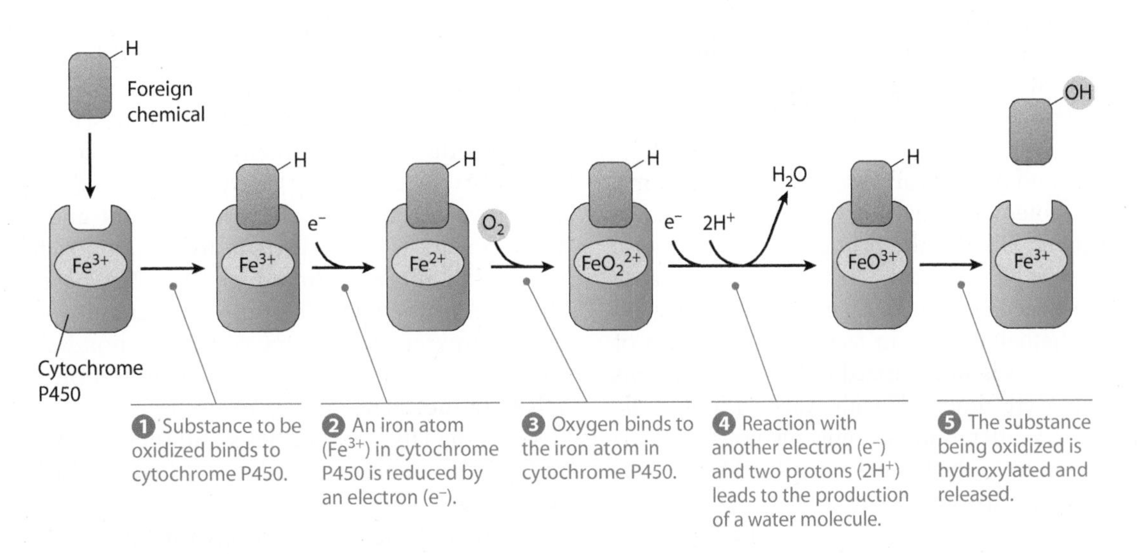

Figure 5-8 Hydroxylation Reaction Catalyzed by Cytochrome P450. Cytochrome P450 oxidizes chemicals by linking them to a hydroxyl group in a five-step oxidation reaction.

cytochrome P450 1A1 were involved in carcinogen activation, cancer rates are elevated in the mutant mice that produce large amounts of this enzyme. Cancer rates can be reduced in these same animals by using inhibitors that block the action of cytochrome P450 1A1.

Elevated amounts of cytochrome P450 1A1 are found in the livers of people who smoke cigarettes, apparently because tobacco smoke stimulates production of the enzyme by the liver. This means that in addition to containing dozens of chemicals that cause cancer, cigarette smoke also induces the production of liver enzymes that make the situation worse by activating carcinogenic activity in chemicals that might not otherwise cause cancer. About one person in ten inherits a form of cytochrome P450 1A1 that is produced in especially large amounts in response to tobacco smoke. If such a person smokes cigarettes, he or she has an even higher risk of developing lung cancer than other smokers.

The role played by liver enzymes in carcinogen activation explains why chemicals being assayed for mutagenic activity in the Ames test (p. 66) are first incubated with a liver homogenate to mimic any reactions that might take place in the body. The requirement for metabolic activation also helps explain why some chemicals only cause cancer in certain organisms. For example, we saw earlier in the chapter that 2-acetylaminofluorene (AAF) is carcinogenic in rats but not in guinea pigs because guinea pigs lack the enzyme needed to convert AAF into an active carcinogen. Because of such differences in liver enzymes between organisms, it is important that suspected carcinogens be tested in more than one animal species.

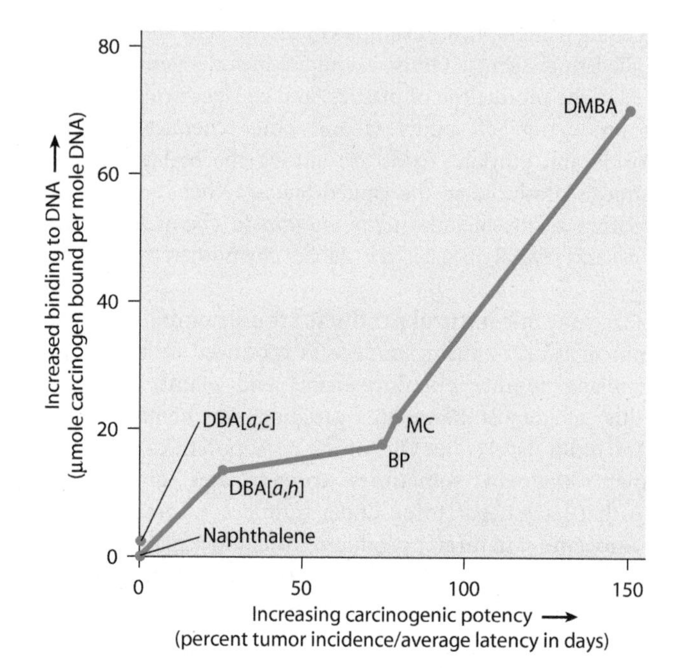

Figure 5-9 Relationship Between Carcinogenic Potency and DNA-Binding Ability. Six polycyclic hydrocarbons of varying carcinogenic potency were injected into animals and measurements were then made to identify the intracellular molecules to which they had become bound. The data show that the more potent the carcinogen, the more extensively it binds to DNA. *Abbreviations*: DBA[*a,c*] = dibenz[*a,c*]anthracene, DBA[*a,h*] = dibenz[*a,h*]anthracene, BP = benzo[*a*]pyrene, MC = 3-methylcholanthrene, DMBA = dimethylbenz[*a*]anthracene. [Data from P. Brookes and P. D. Lawley, *Nature* 202 (1964): 781 (Figure 5).]

Many Carcinogens Are Electrophilic Molecules That React Directly with DNA

Despite the variations in molecular structure exhibited by the carcinogens illustrated in Figure 5-7, many share the same property: When metabolized in the liver, they are converted into highly unstable compounds with electron-deficient atoms. Such molecules are said to be **electrophilic** ("electron-loving") because they readily react with substances possessing atoms that are rich in electrons.

DNA, RNA, and proteins all have electron-rich atoms, making each a potential target for electrophilic carcinogens. Of the three, DNA is the prime candidate because the Ames test has shown that most carcinogens cause DNA mutations (see Figure 4-3). An experiment designed to determine whether DNA is in fact the direct target of chemical carcinogens is summarized in Figure 5-9. In this study, animals were injected with various polycyclic hydrocarbons that differed in carcinogenic potency. Cells were then isolated from the treated animals and measurements were made to determine which intracellular molecules (if any) had become bound to the polycyclic hydrocarbons. The data revealed a direct relationship between the carcinogenic potency of different polycyclic hydrocarbons and their ability to become covalently linked to DNA; in other words, those polycyclic hydrocarbons that became extensively bound to DNA were the most effective at causing cancer.

Before a polycyclic hydrocarbon can interact with DNA, it must be activated. For example, consider the behavior of benzo[*a*]pyrene, which is normally a nonreactive, nonmutagenic compound. After entering the body, metabolic reactions catalyzed by cytochrome P450 in the liver convert benzo[*a*]pyrene into activated derivatives containing an **epoxide** group (Figure 5-10). An epoxide is a three-membered ring containing an oxygen atom covalently bonded to two carbon atoms; these two carbons are electron deficient and therefore tend to react with atoms that are electron rich, such as the amino nitrogen found in the DNA base *guanine*. Reaction of the epoxide group with guanine causes the benzo[*a*]pyrene to become covalently bonded to DNA, thereby forming a DNA-carcinogen complex called a **DNA adduct**. The presence of the bound carcinogen distorts the DNA double helix and thereby causes errors in base sequence (mutations) to arise during DNA replication.

Epoxide formation is also involved in the activation of other classes of chemical carcinogens. For example, *aflatoxin* and *vinyl chloride*, which differ significantly from

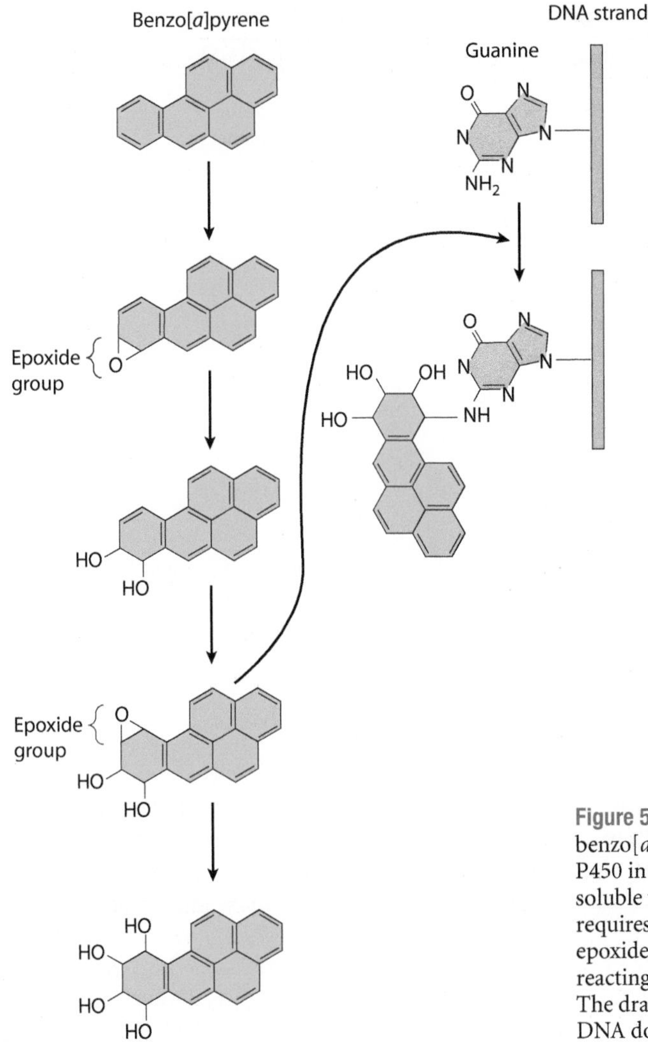

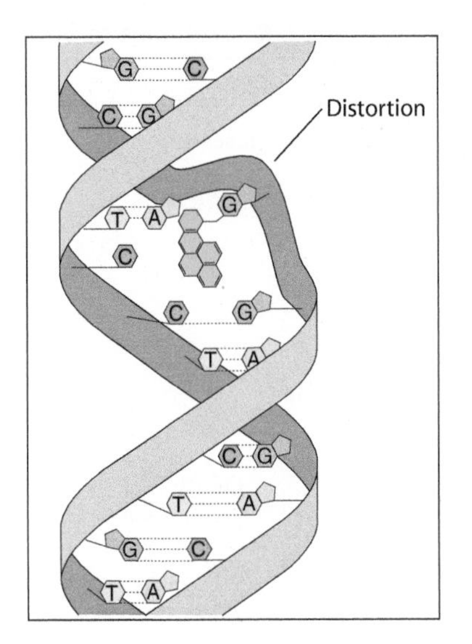

Figure 5-10 Metabolic Activation of Benzo[*a*]pyrene. When benzo[*a*]pyrene enters the body, reactions catalyzed by cytochrome P450 in the liver make the molecule progressively more water soluble to facilitate its excretion from the body. This process requires several steps and creates intermediates containing an epoxide group. One of these intermediates is capable of reacting with the base guanine in DNA to form a DNA adduct. The drawing on the right shows how such an adduct distorts the DNA double helix.

polycyclic hydrocarbons in chemical structure, are both oxidized by cytochrome P450 into epoxides that, like benzo[*a*]pyrene, react with DNA bases to form DNA adducts (Table 5-3). However, the various epoxides do not all react with the same DNA bases. In fact, depending on the carcinogen involved, almost every electron-rich site in the various DNA bases can serve as a target for carcinogen attachment (Figure 5-11). And epoxides are not the only electrophilic groups that react with DNA. Some carcinogens are activated by reactions that create other types of electrophilic groups, such as positively charged nitrogen atoms (*nitrenium ions*) or carbon atoms (*carbonium ions*), or compounds containing an unpaired electron (*free radicals*). Like epoxides, these electrophilic groups also attack electron-rich atoms in DNA.

The preceding mechanisms illustrate that, despite their chemical diversity, many carcinogens share the property of being converted into electrophilic molecules that in turn become linked to DNA. This ability to form DNA adducts is one of the best predictors of a molecule's capacity to cause cancer. In addition, carcinogens can inflict DNA damage in several other ways; for example, they may generate crosslinks between the two strands of the double helix, create chemical linkages between adjacent bases, hydroxylate or remove individual DNA bases, or cause breaks in one or both DNA strands (Figure 5-12).

Chemical Carcinogenesis Is a Multistep Process

An attack on DNA by an activated carcinogen is just the first of several steps involved in creating a cancer cell. The idea that cancer arises through a multistep process was first proposed in the early 1940s by Peyton Rous to explain a phenomenon he encountered when studying the ability of coal tar to cause cancer in rabbits. Rous had observed that repeated application of coal tar to rabbit skin caused tumors to develop, but the tumors disappeared when application of the coal tar was stopped. Subsequent application of an irritant such as turpentine, which does not induce cancer by itself, caused the tumors to reappear. This pattern suggested to Rous that coal tar and turpentine play two

Table 5-3 Examples of Several Carcinogens Activated by Epoxide Formation

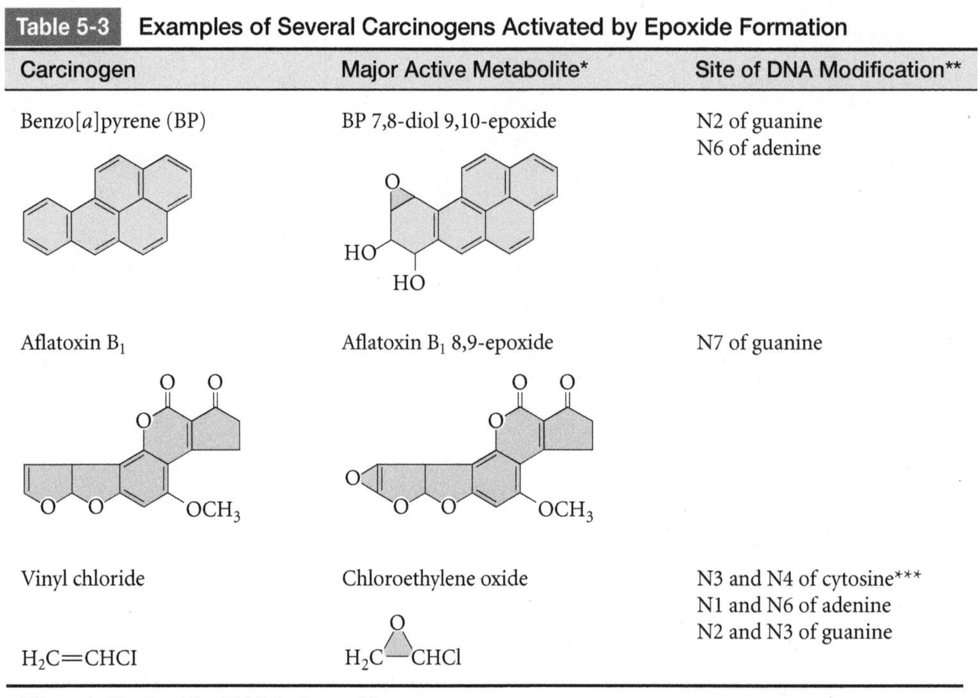

Carcinogen	Major Active Metabolite*	Site of DNA Modification**
Benzo[*a*]pyrene (BP)	BP 7,8-diol 9,10-epoxide	N2 of guanine N6 of adenine
Aflatoxin B_1	Aflatoxin B_1 8,9-epoxide	N7 of guanine
Vinyl chloride $H_2C{=}CHCl$	Chloroethylene oxide $H_2C{-}CHCl$ (O bridge)	N3 and N4 of cytosine*** N1 and N6 of adenine N2 and N3 of guanine

*Green shading is used to highlight the epoxide group.
**The numbers in the third column refer to the numbered positions of nitrogen atoms illustrated in Figure 5-11.
***Vinyl chloride simultaneously attacks two positions on the same base, forming a cyclic adduct.

different roles, which he called **initiation** and **promotion**. According to his theory, initiation converts normal cells to a precancerous state and promotion then stimulates the precancerous cells to divide and form tumors.

Because coal tar is a mixture of various chemicals, clarification of the initiation/promotion hypothesis required the isolation and study of individual coal tar components. One such chemical is the polycyclic hydrocarbon *dimethylbenz[a]anthracene (DMBA)*. DMBA is a potent carcinogen, but feeding mice a single dose rarely causes tumors to arise. However, if the skin of a mouse fed a single dose of DMBA is later painted with a substance that causes skin irritation, cancer develops in the treated area (Figure 5-13). Besides turpentine, the irritant most commonly used for triggering tumor formation in such experiments is *croton oil*, a substance derived from seeds of the tropical plant *Croton tiglium*. Croton oil does not cause cancer in the absence of prior exposure to a carcinogen such as DMBA, nor will cancer arise if DMBA is administered *after* the croton oil. These observations support the concept that chemical carcinogenesis is a multistep process in which an *initiator* (in this case, DMBA) first creates an altered, precancerous state and then a *promoting agent* (in this case, croton oil) stimulates the development of tumors.

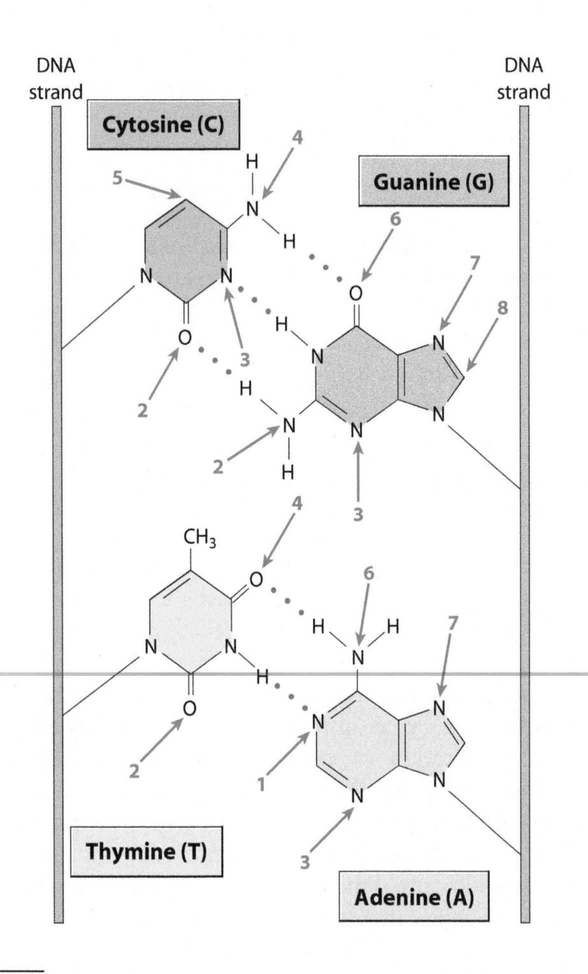

Figure 5-11 Sites of Carcinogen Attack in DNA Bases. The red arrows indicate the main sites in the four DNA bases that serve as targets for carcinogen attachment. The red numbers refer to the numbering system used to identify the atoms located at different positions within each base. The dotted blue lines represent the hydrogen bonds that normally hold the complementary base pairs together in the DNA double helix. Attachment of carcinogen molecules to the bases tends to distort the double helix and interfere with this hydrogen bonding, thereby leading to errors in DNA replication.

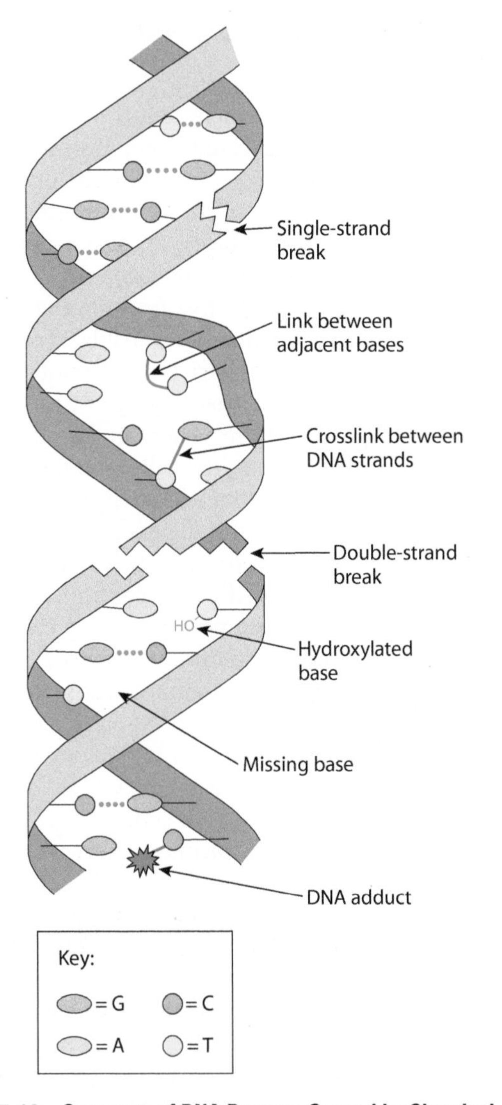

Figure 5-12 **Summary of DNA Damage Caused by Chemical Carcinogens.** Chemical carcinogens inflict DNA damage in a variety of ways, altering or removing individual bases and triggering breaks in one or both DNA strands.

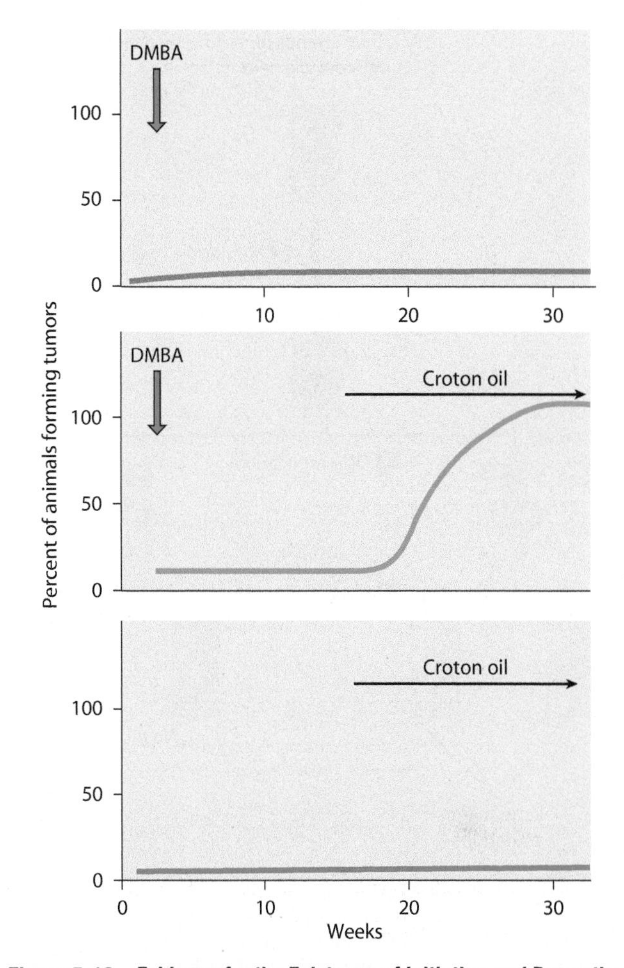

Figure 5-13 **Evidence for the Existence of Initiation and Promotion Stages in Chemical Carcinogenesis.** (*Top*) Mice treated with a single dose of DMBA (dimethylbenz[*a*]anthracene) do not form tumors. (*Middle*) Painting the skin of such animals twice a week with croton oil after the DMBA treatment leads to the appearance of skin tumors. If the croton oil application is stopped a few weeks into the treatment (data not shown), the tumors will regress. (*Bottom*) Croton oil alone does not produce skin tumors. These data are consistent with the conclusion that DMBA is an initiator and croton oil is a promoter. [Adapted from R. K. Boutwell, *Prog. Exp. Tumor Res.* 4 (1963): 207.]

The Initiation Stage of Carcinogenesis Is Based on DNA Mutation

A year or more can transpire after feeding animals a single dose of DMBA and yet tumors will still arise if an animal's skin is then irritated with croton oil. Thus a single DMBA treatment creates a permanently altered, initiated state in cells located throughout the body, and a promoting agent (croton oil) can then act on these altered cells to promote tumor development. Because the carcinogenic potency of most chemicals correlates with their ability to bind to DNA and cause mutations (see Figures 4-3 and 5-9), the permanent alteration is thought to be a mutation. Carcinogens that act in this way are said to be **genotoxic** because they cause gene damage. The ability to cause gene mutations explains how a single exposure to an initiating carcinogen can create a permanent, inheritable change in a cell's properties.

Referring to carcinogen-induced mutations as "permanent," however, implies that DNA damage cannot be repaired, which seems to contradict what we know about the existence of DNA repair mechanisms (see Chapter 2). Such mechanisms are in fact capable of repairing mutations created by initiating carcinogens as long as the damage is repaired in a timely fashion. Once a damaged DNA molecule has been replicated, as occurs each time a cell divides, mutations become very difficult, if not impossible, to repair and the initiated state therefore becomes permanent.

Figure 5-14 illustrates why this is the case, using the carcinogen *methylnitrosourea* as an example. Methylnitrosourea attacks the base guanine (G) in DNA, creating a methylated guanine derivative. If the cell's DNA is replicated before repair mechanisms correct the defect, the methylated guanine tends to form an incorrect base

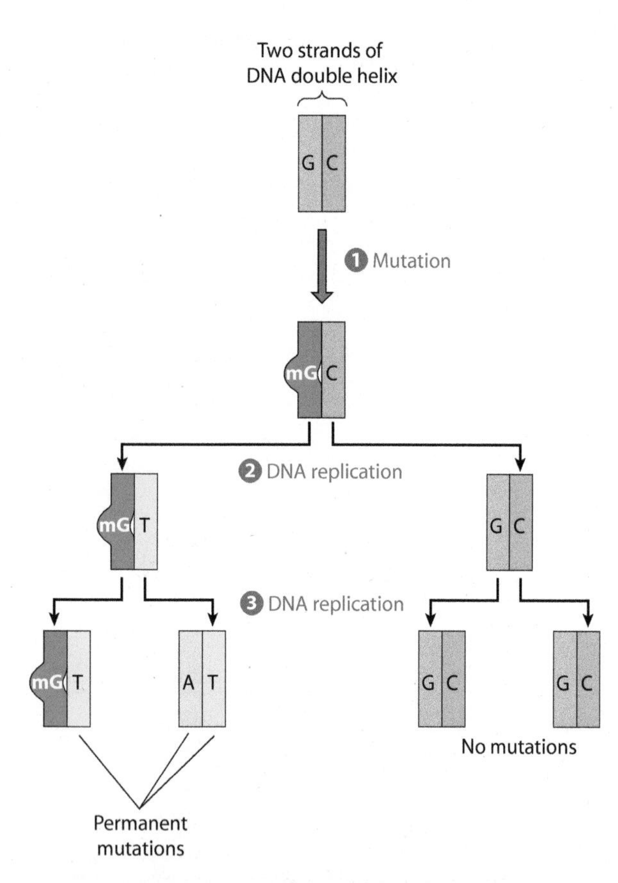

Figure 5-14 Process by Which a DNA Mutation Becomes Permanent. This hypothetical illustration involves a mutation triggered by exposure to the carcinogen *methylnitrosourea*, but a similar principle applies to many kinds of base mutations caused by chemical carcinogens. ① Methylnitrosourea causes a mutation by adding a methyl group to the base G (the methylated base is designated *mG*). ② If the DNA is replicated before the mG mutation can be repaired, the mG will form an incorrect base pair with T during DNA replication. ③ When the DNA is again replicated, the T forms a normal base pair with A. Hence, the DNA molecule now contains an AT base pair where a GC base pair had originally been located. Because the cell would not recognize anything abnormal about an AT base pair, the mutation is permanent.

pair with thymine (T) during DNA replication rather than pairing with its correct partner, cytosine (C). During the next round of DNA replication, the incorrectly inserted T will form a base pair with its normal complementary base, adenine (A), creating an AT base pair. The DNA molecule now contains an AT base pair where a GC base pair had originally existed. Since DNA repair mechanisms would not recognize anything abnormal about an AT base pair, the error will persist.

The preceding scenario demonstrates an important principle that applies to many mutations: If DNA replication occurs before mutations are repaired, base-pair alterations tend to arise during replication that cannot be subsequently detected as mutations by cellular repair mechanisms. For this reason it is crucial that mutations be repaired swiftly, before subsequent rounds of DNA replication create a permanent mutation.

Tumor Promotion Involves a Prolonged Period of Cell Proliferation

In contrast to initiation, which requires only a single exposure to an initiating carcinogen, promotion is a gradual process that depends on prolonged or repeated exposure to a promoting agent. If the promoting agent is removed during the early stages of tumor formation, tumors stop growing and may even disappear.

How do we explain the ability of promoting agents to trigger an event that is potentially reversible, at least in its early stages? Studies of a wide variety of promoting agents have revealed that their main shared property is the ability to stimulate cell proliferation. When an initiated cell is exposed to a promoting agent, the cell starts dividing and the number of initiated cells increases. In the early stages of this process, cell proliferation depends on the presence of the promoting agent, and the cells will stop dividing if the agent is removed. As cell division continues, however, natural selection favors those newly forming cells whose proliferation is faster and autonomous, eventually leading to the formation of a malignant tumor whose growth no longer depends on external promoting agents. The time required for promotion contributes to the long delay that often transpires between exposure to an initiating carcinogen and the development of cancer.

The way in which specific promoting agents stimulate cell proliferation was first established for **phorbol esters**, the class of tumor promoters found in croton oil. In terms of its tumor promoting activity, the most potent phorbol ester is *tetradecanoyl phorbol acetate* (*TPA*). TPA binds to and activates an enzyme called **protein kinase C**, which plays a key role in one of the cell's normal pathways for controlling cell proliferation (Figure 5-15). In the normal operation of this pathway, external signaling molecules bind to cell surface receptors whose activation leads to the production of an intracellular signaling molecule called *diacylglycerol* (*DAG*). DAG in turn activates protein kinase C, which triggers events leading to cell division. Phorbol esters mimic the action of DAG, binding to and activating protein kinase C. Unlike DAG, however, which is converted to inactive forms, phorbol esters continually activate the protein kinase C molecule. Activation of protein kinase C by TPA is a highly selective interaction; small changes in the chemical structure of TPA yield derivatives that exhibit diminished ability to bind to protein kinase C and, as a result, decreased ability to function as tumor promoters (Figure 5-16).

In addition to phorbol esters, a variety of other foreign substances stimulate cell proliferation and thereby act as promoting agents. Some of these molecules resemble phorbol esters in being able to interact with protein kinase C. The fungal toxin *teleocidin* and the algal toxin *aplysiatoxin* are two such agents that function by activating protein kinase C, even though their chemical structures differ significantly from those of phorbol esters. Other promoting agents stimulate cell proliferation indirectly, causing tissue damage and cell destruction that

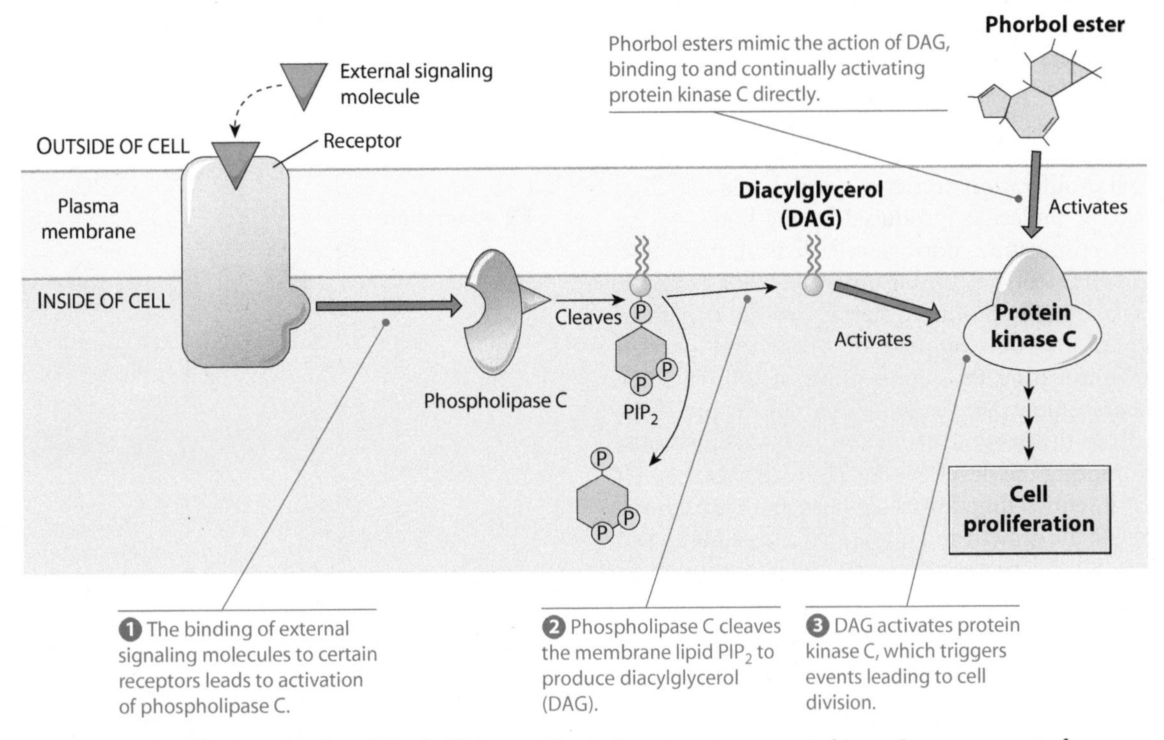

Figure 5-15 Mechanism of Action of Phorbol Esters. Phorbol esters activate protein kinase C, a component of a signaling pathway that stimulates cell proliferation. In the normal operation of this pathway, external signaling molecules bind to cell surface receptors whose activation leads to the production of diacylglycerol (DAG). The DAG then activates protein kinase C, which triggers events leading to cell division. Phorbol esters mimic the action of DAG, binding to and activating protein kinase C directly. (Green arrows represent activation reactions.)

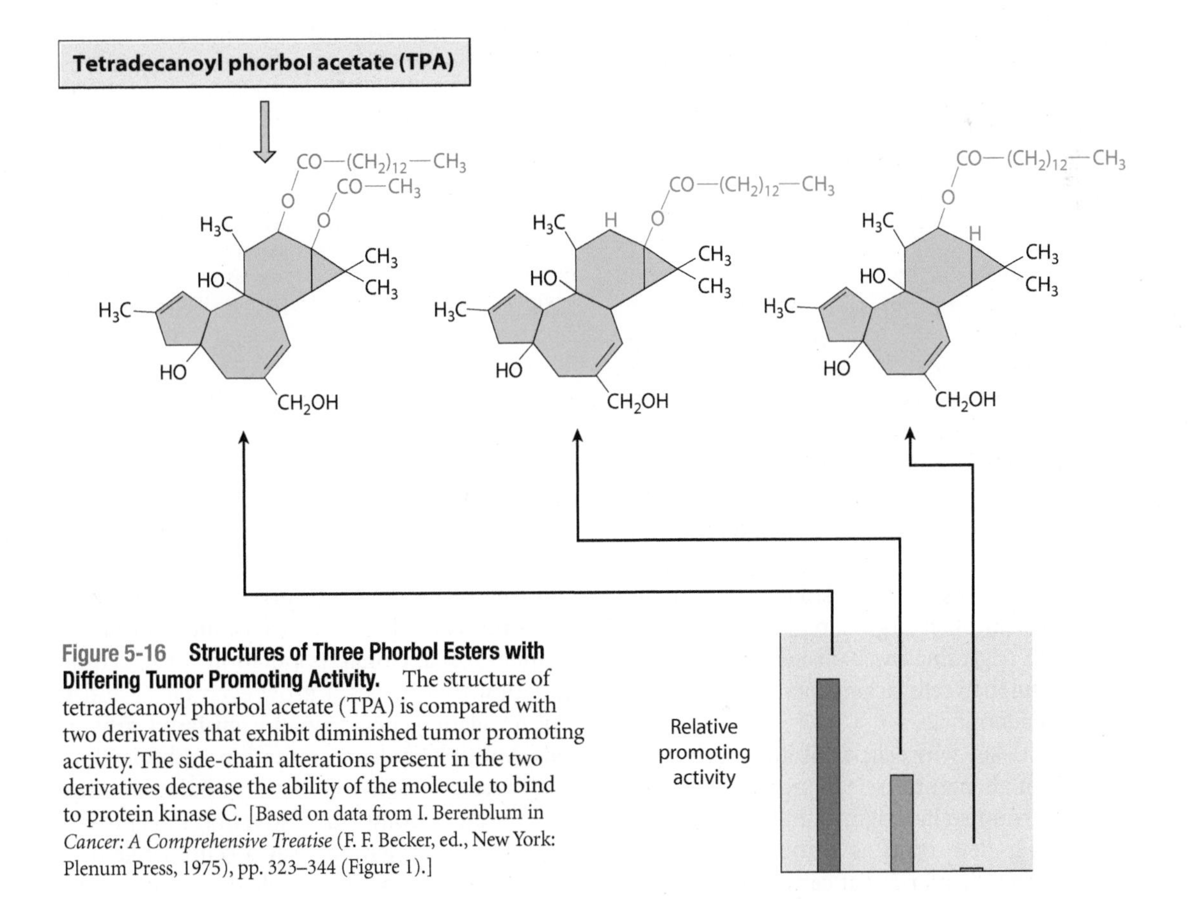

Figure 5-16 Structures of Three Phorbol Esters with Differing Tumor Promoting Activity. The structure of tetradecanoyl phorbol acetate (TPA) is compared with two derivatives that exhibit diminished tumor promoting activity. The side-chain alterations present in the two derivatives decrease the ability of the molecule to bind to protein kinase C. [Based on data from I. Berenblum in *Cancer: A Comprehensive Treatise* (F. F. Becker, ed., New York: Plenum Press, 1975), pp. 323–344 (Figure 1).]

make it necessary for the remaining cells of the affected tissue to proliferate to replace the cells that have been damaged and destroyed. Asbestos and alcohol are two previously discussed substances that function in this way.

Not all tumor promoters are foreign substances. Because cell proliferation occurs in normal cells as well as in tumor cells, molecules produced by the body for the purpose of stimulating normal cell division may also function inadvertently as tumor promoters. For example, *estrogen* is a naturally produced steroid hormone that can contribute to the development of breast and ovarian cancer by stimulating the proliferation of cells in these tissues. The hormone *testosterone* stimulates the proliferation of cells in the prostate gland and plays a comparable role in promoting the development of prostate cancer. Of course, the intended function of estrogen and testosterone is to stimulate the growth and division of normal cells, not cancer cells. But if a breast or prostate cell has acquired an initiating mutation caused by a carcinogen or by an error in DNA replication, any normal hormone or growth factor that stimulates the proliferation of the mutant cell will act inadvertently as a tumor promoter.

In addition to foreign chemicals and natural hormones, certain components of the diet may also act as promoting agents—that is, agents that increase cancer risk by stimulating cell proliferation rather than by creating mutations. Dietary fat and alcohol, whose ability to increase cancer risk were discussed in Chapter 4, are two examples that fit this category. In general, any chemical associated with an increased cancer risk that is found not to be genotoxic can be suspected of acting as a promoting agent.

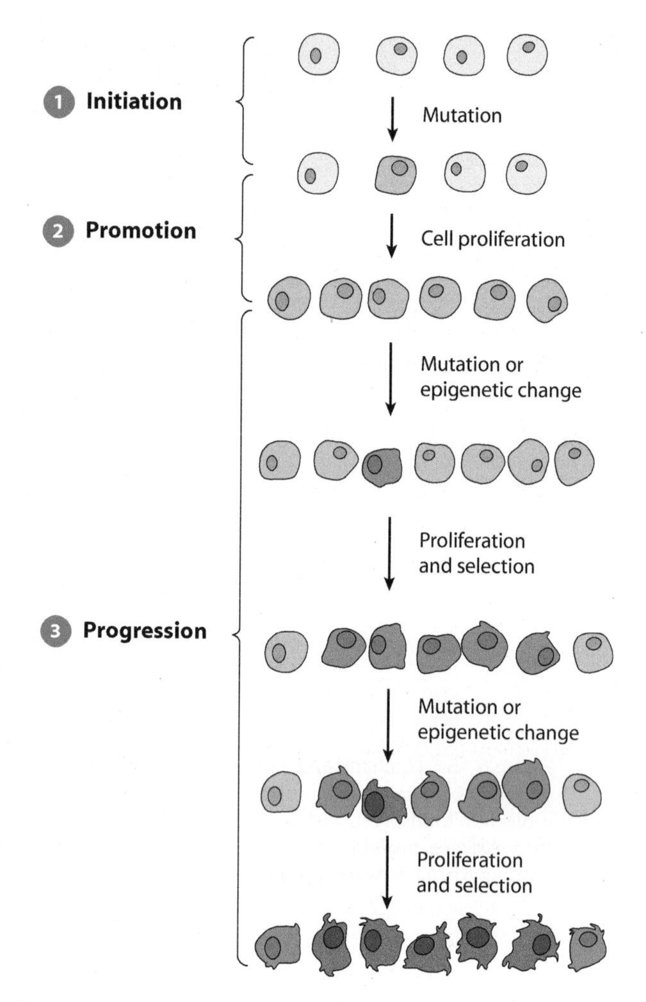

Figure 5-17 Main Stages of Carcinogenesis. Cancer arises by a complex process involving three main stages. ① The first stage, initiation, is based on DNA mutation. Initiation is followed by a ② promotion stage in which the initiated cell is stimulated to proliferate. ③ During tumor progression, further mutations and epigenetic changes in gene expression create variant cells exhibiting enhanced growth rates or other aggressive properties that give certain cells a selective advantage. Such cells tend to outgrow their companions and become the predominant cell population in the tumor. Repeated cycles of this process, called clonal selection, create a population of cells whose properties change over time.

Tumor Progression Involves Repeated Cycles of Selection for Rapid Growth and Other Advantageous Properties

When Rous first proposed that chemical carcinogenesis is a multistep process, he identified only two stages: initiation and promotion. It has gradually become apparent that a third stage, known as **tumor progression**, follows initiation and promotion (Figure 5-17). The concept of tumor progression, briefly introduced in Chapter 2, refers to the gradual changes in the properties of proliferating tumor cell populations that occur over time as cells acquire more aberrant traits and become increasingly aggressive. The underlying explanation for tumor progression is that cells exhibiting traits that confer a selective advantage—for example, increased growth rate, increased invasiveness, ability to survive in the bloodstream, resistance to immune attack, ability to grow in other organs, resistance to drugs, and evasion of death-triggering mechanisms (apoptosis)—will be more successful than cells lacking these traits and will gradually come to predominate.

While it is easy to see why cells exhibiting such traits tend to prevail through natural selection, that does not explain how the aberrant traits originate in the first place. One way of creating new traits is through additional mutations (recall from Chapter 2 that cancer cells tend to accumulate mutations at higher-than-normal rates). If a particular mutation causes a cell to divide more rapidly, cells produced by the proliferation of this mutant cell will outgrow their companions and become the predominant cell population in the tumor. Such a process is called *clonal selection* because the cells that predominate represent a **clone**—that is, a population of cells derived from a single initial cell by successive rounds of cell division. One member of a clonal population may acquire another mutation that makes it grow even faster and the process repeats again, generating an even faster growing clone of cells. Multiple cycles of mutation and selection can occur in succession, each creating a population with enhanced growth rate or some other advantageous property.

Although mutations play a central role in tumor progression, they are not the whole story. Cancer cell properties are also influenced by alterations in the expression of normal genes. The term **epigenetic change** is used to refer to any such alteration in gene expression that does not involve mutating the structure of a gene itself. Cells possess a variety of mechanisms for altering gene expression. Among them, activating or inhibiting the transcription of individual genes into messenger RNA is especially important in cancer cells. For example, many of the traits required for cancer cell invasion and metastasis are produced by epigenetically turning on or turning off the transcription of normal genes rather than by gene mutation.

Because the DNA base sequence is not being altered, epigenetic changes are easier to reverse than mutations. The question therefore arises as to whether a cancer cell can be epigenetically reprogrammed to reverse some of the changes responsible for malignant behavior. One way of addressing this question experimentally is to transfer the nucleus of a cancer cell into a different cytoplasmic environment to see if its gene expression patterns can be converted to a more normal state. When nuclei are taken from mouse cancer cells and transplanted into mouse eggs whose own nuclei have been removed, the eggs divide and proceed through the early stages of embryonic development, even though the cells possess cancer cell nuclei. Especially striking results have been reported when mouse melanoma cells (a cancer of pigment cells) are used as a source of nuclei for transplantation. Eggs receiving melanoma nuclei divide and produce embryonic cells that give rise to normal-appearing cells and tissues of adult mice (Figure 5-18). Nonetheless, mice containing such cells are not completely normal; the mice still exhibit an increased susceptibility to developing cancer. Such results indicate that the DNA of a cancer cell nucleus can be reprogrammed to a more normal state, but a propensity for cancer to arise still remains. In other words, epigenetic and genetic changes both play important roles in tumor development.

To sum up, tumor progression is a phase of carcinogenesis that involves the gradual acquisition of DNA mutations and epigenetic changes in gene expression, accompanied by natural selection of cells that have acquired advantageous properties generated by these mechanisms. The net result is a population of cells whose properties, including growth rate and the ability to invade and metastasize, slowly change over time. The time required for tumor progression contributes to the lengthy delay commonly observed between exposure to carcinogenic chemicals and the development of cancer. These principles, derived largely from studies of chemical carcinogenesis, apply to cancers triggered by other cancer-causing agents as well.

Carcinogenesis Is a Probabilistic Event That Depends on Carcinogen Dose and Potency

The realization that chemical carcinogenesis is a multistep process involving several distinct stages and mechanisms can cause some confusion about the meaning of the term *carcinogen*. In common usage, any agent that increases the risk of developing cancer in animals or humans is considered to be a carcinogen. In this sense, either an initiating or a promoting agent would qualify as a carcinogen. For clarification, the term **incomplete carcinogen** is sometimes employed when referring to a chemical that exerts only one of these two actions. Some chemicals possess both initiating and promoting activities, and can therefore cause cancer by themselves; such chemicals are called **complete carcinogens**. The ability to function as a complete carcinogen may be dose dependent. For example, certain polycyclic hydrocarbons act as initiating agents at

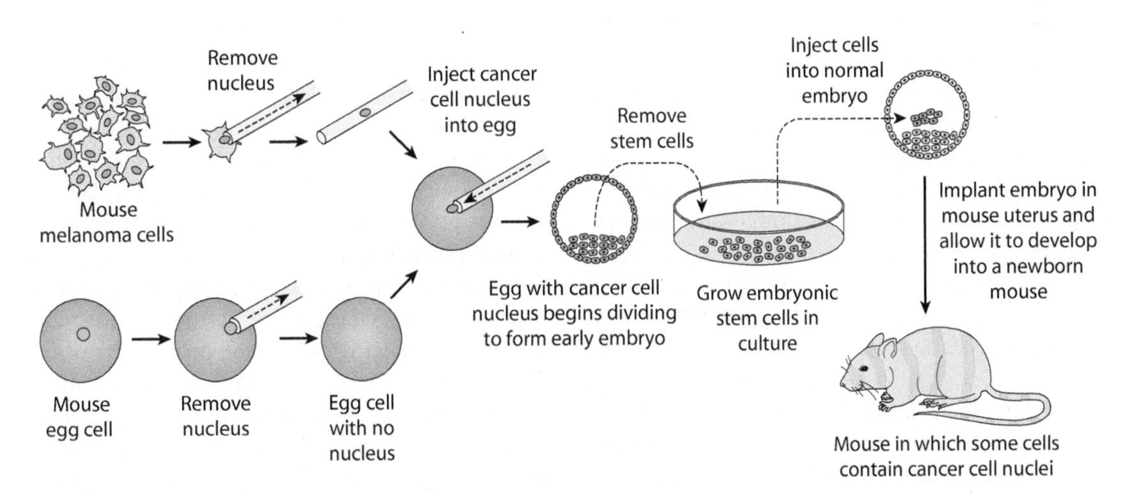

Figure 5-18 Transplanting a Cancer Cell Nucleus into an Egg Cell. When the nucleus of a mouse melanoma cell is transferred into a mouse egg whose own nucleus has been removed, the egg divides and proceeds through the early stages of embryonic development, although it cannot complete the process to form a new mouse. If *stem cells* (undifferentiated cells whose division gives rise to specialized cells) are removed at this stage and injected into a normal early embryo, these embryonic stem cells, with their cancer cell nuclei, will participate in the development of normal-appearing cells and tissues of a new mouse. Nonetheless, mice containing such cells are not completely normal; they still exhibit an increased susceptibility to developing cancer. [Based on experiments of K. Hochedlinger et al., *Genes Dev.* 18 (2004): 1875.]

lower doses but are complete carcinogens at higher doses. In normal human experience, people are exposed to chemical mixtures, such as tobacco smoke or coal tar, that contain both initiating and promoting carcinogens. In such cases, the mixture acts as a complete carcinogen.

The multistep nature of chemical carcinogenesis also complicates the question of what scientists mean when they say that something "causes" cancer. For example, exposure to an incomplete carcinogen (i.e., an initiating or promoting agent) will not, by itself, cause cancer. Even a complete carcinogen rarely causes cancer in every exposed person or animal. When it is stated that a particular carcinogen causes cancer, what is really meant is that the agent in question increases the *probability* that cancer will arise. The magnitude of the increased risk depends on several factors, including the dose and potency of the agent involved and the issue of whether it is acting as an incomplete or complete carcinogen (complete carcinogens obviously carry a greater risk).

The fact that cancer risk is related to carcinogen dose was already introduced in Chapter 4, but the reason for this dose dependence should now be more apparent. As the dose of an initiating carcinogen is increased, more DNA adducts and other types of DNA damage accumulate. To initiate the development of cancer, this damage must affect certain critical genes (the identity and properties of these cancer-related genes will be described in Chapters 9 and 10). The probability that one of these cancer-related genes will happen to undergo mutation is quite small because carcinogens trigger random DNA damage and the critical genes constitute only a tiny fraction of the total DNA. The higher the dose of carcinogen, however, the greater the overall DNA damage and hence the greater the chance that a critical gene will be affected by a random mutation.

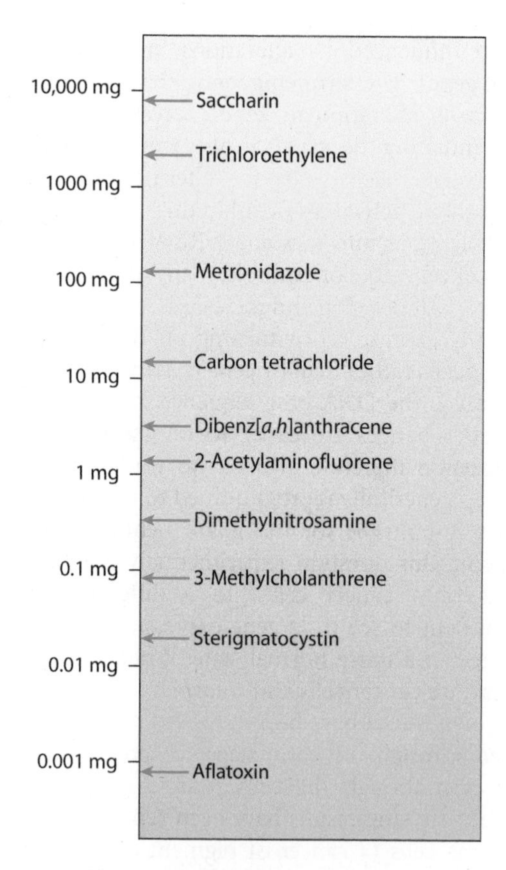

Figure 5-19 Differences in Carcinogenic Potency. The relative potency of several carcinogenic chemicals is compared on a scale that indicates how large a dose is needed to cause cancer in 50% of the animals tested. Note that the required dose of the weakest carcinogen shown (saccharin) is about ten-million-fold higher than the dose of the strongest carcinogen shown (aflatoxin). [Data from T. H. Maugh, *Science* 202 (1978): 38.]

The likelihood that a particular carcinogen will cause cancer also depends on its potency. Carcinogen potency is generally assessed in animals by determining how large a dose is needed to cause cancer in 50% of the animals tested. Such testing has revealed that a ten-million-fold difference in strength separates the strongest carcinogens from the weakest (Figure 5-19). Two properties are especially important in explaining this enormous variation in potency. The first involves the activation reactions catalyzed by cytochrome P450, which are more effective in converting certain types of chemicals into active carcinogens than they are for other chemicals. The second factor is related to the electrophilic strength of different carcinogens. Some substances are strongly electrophilic and are highly reactive with DNA, whereas others are weaker electrophiles and are less reactive with DNA. The probability that a carcinogen will happen to mutate a critical gene randomly is much greater for carcinogens that are stronger electrophiles because they trigger more mutations.

The random nature of mutation helps explain why everyone who is exposed to carcinogens does not develop cancer. For many years, tobacco companies tried to cast doubt on the relationship between cigarette smoking and lung cancer by pointing out that some cigarette smokers live long lives without ever developing cancer. The ability of carcinogens to trigger random DNA mutations provides a simple explanation for such observations: It is largely a matter of chance. Tobacco smoke contains numerous carcinogens that cause random DNA damage, but for cancer to develop, a mutation must arise in a critical cancer-related gene. An apt metaphor is the game of Russian roulette, in which a single bullet is placed in a gun containing six chambers and the cylinder is then spun. When the trigger is pulled, the probability of firing a bullet that can kill is 1 in 6. A similar principle applies to smoking cigarettes, a potentially lethal practice governed by the laws of probability. Each cigarette has a small but finite probability of randomly creating a mutation that can cause cancer. Like Russian roulette, the more the game is played, the greater the chance that lethal damage will occur. So when it is stated that smoking cigarettes (or exposure to any other carcinogen) "causes" cancer, it simply means that a person's risk of developing cancer is increased. For this reason, agents exhibiting the potential to cause cancer are sometimes referred to as cancer **risk factors**.

In concluding, brief mention should be made of the fact that immunosuppressive drugs increase cancer risk in a fundamentally different way from the chemical carcinogens we have been discussing. As mentioned earlier in this chapter, immunosuppressive drugs are given to organ transplant patients to inhibit the immune system and thereby minimize the possibility that a transplanted organ will be rejected. Because they inhibit immune function, some immunosuppressive drugs increase cancer risk by diminishing the likelihood that immune surveillance will destroy newly forming cancer cells. These immunosuppressive drugs differ from typical carcinogens in that they increase cancer risk indirectly, targeting the immune system rather than acting directly on the cells destined to become cancerous.

Summary of Main Concepts

- Everyone is exposed to cancer-causing chemicals, but most of these molecules are weakly carcinogenic and the more potent ones can largely be avoided by the general public.
- Bladder cancer arising in workers who were exposed to 2-naphthylamine in the clothing dye industry was the first human cancer to be linked to a specific chemical. Like many chemically induced cancers, the relationship is dose dependent, it involves a long delay after exposure, hereditary susceptibility does not play a significant role, and one particular type of cancer (bladder cancer) predominates.
- Asbestos is the second most lethal commercial product (after tobacco) in causing cancer deaths. In addition to triggering the development of lung cancer, microscopic fibers of inhaled asbestos penetrate through the lung and enter the chest cavity, where they cause a rare type of cancer known as mesothelioma.
- Several dozen chemical carcinogens were identified as significant workplace hazards during the twentieth century, but the most dangerous ones have been eliminated, exposure to others has been decreased, and relatively few new ones have been identified in recent years.
- Most pollution of the environment by chemical carcinogens involves substances that are present in small concentrations. Age-adjusted cancer rates did not increase significantly in response to the explosive growth in such pollutants that occurred during the twentieth century.
- The cancer risk associated with low-dose exposures to chemical carcinogens is difficult to quantify because the epidemiological approach cannot reliably detect small increases in cancer rates and because data from high-dose animal testing cannot be reliably extrapolated to low-dose human exposures.
- Animal testing often involves exposing animals to the maximum tolerated dose of a suspected carcinogen, which can lead to cancer via mechanisms that are not relevant to low-dose exposures. Under such conditions the dose-response curve may exhibit a threshold—that is, a dose that must be exceeded before cancer rates begin to rise. Repair pathways that correct DNA damage after low-dose exposures but become overwhelmed at higher doses may also contribute to threshold effects. In addition to the threshold problem, humans and animals differ in their susceptibilities to certain carcinogens. All these factors make it difficult to extrapolate from high-dose effects in animals to low-dose effects in humans.
- Chemical carcinogens are a structurally diverse group of substances that include polycyclic hydrocarbons, aromatic amines, N-nitroso compounds, alkylating agents, natural products, and inorganic elements. Many of these substances must be metabolically activated by enzymes in the liver before they can cause cancer. Metabolic activation creates highly unstable electrophilic molecules that react directly with DNA, creating DNA adducts that distort the double helix and cause mutations to arise during DNA replication.
- Chemical carcinogenesis is a multistep process that involves initiation, promotion, and tumor progression. Initiation is based on exposure to chemicals that create permanently altered, precancerous cells through their ability to cause DNA mutation. Promotion involves a prolonged period of proliferation of the initiated cells, accompanied by a gradual selection of cells exhibiting enhanced growth rate. During tumor progression, cells accumulate additional mutations and undergo epigenetic changes in gene expression that allow for the progressive selection of cells with increasingly aggressive and aberrant traits.
- A complete carcinogen possesses both initiating and promoting activities; an incomplete carcinogen exerts only one of these two actions. Saying that a carcinogen *causes* cancer really means that it increases the probability that cancer will arise. The magnitude of the increased risk depends on carcinogen dose as well as potency. A ten-million-fold difference in potency separates the strongest carcinogens from the weakest.

Key Terms for Self-Testing

Identifying Chemicals That Cause Cancer

asbestos (p. 82)
mesothelioma (p. 82)
maximum tolerated dose (MTD) (p. 86)
threshold (p. 86)
linear model (p. 86)
threshold model (p. 86)
hormetic model (p. 86)
immunosuppressive drug (p. 88)

Mechanisms of Chemical Carcinogenesis

organic chemical (p. 89)
polycyclic aromatic hydrocarbon (polycyclic hydrocarbon) (p. 89)
aromatic amine (p. 89)
N-nitroso compound (p. 89)
alkylating agent (p. 89)
natural products (p. 91)
inorganic substance (p. 91)
precarcinogen (p. 91)
cytochrome P450 (p. 91)
carcinogen activation (p. 91)
electrophilic (p. 92)
epoxide (p. 92)
DNA adduct (p. 92)
initiation (p. 94)
promotion (p. 94)
genotoxic (p. 95)
phorbol ester (p. 96)
protein kinase C (p. 96)
tumor progression (p. 98)
clone (p.98)
epigenetic change (p. 99)
incomplete carcinogen (p. 99)
complete carcinogen (p. 99)
risk factor (p. 100)

Suggested Reading

Identifying Chemicals That Cause Cancer

Abelson, P. H. Risk assessment of low-level exposures. *Science* 265 (1994): 1507.

Calabrese, E. J., and L A. Baldwin. Can the concept of hormesis be generalized to carcinogenesis? *Regulatory Toxicol. Pharmacol.* 28 (1998): 230.

Gammon, M. D., et al. Environmental toxins and breast cancer on Long Island. II. Organochlorine compound levels in blood. *Cancer Epidemiol. Biomarkers Prevention* 11 (2002): 686.

Guba, M., et al. Rapamycin inhibits primary and metastatic tumor growth by antiangiogenesis: Involvement of vascular endothelial growth factor. *Nature Medicine* 8 (2002): 128.

Kaiser, J. Just how bad is dioxin? *Science* 288 (2000): 1942.

Ott, W. R., and J. W. Roberts. Everyday exposure to toxic pollutants. *Sci. Amer.* 278 (February 1998): 86.

Pope, C. A., III, et al. Lung cancer, cardiopulmonary mortality, and long-term exposure to fine particulate air pollution. *JAMA* 287 (2002): 1132.

Stallone, G., et al. Sirolimus for Kaposi's sarcoma in renal-transplant recipients. *New England J. Med.* 352 (2005): 1317.

Tweedale, G. Asbestos and its lethal legacy. *Nature Reviews Cancer* 2 (2002): 311.

Zurlo, J., and R. A. Squire. Is saccharin safe? Animal testing revisited. *J. Natl. Cancer Inst.* 90 (1998): 2.

Mechanisms of Chemical Carcinogenesis

Cohen, S. M., and L. B. Ellwein. Genetic errors, cell proliferation, and carcinogenesis. *Cancer Res.* 51 (1991): 6493.

Gooderham, N. J. Mechanisms of Chemical Carcinogenesis. In *The Cancer Handbook* (M. Alison, ed.). London: Nature Publishing Group, 2002, Chapter 20.

Guengerich, F. P. Metabolism of chemical carcinogens. *Carcinogenesis* 21 (2000): 345.

Hochedlinger, K., et al. Reprogramming of a melanoma genome by nuclear transplantation. *Genes Dev.* 18 (2004): 1875.

Luch, A. Nature and nurture—lessons from chemical carcinogenesis. *Nature Reviews Cancer* 5 (2005): 113.

Philips, D. H. The Formation of DNA Adducts. In *The Cancer Handbook* (M. Alison, ed.) London: Nature Publishing Group, 2002, Chapter 21.

Poirier, M. C. Chemical-induced DNA damage and human cancer risk. *Nature Reviews Cancer* 4 (2004): 630.

Radiation and Cancer

In its broadest sense, the term **radiation** refers to any type of energy traveling through space. Radiation is a natural component of the environment and comes in a variety of forms, including heat, sound, radio waves, microwaves, visible light, ultraviolet light, X-rays, and nuclear radiation. While people often associate the word "radiation" with risk and danger, many types of radiation serve useful functions with little or no risk. For example, life could not exist without visible light from the sun, whose energy is trapped by photosynthetic organisms and used to power the synthesis of molecules we consume as food. Sound and heat are other forms of radiation that serve useful functions with little risk from moderate exposures.

Certain kinds of radiation, however, are associated with significant health risks, including the ability to cause cancer. The greatest cancer risks are associated with exposure to high-energy forms of radiation, which interact with biological molecules and impart enough energy to alter the chemical bonds that hold molecules together. Natural sources of such radiation include ultraviolet rays from the sun, cosmic rays from outer space, and nuclear radiation emitted by naturally occurring radioactive elements. Medical, industrial, and military activities have created additional sources of high-energy radiation, mainly in the form of X-rays and radioactivity. In this chapter, we will explore the roles played by these various types of radiation in causing cancer.

SUNLIGHT AND ULTRAVIOLET RADIATION

Several types of radiation, differing from one another in source, energy, and wavelength, are known to cause cancer. Among these, sunlight is responsible for the most cancer cases. In the United States, about half of all newly diagnosed cancers—roughly 1.3 million new cancers each year—are caused by the sun, giving sunlight the dubious distinction of causing more people to develop cancer than all other causes of cancer combined.

Skin Cancer Risk Is Related to Sunlight Exposure and Intensity

The idea that sunlight can trigger cancer initially came from the discovery that skin cancer rates are elevated in people who spend long hours in the sun, especially in areas of the world where the sunlight is intense. A striking example involves the population of Australia, where skin cancer rates are high for reasons based on a quirk of eighteenth-century history. During the 1780s, the British House of Commons decided to deal with overcrowding in British jails by banishing criminals to the (then) remote island of Australia. Within a few decades, the east coast of Australia came to be inhabited by light-skinned British men and women whose descendants now represent a large part of the Australian population. The white skin and fair complexion of these people makes them particularly vulnerable to the intense Australian sunlight, and as a result, the white population of Australia has the highest skin cancer rate of any people in the world. Such high rates cannot by explained by hereditary factors because in England, where the sun is weaker and often covered by clouds, this same group of people had low skin cancer rates.

The Australian episode is particularly striking because it created the world's biggest skin cancer epidemic, but similar associations between sunlight and skin cancer have been observed in other contexts. For example, skin cancer rates vary in different regions of the United States, depending on the average amount of sunlight received. In southern states with intense sunlight, such as Texas and New Mexico, skin cancer is more frequent than in northern states such as Iowa and Michigan, where it is cloudier and the sunlight is less intense (Figure 6-1). As would be expected if sunlight were responsible, skin cancers arise most frequently on parts of the body that are routinely exposed to the sun, such as the face, neck, arms, and hands. Skin cancer rarely affects areas of the body that are normally covered by clothing and the sun cannot reach.

Because skin cancers are located on the outside of the body, they tend to be noticed and diagnosed at earlier stages than other cancers. As a result, most skin cancers can be removed before invasion and metastasis have occurred and cure rates for skin cancer are very high (around 99%). Nonetheless, some forms of skin cancer do represent a significant hazard. Of the three forms of skin cancer that are routinely distinguished from one another, **basal cell carcinomas** are the most common. Basal cell carcinomas account for 75% of all skin cancers but cause few deaths because the tumors metastasize in less than one patient per thousand. The next most frequent type of skin cancer is **squamous cell carcinoma**, which accounts for about 20% of all skin cancers. Squamous cell carcinomas are more serious than basal cell carcinoma because they metastasize more frequently, although metastasis still occurs in only one out of twenty patients.

The most serious form of skin cancer is **melanoma**, a type of cancer that arises from pigment cells called *melanocytes*. Melanomas account for only 5% of all skin cancers but are the most dangerous because they frequently metastasize, often before the tumor has even been noticed. For this reason, melanomas account for the vast majority of skin cancer fatalities (Figure 6-2). Basal and squamous cell carcinomas, being less hazardous than melanomas, are often lumped together and referred to as *nonmelanoma skin cancers* to distinguish them from melanomas.

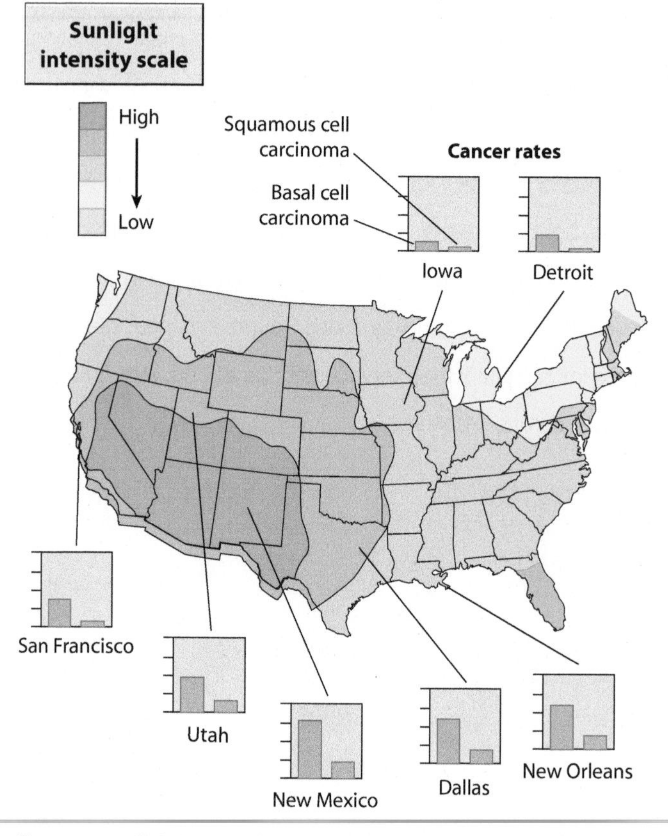

Figure 6-1 Relationship Between Skin Cancer and Sunlight Exposure. The prevalence of skin cancer varies across the United States, depending on the amount of sunlight received. In southern states with intense sunlight exposure, skin cancer is more frequent than in northern states, where it is cloudier and the sunlight is less intense. The bar graphs illustrate the relative rates of basal cell and squamous cell carcinoma among whites. [Based on cancer data from D. J. Leffell and D. E. Brash, *Sci. Amer.* 275 (July 1996): p. 58; solar map from National Renewable Energy Laboratory, U.S. Department of Energy.]

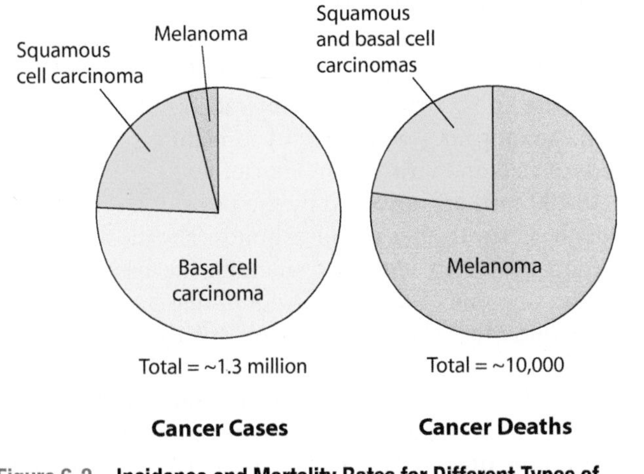

Figure 6-2 Incidence and Mortality Rates for Different Types of Skin Cancer. Basal cell carcinoma is the most frequent type of skin cancer, but melanoma accounts for most skin cancer fatalities. Data are for the United States. [Based on data from *Cancer Facts & Figures 2004* (Atlanta, GA: American Cancer Society, 2004), p. 4.]

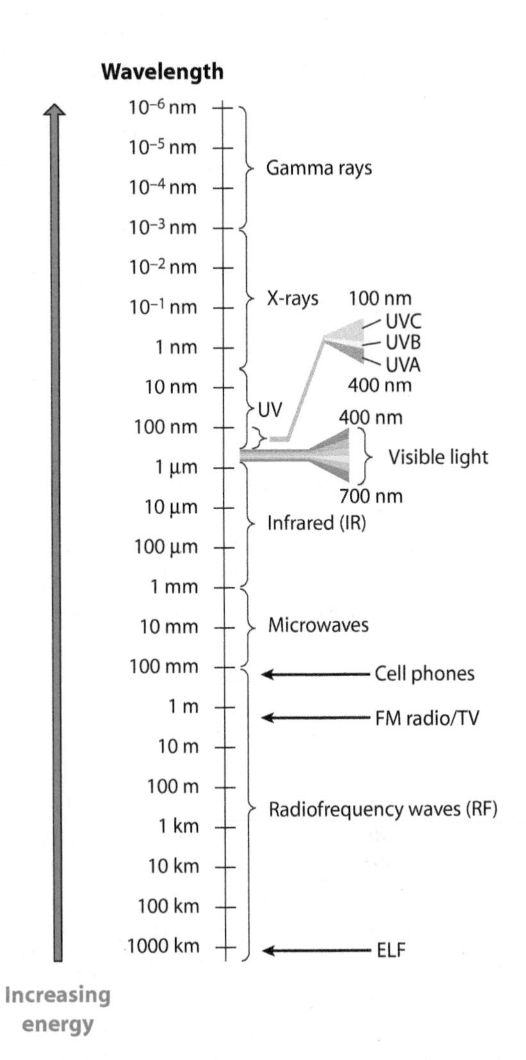

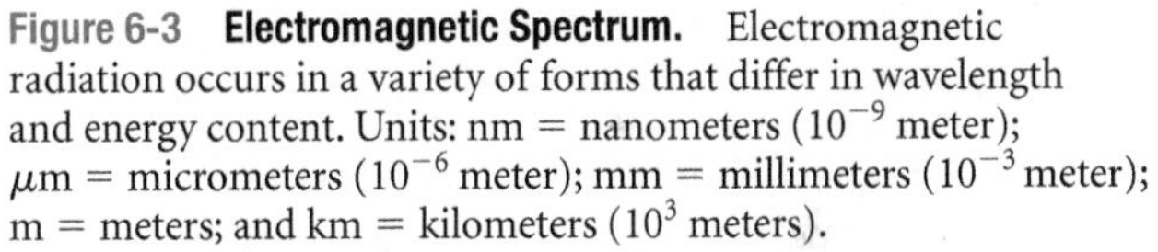

Figure 6-3 Electromagnetic Spectrum. Electromagnetic radiation occurs in a variety of forms that differ in wavelength and energy content. Units: nm = nanometers (10^{-9} meter); μm = micrometers (10^{-6} meter); mm = millimeters (10^{-3} meter); m = meters; and km = kilometers (10^{3} meters).

Sunlight has been implicated in causing both melanoma and nonmelanoma skin cancers, although the nature of the connection differs for the two groups of cancers. Nonmelanoma skin cancers arise mainly in individuals who are exposed to the sun on a regular basis and cancer tends to appear in locations that receive the most sunlight, such as the face, arms, and hands. In contrast, melanomas often arise in areas of the body that are not routinely exposed to the sun, such as the legs and back, and tend to occur in people who work indoors but have intense periodic exposures to sunlight on weekends or vacations or who had intense sunburn episodes when they were young.

Sunlight Contains Several Classes of UV Radiation

To explain how sunlight causes cancer, we need to describe the types of radiation given off by the sun. The sunlight that reaches the earth contains several forms of **electromagnetic radiation**, which is defined as waves of electric and magnetic fields that are propagated through space at the speed of light. Electromagnetic radiation occurs in a variety of forms that differ in wavelength and energy content (Figure 6-3). Wavelength and energy are inversely related to each other—that is, radiation of shorter wavelength possesses more energy than radiation of longer wavelength. The longest-wavelength component of sunlight is *infrared radiation*, which creates the warmth we feel from the sun. Next comes *visible light*, which is of shorter wavelength than infrared radiation and provides the illumination that allows us to see colors. Finally, **ultraviolet radiation** (UV) is the shortest-wavelength component of sunlight and possesses the greatest energy, making it capable of inflicting damage on human tissues.

The ultraviolet radiation in sunlight is in turn subdivided into three classes—A, B, and C—in order of decreasing wavelength (Table 6-1). **UVA** has the longest wavelength and the least energy. Defined as the portion of the UV spectrum whose wavelength falls between 315 and 400 nanometers (nm), UVA is the predominant type of ultraviolet radiation to reach the earth because it is not filtered out by the earth's atmosphere. UVA was once thought to be harmless because of its lower energy content, but long-term exposure to UVA is now known to cause aging of the skin and to act as a promoting agent for skin cancer by stimulating cell proliferation.

UVB radiation is of higher energy than UVA, falling in the wavelength range of 280 to 315 nm. Animal studies have shown the UVB is largely responsible for the carcinogenic properties of sunlight. More than 90% of the UVB radiation emitted by the sun is absorbed by **ozone** molecules present in the upper atmosphere, but enough UVB passes through to the earth's surface to cause sunburn, tanning, aging of the skin, and skin cancer. During the latter half of the twentieth century, the earth's ozone layer was partially destroyed by chemicals

Table 6-1 Types of Ultraviolet Radiation in Sunlight

Type	Wavelength	Properties
UVA	315–400 nm	• Not filtered by ozone layer • Causes skin aging • Stimulates cell proliferation
UVB	280–315 nm	• Partially filtered by ozone layer • Causes sunburn, tanning, skin cancer
UVC	100–280 nm	• Filtered out by ozone layer • Artificial sources cause skin burns and skin cancer

called **chlorofluorocarbons** (**CFCs**), which were used as refrigerants and for several industrial purposes. When they escape into the atmosphere, CFCs react with and destroy ozone molecules. Although the production of CFCs is now being phased out, the ozone layer will take several decades to recover, and ozone depletion may be causing skin cancer rates to rise because more UVB radiation is currently reaching the earth.

Finally, **UVC** falls in the wavelength range of 100 to 280 nm and is the most energetic type of UV radiation emitted by the sun. This high-energy, short-wavelength form of UV radiation can cause severe burns, but it is completely absorbed by the upper layers of the atmosphere before reaching the earth. UVC radiation is generally encountered only from artificial light sources, such as the germicidal lamps that use UVC to destroy bacteria when sterilizing medical and scientific equipment.

UVB Radiation Creates Pyrimidine Dimers in DNA

UVB is the highest-energy component of sunlight to reach the earth, but its energy level is still relatively low and thus it cannot penetrate very far into the body. Instead, UVB is absorbed by cells located in the outer layers of the skin, which explains why sunlight rarely causes any type of malignancy other than skin cancer. The damaging effects of UVB on skin cells often precede the development of cancer by many years. For example, consider what happens to people who move from England, with its weak sunlight and cloudy skies, to the intensely sunny climate of Australia. Those who move to Australia when they are young develop skin cancer at high rates when they reach middle age, whereas those who move to Australia later in life retain the low skin cancer rates that are typical of people who remain in England. Such observations suggest that skin cancers observed later in life are the result of sunlight damage that occurred many years earlier.

Such a pattern is reminiscent of the initiation phase of chemical carcinogenesis, in which carcinogens trigger DNA mutations that persist for many years, passed from one cell generation to the next as genetically damaged cells proliferate and give rise to tumors. By analogy, researchers have looked to see whether sunlight causes skin cell mutations early in life that can be linked to the later development of cancer. This is a complicated task because even if mutations are discovered in skin cancer cells, how can you be certain that sunlight caused them? A useful clue has come from studying the interactions of UVB—the main carcinogenic component of sunlight—with different kinds of cells and viruses. The shorter wavelengths of UVB (near 280 nm) are absorbed by the DNA bases, imparting enough energy to alter chemical bonds. The most common reaction occurs in regions containing the bases cytosine (C) and thymine (T), a class of single-ring bases known as *pyrimidines*. In locations where two of these pyrimidine bases lie next to each other, absorption of UVB radiation triggers the formation of covalent bonds between the adjacent bases, creating a unique type of mutation called a **pyrimidine dimer** (see Figure 2-15, *bottom*). All four combinations of two adjacent pyrimidines—that is, CC, CT, TC, and TT—are frequently converted into covalently linked dimers by UVB radiation.

Although cells can repair pyrimidine dimers, repair needs to occur before DNA replication creates a permanent, noncorrectable mutation. Figure 6-4 illustrates how such a permanent mutation could arise, using a CC dimer as an example. During DNA replication, the DNA strand in the region of the CC dimer is distorted and therefore tends to pair improperly with bases in the newly forming DNA strand. Instead of pairing correctly with its complementary base G, the base C in a CC dimer often pairs incorrectly with the base A (Figure 6-4, step ②). During the next round of DNA replication, the incorrectly inserted A will then form a base pair with its normal complementary base, T, creating an AT base pair (Figure 6-4, step ③). This AT base pair now looks normal to the cellular DNA repair machinery and so will continue to be replicated as if no error had been introduced.

Because the base T resides where the base C had been located in the original DNA molecule, the preceding type of mutation is called a C → T substitution. In some cases both C's of the dimer are replaced by the same mechanism, creating a CC → TT mutation. At this point the initial CC dimer in the original DNA strand could be repaired, but the C → T or CC → TT substitution in the newly replicated DNA will be permanent. Such base substitution patterns involving adjacent pyrimidines are unique to UV radiation and are therefore used as a distinctive "signature" to identify mutations caused by sunlight.

Mutations in the *p53* Gene Triggered by UVB Radiation Can Lead to Skin Cancer

After scientists discovered that UV radiation selectively induces the formation of pyrimidine dimers, the next task was to determine whether these mutations are associated with skin cancer. Among the first genes to be examined for the presence of pyrimidine dimers was the ***p53* gene**, a gene chosen for study because it is known to be mutated in many kinds of human cancer. (The *p53* gene codes for the p53 protein, whose role in preventing cells with damaged DNA from proliferating was introduced in Chapter 2 and will be

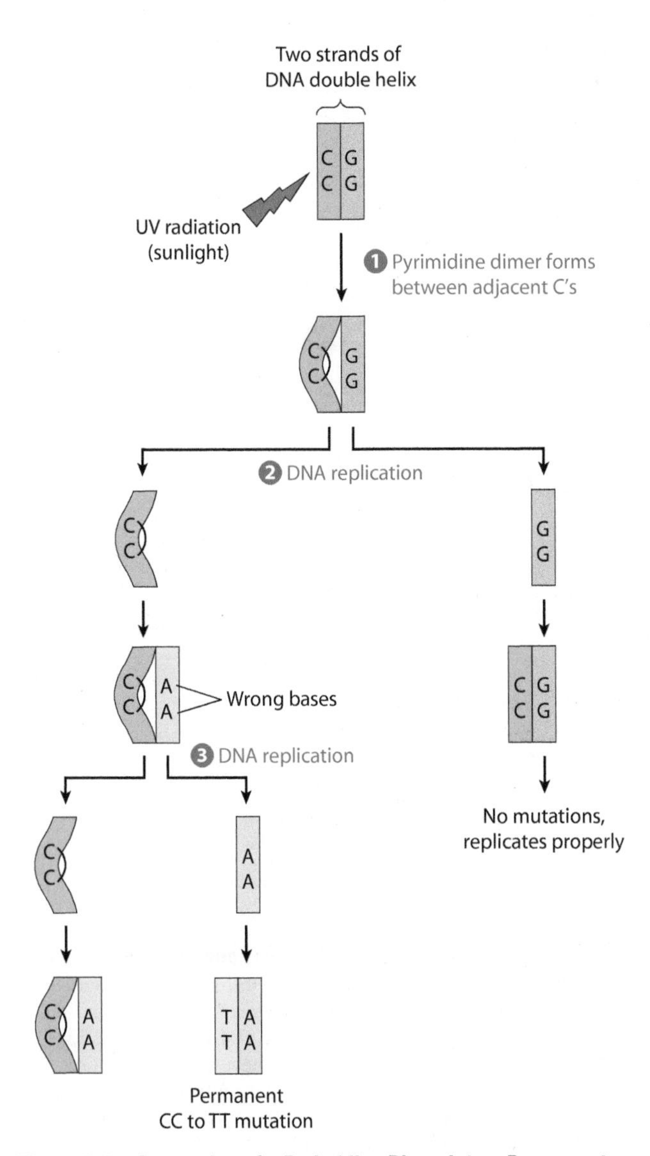

Figure 6-4 Conversion of a Pyrimidine Dimer into a Permanent Mutation. Ultraviolet radiation triggers the formation of covalent bonds between adjacent pyrimidine bases, creating dimers such as the CC dimer illustrated here (①). If the dimer is not repaired, the distortion in the DNA double helix causes the C's in the dimer to pair incorrectly with the base A during the next round of DNA replication (②). During the subsequent round of DNA replication, these A's will form base pairs with their normal complementary base, T, generating AT base pairs (③). Since DNA repair mechanisms will not recognize anything abnormal about AT base pairs, the mutation will now be replicated indefinitely. (Recall from Figure 5-14 that mutations triggered by chemical carcinogens become permanent in a similar way.)

described in detail in Chapter 10.) When skin cancer cells are examined for the presence of *p53* mutations, nonmelanoma skin cancers are routinely found to exhibit *p53* mutations with the distinctive UV "signature"—that is, C → T or CC → TT substitutions at dipyrimidine sites. In contrast, the *p53* mutations arising in cancers of internal body organs do not generally exhibit this UV-specific pattern (Figure 6-5).

The preceding observations indicate that the *p53* mutations seen in nonmelanoma skin cancer cells are triggered by sunlight, but do these mutations actually cause cancer to arise, or are they simply an irrelevant byproduct of long-term exposure to sunlight? This question can be resolved by looking at the precise location of UV-induced mutations within the *p53* gene. The DNA base sequence of most genes is arranged in a series of three-base units called *codons*, each of which specifies a particular amino acid in the protein encoded by the gene. Typically, the first two bases of a codon are more important in determining the amino acid than is the third. For example, the codons GAA and GAG both specify the same amino acid (glutamine), so changing the third base from A to G in this codon does not change the amino acid. A similar principle applies to the codons for many other amino acids.

If the *p53* mutations seen in nonmelanoma skin cancers were simply a random by-product of sunlight exposure, mutations in a codon's third base (which do not change an amino acid) should be as frequent as mutations in the first or second base (which do change an amino acid). In fact, DNA sequencing has revealed that *p53* mutations are not randomly distributed but instead involve base changes that alter amino acids. In other words, the *p53* mutations seen in nonmelanoma skin cancers alter the amino acid sequence of the protein encoded by the *p53* gene, as would be expected if these mutations are involved in the mechanism by which sunlight causes cancer. (The *p53* gene is not, however, the only mutant gene to be involved in nonmelanoma skin cancers, nor is it as frequently mutated in melanomas. Two genes commonly mutated in melanoma, called *BRAF* and *CDKN2A*, will be discussed in Chapters 9 and 10, respectively.)

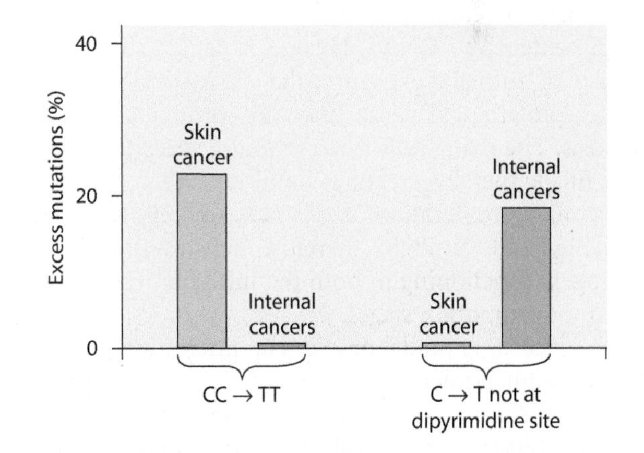

Figure 6-5 Incidence of Two Types of *p53* Mutations in Skin Cancer and Internal Cancers. The two bars on the left represent the frequency of CC → TT mutations, which are triggered by UV radiation. The two bars on the right represent the frequency of C → T mutations not located at dipyrimidine sites, which are not caused by UV radiation. Note that the UV-triggered type of mutation is found in the *p53* gene of skin cancers (squamous cell carcinomas) but not in cancers of internal organs. Mutation frequencies are plotted relative to what would be expected to occur randomly. [Based on data from D. E. Brash et al., *Proc. Natl. Acad. Sci. USA* 88 (1991): 10124 (Figure 2).]

The p53 Protein Protects Against Skin Cancer by Preventing Cells with Damaged DNA from Proliferating

Given that UV radiation triggers mutations that alter the p53 protein of skin cells, the question arises as to how these p53 abnormalities lead to cancer. We saw in Chapter 2 that the normal function of the p53 protein is to stop cells with damaged DNA from proliferating. In the presence of damaged DNA, the p53 protein accumulates and activates a pathway that halts the cell cycle, thereby allowing time for DNA repair. If the DNA damage is too severe to be repaired, the p53 protein eventually triggers cell suicide by *apoptosis* (p. 28). The net result is that cells containing mutations that might lead to cancer are not allowed to proliferate.

But what happens if sunlight happens to trigger a mutation that renders the p53 protein nonfunctional? In such a case, DNA damage caused by subsequent exposure to sunlight will not be able to trigger p53-mediated apoptosis, even if the DNA is damaged beyond repair. Damaged DNA will therefore be passed on to future cell generations during ensuing cell divisions, creating conditions that can lead to the development of cancer (Figure 6-6). Sunlight-induced mutation of the *p53* gene is thus comparable to the initiation stage of chemical carcinogenesis described in Chapter 5, in which an initial mutation creates a precancerous cell that is later converted into a tumor by a promotion phase involving sustained cell proliferation.

In addition to initiating carcinogenesis by mutating the *p53* gene, sunlight also affects tumor promotion through an indirect mechanism involving neighboring cells. An initiated, precancerous skin cell that has incurred a *p53* mutation will be surrounded by neighboring cells in which the *p53* gene is normal. If these surrounding cells sustain extensive DNA damage during subsequent episodes of sunlight exposure, the normal operation of the p53 protein will cause them to commit suicide by apoptosis. The dying cells leave behind a space that needs to be filled, thereby creating conditions that allow the uncontrolled proliferation of *p53*-mutated cells to replace the dying cells. Sunlight therefore acts as a complete carcinogen, functioning in both the initiation (mutation) and tumor promotion stages.

The preceding model provides an interesting perspective on the phenomenon of **sunburn**, the reddening and peeling of the skin that is commonly observed after intense sunlight exposure. Microscopic and biochemical examination of sunburned, peeling skin cells reveals that they look like cells being destroyed by apoptosis. Does that mean that a sunburn is simply the reflection of p53-mediated apoptosis triggered by sunlight-induced DNA damage? An answer has come from studies involving a mutant strain of mice possessing a defective *p53* gene. When the skin of such mice is exposed to intense UV radiation, fewer sunburned, apoptotic cells are observed than in normal mice exposed to the same radiation. It therefore appears

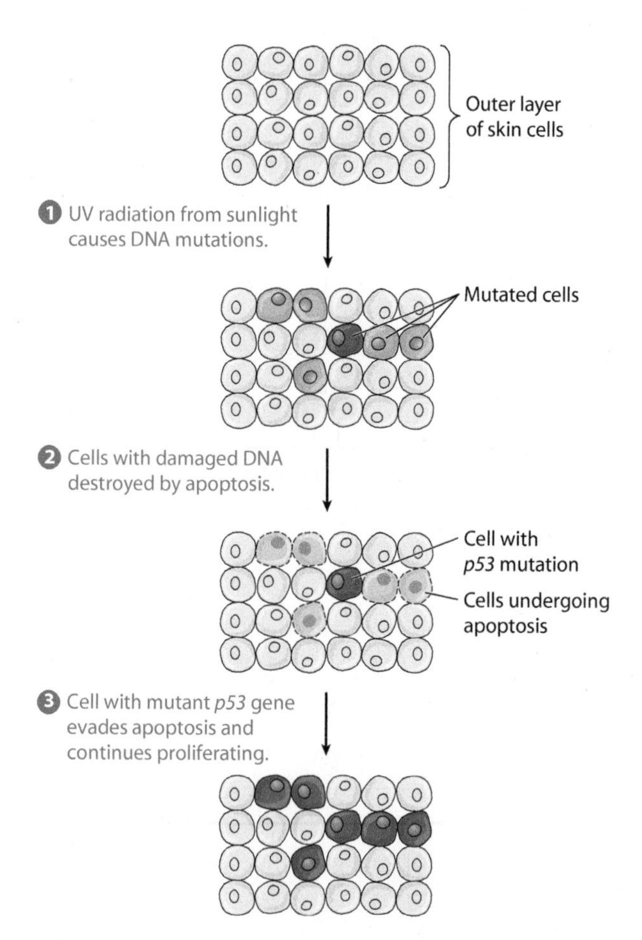

Figure 6-6 Sunlight-Induced *p53* Mutations and the Development of Skin Cancer. When skin is exposed to sunlight, the ultraviolet radiation causes DNA damage in the outer regions of the skin (①). In response, cells with extensively damaged DNA commit suicide by p53-mediated apoptosis (②). Cells that have acquired mutations in the *p53* gene may not be able to commit apoptosis. As a result they continue to proliferate, filling in the spaces left behind by those cells that do commit suicide (③).

that sunburn merely reflects the p53-mediated destruction of cells containing damaged DNA. So despite the pain, a sunburn is a protective mechanism, a deliberate effort to avert the development of cancer by destroying cells that have been damaged by UV radiation.

Clothing, Sunscreens, Skin Pigmentation, and Avoiding Strong Sunlight Are All Helpful in Decreasing Skin Cancer Risk

Because sunlight is its main cause, skin cancer is one of the easiest kinds of cancer to prevent. The best protection is afforded by simply staying out of the sun, especially at midday when sunlight is most intense. People living in tropical climates have traditionally taken a siesta at this time of day, a practice born of common sense and immortalized in a Noel Coward song whose lyrics state that only "mad dogs and Englishmen go out in the midday sun." (The English, of course, can safely go outdoors at midday when living in England because the sun is relatively weak and often covered by clouds; going out at midday only creates a significant cancer problem for the English when they travel to tropical climates.)

Although staying out of the sun provides the best protection, it is often impractical advice because occupational or recreational activities may require people to spend prolonged hours outdoors. Since only exposed areas of the body tend to be affected, the risk of developing skin cancer can be minimized by wearing protective clothing such as hats, long-sleeved shirts, and pants. Recently developed lightweight fabrics that exhibit enhanced sun-blocking properties, even when wet, are especially useful for this purpose.

Protection can also be provided by **sunscreen** lotions, which contain substances that prevent UV radiation from reaching skin cells. Sunscreens are of two different types: *physical sunscreens* that reflect UV radiation and *chemical sunscreens* that absorb it. The most popular physical sunscreens, zinc oxide and titanium dioxide, reflect both UVA and UVB radiation. Early lotions containing zinc oxide or titanium dioxide had the disadvantage of coloring the skin white, but these substances are now manufactured with particles so small that the white coloring is almost invisible. Unlike physical sunscreens, chemical sunscreens preferentially absorb either UVA or UVB radiation. Para-aminobenzoic acid (PABA), which absorbs UVB radiation, was introduced in the early 1970s and became the first widely used chemical sunscreen. However, PABA was subsequently found to elicit allergic reactions and has been largely replaced with other UVB-absorbing molecules, such as cinnamates, salicylates, and octocrylene. For absorbing UVA radiation, benzophenones and avobenzone are the most common chemical sunscreens.

Modern "broad-spectrum" sunscreen lotions usually contain a mixture of the preceding ingredients to provide maximum protection against both UVB radiation, the main cause of DNA mutation and skin cancer, and UVA radiation, which stimulates cell proliferation and can therefore promote tumor development. Sunscreens clearly diminish the risk of sunburn, but their effectiveness in decreasing skin cancer risk has been more difficult to assess. One complication is that skin cancer can take 15 to 30 years to develop. Because the use of broad-spectrum sunscreens is a relatively recent phenomenon, it may take several decades before epidemiological studies will detect any effects on skin cancer rates. In addition, the use of sunscreen lotions often lets people stay outdoors longer, so the protective effects of sunscreens might be offset by extra time spent in the sun. Although it may therefore take many years before the effects of sunscreens on skin cancer rates are precisely quantified, their ability to block UV radiation and prevent sunburning justifies their continued use. (Note: The same cannot be said for so-called sunless tanning lotions, which contain a chemical that directly imparts a bronze-like color to the skin; such products do not absorb UV radiation and thus afford no protection against skin cancer.)

Like broad-spectrum sunscreens, the natural pigments found in the skin also absorb UV radiation. As a result, people with darkly pigmented skin (e.g., Africans, Latinos, Hispanics, and Australian Aborigines) have lower skin cancer rates than people with lightly pigmented skin. Skin pigmentation comes from **melanin**, a family of brown pigments that are synthesized by melanocyte cells present in the skin. Regular exposure to sunlight stimulates melanin production, thereby causing a darkening of the skin or "suntan." A suntan helps protect the skin during subsequent exposures to sunlight because the melanin molecules absorb UV radiation. For this reason, some people use artificial tanning devices—such as sunlamps or commercial tanning beds—to start a suntan before vacationing in a sunny climate. Such tanning devices have even been claimed to be "safer" than sunlight because most of the radiation is UVA rather than UVB. However, people who use such devices develop skin cancer at higher rates than people who do not, even when their total sun exposure is equivalent. The popularity of artificial tanning devices is therefore of considerable concern, especially in view of reports that 50% of high school-aged girls in certain areas of the northern United States use commercial tanning beds on a regular basis.

IONIZING RADIATION

Although UV radiation is responsible for more cases of cancer than all other carcinogens combined, its inability to penetrate very far into the body means that it only causes skin cancer, which is often easy to cure. We now turn our attention to higher-energy forms of radiation that penetrate into the body and can therefore cause cancer to arise in internal organs. This type of radiation is called **ionizing radiation** because it removes electrons from biological molecules, thereby generating highly reactive ions that damage DNA in various ways.

X-rays Penetrate Through Body Tissues and Cause Cancers of Internal Cells and Organs

In 1895, the first form of ionizing radiation that would turn out to cause cancer in humans was accidentally discovered by Wilhelm Roentgen, a German physicist. Roentgen was passing an electric current through a partially evacuated glass tube, called a *cathode-ray tube*, when he noticed that a fluorescent screen located across the room began to glow. Even after he covered the cathode-ray tube with black paper and moved it to another room, the screen glowed when the cathode-ray tube was turned on. Most astonishing, however, was the discovery that an image of the bones in Roentgen's hand appeared on the screen when he placed his hand between the cathode-ray tube and the screen. Radiation exhibiting such unusual properties was completely unknown at that time, so Roentgen named it **X-rays.** In recognition of the importance of this discovery, Roentgen was awarded the first Nobel Prize in Physics in 1901.

X-rays are a type of electromagnetic radiation exhibiting a wavelength shorter than that of UV radiation (see Figure 6-3). Because of their short wavelength, X-rays are highly energetic and will pass through many materials that cannot be penetrated by UV radiation, visible light, or other weaker forms of electromagnetic radiation. This is the property that allows X-rays to be used for viewing the inside of objects such as the human body. Shortly after X-rays were discovered in 1895, newspaper headlines proclaimed "new light sees through flesh to bones!" and X-ray studios were opened around the country so that people could have "bone portraits" taken of themselves, even if they had no health problems! And doctors, of course, quickly embraced the new tool, which was to revolutionize many aspects of medical diagnosis and treatment.

Unfortunately, medical practitioners and researchers were slow to recognize the hazards of X-rays. An early danger signal came from the laboratory of Thomas Edison, whose research technician routinely tested X-ray equipment by using it to take pictures of his own hands. The technician soon developed severe radiation burns and cancer arose in the burned tissue. Although both his arms were subsequently amputated, he died of metastatic cancer in 1904, the first cancer fatality attributed to X-rays.

More cancers appeared in the next few decades as doctors specializing in the use of X-rays (radiologists) began to develop leukemia at rates several times higher than normal. The suspicion that X-rays were causing these cancers was eventually confirmed by animal studies, which showed that animals exposed to X-rays develop cancer at rates that are directly proportional to the dose of radiation received (Figure 6-7). The risk of leukemia is especially elevated, but X-rays pose a cancer threat to almost every tissue of the body.

Human exposure to such hazards could be virtually eliminated if X-rays served no useful purpose, but in many situations the health benefits to be gained from medical X-rays far outweigh the risk of inducing cancer.

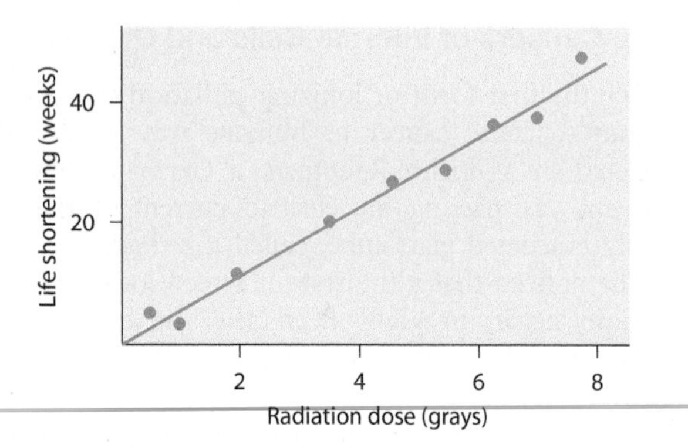

Figure 6-7 Effect of X-ray Exposure on Lifespan in Mice. A direct relationship exists between the dose of radiation administered to mice and a resulting reduction in their lifespan. Most of the lifespan reduction is caused by the development of cancer. Radiation dose is expressed in grays, which is a measurement unit for quantifying how much radiation is absorbed by biological tissues. [Data from P. J. Lindop and J. Rotblat, *Proc. Roy. Soc. London Ser. B* 154 (1961): 332 (Figure 10).]

That is not the case in all situations, however. Between the 1920s and 1950s, some doctors used high-dose X-rays to treat children with superficial skin conditions of the head and neck, such as ringworm and acne. Later in life, these individuals developed thyroid cancer at much higher rates than normal. Thus, the medical benefits to be gained from every X-ray procedure need to be prudently weighed against the increased cancer risk.

Radioactive Elements Emit Alpha, Beta, and Gamma Radiation

A year after the discovery of X-rays by Roentgen in 1895, the French physicist Henri Becquerel discovered another form of radiation called **radioactivity**, which is emitted by chemical elements that are intrinsically unstable. Some elements are naturally radioactive, whereas others are created artificially. The radiation emitted by a radioactive element emanates from an unstable atomic nucleus and is therefore referred to as **nuclear radiation**.

There are three main forms of nuclear radiation, known as alpha (α), beta (β), and gamma (γ) radiation. Alpha and beta radiation both involve streams of charged particles of matter. **Alpha particles** are positively charged entities composed of two neutrons plus two protons (the nucleus of one helium atom); **beta particles** are electrons and therefore exhibit a negative charge. In addition to these particulate forms of nuclear radiation, some radioactive elements emit **gamma rays**, which are a type of electromagnetic radiation and therefore exhibit no mass or charge. The wavelength of gamma rays is shorter than that of X-rays, making them the most energetic form of electromagnetic radiation (see Figure 6-3).

One of the first scientists to work with radioactivity was Marie Curie, codiscoverer of two naturally radioactive elements, polonium and radium. Awarded two Nobel Prizes (one in physics and one in chemistry) for her pioneering work on radioactivity, she suffered severe radiation burns and eventually died of leukemia, presumably caused by extensive exposure to nuclear radiation. Marie Curie's daughter, Irène Joliot-Curie, followed in her mother's footsteps and eventually received a Nobel Prize in Chemistry for showing that radioactive elements can be artificially created in the laboratory by bombarding chemically stable elements with high-energy radiation. Like dozens of other early workers who handled radioactive materials, Irène Joliot-Curie also died of cancer.

Radiation Dose Is Measured in Grays and Sieverts to Account for Differences in Tissue Absorption and Damage

As with other carcinogens, the cancer risk associated with nuclear radiation is directly related to the dose received. Measuring radiation dosage is complicated because various types of radiation differ both in their energy content and in their ability to penetrate and damage biological tissues. To take these variables into account,

Table 6-2 Some Radiation Measurement Concepts

Concept	Unit of Measurement
Radiation energy	electronvolt (eV)
Absorbed dose	gray (Gy)
historical unit (obsolete)	rad (= 0.01 Gy)
Biologically equivalent dose*	sievert (Sv)
historical unit (obsolete)	rem (= 0.01 Sv)
Type of Radiation	**RBE Value**
X-rays	1
Alpha particles	20
Beta particles	1
Gamma radiation	1
Protons	10

*Calculation: Biologically equivalent dose = absorbed dose × RBE.

several measurement units are used when describing radiation doses (Table 6-2).

The basic unit of radiation energy is the *electron volt* (*eV*), which expresses the total amount of energy present. However, such a measurement does not reflect how much energy is absorbed when a given type of radiation interacts with biological tissue. An additional unit of measurement, called the **gray** (**Gy**), is therefore used to describe how much energy is actually absorbed. Even when two types of radiation are absorbed in equal amounts, they may differ in their ability to damage biological tissue. To account for such differences, the absorbed dose (measured in grays) is commonly multiplied by a correction factor called the **relative biological effectiveness** (**RBE**) to yield a *biologically equivalent dose* (or simply *equivalent dose*) measured in units called **sieverts** (**Sv**). As a standard of reference, X-rays are defined as having an RBE = 1; the tissue-damaging potencies of other forms of radiation are then compared with that of X-rays, and each type of radiation is assigned an RBE value that reflects its relative effectiveness.

For example, alpha particles have an RBE = 20, which means that alpha particles are 20 times more effective than X-rays in causing tissue damage. Therefore, exposure to 1 Gy of alpha particles corresponds to an equivalent dose of 20 Sv (the absorbed dose in grays multiplied by RBE = 20). In contrast, exposure to 1 Gy of X-rays corresponds to an equivalent dose of only 1 Sv (the absorbed dose in grays multiplied by RBE = 1). The reason for expressing radiation dose in sieverts rather than grays is that it provides a better indication of how much biological damage a given exposure to radiation will cause. (Note: The gray and sievert replace older historical units of radiation exposure called *rads* [= 0.01 Gy] and *rems* [= 0.01 Sv], respectively.)

Radon, Polonium, and Radium Emit Alpha Particles That Can Cause Cancer in Humans

Of the three main types of nuclear radiation (alpha, beta, and gamma), alpha particles are potentially the most hazardous because they are highly damaging to biological tissues and many radioactive elements emit alpha particles. Nonetheless, *external* exposure of the body to alpha emitters involves little danger because alpha particles are relatively large and are therefore easily blocked by most materials, making it difficult for them to penetrate very far into biological tissue. From outside the body they do not infiltrate more than 50 micrometers into the skin, which does not get them beyond the outermost layer of dead skin cells.

The situation is different when an alpha-emitting radioactive substance is inhaled or ingested, bringing the radioactivity into direct contact with living cells. That is what happens with **radon**, a radioactive gas produced during the spontaneous breakdown of radium (which is itself produced from the spontaneous breakdown of uranium). In regions of the country where large amounts of radium are present in underground rock formations (Figure 6-8), the radium gives rise to radioactive radon gas that seeps out of the earth and can accumulate in buildings if the ventilation is inadequate. When a radioactive atom of radon emits an alpha particle, the atom is converted into **polonium**, a radioactive metal that forms tiny particles that may be inhaled and become lodged in a person's lungs. The subsequent radioactive decay of polonium produces more alpha particles and a series of additional radioactive elements, which emit yet more radiation. The alpha particles released during these events enter the cells lining the inner surface of the lung, causing DNA mutations that can initiate the development of cancer. As a consequence, increased lung cancer rates are observed in people who have been exposed to high levels of environmental radon, especially in those individuals who also smoke cigarettes.

Exposure to environmental radon is not the only explanation for radioactivity that might be present in a person's lungs. Cigarette smoke is also radioactive, in this case because of the fertilizers that are used for growing commercial

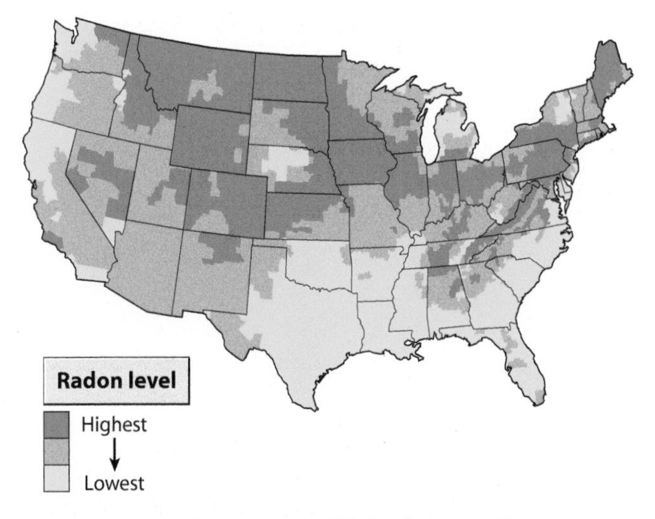

Figure 6-8 Map of Environmental Radon Concentrations. The amount of radon released into the atmosphere varies in different regions of the United States, depending on the uranium and radium concentration in underground rock formations. [Data from U.S. Environmental Protection Agency, 2005, http://www.epa.gov/radon/zonemap.html.]

tobacco. Most of these fertilizers contain phosphate derived from crushed rocks that contain trapped radon gas, which decays to form radioactive polonium; the radioactive polonium in turn appears both in cigarette smoke and in high concentration in the lungs of cigarette smokers. Animal studies indicate that radioactive polonium is one of the more potent carcinogens present in tobacco smoke.

When radioactive radon or polonium is inhaled, the lung is the initial organ encountered by the inhaled material and is therefore the main site where cancer appears. For other radioactive elements, specific chemical properties may cause the inhaled or ingested radioactivity to become concentrated in a different location. A striking illustration of this principle dates back to the 1920s and involves a group of women working in a New Jersey factory that produced watch dials that glow in the dark. The luminescent paint used in painting the dials contained **radium**, a radioactive element that resembles calcium in some of its chemical properties. The radium paint was applied using a fine-tipped brush that the employees frequently wetted with their tongues. As a result, minute quantities of radium were inadvertently ingested and, like calcium, the radium became concentrated in bone tissue. Many of these women subsequently developed bone cancer caused by the radioactive radium that had become concentrated in their bones.

Nuclear Explosions Have Exposed People to Massive Doses of Ionizing Radiation

The most dramatic episode of large-dose exposure to ionizing radiation involved the atomic bombs that were exploded over the Japanese cities of Hiroshima and Nagasaki at the end of World War II. People who survived the initial blast and the short-term toxic effects of the massive amounts of ionizing radiation released by the explosions later developed cancer at higher-than-normal rates. Leukemia initially appeared to be the predominant cancer, but it is now clear that many other kinds of cancer were caused as well.

The ability of ionizing radiation to cause leukemia versus other types of cancer tends to be overestimated for two reasons. First, leukemia is not as common as many other kinds of cancer, so increases in leukemia rates caused by radiation exposure are easier to spot. Second, the latent period is generally shorter for leukemia than for other cancers, so leukemias are the first cancers to be seen. Leukemia rates began to increase within the first 2 years after the atomic explosions in Japan, whereas other cancer rates did not rise until 10 to 15 years later. As a result, initial reports tended to emphasize the effects of radiation on leukemia rates. The long-term data now make it clear that leukemias accounted for only about 15% of the total number of cancer deaths caused by the two atomic explosions (Figure 6-9). Many of the more common types of

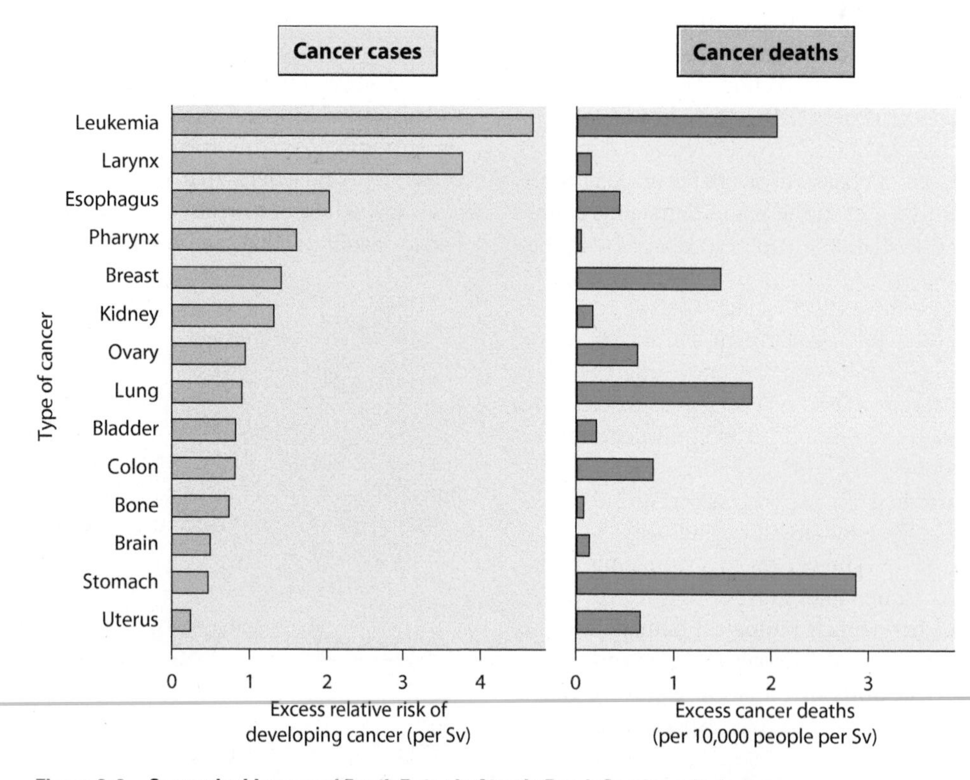

Figure 6-9 Cancer Incidence and Death Rates in Atomic Bomb Survivors in Japan. (*Left*) Cancer rates in atomic bomb survivors are shown relative to the rates observed in individuals who were not exposed to the atomic blast. Leukemia exhibited the greatest percentage increase in incidence relative to its normal rate, but the rates for many other cancers increased as well. (*Right*) In terms of the actual number of cancer deaths caused, the nuclear explosions were responsible for as many stomach, lung, and breast cancer deaths as leukemia deaths. [Based on data from D. A. Pierce et al., *Radiation Res.* 146 (1996): 1 (Table AIV).]

cancer—including those of the breast, lung, colon, stomach, ovary, and uterus—also increased in response to the massive doses of radiation received by the citizens of Hiroshima and Nagasaki.

Individuals who survived the initial impact of the atomic bombs were located at varying distances from the explosions and therefore received differing doses of radiation. Using such information to estimate the doses received by different individuals, scientists have constructed dose-response curves for radiation exposure versus subsequent cancer risk (Figure 6-10). As expected, the data reveal that cancer rates go up as the radiation dose is increased, both for leukemia and for other types of cancer. However, a transient decline in cancer rates occurs at very high doses of radiation, which is thought to reflect the fact that at higher doses, radiation kills cells in addition to causing mutations that can lead to cancer. The pattern of increasing cancer risk may therefore be disrupted at high radiation doses because the probability that a cell will be killed by radiation may be as great as (or greater than) the probability that it will sustain a cancer-inducing mutation.

Japan is not the only place where nuclear explosions have triggered cancer outbreaks. During the 1950s, the United States tested its nuclear weapons by exploding them in the open deserts of Nevada. In contrast to the situation in Japan, where a large population was exposed to whole-body radiation from gamma rays and neutrons emitted by the initial explosion, relatively few people were close enough to the Nevada test sites to receive much direct radiation. However, radioactive particles created during the explosions were carried by the prevailing winds into southwestern Utah, where individuals exposed to the radioactive fallout suffered an increased incidence of leukemia several decades later. Recent reports suggest that most people living in the United States have been exposed to at least a little bit of the radioactive fallout produced by these tests, although the exact impact on cancer rates has been difficult to estimate.

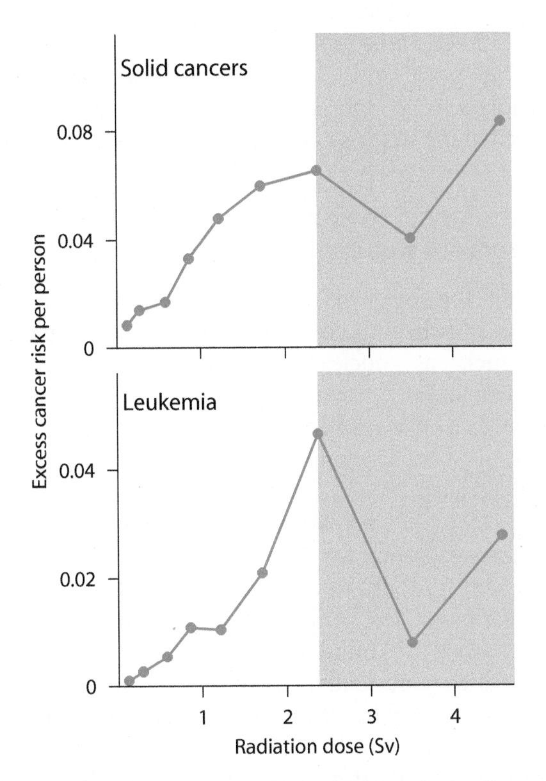

Figure 6-10 Dose-Response Curves Relating Radiation Exposure to Cancer Rates in Atomic Bomb Survivors in Japan. The shapes of the curves for both leukemias and solid cancers (tumors other than leukemias) show that the chance of developing cancer increased with increasing exposure to nuclear radiation. At high doses of radiation, however, the dose-response curves exhibit an area of decline (shaded area). Such patterns are thought to occur because high doses of radiation kill cells as well as causing cancer-inducing mutations, thereby diminishing the number of cells that might form cancers. [Based on data from D. A. Pierce et al., *Radiation Res.* 146 (1996): 1 (Figure 5 and 6).]

Another nuclear radiation incident happened in 1986 at the Chernobyl nuclear power plant in the former Soviet Union (now Ukraine). During a routine test, one of the plant's reactors went out of control and exploded, discharging several hundred Hiroshima bombs' worth of radioactive fallout across a large portion of Eastern Europe. Although many radioactive chemicals were released by the Chernobyl explosion, beta-particle emitting forms of iodine represented a large fraction. When iodine is ingested, it becomes concentrated in the thyroid gland so efficiently that the radiation dose experienced by the thyroid is 1000 to 2000 times higher than the average body dose. It is therefore not surprising that a few years after the accident, thyroid cancer rates in children were almost one hundred times higher than normal in regions receiving the largest amounts of radioactive fallout. These thyroid cancers represent the largest number of cancers of one particular type ever triggered by a single event.

Ionizing Radiation Initiates Carcinogenesis by Causing DNA Damage

As was the case for carcinogenic chemicals and UV radiation, DNA damage lies at the heart of the mechanism by which ionizing radiation causes cancer. The ability of ionizing radiation to trigger mutations was first described in the 1920s by Hermann Muller in studies involving fruit flies. When the mutation rate is plotted against the dose of ionizing radiation, the dose-response curve appears to be linear over a wide range of radiation doses.

In contrast to UV radiation, which creates a distinctive type of DNA mutation (pyrimidine dimers), ionizing radiation damages DNA in a variety of ways (Figure 6-11). By definition, ionizing radiation strips away electrons from molecules, generating highly unstable ions that rapidly undergo chemical changes and break chemical bonds. Because roughly 80% of the mass of a typical cell is accounted for by water molecules, many of the bonds broken by ionizing radiation reside in water. The disruption of water molecules produces highly reactive fragments called **free radicals**, a general term that refers to any atom or molecule containing an unpaired electron. The presence of an unpaired electron makes free radicals extremely reactive. One of the free radicals produced

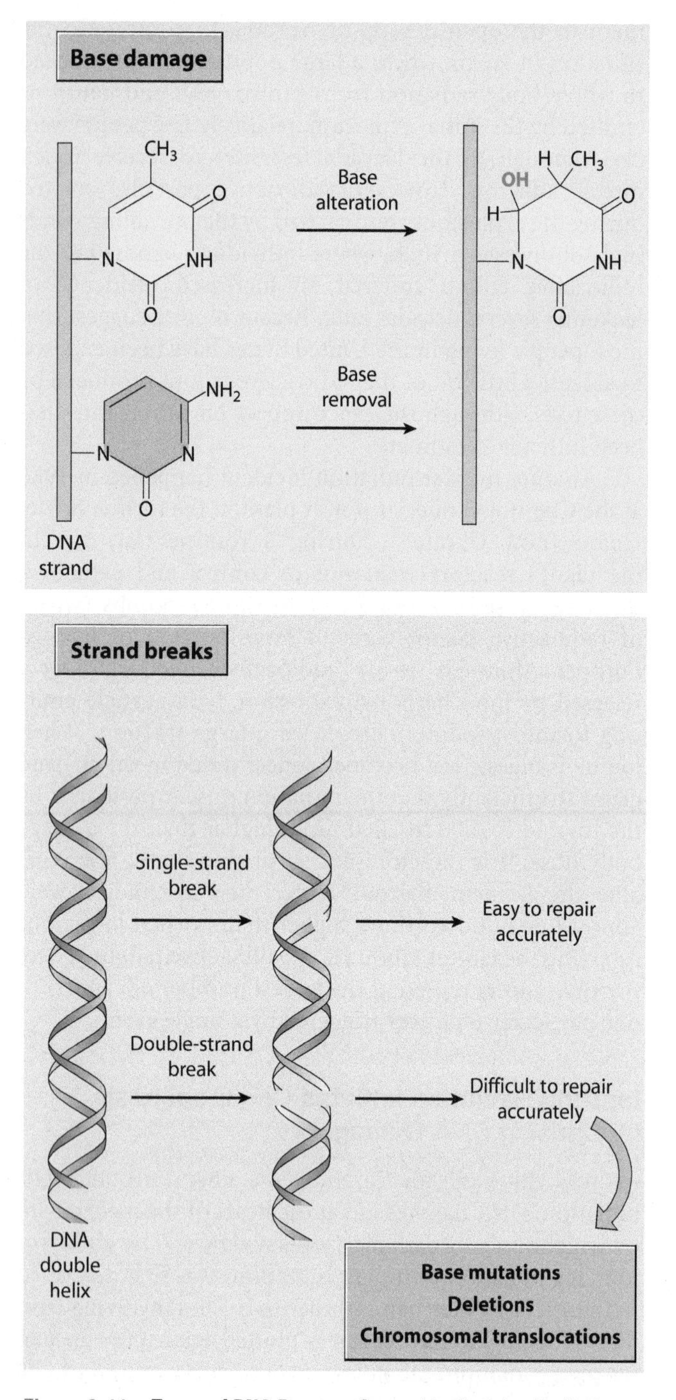

Figure 6-11 Types of DNA Damage Caused by Ionizing Radiation. Ionizing radiation interacts with DNA in a variety of ways, causing damage to individual bases (*top*) and triggering DNA strand breaks (*bottom*).

when ionizing radiation interacts with water is the hydroxyl radical (OH•), which readily attaches itself to DNA bases. The presence of these added hydroxyl groups alters the base-pairing properties of the bases during DNA replication, leading to various mutations.

In addition to generating water-derived free radicals, ionizing radiation also attacks DNA directly, stripping away electrons and breaking bonds. Such reactions cleave the bonds that join bases to the DNA backbone, thereby causing individual bases to be lost; ionizing radiation also attacks the DNA backbone itself, creating single- or double-strand breaks in the DNA double helix. Fortunately, it is relatively easy to repair single-strand breaks or the loss of individual bases because the opposite DNA strand of the double helix remains intact and serves as a template for fixing the defective strand by normal repair mechanisms. Double-strand breaks are more difficult to fix, and imperfect attempts at repair may create localized mutations in the region of the break or larger-scale alterations, such as major deletions or sequence rearrangements. If double-strand breaks occur in more than one chromosome, DNA derived from two different chromosomes may be mistakenly joined together. The result is a *chromosomal translocation* in which a segment of one chromosome is physically joined to another chromosome.

It usually takes many years for cancer to arise following radiation-induced DNA damage. Radiation is thus acting in the initiation phase of carcinogenesis, playing a role comparable to that of mutagenic chemicals in the initiation of chemical carcinogenesis. As would be expected, treating radiation-exposed cells with promoting agents, such as phorbol esters, increases the rate at which tumors appear. Cells that have been initiated by exposure to ionizing radiation often exhibit a persistent elevation in the rate at which new mutations and chromosomal abnormalities arise. This condition, called *genetic instability* (pp. 35 and 187), creates conditions favorable for accumulation of the subsequent mutations that are required in the stepwise progression toward malignancy.

Most Human Exposure to Ionizing Radiation Comes from Natural Background Sources

Although the ability of ionizing radiation to cause cancer is well established, many of the examples we have considered—such as nuclear explosions or occupational exposures to X-rays or radioactivity—are not particularly relevant to most people. So how much of a hazard do the various types of ionizing radiation actually pose for the typical citizen?

Figure 6-12 summarizes the main sources of exposure to ionizing radiation for the average person in the United States, who typically receives an annual radiation dose of about 3.6 mSv. Most of this exposure comes from natural sources of background radiation such as radioactive radon, which continually seeps out of the earth's crust and accounts for roughly 2.0 mSv per year. Additional sources of natural radiation include other radioactive elements present in the earth's crust and in our bodies, plus *cosmic rays*, which are high-energy charged particles that bombard the earth from outer space. Exposure to cosmic rays is greater at higher altitudes where the atmosphere is thinner. For example, a round-trip transatlantic airplane flight at an altitude of 35,000 feet exposes passengers to almost as much ionizing radiation as a chest X-ray.

Taken together, the various sources of natural radiation account for about 80% of our annual exposure to ionizing radiation. The remaining 20% comes from

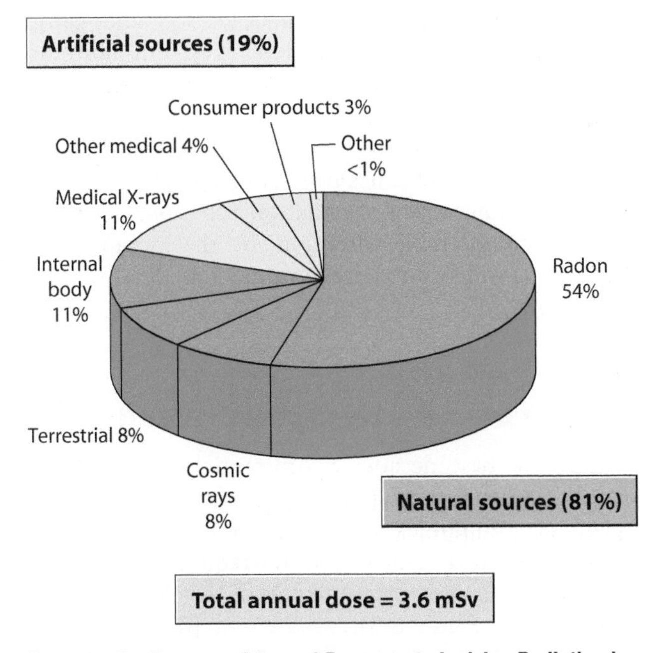

Figure 6-12 Sources of Annual Exposure to Ionizing Radiation in the United States. For the average person, most exposure to ionizing radiation comes from natural background sources, mainly radon. Medical X-rays are the largest artificial source of exposure to ionizing radiation. [Based on data from J. L. Schwartz in *The Cancer Handbook* (M. Alison, ed., London: Nature Publishing Group, 2002), Chapter 22 (Table 3).]

artificial sources of radiation generated by human activities. Some of this ionizing radiation is emitted by consumer products, such as television sets and smoke detectors, but most comes from medical procedures, mainly diagnostic X-rays. The benefits of medical X-rays almost always outweigh the small risks involved, but people sometimes express concerns about being X-rayed because they have heard that radiation causes cancer. Such individuals usually lack scientific training and are not likely to be reassured when told that a chest X-ray involves only 0.08 mSv of radiation.

It would be easier for the average person to understand how much radiation is involved in a medical X-ray if it were compared to the dose of radiation we all receive every day from natural sources of background radiation. For this reason, scientists have developed an alternative measurement unit for expressing radiation exposure known as the **BERT** (for *Background Equivalent Radiation Time*) value. The BERT value converts a given dose of ionizing radiation into the amount of time it would take a person to receive that same dose from the natural background radiation that we are all exposed to every day. Table 6-3 lists the BERT values for several types of radiation. It shows, for example, that the BERT value of a typical chest X-ray is roughly ten days. In other words, the amount of radiation received during a routine chest X-ray is equivalent to the amount of natural background radiation we all receive every ten days. Thus a chest X-ray represents a small fraction of the natural background radiation we are exposed to each year. Living near a nuclear power plant for an entire year has a BERT value of less than one day, indicating that it would add an extremely small amount to the annual radiation dose we already receive. In contrast, smoking a pack of cigarettes per day for one year has a BERT value of about ten years, which means that such a smoker would receive ten times more ionizing radiation each year from the radioactive polonium in tobacco smoke than they do from natural background radiation.

Table 6-3 BERT Values for Some Common Exposures to Ionizing Radiation

Radiation Source	Dose (mSv)	BERT
Living near nuclear power plant (all year long)	< 0.01	< 1 day
Transatlantic flight (round trip)	0.06	1 week
Dental X-ray	0.06	1 week
Chest X-ray	0.08	10 days
Mammogram	0.75	3 months
Polonium in cigarettes (500 packs/year)	30	10 years

The Cancer Dangers Associated with Typical Exposures to Ionizing Radiation Are Relatively Small

The BERT concept is useful in helping people understand how much additional radiation exposure (compared to background levels) is associated with various activities, but it does not provide any direct information about cancer risk. Assessing cancer risk is a difficult task because epidemiological data cannot reliably quantify the small numbers of cancer cases that would be expected when individuals are exposed to low levels of ionizing radiation. Scientists must therefore extrapolate from higher-dose human exposures, such as those experienced by atomic bomb survivors and radiation workers, to estimate human cancer risk at lower exposures.

Although such extrapolations are inherently imprecise, the data suggest that a person receiving a single whole-body exposure to 100 mSv of X-rays would incur about a 1% lifetime risk of developing a fatal cancer as a result of this exposure. Since 100 mSv is roughly one thousand times higher than the dose from a routine chest X-ray and 30 times higher than a person's typical annual exposure to all sources of ionizing radiation (3.6 mSv), it can be concluded that the cancer risks associated with typical exposures to ionizing radiation are quite small. Such calculations, however, are for whole-body radiation. If a radioactive element is inhaled and becomes permanently lodged in the lungs, as is the case for the radioactive polonium in tobacco smoke, the cells lining each lung are subjected to radiation levels that are much higher than those experienced by the body as a whole.

Another variable to be considered is the effect of ongoing exposure to low doses of radiation spread over long periods of time. For example, consider a workplace environment that exposes workers to an annual radiation dose of 50 mSv, which is just within acceptable governmental guidelines. A 30-year employee would acquire an accumulated dose of 30 yrs × 15 mSv/yr = 450 mSv of radiation. Since it was indicated in the preceding paragraph that 100 mSv increases the risk of developing a fatal cancer by about 1%, a worker with an accumulated dose of 450 mSv would have incurred a 4.5% added risk of dying of cancer.

Even this modest estimate of increased risk is probably an overestimate. First of all, radiation exposures that are accumulated gradually over a period of many years are less damaging to tissue than the same total dose of radiation administered all at once, so the preceding calculation for workplace exposure is likely to overestimate the long-term cancer risk. In addition, if radiation-induced mutations only begin to accumulate after the capacity of DNA repair pathways has been exceeded by higher radiation doses, the dose-response curve for radiation-induced cancers may exhibit a threshold (see p. 86 for a more detailed consideration of the threshold concept). If such a threshold were to exist, the actual cancer risk from low-dose exposures to radiation would be even smaller than suggested by current estimates, which are already quite low.

Given the relatively small cancer threat likely to be associated with typical human exposures to ionizing radiation, it is worth noting that people's perceptions of risk are often quite different from actual risks. When individuals are asked to rank their perceived risk of dying from various activities, nuclear power tends to be ranked ahead of much riskier activities, such as smoking cigarettes or driving a car. In reality, the risk of dying from a radiation-induced cancer is quite small compared with most of the other risks associated with everyday life (Table 6-4).

OTHER FORMS OF RADIATION

Of the various types of radiation introduced at the beginning of the chapter, we have seen that at least two can cause cancer. The first is *ultraviolet radiation*, a nonionizing type of electromagnetic radiation that lies between visible light and X-rays in the electromagnetic spectrum. The other is *ionizing radiation*, which includes X-rays and gamma radiation from the electromagnetic spectrum along with particulate forms of nuclear radiation, such as alpha and beta particles.

While these types of radiation are the only ones for which a cancer hazard has been convincingly established, claims that other forms of radiation also cause cancer have received considerable attention in the popular news media. Two widely publicized examples are briefly considered below.

There Is Little Evidence That Cell Phones or Power Lines Are Significant Cancer Hazards

Within the past decade, a substantial portion of the world's industrialized population has adopted a new means of communication: the cell phone. This familiar device emits and transmits **radiofrequency (RF) waves** located near the microwave region of the electromagnetic spectrum (see Figure 6-3). A cell phone RF wave, which is a bit shorter than an FM radio wave and somewhat longer than the microwaves used in a microwave oven, contains billions of times less energy than an X-ray and is therefore a *nonionizing* type of radiation.

Because cell phones are placed directly against a person's head, concerns have been expressed that the emitted RF waves might cause brain cancer. Support for this idea seemed to come from early reports of DNA damage in animals exposed to RF waves, but most studies have failed to confirm such results or demonstrate any carcinogenic activity of RF waves in laboratory animals. In humans, lag times of 20 to 30 years between exposure to a carcinogen and the development of cancer are not uncommon, so it is too early to know whether any long-term cancer hazards exist. However, the epidemiological evidence has failed to reveal any consistent relationship between cell phone usage and brain cancer rates thus far. The low-energy nature of RF waves and the failure to detect any consistent carcinogenic effects suggest that if any cancer risk exists at all, it will probably turn out to be rather small. At present the only well-established health risk associated with cell phone usage is an increased risk

Table 6-4 Risk of Dying from Various Activities

Activity	Cause of Death	Risk of Death (per million per year)
Smoking one pack of cigarettes per day	All causes	3500
Canoeing for 20 hours	Drowning	200
Traveling 1500 miles by car	Accident	40
Fishing	Drowning	10
Eating	Choking	8
Traveling 5000 miles by airplane	Accident	5
Having a chest X-ray	Cancer (radiation induced)	1
Living near a nuclear power plant	Cancer (radiation induced)	< 0.1

Based on data from J. B. Little in *Holland-Frei Cancer Medicine*, 5th ed. (R. C. Bast et al., eds., Lewiston, NY: Decker, 2000), Chapter 14, Table 14.3.

of dying in a car accident when the driver is talking on the phone!

High-voltage power lines, which emit electric and magnetic fields of extremely long wavelength, are another source of electromagnetic energy that has been claimed to pose a cancer risk. The electromagnetic energy emitted by power lines, called **ELF** (for Extremely Low Frequency), is at the low-energy end of the radiofrequency region of the electromagnetic spectrum (see Figure 6-3). Concerns about the possible health hazards of ELF fields were initially raised by several reports suggesting that children living near high-voltage power lines have elevated rates of leukemia, brain cancer, and other childhood cancers. The observed increase in cancer rates is rather small, however, and subsequent studies have failed to consistently detect such effects. Moreover, ELF fields have insufficient energy to damage DNA directly.

Summary of Main Concepts

- Radiation is energy traveling through space. Of the many forms of radiation, only a few cause cancer.
- Ultraviolet radiation is the main cause of skin cancer, which arises most frequently on areas of the body that are routinely exposed to sunlight. The risk of developing skin cancer is related to the length and intensity of sunlight exposure.
- Because skin cancers are located on the outer surface of the body, they are relatively easy to detect in their early stages, before invasion and metastasis have occurred. Basal cell and squamous cell carcinomas, which together account for about 95% of all skin cancers, rarely metastasize and are almost always curable. Melanomas account for only 5% of all skin cancers but are responsible for most skin cancer fatalities because they frequently metastasize.
- The ultraviolet radiation in sunlight is subdivided into UVA, UVB, and UVC based on differences in wavelength and energy content. Animal studies have shown that UVB is largely responsible for the carcinogenic effects of sunlight. Although more than 90% of the sun's UVB radiation is absorbed by the ozone layer, enough reaches the earth's surface to cause sunburn, tanning, aging of the skin, and skin cancer. The ozone layer was partially destroyed by CFCs during the latter half of the twentieth century, which may be causing skin cancer rates to rise because more UVB radiation is now reaching the earth. Clothing, sunscreens, skin pigmentation, and avoiding strong sunlight all help decrease skin cancer risk.
- UVB radiation creates unique mutations called pyrimidine dimers in the DNA of skin cells. In response to the DNA damage caused by UVB, the p53 protein accumulates and triggers cell suicide by apoptosis, which accounts for the peeling of dead skin that we call a "sunburn." If a UV-induced mutation happens to randomly inactivate the *p53* gene, however, cells with damaged DNA can no longer self-destruct by apoptosis and cancer may arise from the uncontrolled proliferation of these mutant cells.
- Ionizing radiation includes X-rays and several forms of nuclear radiation emitted by radioactive elements. These types of radiation have more energy than UV radiation and can cause cancers in almost any tissue of the body. Ionizing radiation removes electrons from molecules, thereby generating highly reactive free radicals. In addition to generating water-derived free radicals that react with DNA, ionizing radiation also attacks DNA directly, stripping away electrons and breaking bonds. Such reactions cause individual bases to be lost and trigger single- or double-strand breaks in the DNA double helix.
- Of the three main types of nuclear radiation (alpha, beta, and gamma), alpha particles are the most damaging to biological tissues. External exposure to alpha-emitting radiation involves little danger because alpha particles are too large to penetrate beyond the outermost layer of dead skin cells. But when radioactive alpha emitters are inhaled or ingested, as occurs with the radioactive radon gas that seeps out of the earth, alpha particles can cause mutations and cancer inside the body.
- Large-scale exposures of humans to ionizing radiation have come from the atomic bombs dropped over Japan, the open-air testing of nuclear weapons in the United States, and the explosion of the Chernobyl nuclear power plant in the former Soviet Union. All three events were followed by increased cancer rates in surrounding populations.
- Today most human exposure to ionizing radiation comes from natural background sources and, to a lesser extent, medical procedures. Taken together, the lifetime cancer risk associated with exposure to these sources of radiation is quite small. For cigarette smokers, the largest exposure to ionizing radiation comes from the radioactive polonium found in tobacco smoke.
- There is little evidence to support claims that radiation emitted by cell phones or high-voltage power lines represents a significant cancer hazard.

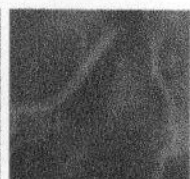

Key Terms for Self-Testing

radiation (p. 103)

Sunlight and Ultraviolet Radiation

basal cell carcinoma (p. 104)
squamous cell carcinoma (p. 104)
melanoma (p. 104)
electromagnetic radiation (p. 105)
ultraviolet radiation (UV) (p. 105)
UVA (p. 105)
UVB (p. 105)
ozone (p. 105)
chlorofluorocarbons (CFCs) (p. 106)
UVC (p. 106)
pyrimidine dimer (p. 106)
p53 gene (p. 106)
sunburn (p. 108)
sunscreen (p. 109)
melanin (p. 109)

Ionizing Radiation

ionizing radiation (p. 109)
X-rays (p. 109)
radioactivity (p. 110)
nuclear radiation (p. 110)
alpha particles (p. 110)
beta particles (p. 110)
gamma rays (p. 110)
gray (Gy) (p. 111)
relative biological effectiveness (RBE) (p. 111)
sievert (Sv) (p. 111)
radon (p. 111)
polonium (p. 111)
radium (p. 112)
free radical (p. 113)
BERT (Background Equivalent Radiation Time) (p. 115)

Other Forms of Radiation

radiofrequency (RF) waves (p. 116)
ELF (Extremely Low Frequency) (p. 117)

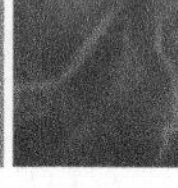

Suggested Reading

Sunlight and Ultraviolet Radiation

Armstrong, B. K., and A. Kricker. The epidemiology of UV induced skin cancer. *J. Photochem. Photobiol B: Biology* 63 (2001): 8.

Autier, P., et al. Sunscreen use and duration of sun exposure: a double-blind, randomized trial. *J. Natl. Cancer Inst.* 91 (1999): 1304.

Brash, D. E., et al. A role for sunlight in skin cancer: UV-induced p53 mutations in squamous cell carcinoma. *Proc. Natl. Acad. Sci. USA* 88 (1991): 10124.

Gilchrest, B. A., M. S. Eller, A. C. Geller, and M. Yaar. The pathogenesis of melanoma induced by ultraviolet radiation. *New England J. Med.* 340 (1999): 1341.

Leffell, D. J., and D. E. Brash. Sunlight and skin cancer. *Sci. Amer.* 275 (July 1996): 52.

Robinson, J. K., D. S. Rigel, and R. A. Amonette. Trends in sun exposure knowledge, attitudes, and behaviors: 1986 to 1996. *J. Amer. Acad. Dermatol.* 37 (1997): 179.

Ziegler, A., et al. Sunburn and p53 in the onset of skin cancer. *Nature* 372 (1994): 773.

Ionizing Radiation

Boice, J. D., Jr., and J. H. Lubin. Occupational and environmental radiation and cancer. *Cancer Causes Control* 8 (1997): 309.

Field, R. W. A review of residential radon case-control epidemiologic studies performed in the United States. *Reviews Environ. Health* 16 (2001): 151.

Grosovsky, A. J., and J. B. Little. Evidence for linear response for the induction of mutations in human cells by x-ray exposures below 10 rads. *Proc. Natl. Acad. Sci. USA* 82 (1985): 2092.

Little, J. B. Radiation carcinogenesis. *Carcinogenesis* 21(2000): 397.

Pierce, D. A., et al. Studies of the mortality of atomic bomb survivors. Report 12, Part I. Cancer: 1950–1990. *Radiation Res.* 146 (1996): 1.

Tubiana, M. Radiation risks in perspective: radiation-induced cancer among cancer risks. *Radiation Environmental Biophys.* 39 (2000): 3.

Ullrich, R. L., and B. Ponnaiya. Radiation-induced instability and its relation to radiation carcinogenesis. *Int. J. Radiation Biol.* 74 (1998): 747.

Williams, D. Cancer after nuclear fallout: lessons from the Chernobyl accident. *Nature Reviews Cancer* 2 (2002): 543.

Winters, T. H., and J. R. Di Franza. Radioactivity in cigarette smoking. *New England J. Med.* 306 (1982): 364.

Other Forms of Radiation

Ahlbom, I. C., et al. Review of the epidemiologic literature on EMF and health. *Environ. Health Perspect.* 109 Suppl. 6 (2001): 911.

Frumkin, H., A. Jacobson, T. Gansler, and M. J. Thun. Cellular phones and risk of brain tumors. *CA Cancer J. Clin.* 51 (2001): 137.

Moulder, J. E., et al. Cell phones and cancer: What is the evidence for a connection? *Radiation Res.* 151 (1999): 513.

7 Infectious Agents and Cancer

Microscopic infectious agents that can spread from person to person are responsible for a large number of human diseases. Some of these diseases—including influenza, hepatitis, measles, mumps, polio, rabies, chickenpox, and smallpox—come from exposure to *viruses*, which are nonliving DNA- or RNA-containing particles that only reproduce inside living cells. A second group of diseases arise from infection with *bacteria*, which are single-cell microorganisms lacking a nuclear membrane. Diseases triggered by bacteria include anthrax, cholera, tetanus, whooping cough, diphtheria, and certain forms of pneumonia and meningitis. Finally, some infectious diseases are caused by tiny *parasitic organisms*, such as the protozoan responsible for malaria or the flatworm responsible for schistosomiasis.

While cancer does not generally behave like a contagious disease, it has become increasingly apparent that infectious agents trigger a small but growing list of human cancers. Taken together, about 15% of all cancers worldwide (close to 1.5 million cases per year) are now thought to originate from viral (11%), bacterial (4%), or parasitic (0.1%) infections. Since such infections are theoretically preventable by either vaccination or by early treatment to eliminate the infectious agent, a better understanding of the role played by infectious agents in causing cancer could lead to a significant reduction in cancer deaths. ■

INFECTIOUS AGENTS IN HUMAN CANCERS

Cancer does not usually spread between people like a contagious disease, so it has taken many years to establish the underlying role that is sometimes played by viruses, bacteria, and parasites. To prove that a disease is caused by a specific **pathogen** (disease-producing agent), scientists normally require that four conditions known as *Koch's postulates* be met: (1) the suspected pathogen must be detected in the diseased tissue, (2) the suspected pathogen must be isolated from the diseased host and grown in the laboratory, (3) the laboratory-grown pathogen must cause the disease when administered to a healthy susceptible organism, and (4) the pathogen isolated from the newly infected host must be identical to the original pathogen.

While it has been relatively easy to satisfy these criteria for the dozens of viruses that cause cancer in animals, the third and fourth criteria cannot be routinely met in human studies because it would be unethical to deliberately inject people with a pathogen suspected of causing cancer. Indirect, and often less rigorous, evidence must therefore be used when investigating agents suspected of causing cancer in humans. Such indirect evidence might include microscopic or biochemical data showing the presence of the pathogen (or its genes) in human cancer tissue, epidemiological data demonstrating a linkage between exposure to the pathogen and increased cancer rates, clinical data showing that antibodies against the pathogen are present in a person's bloodstream, and laboratory data showing that cultured human cells acquire the traits of cancer cells after being exposed to the pathogen. The need to accumulate evidence from such diverse experimental approaches makes it a time-consuming and complex task to establish that a particular kind of human cancer is caused by a specific pathogen.

Cancer Viruses Were Initially Discovered in Animals

The idea that viruses can cause cancer was first proposed in the early 1900s. Although the role of chemicals and radiation in causing cancer was already recognized at the time, the possibility that infectious agents might also be involved was not seriously considered because cancer did not seem to behave like a contagious disease. However, in 1908 Danish scientists Wilhelm Ellerman and Olaf Bang reported that leukemia could be transmitted to healthy chickens by injecting them with blood extracts obtained from chickens that already had the disease. Shortly thereafter Peyton Rous discovered that sarcomas could likewise be transmitted between chickens by injection of filtered tumor extracts. The latter experiments, described in more detail in Chapter 4 (see Figure 4-12), led Rous to conclude that chicken sarcomas are transmitted by an infectious pathogen, and he eventually isolated several cancer viruses from chickens brought to him by local farmers.

Despite the clarity of Rous's data, his work was initially greeted with skepticism and many years passed before the existence of **oncogenic viruses** (cancer-causing viruses) came to be widely accepted. Convincing proof required researchers to demonstrate the existence of other viruses that can cause cancer. One of the first successes was reported in 1933, when Richard Shope showed that he could transmit skin cancer between rabbits using tumor extracts containing no intact cells. Such extracts contained an oncogenic virus subsequently named the *Shope papillomavirus*. In the following year, Baldwin Lucké observed that kidney tumors in New England frogs could be transmitted in a similar fashion. And a few years later, John Bittner reported that breast cancer in mice is transmitted from mother to offspring by a virus present in breast milk.

In the following decades, dozens of additional oncogenic DNA and RNA viruses were discovered (Table 7-1). Most are selective in the hosts they infect and in the types of tumors they cause, although there are exceptions to this rule. The *murine polyomavirus*, for example, infects a variety of mammals and causes more than 20 kinds of tumors, including cancers of the liver, kidney, lung, skin, bone, blood vessels, nervous tissues, and connective tissues.

A common feature shared by many oncogenic viruses is the ability to conceal themselves inside cells in a hidden or *latent* form in which no new virus particles are produced or released. Some latent viruses do not become active until the cell in which they reside is exposed to an appropriate stimulus, such as radiation, toxic chemicals, hormones, or even other viruses. An example is provided by the *feline leukemia virus*, which can be harbored by otherwise healthy cats for many years. Only upon exposure to a physically stressful situation, such as a respiratory infection, does the latent virus become activated and cause cancer. At the same time, large numbers of new virus particles are produced and released, thereby triggering a potential cancer epidemic among neighboring cats.

The Epstein-Barr Virus Is Associated with Burkitt's Lymphoma and Several Other Proliferative Disorders of Lymphocytes

By the late 1950s many viruses had been shown to cause cancer in animals, but no human examples were yet known. The situation was about to be changed, however, by the astute observations of a British surgeon named Denis Burkitt, who was running a medical clinic in Africa. Periodically, large numbers of children came to Burkitt's clinic with massive swellings of the jaw, which he diagnosed as a lymphocytic cancer now called **Burkitt's lymphoma**. When Burkitt mapped the geographical distribution of the disease, he discovered that it only appeared in a region of Africa stretching between

Table 7-1 Some Examples of Cancer Viruses

Family / Examples	Tumors Induced
DNA Viruses	
Adenoviruses	
Human adenoviruses (many types)	Subcutaneous, intraperitoneal, and brain tumors (hamsters)
Hepadnaviruses	
→ Hepatitis B virus (HBV)	Liver cancer (humans)
Herpesviruses	
→ Epstein-Barr virus (EBV)	Burkitt's lymphoma, nasopharyngeal carcinoma, Hodgkin's disease (humans)
→ Kaposi's sarcoma-associated herpesvirus (KSHV)	Kaposi's sarcoma (humans)
Lucké virus	Kidney adenocarcinoma (frog)
Marek's disease virus	Lymphoma (chickens)
Papillomaviruses	
→ Human papillomavirus (HPV)*	Cervical cancer and cancer of the penis (humans)
Shope papillomavirus	Papillomas (rabbits)
Polyomaviruses	
Murine polyomavirus	Cancers of liver, kidney, lung, bone, vessels, nerve, others (mice)
SV40	Lymphoma, mesothelioma, bone, brain cancers (hamsters); also humans?
RNA Viruses	
Retroviruses	
Mouse mammary tumor virus (Bittner)	Mammary carcinoma (mice)
Rous sarcoma virus	Sarcomas (chickens)
Murine leukemia viruses	Leukemia (mice)
Feline leukemia virus	Leukemia (cats)
Murine sarcoma virus	Sarcoma (mice)
Feline sarcoma virus	Sarcoma (cats)
Avian leukosis virus (ALV)	Leukemia (chickens)
Avian myelocytomatosis virus	Leukemia (chickens)
→ Human T-cell lymphotropic virus-I (HTLV-I)	Adult T-cell leukemia and lymphoma (humans)
Flaviviruses	
→ Hepatitis C virus (HCV)	Liver cancer (humans)
Reoviruses	
Wound tumor virus	Roots and stems (plants)

Note: Red arrows indicate viruses associated with human cancers.
*There are more than 100 different types of HPV, but only a small number are linked to cancer (e.g., HPV types 16, 18, 45, and 31). Some other forms of HPV cause genital warts.

10 degrees north of the equator and 10 degrees south, with a tail down the east coast (Figure 7-1). Because the rainfall and temperature patterns in this area make mosquito-transmitted infections exceptionally frequent, Burkitt proposed that the lymphoma was being transmitted by a mosquito-borne infectious pathogen.

Burkitt's ideas soon attracted the attention of two virologists, Michael Anthony Epstein and Yvonne Barr, whose electron microscopic studies revealed virus particles in tumor cells isolated from patients with Burkitt's lymphoma (Figure 7-2). The virus, identified as a member of the DNA-containing herpesvirus group, is now called the **Epstein-Barr virus (EBV)** in recognition of the scientists who discovered it. Unlike the situation with animal viruses, it is difficult to prove that a virus such as EBV causes cancer because ethical constraints prevent direct testing of the hypothesis by injecting the virus into healthy individuals. Nonetheless, several lines of evidence support the idea that EBV causes Burkitt's lymphoma. First, DNA sequences and proteins encoded by EBV have been found in tumor cells obtained from patients with Burkitt's lymphoma but not in normal cells from the same individuals. Second, adding EBV to cultures of normal human lymphocytes stimulates cell proliferation and causes the lymphocytes to acquire some of the traits of cancer cells. Finally, injecting EBV into monkeys triggers the development of lymphomas.

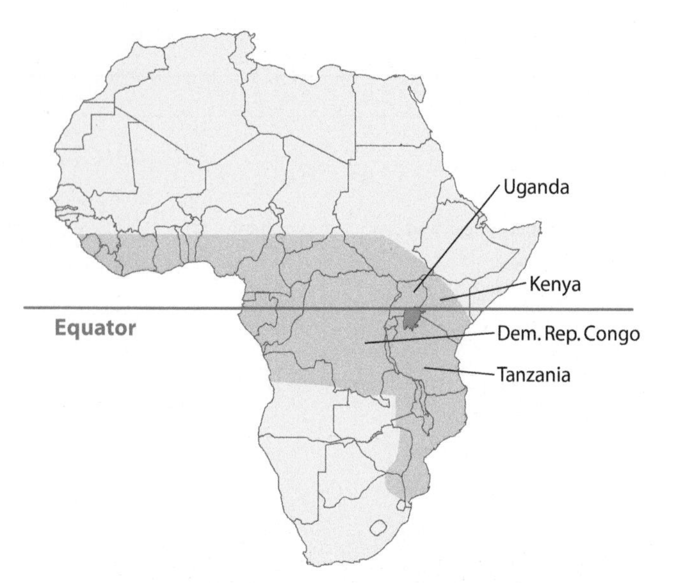

Figure 7-1 Distribution of Burkitt's Lymphoma in Africa. The shaded area indicates the region of Africa in which Burkitt's lymphoma is prevalent. The heat and rainfall in this part of Africa make mosquito-transmitted infections, such as malaria, exceptionally frequent. By suppressing childhood immunity, malarial infections are thought to make EBV-infected individuals more susceptible to developing Burkitt's lymphoma.

Although the evidence linking EBV to Burkitt's lymphoma is strong, the story is not quite so simple. We now know that EBV is a ubiquitous virus that has infected more than 90% of the world's population. If the virus is so common, why is Burkitt's lymphoma concentrated in a few geographical locations, such as central Africa, and infrequent elsewhere? A possible answer is suggested by research showing that Burkitt's lymphoma only appears in regions of Africa where the disease *malaria* is prevalent and childhood immunity has been depressed by malarial infection. The role of mosquitoes in Burkitt's lymphoma is thus indirect; by spreading malaria, mosquitoes cause a disease that depresses immune function and thereby allows EBV infections to proceed unchecked by an effective immune response. The idea that a deficient immune response plays a permissive role in allowing EBV-induced lymphomas to arise has received independent support from studies involving AIDS patients, whose immune systems are also debilitated. As we will see shortly, individuals with AIDS have abnormally high rates of several kinds of cancer, including lymphomas.

Burkitt's lymphoma is not the only kind of lymphocytic cancer to be linked to EBV infection. Based on differences in their appearance, lymphomas are commonly subdivided into a dozen or more subtypes. One prominent form, called **Hodgkin's disease**, is a lymphoma characterized by the presence of *Reed-Sternberg cells,* which possess two nuclei and have a uniquely distinctive appearance. A role for EBV in Hodgkin's disease is suggested by the finding that about half the patients with this cancer have EBV DNA sequences in their tumor cells.

In addition to its involvement in lymphocytic cancers, EBV infection can also trigger a nonmalignant proliferation of lymphocytes known as **infectious mononucleosis** (or simply "mono"). Infectious mononucleosis produces temporary flu-like symptoms and occurs mainly in individuals whose first encounter with EBV takes place when they are teenagers. If a person's initial exposure to EBV happens earlier in life, it rarely produces disease symptoms. Because EBV is transmitted mainly through exchange of saliva, mononucleosis has been nicknamed the "kissing disease." Roughly half the individuals who are first exposed to EBV as teenagers develop the symptoms of mononucleosis, which include swollen lymph glands, fever, headache, muscle aches, and fatigue. While the condition is almost always self-limiting, individuals who have had mononucleosis exhibit a slightly elevated risk for developing Hodgkin's disease or other lymphomas.

EBV Infection Is Also Associated with Nasopharyngeal Carcinoma and a Few Other Epithelial Cancers

Lymphocytes are not the only cells targeted for attack by EBV; the virus can also infect and trigger abnormal proliferation of epithelial cells. The clearest example of an epithelial cancer induced by EBV is **nasopharyngeal carcinoma**, a tumor of the nasal passages and throat that is frequent in Southeast Asia but rare elsewhere in the world. In patients with nasopharyngeal carcinoma, EBV DNA is detected in their tumor cells but not in the cells of the surrounding normal epithelium, suggesting that the cancer arises from EBV-infected cells. While it is unclear why nasopharyngeal carcinoma is common in Southeast Asia but not elsewhere, hereditary or dietary factors unique to that part of the world may play a role in determining susceptibility.

Even though 90% of the world's population has been exposed to EBV, cells infected by the virus rarely become malignant and little evidence implicates EBV in most of the common epithelial cancers. A small percentage of individuals with stomach or breast cancer do have EBV DNA sequences in their cancer cells, but the precise role played by EBV in such cases is not clear. Some evidence suggests that the presence of EBV may influence the behavior of cancer cells even when it is not the underlying cause of the disease. For example, one protein produced by EBV binds to and inhibits the activity of a normal cellular protein called Nm23-H1. The Nm23-H1 protein, which inhibits cell migration, is produced by one of the cell's *metastasis suppressor genes* (p. 58). The ability of EBV to bind to and inhibit Nm23-H1 may enhance cancer cell motility and thereby facilitate the ability of cancer cells to invade and metastasize.

Human Papillomavirus Is the Main Cause of Cervical Cancer

EBV was the first oncogenic virus to be identified in humans, but it is not the virus that causes the most human cancers. That distinction belongs to **human papillomavirus** (**HPV**), a DNA virus implicated in cancer

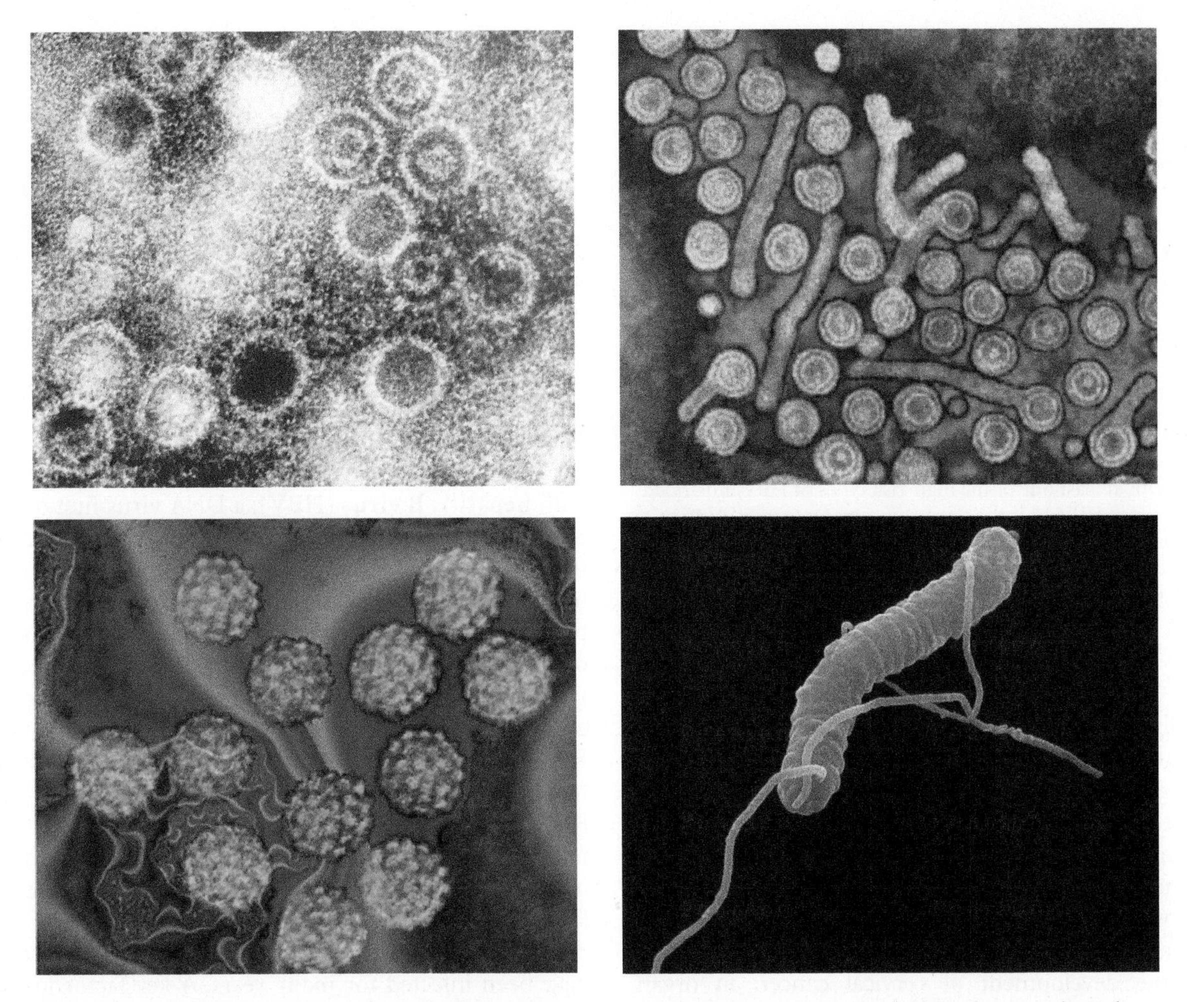

Figure 7-2 Some Infectious Agents That Can Cause Cancer. (*Top left*) Epstein-Barr virus, which has been implicated as a cause of Burkitt's lymphoma, nasopharyngeal carcinoma, and several other cancers. (*Top right*) Hepatitis B virus, which causes hepatitis that can lead to liver cancer. (*Bottom left*) Human papillomavirus, which causes cervical cancer and cancer of the penis. (*Bottom right*) *Helicobacter pylori*, a bacterium linked to the development of ulcers and stomach cancer. [Courtesy of M. A. Epstein (Epstein-Barr virus), Linda Stannard, UCT/Science Photo Library (hepatitis B virus and human papillomavirus), and SPL/Photo Researchers (*Helicobacter pylori*).]

of the uterine cervix (*cervical cancer*) and in several other cancers of the anal and genital area, including cancer of the penis. The idea that cervical cancer is triggered by an infectious agent can be traced back to the nineteenth century, when epidemiologists first noted that cervical cancer is relatively common in prostitutes but unknown in nuns. Such observations led to the suggestion that cervical cancer is a sexually transmitted disease, an idea now supported by much additional evidence. For example, cervical cancer rates are elevated in women who have had multiple sexual partners and in women who have sexual relationships with men who were previously married to women who developed cervical cancer.

Although the idea that cervical cancer is caused by a sexually transmitted pathogen was first proposed more than one hundred years ago, not until the 1980s did evidence begin to point to HPV as the responsible agent. The situation was initially complicated by the fact that HPV is not a single virus but a heterogeneous family of related viruses consisting of more than 100 different types (designated HPV 1, HPV 2, HPV 3, and so forth). As methods for identifying the various types of HPV improved, some striking patterns began to emerge. Epidemiological studies revealed that certain forms of HPV, designated *high-risk* types, are consistently associated with cervical cancer. The most prevalent member of this group, HPV 16, is detected in roughly half of all cervical cancers, followed in frequency by HPV 18, HPV 45, HPV 31, and a small group of others (Figure 7-3). In worldwide studies of cervical cancer, about 90% of the cases have been found to involve at least one of these high-risk types of HPV. Other forms of HPV are considered to be *low-risk* types because they are associated with nonmalignant conditions; for example, HPV 6 and HPV 11 are a common cause of tiny skin-colored bumps, called *genital warts*, in the male or female genital area.

Roughly 50% of all women become infected with HPV sometime during their life. Most infections are transient and disappear within a year or two, but a small fraction of the infections involving high-risk forms of HPV become persistent. Over a period of many years, persistent infections can trigger cervical dysplasia that, if untreated,

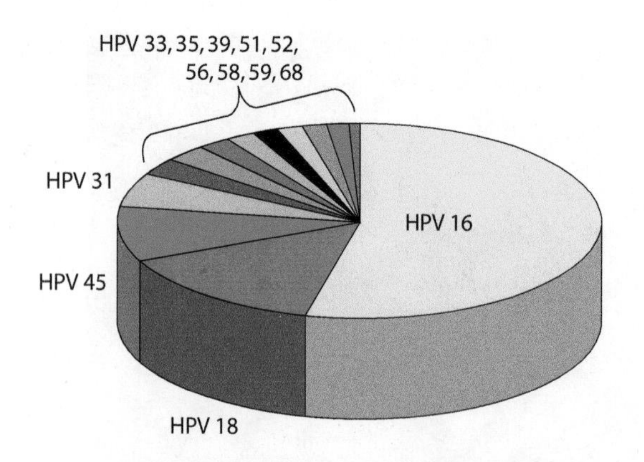

Figure 7-3 Prevalence of Various Types of HPV in Cervical Cancers. About 90% of all cancers of the uterine cervix involve infection with at least one of the high-risk types of HPV illustrated in this pie chart. HPV 16 is the most common form of HPV to be detected in cervical cancers, occurring in slightly more than 50% of all cases. [Based on data from F. X. Bosch et al., *J. Natl. Cancer Inst.* 87 (1995): 796 (Table 2).]

eventually develops into cancer. Of the various risk factors that influence the likelihood that an infection will persist and progress to cervical cancer, cigarette smoking is the strongest. Epidemiological studies of HPV-infected women have revealed that cigarette smokers are up to four times more likely to develop cervical cancer than are women who have never smoked. The evidence also suggests that normal immune function can help protect against the development of cervical cancer; in organ transplant patients treated with immunosuppressive drugs to decrease the risk of immune rejection of the transplanted organ, cervical cancer rates are ten times higher than normal.

The Hepatitis B and Hepatitis C Viruses Are Responsible for Most Liver Cancers

The disease known as **hepatitis**—inflammation of the liver—was first recognized thousands of years ago based on the distinctive yellow color imparted to a person's skin when the liver fails to metabolize and excrete pigments properly. In recent years, diagnosis of hepatitis has become more refined and doctors can now distinguish between several forms of the disease; some are triggered by the toxic effects of alcohol or drug consumption, whereas others are caused by viruses. At least six hepatitis viruses are known to exist, two of which are strongly linked to the development of liver cancer.

The largest number of cancers stem from infection with **hepatitis B virus** (**HBV**), a DNA virus first detected in hepatitis patients around 1970. HBV is usually transmitted by exchange of bodily fluids, such as semen or blood. The virus has infected more than 300 million people worldwide and is responsible for more than 75% of the world's cases of liver cancer. People infected with HBV are about 100 times more likely to develop liver cancer than are those who have never been infected. Infections are especially prevalent in Southeast Asia, China, Africa, Alaska, northern Canada, and the Amazon River basin of South America (Figure 7-4). In these regions of the world, close to 10% of the population is infected with HBV, and liver cancer is one of the most frequently encountered cancers.

HBV-induced liver cancers only arise after individuals have been infected for many years. A key factor in determining whether a person will develop such a persistent infection is the age when initially infected. Roughly 90% of all children infected at birth become chronically infected, but the value falls to 25% for infections occurring up to the

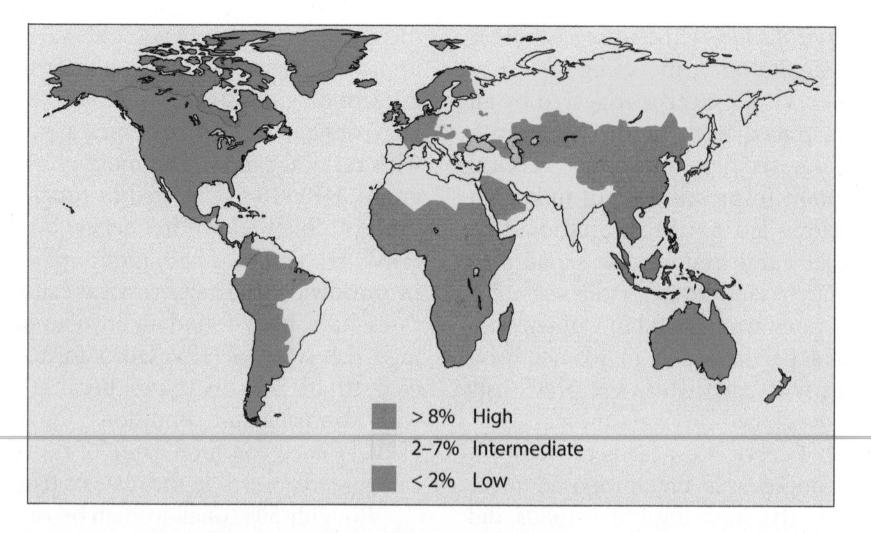

Figure 7-4 Worldwide Infection Rates for Hepatitis B Virus. The map is color coded to reflect the percentage of the population that is infected with hepatitis B virus (HBV) in various regions of the world. Infections are especially common in Southeast Asia, China, Africa, Alaska, northern Canada, and the Amazon River basin of South America, making liver cancer one of the most common cancers in these parts of the world. [Data reported in 2003 by the *United States Centers for Disease Control and Prevention* (http://www.cdc.gov).]

age of 5 years and continues to decline as infections occur later in life. This gradual improvement in a person's resistance to chronic infection may result from changes in the ability of the immune system to restrain the virus, but the exact mechanism is not clear.

The other virus linked to liver cancer is **hepatitis C virus** (**HCV**), an RNA-containing virus discovered in 1988. In contrast to HBV, which is easily spread through sexual activity, HCV is difficult to transmit by mechanisms other than direct contact with contaminated blood. Because blood used for medical transfusions is routinely screened for HCV (and HBV) contamination, the main route for acquiring HCV is through the sharing of dirty needles by intravenous drug users. Blood-screening techniques for identifying HCV were not introduced until 1990, however, which means that millions of people who received blood transfusions prior to that time were at risk of being infected with HCV. As a consequence, HCV is being carried by several million people in the United States, most of whom do not realize that they are infected.

Although HCV is more difficult to transmit than HBV, a larger number of people in the United States are infected with HCV (about 1.8% of the population) than with HBV (about 0.5% of the population). The explanation for this apparent contradiction is that HCV has a greater tendency to trigger long-term, chronic infections than does HBV. Less than 25% of the individuals infected with HBV develop chronic infections, whereas 75% of those infected with HCV will do so. More people therefore end up being chronic carriers of HCV. Roughly 5% of the people with chronic HCV infections eventually develop liver cancer, although it usually requires several decades to appear.

The HTLV-I Retrovirus Is Associated with Adult T-Cell Leukemia and Lymphoma

In animals, many of the viruses that cause cancer are members of the *retrovirus* family. Retroviruses contain RNA as their genetic material packaged along with *reverse transcriptase*, an enzyme that synthesizes a DNA copy from the RNA's base sequence. As we will see later in the chapter, making a DNA copy allows the viral genes to become integrated into the chromosomal DNA of an infected host cell.

Dozens of retroviruses have been shown to cause cancer in animals, but only one has been clearly linked to human cancer. This virus, called **human T-cell lymphotropic virus-I** (**HTLV-I**), was first identified in 1980 in lymphocytes obtained from patients with *adult T-cell leukemia/lymphoma*, a particularly aggressive type of cancer that is rare in the United States but prevalent in certain parts of Japan, Africa, and the Caribbean. Infection with HTLV-I is so tightly linked to adult T-cell leukemia/lymphoma that the presence of the virus is one of the diagnostic criteria for identifying this particular kind of cancer. HTLV-I is transmitted mainly through sexual contact, through blood products, and from mother to child during breast-feeding. In regions of the world where HTLV-I is prevalent, up to 10% of the population may be infected. Cancer develops in only a few percent of infected individuals; it usually arises many decades after initial infection and is more common among those infected early in life.

HIV-Infected Individuals Are at Increased Risk for Kaposi's Sarcoma and Several Other Viral Cancers

HTLV-I is the only retrovirus to be directly implicated in causing a human cancer, but the well-known retrovirus **HIV** (**human immunodeficiency virus**) represents an *indirect* cancer risk because of its destructive effects on the immune system. HIV-infected individuals who develop the symptoms of **acquired immunodeficiency syndrome** (**AIDS**) have a debilitated immune system that puts them at increased risk for several types of cancer (Figure 7-5). The greatest risk is for **Kaposi's sarcoma**, a cancer arising from blood vessels in the skin. Kaposi's sarcoma is generally quite rare in the United States, but rates are increased 100-fold in people infected with HIV. Reports of a sudden increase in these unusual, reddish purple skin tumors in Los Angeles and New York City in 1981 were one of the earliest signs heralding the onset of the AIDS epidemic. By 2004 it was estimated that 40 million people worldwide were infected with HIV, and in some regions of Africa, where more than 20% of the adult population is infected, Kaposi's sarcoma has become the most common type of cancer.

In 1994, a DNA virus called **Kaposi's sarcoma-associated herpesvirus** or **KSHV** (also known as *human herpesvirus-8* or *HHV-8*) was discovered in specimens of Kaposi's sarcoma tissue. KSHV is a sexually transmitted virus that exhibits a number of properties suggesting a causative role in Kaposi's sarcoma. For example, KSHV is found in virtually all Kaposi's sarcoma tumors, it specifically targets the cell type that becomes cancerous, and infection with KSHV precedes tumor development. KSHV infection by itself, however, is not sufficient to cause cancer; some degree of immune deficiency, as occurs in AIDS, is also required. Evidence for the role of immune deficiency in permitting Kaposi's sarcoma to develop has come from studies of AIDS patients treated with drug cocktails known as *highly active antiretroviral therapy* (*HAART*). In those individuals whose immune function improves in response to HAART treatment, Kaposi's sarcoma often disappears.

While the increase in cancer risk is most dramatic for Kaposi's sarcoma, this is only one of several viral cancers to occur at higher-than-normal rates in people with HIV/AIDS. Lymphomas triggered by infection with EBV or KSHV, cervical and skin cancers induced by HPV, and liver cancer triggered by HBV or HCV are other viral cancers whose incidence increases in individuals whose immune systems have been debilitated by HIV.

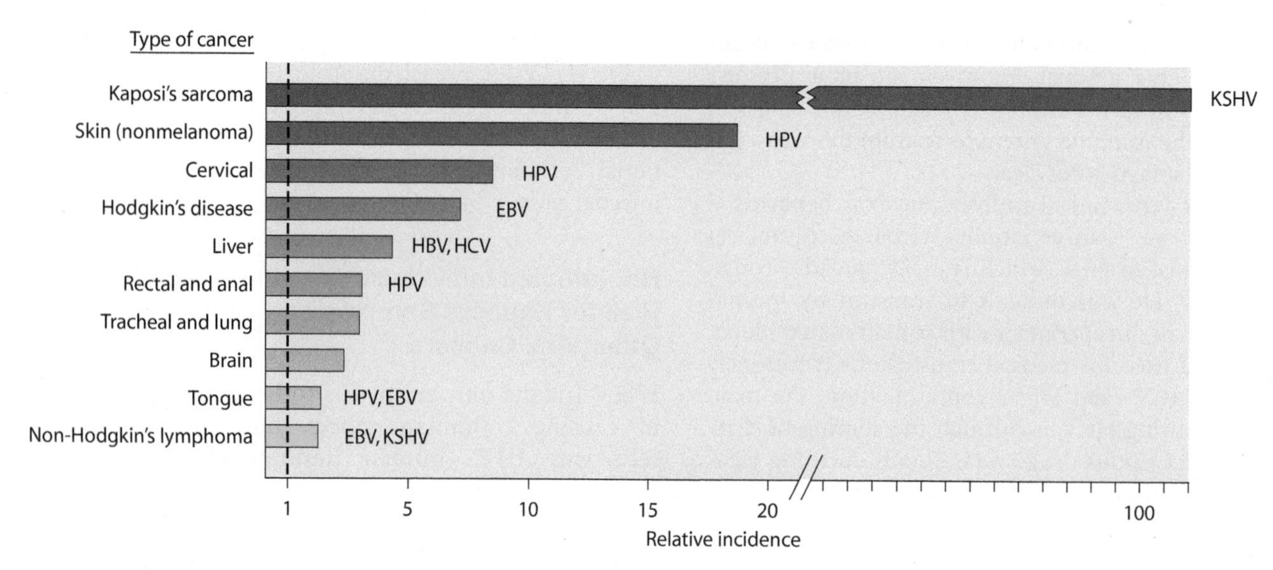

Figure 7-5 Increased Cancer Risk in Individuals Infected with HIV. People infected with HIV who develop AIDS have a debilitated immune system that puts them at increased risk for several types of cancer, especially Kaposi's sarcoma. Most of these cancers are virally induced; the names of the viruses that have been implicated are included for such cancers. "Relative incidence" is the rate for each type of cancer in HIV-infected individuals divided by the rate in the general population. The dashed line represents a value of "1", which would indicate no increase in cancer risk. [Based on data from C. Boshoff and R. Weissl, *Nature Reviews Cancer* 2 (2002): 373 (Table 1).]

SV40 Is an Animal Cancer Virus That Contaminated Early Batches of Polio Vaccine

Between 1955 and 1963, batches of polio vaccine that were used to immunize hundreds of millions of people around the world, including almost 100 million people in the United States, were unknowingly contaminated with a monkey virus called **SV40** (an abbreviation for *simian virus 40*). SV40 was present in the monkey kidney cells that had been used to grow the polio virus, but scientists did not know it was there and so had not taken steps to remove it. When the polio vaccine was manufactured, the procedures that were designed to inactivate the polio virus did not fully inactivate the SV40 that contaminated the vaccine.

SV40 was removed from the vaccine as soon as the virus was discovered, but concerns have persisted about the possible health consequences for the millions of people who were inadvertently exposed to the monkey virus. Animal and laboratory testing has not been reassuring: SV40 triggers the development of several types of cancer in animals and causes cultured human cells to develop some of the traits of cancer cells. Such findings raise the alarming possibility that the vaccine that eradicated polio in the United States might have inadvertently caused an outbreak of cancer.

Fortunately, four decades of epidemiological studies have failed to reveal any significant increases in cancer rates among people exposed to the contaminated polio vaccine. It is still possible, of course, that SV40 caused smaller numbers of cancers that cannot be reliably detected using epidemiological approaches. Investigators using an alternative tactic have discovered that SV40 DNA sequences are sometimes detectable in the cancer cells of individuals with non-Hodgkin's lymphoma, mesothelioma, brain cancer, or bone cancer. Although such data do not prove that the virus actually caused these tumors, it is interesting to note that the same four kinds of cancer are triggered by SV40 in animals. Additional concerns are raised by the discovery that many people with SV40 DNA in their tumor cells were never vaccinated with the contaminated polio vaccine, suggesting that SV40 might be spreading from person to person. The methods used for detecting tiny amounts of SV40 in cancer cells are fraught with difficulties, however, and the reliability of the data is open to question. For the moment, we do not really know how much (if any) cancer risk was created by the inadvertent exposure of humans to SV40. Nevertheless, the failure of the epidemiological data to demonstrate a relationship between exposure to the contaminated vaccine and cancer risk suggests that any hazards that might have been created are relatively small.

Infection with the Bacterium *Helicobacter pylori* Is Linked to Stomach Cancer

In contrast to viruses, whose roles in causing cancer have been studied for almost one hundred years, the possible cancer risks associated with chronic bacterial infections were not recognized until quite recently. One bacterium that has now been clearly linked to a human cancer is ***Helicobacter pylori***, also known as ***H. pylori***. This spiral-shaped bacterium was initially isolated in 1982 from the stomachs of people suffering from *gastritis*, an inflammation of the stomach lining that represents the first step in the development of stomach ulcers. To determine whether *H. pylori* is the underlying cause of gastritis, one of the investigators who discovered the organism, Barry Marshall, deliberately swallowed a sample of *H. pylori* himself. Marshall had been healthy prior to ingesting the bacteria, but he soon developed gastritis.

During the following decade, epidemiological studies began to reveal a strong connection between stomach cancer and exposure to *H. pylori*. It was found that *H. pylori* infections are present prior to the appearance of stomach cancer and that stomach cancer only develops in people who have had such an infection. Laboratory studies eventually confirmed the relationship by showing that mice deliberately infected with *H. pylori* develop stomach cancer. Although the way in which *H. pylori* contributes to cancer development is not completely understood, the bacterium is known to secrete toxins and cause inflammation in the stomach, which in turn stimulates cell proliferation and triggers DNA damage. Over a period of many years the proliferating cells become progressively more abnormal, creating areas of dysplasia that eventually develop into cancer. Because it is linked to a bacterial infection, stomach cancer can usually be prevented by treating infected individuals with antibiotics to rid them of *H. pylori*. Antibiotics even reduce the rate of cancer recurrence in stomach cancer patients who are treated after their tumors have been removed.

H. pylori has infected at least half the world's population, making stomach cancer one of the most common cancer killers worldwide, second only to lung cancer. Nevertheless, significant global variations exist in the prevalence of both *H. pylori* and stomach cancer. In the United States, stomach cancer rates have been declining for many decades, making the disease less common than in most other countries (see Figure 1-2). One factor contributing to these low stomach cancer rates is the widespread use of antibiotics, which decreases the prevalence of *H. pylori* infections. Public health conditions also play an important role because transmission of *H. pylori* from person to person is facilitated by poor sanitation and crowded living environments, which are more frequently encountered in developing countries than in the United States. As a result, *H. pylori* infections are less frequent in the United States than in many other parts of the world, and infections tend to occur later in life, when they are associated with a lower risk of causing stomach cancer (Figure 7-6).

Of the millions of people infected with *H. pylori* worldwide, only a few percent develop stomach cancer. It is not clear why such a small percentage of *H. pylori* infections lead to cancer, but several factors are thought to play a role. One variable is the type of bacterial strain involved. *H. pylori* is a heterogeneous group of bacteria, and a single infected person often carries multiple bacterial strains that change with time. The risk of developing cancer may depend to a significant extent on the type of *H. pylori* that prevails. Differing immune responses to the various bacterial strains can influence the course of infection, and blood type also plays a role because *H. pylori* adheres to the stomach lining by binding to cell surface receptors that are related to a person's blood type. Some strains of

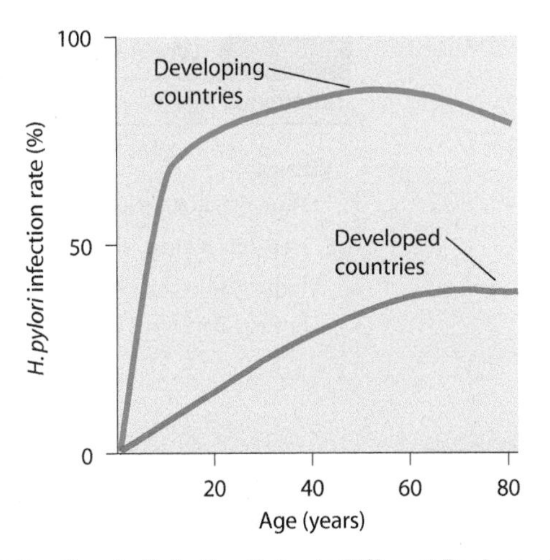

Figure 7-6 *H. pylori* Infection Rates in Different Regions of the World. In economically developed regions of the world, infection with *H. pylori* is less prevalent than in developing countries and tends to occur later in life, when it is less likely to cause stomach cancer. [Based on data from M. J. Blaser, *Sci. Am* 274 (February 1996): 104.]

H. pylori adhere preferentially to receptors associated with blood type O, making individuals with this blood type more susceptible to chronic *H. pylori* infections than individuals with other blood types.

Diet is another factor that influences the likelihood that exposure to *H. pylori* will lead to chronic infection and cancer. Stomach cancer rates are highest in those regions of the world, such as Japan, China, and Latin America, where diets are very salty because meat, fish, and vegetables preserved by salting are traditionally eaten on a regular basis. In animal studies, high-salt diets have been shown to enhance the ability of *H. pylori* to infect the stomach lining and stimulate cell proliferation.

Flatworm Infections Can Lead to Cancers of the Bladder and Bile Ducts

The final group of infections that pose a cancer hazard involve parasitic flatworms. Compared with viral and bacterial infections, which are linked to more than a million cases of cancer worldwide each year (Table 7-2), the cancer burden associated with parasitic infections is quite small, although it still involves about 10,000 cases of cancer annually. The biggest risk is posed by **blood flukes**, which are tiny flatworms that cause the disease *schistosomiasis* in tropical and subtropical regions of the world, mainly in Africa. Blood flukes are carried by freshwater snails, which harbor the larvae of the parasite. When people bathe in water contaminated by infected snails, the larvae penetrate the skin and, after maturing, settle in the blood vessels of the intestine or bladder, where they cause chronic inflammation. Most of the fatalities associated with schistosomiasis come from kidney or liver failure, but chronic inflammation of the bladder occasionally leads to bladder cancer.

Another type of parasitic flatworm, called a **liver fluke**, resides mainly in Asia and is acquired by eating raw or

Table 7-2 Worldwide Incidence of Cancers Attributed to Infections

Infectious Agent	Cancer	Total Cases
Viruses		
Human papillomavirus (HPV)	Cervical, other genital cancers	353,400
Hepatitis B virus (HBV)	Liver cancer	228,900
Hepatitis C virus (HCV)	Liver cancer	109,700
Epstein-Barr virus (EBV)	Burkitt's lymphoma	6,100
	Nasopharyngeal carcinoma	56,200
	Hodgkin's disease	26,200
	Non-Hodgkin's lymphoma*	8,800
KSHV	Kaposi's sarcoma*	43,600
HTLV-I	T-cell leukemia/lymphoma	2,600
Bacteria		
Helicobacter pylori	Stomach cancer	346,100
Parasites		
Blood flukes	Bladder cancer	9,500
Liver flukes	Bile duct cancer	800

*Cancers marked with an asterisk are in HIV-infected individuals.
Data from D. M. Parkin et al., The global health burden of infection associated cancers, in *Infections and Human Cancer* (R. Newton et al., eds., Cold Spring Harbor, NY: Cold Spring Harbor Press, 1999).

undercooked fish. The ingested parasite moves from the intestines into the bile ducts, where it becomes lodged in the smaller ducts of the liver and triggers local inflammation. Such infections can eventually lead to the development of bile duct cancer (*cholangiocarcinoma*), which is very rare in most parts of the world but occurs more frequently in regions of Asia where liver fluke infections are common.

HOW INFECTIONS CAUSE CANCER

Given the diverse group of viruses, bacteria, and parasites that have been linked to the development of cancer, it is perhaps not surprising to learn that these agents trigger the disease in a variety of different ways. Some of them function indirectly, creating conditions that are favorable for the development of cancer, whereas others alter the behavior of infected cells in ways that lead directly to malignancy. In the remainder of the chapter, we will explore the main types of mechanisms involved.

Chronic Infections Trigger the Development of Cancer Through Indirect as Well as Direct Mechanisms

The mechanisms by which infectious agents cause cancer can be subdivided into three broad categories. The first involves agents that increase cancer risk indirectly by interfering with immune function. The most dramatic example in this category is HIV, a virus whose ability to debilitate the immune system predisposes individuals to developing cancers that are actually triggered by other viruses, such as KSHV, EBV, HPV, HBV, and HCV. The presumed explanation is that the weakened immune system is unable to limit infection by these viruses, leaving them free to infect tissues and foster the development of cancer. A related phenomenon is seen with malarial infections, which depress immune function and thereby increase the likelihood that EBV infections will proceed to Burkitt's lymphoma.

The second category involves infectious agents that create tissue destruction and chronic inflammation. Pathogens acting in this way include HBV, HCV, *H. pylori*, and parasitic flatworms. By causing tissue destruction, such pathogens make it necessary for the normal cells of the infected tissue to continually proliferate to replace the cells that have been damaged and destroyed. Persistent infections also create chronic inflammatory conditions in which cells of the immune system, mainly *lymphocytes* and *macrophages*, infiltrate the tissue and attempt to kill the pathogen and repair the tissue damage. Unfortunately, the mechanisms used by the infiltrating immune cells to fight infections often produce mutagenic chemicals, such as oxygen *free radicals*. We saw in Chapter 6 that free radicals possess highly reactive, unpaired electrons that trigger various types of DNA damage. This means that proliferation of replacement cells for the injured tissue is taking place under conditions in which DNA damage is likely, thereby increasing the possibility that cancer-causing mutations will arise. Besides producing free radicals, macrophages release substances that enter the injured cells of the surrounding tissue and increase the activity of a protein called **NF-kappa B** (**NF-κB**). NF-κB turns on the transcription of genes coding for proteins that stimulate cell division and make cells resistant to apoptosis, both of which can contribute to unrestrained cell proliferation. The net result is that cells in chronically infected tissues are subjected to conditions that promote persistent cell proliferation as well as the accumulation of mutations, two traits that are central to the development of cancer.

The third mechanism by which infectious agents cause cancer is by directly stimulating the proliferation of infected cells. This tactic is used mainly by viruses, although certain bacteria can also stimulate cell proliferation directly. For example, some strains of *H. pylori* inject a protein called CagA into epithelial cells that line the inner wall of the stomach. After entering a cell, CagA binds to and activates a key protein involved in one of the cell's main pathways for stimulating cell proliferation. Strains of *H. pylori* that produce CagA are more effective in causing disease than those strains that do not produce CagA.

Many cancer viruses can likewise stimulate cell proliferation directly. Some viruses produce proteins that alter the behavior of a cell's normal growth signaling pathways, leading to uncontrolled proliferation; other viruses alter the expression of normal cellular genes that code for components of these same growth signaling pathways. The mechanisms used by viruses to disrupt such pathways illustrate a series of principles that apply not just to viral cancers but to cancers caused by chemicals and radiation as well. In the remainder of this chapter, we will explore some of the tactics used by viruses for disrupting these pathways involved in the control of cell proliferation and survival.

DNA and RNA Viruses Employ Different Mechanisms for Latently Infecting Cells

Before addressing the question of how viruses disrupt the control of cell proliferation, we first need to clarify a few basic issues regarding the nature and behavior of viruses. A **virus** is a tiny particle, too small to be seen with a light microscope, that depends on living cells for its reproduction. Mature virus particles contain a core of nucleic acid (DNA or RNA) surrounded by a protein coat and sometimes an outer envelope as well. After entering a cell, some viruses direct the cell to make multiple copies of the virus's own components, thereby creating new virus particles that are released from the cell. Infections of this type usually kill the cell in which the virus is replicating or expose the cell to destruction by host immune responses.

An alternative mechanism, used by many cancer viruses, is for a virus to conceal itself inside infected cells in a hidden or **latent** form in which no new virus particles are produced or released. The mechanism for establishing such latent infections differs between DNA and RNA viruses. With DNA viruses, entrance of the virus into the cell is usually followed by transcription of its DNA into messenger RNA, which is translated into viral proteins that are involved in establishing and maintaining the latent state. Replication of cellular and viral DNA then ensues, leading to cell division. At this stage, the viral DNA may persist indefinitely as an independently replicating molecule called an **episome** (Figure 7-7, pathway ①). Alternatively, one or more copies of the viral DNA may become integrated into the host chromosomal DNA; the viral genetic information then becomes a permanent part of the cell's genetic material and is replicated by the cell as part of its own DNA (Figure 7-7, pathway ②). Whether the viral DNA is replicated as part of the host chromosome or as an independent episome, no new virus particles are produced or released during a latent infection.

Unlike DNA viruses, RNA viruses cannot insert genes directly into a host cell chromosome because their genes are made of RNA rather than DNA. The mechanism for overcoming this limitation was discovered in 1970 by Howard Temin and David Baltimore, who independently found that some RNA viruses contain an enzyme, called **reverse transcriptase**, that catalyzes the synthesis of DNA using viral RNA as a template. With the aid of another viral protein, called **integrase**, the resulting DNA copy of the viral RNA is integrated into the host's chromosomal DNA and is subsequently replicated along with it (Figure 7-8). When integrated into a host cell chromosome, such a DNA copy of the genes of an RNA virus is referred to as a **provirus**. RNA viruses that employ the reverse transcriptase pathway to integrate their genetic information into a host cell are called **retroviruses**.

Retroviral Oncogenes Are Altered Versions of Normal Cellular Genes

As we have just seen, oncogenic DNA and RNA viruses both possess mechanisms for indefinitely replicating their genes in cells that have been latently infected. The big question, of course, is how do such genes cause cancer? In exploring this question, we will begin by examining the behavior of the RNA retroviruses.

The first cancer gene of any type to be explicitly identified and analyzed was a component of the *Rous sarcoma virus*, the chicken retrovirus discovered by Peyton Rous in the early 1900s. The Rous virus is a small RNA virus whose **genome** (its total genetic information) includes only four genes, making it relatively easy to identify the gene that causes cancer (Figure 7-9). In the early 1970s, Peter Vogt isolated mutant forms of the Rous virus that had lost the ability to cause cancer but could still infect cells and replicate. Examination of the viral RNA revealed that these mutant viruses had lost a single gene whose presence normally allows the Rous virus to cause sarcomas. The gene was therefore named the **v-*src* gene** ("v" for viral and "*src*" for sarcoma). Genes like v-*src*, which trigger the development of cancer, are referred to as **oncogenes**.

Subsequent investigations led to the surprising discovery that genes resembling v-*src* are not unique to cancer viruses. DNA sequences that are very similar to, although not identical with, the Rous v-*src* gene have been detected in the normal cellular DNA of a wide variety of organisms, including salmon, mice, cows, birds, and humans. Because the evolutionary divergence of this group of organisms occurred hundreds of millions of years ago, it can be concluded that genes resembling v-*src*

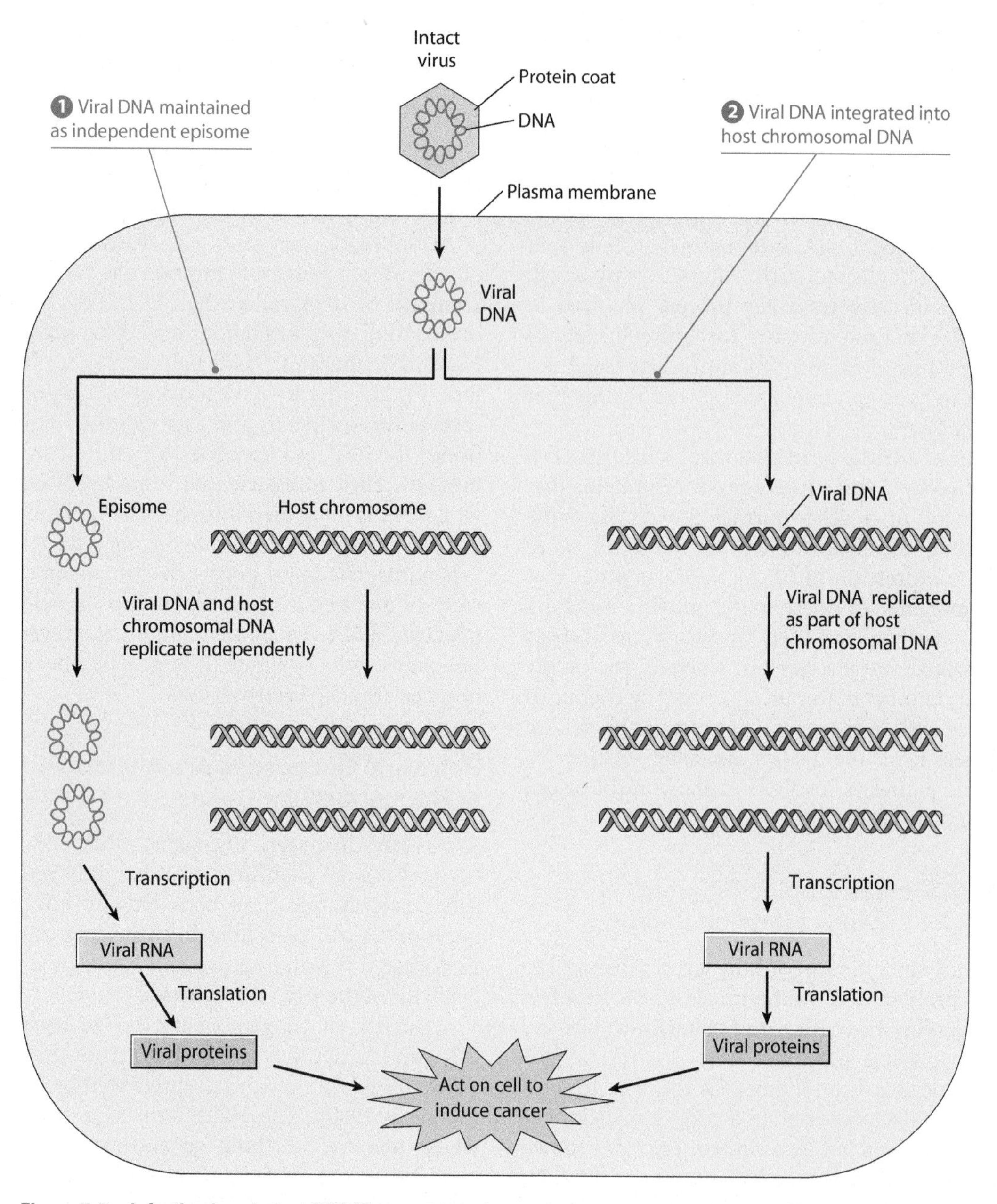

Figure 7-7 Infection by a Latent DNA Virus. DNA viruses that cause latent infections are incorporated into cells in two different ways. ① In some cases, the viral DNA persists indefinitely as an independently replicating DNA molecule called an episome. ② Alternatively, one or more viral DNA copies may become integrated into the host chromosomal DNA, where the viral DNA is then replicated as part of the chromosome. Whether the viral DNA is replicated as part of the host chromosome or as an independent episome, no new virus particles need to be produced during a latent infection.

have been conserved for a large part of evolutionary history and therefore must perform an important function in normal cells. To distinguish it from the v-*src* oncogene of the Rous virus, the normal version of this gene in human cells is called the ***SRC* gene** (sometimes preceded by the letter "c" for "cellular", as in "c-*SRC*").

Following the discovery of v-*src*, dozens of other oncogenes have been identified in different retroviruses. Like v-*src*, these oncogenes are altered versions of genes occurring in normal cells. The term **proto-oncogene** is used to refer to such normal cellular genes that are closely related to oncogenes. For example, v-*src* is a viral oncogene and *SRC* is the corresponding proto-oncogene. In Chapter 9 we will see that cells contain dozens of different proto-oncogenes, each of which plays an important role in normal cellular activities. However, their close resemblance to oncogenes means that alterations in the structure or expression of proto-oncogenes can convert them into oncogenes.

How do we account for the fact that normal cells contain proto-oncogenes that closely resemble the oncogenes of cancer-causing retroviruses? The most likely explanation, illustrated in Figure 7-10, is that retroviral oncogenes were derived from normal cellular genes

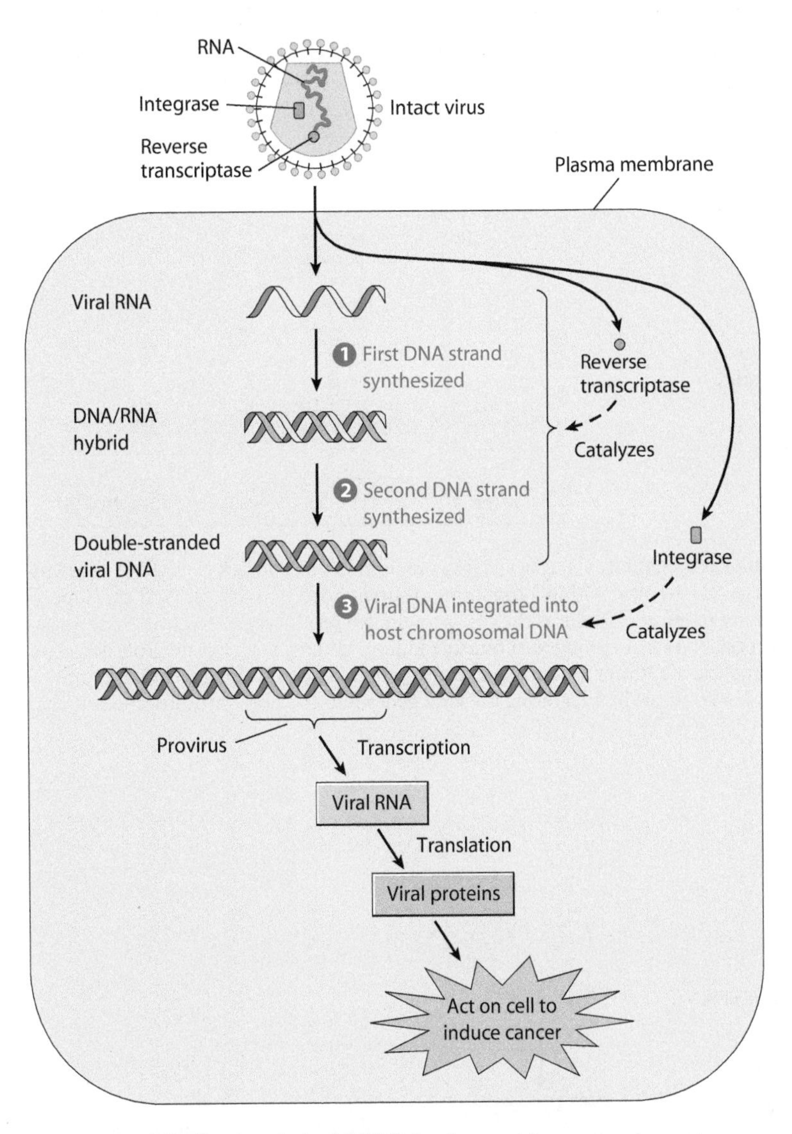

Figure 7-8 Infection by a Latent RNA Retrovirus. After a retrovirus enters a cell, its RNA is ① copied by reverse transcriptase into a complementary strand of DNA, which then ② serves as a template that is copied by reverse transcriptase into a second DNA strand. The resulting DNA double helix is then ③ integrated into the host chromosomal DNA using the viral enzyme integrase, thereby creating an integrated provirus. Only after integration are the viral genes transcribed into proteins that trigger the development of cancer. The proviral DNA is replicated along with the host cell chromosomal DNA every time the cell divides.

millions of years ago. According to this theory, the first step in the creation of retroviral oncogenes occurred when ancient retroviruses infected cells and integrated their proviral DNA into host chromosomes adjacent to normal cellular proto-oncogenes. When the integrated proviral DNA was subsequently transcribed to form new viral RNA molecules, an adjacent proto-oncogene could have been inadvertently copied as well, thereby generating viral RNA molecules containing a normal proto-oncogene alongside the viral genes. Eventually such a proto-oncogene might undergo a mutation that converts it into an oncogene, which would be useful to the retrovirus because of its ability to stimulate the proliferation of infected cells. Such a hypothetical scenario would explain why present-day retroviruses possess oncogenes that are altered versions of normal cellular genes.

Retroviral Oncogenes Code for Proteins That Function in Growth Factor Signaling Pathways

The preceding scenario explains how retroviruses may have come to possess oncogenes that are altered versions of normal cellular genes. But how do these oncogenes cause cancer? An early clue came from studies of the Rous virus and its v-*src* oncogene, which codes for a protein called **v-Src.** (By convention, gene names are usually printed in

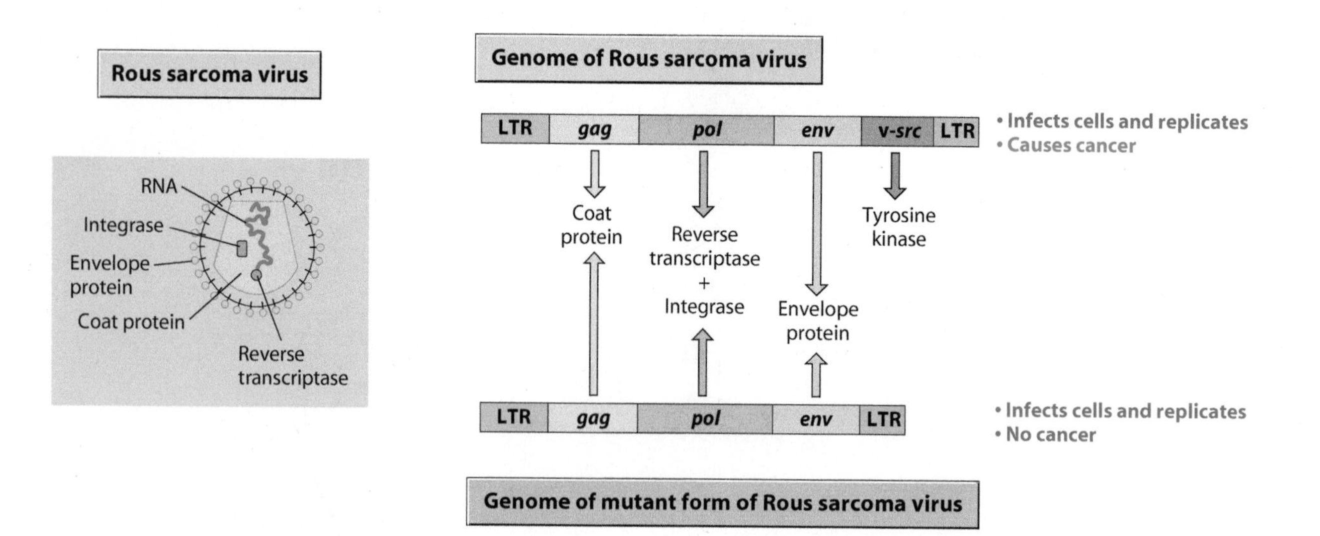

Figure 7-9 Rous Sarcoma Virus Genome. (*Left*) The Rous sarcoma virus consists of an RNA molecule, reverse transcriptase, and integrase, all packaged within a coat protein surrounded by an outer viral envelope. (*Right*) The viral RNA molecule contains four genes: *gag* produces the coat protein, *pol* produces reverse transcriptase and integrase, *env* produces the envelope protein, and v-*src* produces a tyrosine kinase. Mutant forms of the Rous virus that lack the v-*src* gene can infect cells and replicate, but they are unable to cause cancer. The LTR (long terminal repeat) sequences located at both ends of the genome play a role in integrating the viral genes into the host chromosome and in activating gene transcription.

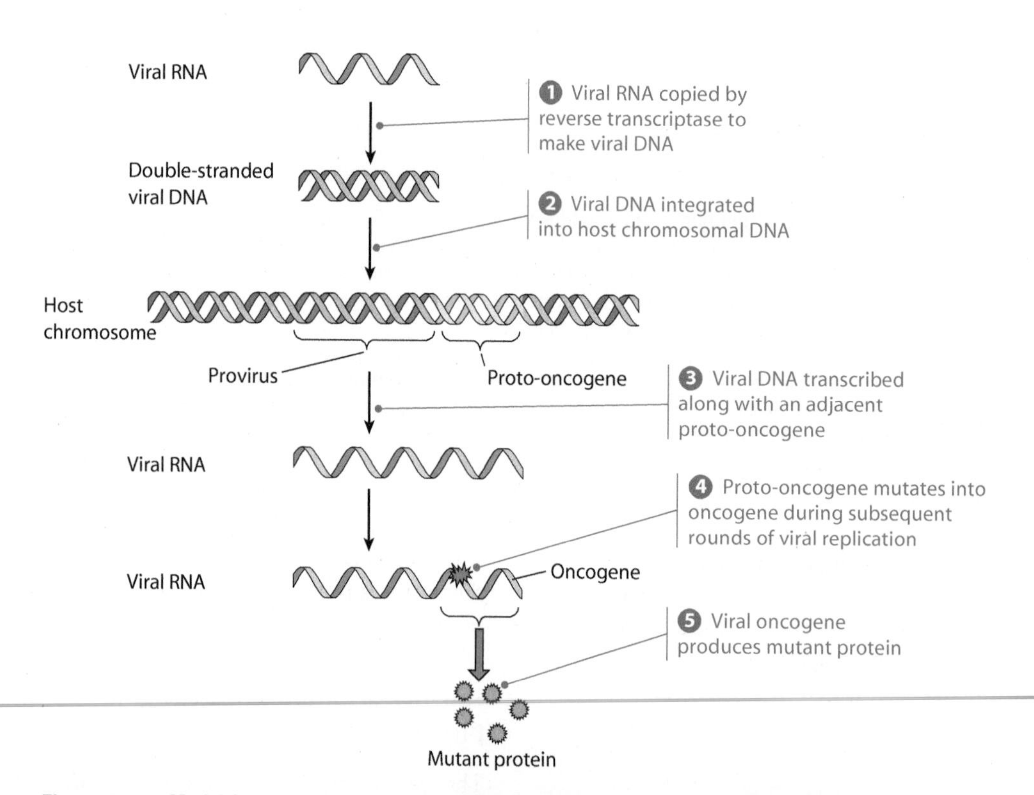

Figure 7-10 Model for the Proposed Origin of Retroviral Oncogenes. According to this model, ancient RNA retroviruses infected normal cells millions of years ago and became integrated in host chromosomes adjacent to proto-oncogenes. When the integrated proviral DNA was later transcribed, neighboring proto-oncogenes might have been copied along with the viral genes. Subsequent mutation of the incorporated proto-oncogene could then convert it into an oncogene.

Table 7-3 Nomenclature Conventions*

Virus	Retroviral Gene		Corresponding Human Gene†	
	Oncogene*	Protein**	Proto-oncogene*	Protein**
Rous sarcoma virus	v-*src*	v-Src	*SRC*	Src
Avian myelocytomatosis virus	v-*myc*	v-Myc	*MYC*	Myc

*Gene names are always in italics (lowercase letters for viral genes and usually all capital letters for human genes).
**Protein names begin with a capital letter and are not in italics.
†Sometimes the names of the human genes and proteins are preceded with a "c" for "cellular" (for example, c-*SRC* gene and c-Src protein, or c-*MYC* gene and c-Myc protein).

italics and the corresponding protein names are printed without italics, as illustrated in Table 7-3.) The v-Src protein, first isolated in the late 1970s, is a *protein kinase.* Chapter 2 introduced the concept that protein kinases are enzymes that regulate the activity of targeted protein molecules by catalyzing their phosphorylation. In the signaling pathways that control cell proliferation, protein kinases play a prominent role in transmitting signals from one molecule to another. The v-Src protein kinase is a member of a particular group of protein kinases that are called **tyrosine kinases** because they phosphorylate the amino acid tyrosine in targeted proteins.

Normal cells possess dozens of their own tyrosine kinases, most of which play roles in the various signaling pathways for controlling cell proliferation and survival. We saw in Chapter 2 that alterations in the proteins involved in growth factor signaling pathways can cause the pathways to become hyperactive, leading to uncontrolled cell proliferation (see Figure 2-6). One protein involved in such pathways, called **Src kinase**, is the normal version of the v-Src kinase produced by the Rous virus. When certain growth factors bind to their corresponding receptors on the cell surface, the activated receptors in turn activate the normal Src kinase. The activated Src kinase then phosphorylates and activates other signaling proteins that cause the cell to divide.

In contrast to this regulated behavior of the normal Src kinase, which is inactive until stimulated by an appropriate signal from a growth factor receptor, the v-Src kinase produced by the Rous virus is **constitutively active**, which means that it remains active whether there is an appropriate signal or not. So when the Rous virus infects a cell, it produces v-Src tyrosine kinase molecules that persistently catalyze the phosphorylation of target proteins that activate cell proliferation. This Rous v-Src kinase is the first example we have encountered of an **oncoprotein**—that is, a protein whose activity contributes to the development of cancer.

Subsequent to the discovery of the Rous v-Src kinase, the oncoproteins produced by dozens of other viral oncogenes have been identified as well. Some of these oncoproteins are also tyrosine kinases, whereas others are abnormal versions of growth factors, receptors, or other proteins involved in growth signaling pathways. The unifying principle to emerge from these discoveries is that many oncoproteins stimulate cell proliferation by functioning as unregulated, hyperactive versions of proteins that are normally used by cells for stimulating cell proliferation. We will return to a detailed consideration of this theme in Chapter 9, which is devoted to the topic of oncogenes.

Insertional Mutagenesis Allows Viruses with No Oncogenes to Cause Cancer by Activating Cellular Proto-oncogenes

Oncogenic retroviruses can be subdivided into two classes based on how quickly and efficiently they trigger cancer in animals. One group consists of *acutely transforming retroviruses* that cause animals to develop tumors rapidly, often within days of injection. The Rous sarcoma virus is a member of this first group. The second group consists of *slow-acting retroviruses* that require months or years to induce cancer. In terms of genetic makeup, the main feature distinguishing the two groups of viruses is that acutely transforming retroviruses possess oncogenes and slow-acting retroviruses do not.

If slow-acting viruses lack oncogenes, how do they cause cancer? The answer is based not on the action of any viral gene but rather on the ability of retroviruses to alter the expression of normal cellular genes. As described earlier, retroviruses use reverse transcriptase to make a DNA copy of their viral RNA, and the DNA copy is then inserted into the host chromosomal DNA. The impact of the integrated viral DNA depends on where it is located. Insertion at some chromosomal locations causes no apparent problems, whereas insertion at other sites alters the expression of nearby host genes. This phenomenon, in which the integration of viral DNA into a host chromosome activates or disrupts a normal cellular gene, is known as **insertional mutagenesis.**

One of the first examples of insertional mutagenesis to be discovered involved a chicken retrovirus called the *avian leukosis virus* (ALV). When ALV infects cells, it inserts its proviral DNA at a variety of different locations in the host chromosomal DNA. Cancer only arises, however, when ALV happens to insert itself near a cellular gene called the ***MYC* gene** (Figure 7-11, *top*). Like *SRC*, the *MYC* gene is a proto-oncogene and thus has the potential to be converted into an oncogene. When ALV integration occurs near the *MYC* gene, the integrated viral DNA increases the rate at which the *MYC* gene is transcribed.

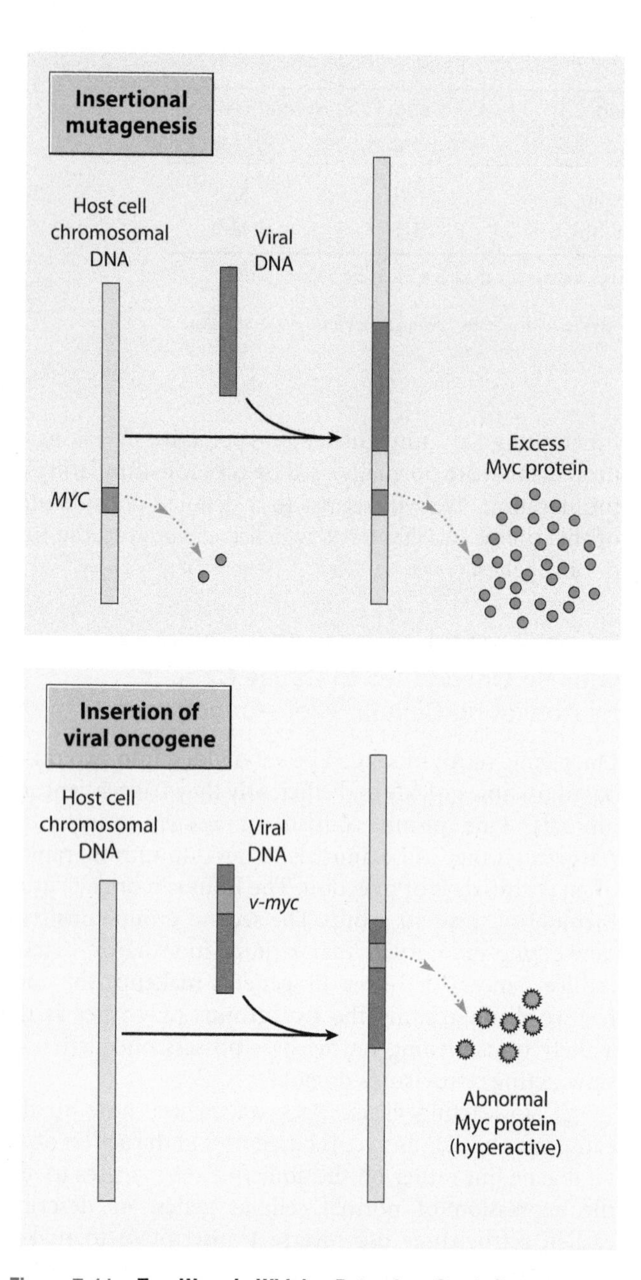

Figure 7-11 Two Ways in Which a Retrovirus Can Influence Production of the Myc Protein. (*Top*) Some retroviruses disrupt regulation of a cell's own *MYC* gene by integrating viral DNA in the vicinity, thereby leading to overproduction of Myc protein. Such an event is an example of insertional mutagenesis. (*Bottom*) Certain retroviruses possess their own v-*myc* oncogene, which is integrated into a host cell chromosome and produces an abnormal form of the Myc protein.

Increased transcription in turn leads to overproduction of **Myc**, the protein encoded by the *MYC* gene. The Myc protein then carries out its normal function, which is to stimulate the transcription of genes required for cell proliferation. There is nothing abnormal about the Myc protein in this particular situation, but its overproduction leads to an excessive stimulation of cell proliferation.

How does the integration of viral DNA into a host chromosome cause the transcription of neighboring genes to be activated? The key lies in special sequences called **long terminal repeats (LTRs)**, which are located at both ends of the genome of retroviruses (see Figure 7-9). LTRs serve two functions for a retrovirus. First, each LTR is bounded at both ends by short repeated sequences that play a role in the mechanism by which proviral DNA is inserted into the host chromosomal DNA. Second, LTRs contain sequences that promote transcription of the adjacent viral genes. LTRs are so efficient at this task, however, that they also activate the transcription of neighboring cellular genes that lie near the inserted viral genes. So if a retrovirus happens to randomly insert its genes near a proto-oncogene, the LTRs may stimulate transcription of the proto-oncogene and trigger overproduction of a normal cellular protein that contributes to cancer development.

Abnormalities Involving the Myc Protein Arise Through Several Mechanisms in Viral Cancers

The preceding mechanism involving the avian leukosis virus is only one of several situations in which the Myc protein has been implicated in the development of cancer. Another example involves the *avian myelocytomatosis virus*, a retrovirus that also causes cancer in chickens. Unlike the avian leukosis virus, which possesses no oncogene but instead enhances expression of a cell's own *MYC* gene, the avian myelocytomatosis virus does possess an oncogene and this oncogene codes for an abnormal version of the Myc protein. Because of its viral origin, the oncogene is called **v-*myc*** to distinguish it from the normal host cell *MYC* gene. Such a relationship is similar to that of v-*src* and *SRC* described earlier. In each case, a retrovirus possesses an oncogene that is an altered version of a normal cellular proto-oncogene.

The behavior of the avian leukosis and avian myelocytomatosis viruses illustrates the principle that oncogenic retroviruses can alter production of the Myc protein in two fundamentally different ways: (1) a virus may disrupt the regulation of a cell's own *MYC* gene by integrating proviral DNA in its vicinity, thereby leading to overproduction of the Myc protein by insertional mutagenesis; or (2) a virus may introduce its own v-*myc* oncogene coding for an abnormal Myc protein (see Figure 7-11, *bottom*). In the first case, cells produce too much normal Myc protein; in the second case, cells produce an additional, abnormal version of the Myc protein encoded by a viral gene, and this abnormal protein is hyperactive. In both cases, excessive activity of the Myc protein disrupts cell signaling mechanisms that control cell proliferation and survival.

Yet another way in which abnormalities involving the Myc protein arise in virally induced cancers is illustrated by Burkitt's lymphoma. As described earlier in the chapter, Burkitt's lymphoma is associated with infection by the Epstein-Barr virus (EBV), which is a DNA virus rather than an RNA virus like the ones we have been discussing. Most people infected with EBV do not develop lymphoma, indicating that other factors in addition to EBV are required. One such factor is disruption of immune function (as occurs with malaria or HIV infection), which permits the unchecked proliferation of EBV-infected cells.

Another event must also occur before Burkitt's lymphoma will arise. This additional step is a chromosomal translocation involving chromosome 8, the chromosome where the *MYC* gene normally resides. In the most common translocation, a piece of chromosome 8 containing the *MYC* gene is translocated to chromosome 14, thereby bringing the normal *MYC* gene into close proximity to genes on chromosome 14 that code for antibody molecules (Figure 7-12). These antibody genes are highly active in the class of lymphocytes that give rise to Burkitt's lymphoma because antibodies account for roughly half of the total protein produced by such cells. Moving the *MYC* gene so close to the highly active antibody genes causes the *MYC* gene to also become activated, thereby leading to an overproduction of Myc protein that in turn stimulates cell proliferation. The mechanism responsible for the chromosome 8 to chromosome 14 translocation is not well understood; it may simply be a random occurrence that is more likely to take place in EBV-infected cells because they are constantly proliferating. If this particular translocation happens to occur by chance in just a single cell, the resulting overproduction of the Myc protein would further enhance proliferation of the affected cell and its progression toward malignancy.

Although we have focused our attention on examples involving viruses, it should be pointed out that abnormalities involving *MYC* genes and the proteins they produce are not restricted to viral cancers. Mutations in the *MYC* gene triggered by other mechanisms—for example, mutagenic chemicals, radiation, or spontaneous replication errors—occur in many cancers that are not caused by viruses.

Several Oncogenic DNA Viruses Produce Oncoproteins That Interfere with p53 and Rb Function

The Epstein-Barr virus, whose role in Burkitt's lymphoma we have been discussing, is just one of several kinds of DNA viruses associated with human or animal cancers. Unlike retroviruses, these DNA viruses do not possess oncogenes that are related to normal cellular genes (proto-oncogenes). Instead, the oncogenes found in DNA cancer viruses are integral parts of the viral genome that facilitate viral replication by disrupting key cellular control mechanisms. At the same time, these viral genes may end up activating the persistent proliferation of the cells they have infected, thereby creating conditions that can lead to cancer.

A good example is provided by the human papillomavirus (HPV), a DNA virus whose association with cervical cancer was discussed earlier in the chapter. HPV contains oncogenes called *E6* and *E7*, which produce proteins that disrupt the activity of two key cellular molecules introduced in Chapter 2: the p53 protein and the Rb protein. As you may recall, the *p53 protein* is a central component of the pathway that halts progression through the cell cycle or triggers cell death in cells that have sustained DNA damage. As a result, cells with damaged DNA are prevented from proliferating and passing the damage on to succeeding generations of cells (see Figure 10-5). The *E6* oncogene of HPV circumvents this protective mechanism by producing the **E6 oncoprotein**, which binds to the p53 protein and promotes its destruction (Figure 7-13). In the absence of p53, HPV-infected cells that incur DNA damage fail to trigger apoptosis by the p53 pathway and are free to proliferate despite numerous mutations.

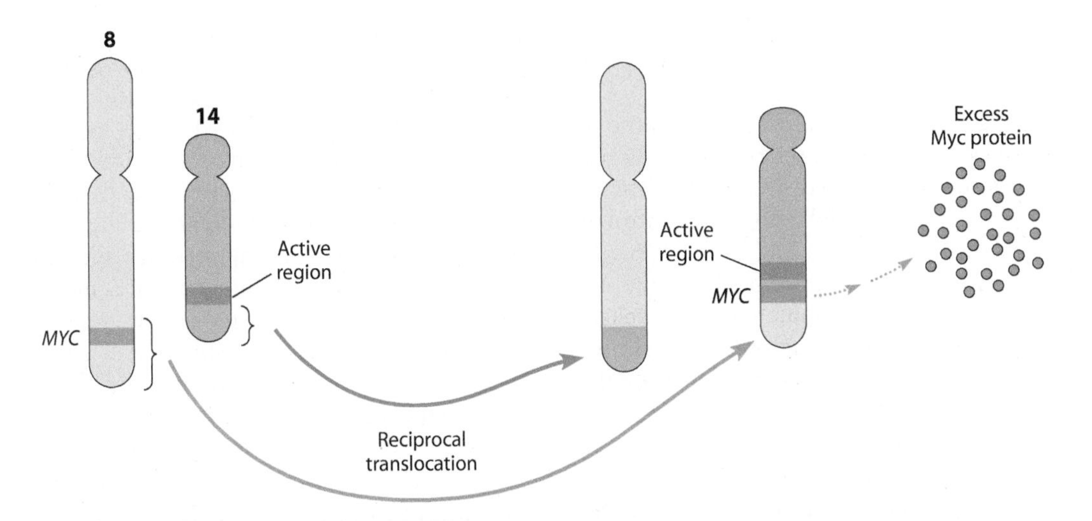

Figure 7-12 Chromosomal Translocation in Burkitt's Lymphoma. In lymphocytes that give rise to Burkitt's lymphoma, a segment of chromosome 8 containing the normal *MYC* gene is frequently exchanged with a segment of chromosome 14. This reciprocal translocation places the normal *MYC* gene adjacent to a very active region of chromosome 14, which contains genes that code for antibody molecules. Moving the *MYC* gene so close to the highly active antibody genes causes the *MYC* gene to also become activated. This activation leads to an overproduction of Myc protein that stimulates cell proliferation.

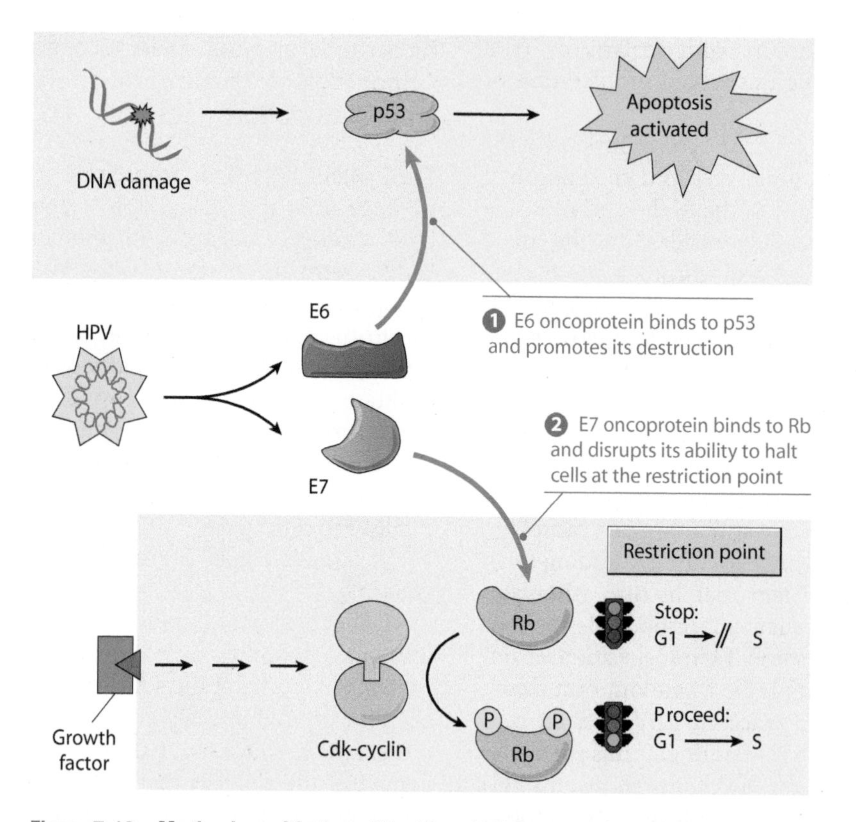

Figure 7-13 Mechanism of Action of the E6 and E7 Oncoproteins of HPV. The human papillomavirus (HPV) produces viral oncoproteins called E6 and E7. ① The E6 oncoprotein binds to the cell's normal p53 protein, thereby promoting its destruction. As a result, p53 can no longer prevent cells with damaged DNA from proliferating. ② The E7 oncoprotein binds to the cell's normal Rb protein and disrupts its ability to stop cells at the restriction point of the cell cycle. Thus the survival of mutant cells is promoted by the E6 oncoprotein and the ability to halt cell proliferation is disrupted by the E7 oncoprotein, both of which contribute to the development of cancer.

The *Rb protein,* which is also targeted by HPV, controls cell proliferation by inhibiting progression through the G1 restriction point in cells that have not been stimulated by an appropriate growth factor (see Figure 2-10). In cells infected with HPV, the virally produced **E7 oncoprotein** binds to the Rb protein and blocks its function. Under such conditions, the Rb protein cannot restrain passage through the restriction point and cell proliferation proceeds unchecked, even in the absence of growth factors. So in cells infected with HPV, control of cell proliferation is disrupted by the E7 oncoprotein and the survival of mutant cells is promoted by the E6 oncoprotein, both of which contribute to the development of cancer. HPV is only one of several DNA viruses that function in this way. The SV40 virus, for example, produces a single protein called the *large T antigen* that accomplishes both tasks, binding to and incapacitating both the p53 protein and the Rb protein.

Although oncogenes coding for proteins that interfere with Rb and p53 play a central role for these viruses, the mere presence of such genes is not necessarily sufficient to cause cancer. The behavior of HPV is an interesting case in point. As was mentioned earlier, only a small number of the more than 100 different types of HPV cause cervical cancer, yet all the known types of HPV have E6 and E7 oncogenes. So why are only some forms of HPV carcinogenic? The answer appears to be related not to the presence or absence of the E6 and E7 oncogenes, but to how these oncogenes are expressed in infected cells. When noncarcinogenic forms of HPV infect a cell, the viral DNA is maintained in the cell nucleus as a separate, independently replicating episome. In contrast, the viral DNA of carcinogenic forms of HPV, such as HPV 16 and 18, integrate their DNA into a host cell chromosome, placing it in a new environment that alters the way in which the viral genes are expressed (Figure 7-14). The net result is the production of slightly different forms of the viral messenger RNAs coding for the E6 and E7 oncoproteins. These altered RNAs are more stable than the RNAs produced by the nonintegrated form of the virus, resulting in the production of larger quantities of the E6 and E7 oncoproteins.

The Ability of Viruses to Cause Cancer Has Created Problems for the Field of Gene Therapy

We have now seen that viruses can cause cancer through a variety of different mechanisms. As a result, viruses with the potential to cause cancer are not always easy to

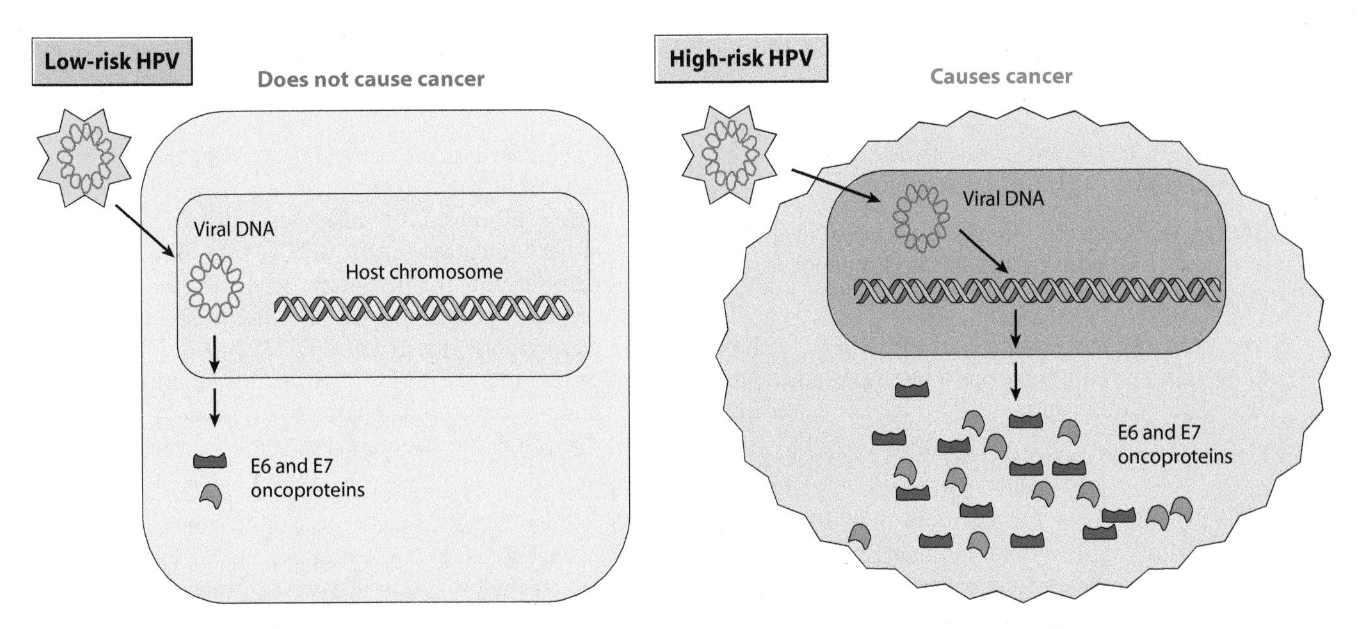

Figure 7-14 Significance of Viral DNA Integration for High-Risk Strains of HPV. (*Left*) When low-risk types of HPV infect a host cell, the viral DNA is maintained as a separate episome and cancer does not arise. (*Right*) High-risk, oncogenic forms of HPV, such as HPV 16 and HPV 18, integrate their viral DNA into a host chromosome, thereby placing the viral genes in an environment that causes them to produce larger amounts of the E6 and E7 oncoproteins.

recognize. An example of the kind of problems this limitation can impose has occurred in the field of **gene therapy**, which involves treating genetic diseases by inserting normal copies of genes into the cells of people who have inherited defective, disease-causing genes. One type of gene therapy uses specially modified viruses to ferry copies of normal genes into cells possessing defective genes. To prevent potentially dangerous infections, the viruses are first modified to remove most of their own genes. A normal copy of the gene to be corrected in the target cells is then inserted into the modified virus, and the virus is used either to infect a patient directly or to infect cells isolated from the patient.

One of the first apparent successes of gene therapy was observed in children suffering from a debilitating immune disorder called *severe combined immunodeficiency (SCID)*. The immune system does not develop properly in babies who inherit one of the defective genes responsible for SCID, leaving children fatally vulnerable to what would otherwise be routine infections, such as cold sores or chicken pox. Children who inherit the disease are sometimes referred to as "bubble babies" because they may be forced to live inside a sterile plastic bubble to avoid threats to their debilitated immune systems.

In 2000, a team of researchers led by French pediatrician Alain Fischer reported dramatic success in using gene therapy to treat children with SCID. These studies employed a specially modified retrovirus that is especially efficient at transferring genes into human cells and integrating them into human chromosomes. The retrovirus, with the correct replacement gene inserted, was used to infect lymphocytes obtained from a small number of SCID patients, and the lymphocytes were then injected back into the bloodstream. The resulting improvement in immune function was so dramatic that, for the first time, some children were able to leave the protective isolation "bubble" that had been used to shield them from infections.

It was therefore a great disappointment when two of the ten children treated in the initial study developed leukemia a few years later. Examination of their cancer cells revealed that the retrovirus had integrated itself into a chromosome near a normal gene called *LMO2*, whose abnormal expression is known to be associated with certain forms of leukemia. Apparently, the retrovirus had triggered an insertional mutagenesis event by integrating in a place where it could activate the *LMO2* gene. Integration at this particular site occurs in no more than 1-in-50,000 treated lymphocytes, but because millions of lymphocytes were injected back into each patient, cells exhibiting viral insertion near *LMO2* were likely to be numerous. One such insertion event, leading to uncontrolled proliferation of a single affected cell, might be all it takes to initiate the development of cancer.

This outcome was an unexpected setback for the field of gene therapy because the retrovirus had no oncogene or other feature that would have allowed scientists to predict its significant cancer risk. One should not, of course, forget that these studies also provided one of the first signs of hope that gene therapy might be able to correct gene defects in children with life-threatening genetic diseases. Nonetheless, the associated cancer risks need to be better understood before further progress can be made.

Summary of Main Concepts

- About 15% of all cancers worldwide are triggered by viral, bacterial, or parasitic infections.
- Cancer viruses were initially discovered in animals, but several human viruses are now known to cause cancer as well.
- Infection with the Epstein-Barr virus (EBV) is linked to Burkitt's lymphoma, nasopharyngeal carcinoma, Hodgkin's disease, and infectious mononucleosis.
- Human papillomavirus (HPV) is a sexually transmitted virus responsible for cervical cancer and cancer of the penis. Of the more than 100 different types of HPV, only a small number (mainly types 16, 18, and a few others) cause cancer.
- Infections involving hepatitis B virus (HBV) or hepatitis C virus (HCV) are responsible for most cases of liver cancer, a disease that is especially prevalent in Southeast Asia, China, and Africa.
- The human T-cell lymphotropic virus-I (HTLV-I) is a retrovirus associated with adult T-cell leukemia/lymphoma, an aggressive cancer that is rare in the United States but common in certain parts of Japan, Africa, and the Caribbean.
- People infected with HIV have a debilitated immune system that puts them at increased risk for certain virally induced cancers. The greatest risk involves Kaposi's sarcoma triggered by KSHV, but increased risk is also exhibited for cancers caused by EBV, HPV, HBV, and HCV.
- SV40 is an animal virus that contaminated early batches of polio vaccine and may have caused a small number of human cancers.
- Infection with the bacterium *H. pylori* is associated with most stomach cancers, and flatworm infections involving blood flukes or liver flukes can cause cancers of the bladder or bile duct, respectively.
- Some infectious agents trigger the development of cancer by indirect mechanisms, such as by disrupting the immune system or by causing a chronic inflammatory state that stimulates cell proliferation. Other infectious agents directly stimulate the proliferation of the cells they infect.
- Cancer viruses can conceal themselves within infected cells in a latent form, with no new virus particles being produced or released. Latent DNA viruses either integrate their DNA into the host cell chromosome or maintain it as an independently replicating episome. RNA retroviruses use reverse transcriptase to make a DNA copy of their RNA genes, and the DNA copy is then integrated into a host chromosome, where it is replicated along with the host chromosomal DNA.
- Retroviral oncogenes are altered versions of normal cellular genes called proto-oncogenes. Retroviral oncogenes code for oncoproteins that are abnormal, hyperactive versions of cellular proteins involved in growth factor signaling pathways.
- Acutely transforming retroviruses cause animals to develop tumors rapidly, whereas slow-acting retroviruses require months or years to induce cancer. The main feature that distinguishes the two groups of viruses is that acutely transforming retroviruses possess oncogenes and slow-acting retroviruses do not. Slow-acting retroviruses cause insertional mutagenesis by integrating their genes into a host chromosomal site where the viral DNA can activate the transcription of a nearby cellular proto-oncogene.
- Abnormalities involving *MYC* genes and the proteins they produce arise in viral cancers by several different mechanisms, including insertional mutagenesis, introduction of a v-*myc* oncogene, and chromosomal translocations that alter the expression of a host *MYC* gene.
- Unlike the oncogenes of RNA retroviruses, the oncogenes of DNA viruses are not related to normal cellular genes but instead appear to be integral parts of the viral genome that facilitate viral replication by disrupting key cellular control mechanisms. Several of these DNA viruses code for oncoproteins that bind to and interfere with the activity of a cell's Rb and p53 proteins, thereby facilitating uncontrolled proliferation and disrupting the pathway that would normally halt the proliferation of cells containing damaged DNA.
- Retroviruses used as tools for gene therapy carry the risk of causing cancer by insertional mutagenesis.

Key Terms for Self-Testing

Infectious Agents in Human Cancers

pathogen (p. 120)
oncogenic virus (p. 120)
Burkitt's lymphoma (p. 120)
Epstein-Barr virus (EBV) (p. 121)
Hodgkin's disease (p. 122)

infectious mononucleosis (p. 122)
nasopharyngeal carcinoma (p. 122)
human papillomavirus (HPV) (p. 122)
hepatitis (p. 124)
hepatitis B virus (HBV) (p. 124)
hepatitis C virus (HCV) (p. 125)
human T-cell lymphotropic virus-I (HTLV-I) (p. 125)
HIV (human immunodeficiency virus) (p. 125)
acquired immunodeficiency syndrome (AIDS) (p. 125)
Kaposi's sarcoma (p. 125)
Kaposi's sarcoma-associated herpesvirus (KSHV) (p. 125)
SV40 (simian virus 40) (p. 126)
Helicobacter pylori (*H. pylori*) (p. 126)
blood fluke (p. 127)
liver fluke (p. 127)

How Infections Cause Cancer

NF-kappa B (NF-κB) (p. 128)
virus (p. 129)
latent (p. 129)
episome (p. 129)
reverse transcriptase (p. 129)
integrase (p. 129)
provirus (p. 129)
retrovirus (p. 129)
genome (p. 129)
v-*src* gene (p. 129)
oncogene (p. 129)
SRC gene (p. 130)
proto-oncogene (p. 130)
v-Src (protein) (p. 131)
tyrosine kinase (p. 133)
Src kinase (p. 133)
constitutively active (p. 133)
oncoprotein (p. 133)
insertional mutagenesis (p. 133)
MYC gene (p. 133)
Myc (protein) (p. 134)
long terminal repeats (LTRs) (p. 134)
v-*myc* gene (p. 134)
E6 oncoprotein (p. 135)
E7 oncoprotein (p. 136)
gene therapy (p. 137)

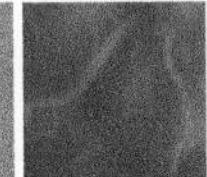

Suggested Reading

Infectious Agents in Human Cancers

Blaser, M. J. The bacteria behind ulcers. *Sci. Amer.* 274 (February 1996): 104.

Boshoff, C., and R. Weiss. AIDS-related malignancies. *Nature Reviews Cancer* 2 (2002): 373.

Castle, P. E. A prospective study of high-grade cervical neoplasia risk among human papillomavirus-infected women. *J. Natl. Cancer Inst.* 94 (2002): 1406.

DiBisceglie, A. M., and B. R. Bacon. The unmet challenges of hepatitis C. *Sci. Amer.* 281 (October 1999): 80.

Ganem, D. KSHV and Kaposi's sarcoma: The end of the beginning? *Cell* 91 (1997): 157.

Gazdar, A. F., J. S. Butel, and M. Carbone. SV40 and human tumours: myth, association or causality? *Nature Reviews Cancer* 2 (2002): 957.

Newton, R. Infectious Agents and Cancer. In *The Cancer Handbook* (M. Alison, ed.). London: Nature Publishing Group, 2002, Chapter 24.

Parsonnet, J., ed. *Microbes and Malignancy: Infection as a Cause of Human Cancers.* New York: Oxford University Press, 1999.

Peek, R. M., Jr., and M. J. Blaser. *Helicobacter pylori* and gastrointestinal tract adenocarcinomas. *Nature Reviews Cancer* 2 (2002): 28.

Subramanian, C., M. A. Cotter II, and E. S. Robertson. Epstein-Barr virus nuclear protein EBNA-3C interacts with the human metastatic suppressor Nm23-H1: A molecular link to cancer metastasis. *Nature Medicine* 7 (2001): 350.

Walboomers, J. N. N., et al. Human papillomavirus is a necessary cause of invasive cervical cancer worldwide. *J. Pathol.* 189 (1999): 12.

zur Hausen, H. Papillomaviruses and cancer: from basic studies to clinical application. *Nature Reviews Cancer* 2 (2002): 342.

How Infections Cause Cancer

Aggarwal, B. B. Nuclear factor-κB: The enemy within. *Cancer Cell* 6 (2004): 203.

Balkwill, F., K. A. Charles, and A. Mantovani. Smoldering and polarized inflammation in the initiation and promotion of malignant disease. *Cancer Cell* 7 (2005): 211.

Cohen, J. I. Herpesviruses. In *Holland-Frei Cancer Medicine*, 5th ed. (R. C. Bast et al., eds.). Lewiston, NY: Decker, 2000, Chapter 19.

Coussens, L. M., and Z. Werb. Inflammation and cancer. *Nature* 420 (2002): 860.

Fine, H. A., and J. G. Sodroski. Tumor Viruses. In *Holland-Frei Cancer Medicine*, 5th ed. (R. C. Bast et al., eds.). Lewiston, NY: Decker, 2000, Chapter 19.

Griffin, B. E. Human DNA Tumor Viruses. In *The Cancer Handbook* (M. Alison, ed.). London: Nature Publishing Group, 2002, Chapter 18.

Hatakeyama, M. Oncogenic mechanisms of the *Helicobacter pylori* CagA protein. *Nature Reviews Cancer* 4 (2004): 688.

Hussain, S. P., L. J. Hofseth, and C. C. Harris. Radical causes of cancer. *Nature Reviews Cancer* 3 (2003): 276.

Keiff, E. Tumor Viruses. In *Clinical Oncology* (M. D. Abeloff, ed.). New York: Churchill Livingstone, 2000, Chapter 14.

Young, L. S., and A. B. Rickinson. Epstein-Barr virus: 40 years on. *Nature Reviews Cancer* 4 (2004): 757.

zur Hausen, H. Papillomavirus infections—a major cause of human cancers. *Biochim. Biophys. Acta* 1288 (1996): F55.

8 Heredity and Cancer

The preceding three chapters have covered the three main causes of cancer, namely chemicals, radiation, and infectious agents. Most of the cancers caused by these agents arise in people who are not otherwise predisposed to developing the disease. Instead, their cancer risk is determined mainly by the potency and dose of the carcinogen involved. Because carcinogen-induced mutations are largely random, luck also plays a role. If two people have identical exposures to the same carcinogens, one person may develop cancer while the other does not simply because random mutations happened to damage critical genes in the unlucky individual.

But there is more to the story than this. Given the exact same set of circumstances, everyone does not have the exact same chance of developing cancer because heredity makes some people more susceptible than others. In Chapter 4 we learned that 80% to 90% of a population's overall cancer risk comes from environmental and lifestyle factors, leaving 10% to 20% to be derived from hereditary makeup. While the impact of heredity is therefore small for populations as a whole, that does not mean that heredity is a minor factor for every person. Some individuals inherit gene mutations that enormously increase their cancer risk. In fact, some inherited mutations increase the risk to almost 100%, making it a virtual certainty that a person will develop cancer within his or her lifetime; other inherited mutations impart smaller, but still significant, increases in cancer risk. The study of such mutations has turned out to illuminate our understanding not just of hereditary cancers, but of nonhereditary forms of cancer as well.

HEREDITARY RISK: GENES INVOLVED IN RESTRAINING CELL PROLIFERATION

When it is said that certain cancers are hereditary, it does not mean that people inherit cancer from their parents. What can be inherited, however, is an increased *susceptibility* to developing cancer. Cancers that arise because of such an inherited predisposition are called **familial** or **hereditary cancers**, whereas the more common cancers that do not exhibit an obvious inherited pattern are referred to as **sporadic** or **nonhereditary cancers**.

Inherited mutations in several different categories of genes can predispose a person to developing cancer. We will start the chapter by considering genes whose normal function is to suppress cell proliferation and survival.

Cancer Risk Is Influenced by the Inheritance of Dominant or Recessive Mutations of Varying Penetrance

To begin our discussion of inherited mutations and cancer risk, it will be useful to review a few of the basic principles that govern inheritance patterns. Human cells other than sperm and eggs possess two copies of each chromosome and are therefore said to be **diploid** (from the Greek word *diplous,* meaning "double"). A normal diploid cell contains two sets of 23 chromosomes, yielding a total of 46 chromosomes. The two members of each chromosome pair are referred to as **homologous chromosomes**; one member of each pair is inherited from a person's father and the other from the mother.

Since each type of chromosome is present in two copies, each gene represented in a pair of homologous chromosomes will be present in two copies. The two versions of each gene—called **alleles**—may be identical or slightly different. If the two alleles for a given gene are identical, a person is said to be **homozygous** for that gene; if the two alleles are different, the person is said to be **heterozygous**. When a cell is heterozygous for a particular gene, one of the two alleles is often **dominant** and the other **recessive**. These terms convey the idea that the dominant allele determines how the trait will appear in a heterozygous individual. In other words, a person with one dominant and one recessive allele for a given trait will express the dominant allele. Although such heterozygous individuals do not express the recessive trait, they can pass that recessive allele on to their children. To actually express a recessive trait, however, a person must be homozygous for the recessive allele (Figure 8-1).

Not all individuals who inherit dominant or recessive mutations express the trait that would be expected. The frequency with which a given dominant or homozygous recessive allele yields the expected trait within a population is known as **penetrance**. For example, we will see shortly

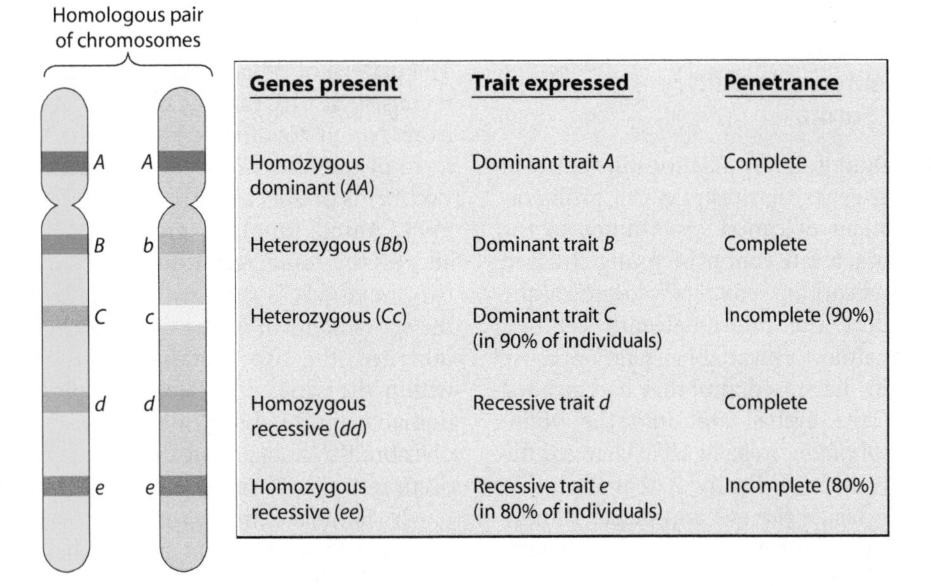

Genes present	Trait expressed	Penetrance
Homozygous dominant (*AA*)	Dominant trait *A*	Complete
Heterozygous (*Bb*)	Dominant trait *B*	Complete
Heterozygous (*Cc*)	Dominant trait *C* (in 90% of individuals)	Incomplete (90%)
Homozygous recessive (*dd*)	Recessive trait *d*	Complete
Homozygous recessive (*ee*)	Recessive trait *e* (in 80% of individuals)	Incomplete (80%)

Figure 8-1 Some Genetic Terminology. This diagram shows a homologous pair of chromosomes from a diploid cell; the chromosomes are the same size and shape and carry genes coding for the same products, arranged in the same linear order. The particular versions of a gene—alleles—found in the two homologous chromosomes may be identical (making a person homozygous for that gene) or different (making a person heterozygous for that gene). If one allele of a gene is dominant and the other recessive, the heterozygous individual exhibits the dominant trait. A recessive trait is only expressed if the individual is homozygous for the recessive allele. Penetrance refers to the frequency with which a given dominant or homozygous recessive allele yields the expected trait within a population of individuals. Five different letters are used to refer to five different genes along the chromosome; capital letters are used to refer to dominant alleles, and lowercase letters are used to refer to recessive alleles.

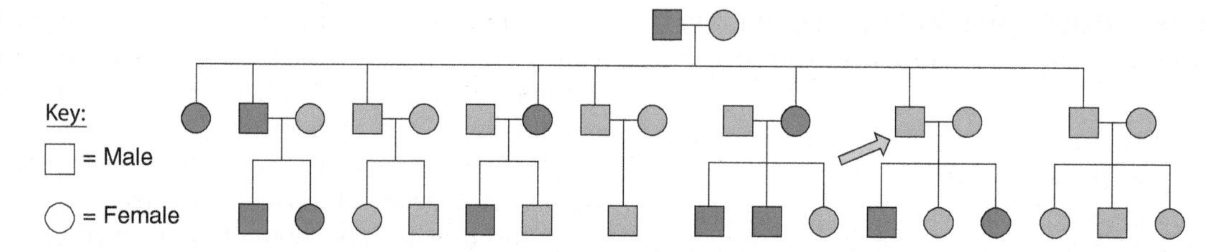

Figure 8-2 Pedigree Typical of the Familial Form of Retinoblastoma. This pedigree summarizes the disease status of 24 children and grandchildren descended from a single male who had retinoblastoma as a child. Roughly half of his descendants also developed the disease (the red symbols represent individuals who developed retinoblastoma). Note that one son (*arrow*) who never developed retinoblastoma nonetheless passed the susceptibility on to two of his children. He represents one of the small fraction (~10%) of individuals who inherit the genetic defect responsible for familial retinoblastoma but, because of incomplete penetrance, do not actually develop cancer.

that certain gene mutations are inherited as dominant traits that dramatically increase a person's risk of developing cancer. In some cases, every person who inherits a mutant copy eventually develops cancer and the penetrance is therefore said to be 100%. More commonly, less than 100% of the individuals who inherit a mutant copy actually develop cancer. In such cases, the mutant form of the gene is said to exhibit *incomplete penetrance*. Individuals who inherit the mutation but do not develop cancer themselves can nonetheless pass the mutation on to their children. Incomplete penetrance arises when environmental conditions or other components of a person's genetic makeup influence whether a particular trait will be expressed.

Retinoblastoma Is a Rare Childhood Cancer of the Eye That Occurs in Hereditary and Nonhereditary Forms

Our current understanding of the relationship between inherited mutations in genes that restrain cell proliferation and the development of cancer owes much to the study of **retinoblastoma**, a rare cancer of young children that arises in the light-absorbing *retinal cells* located at the back of the eye. Before the mid-nineteenth century, retinoblastoma was almost invariably fatal because tumors were not usually diagnosed until they had invaded through the back of the eyeball and into the brain. The invention of the *ophthalmoscope* in 1850 changed the situation dramatically; doctors finally had a tool that permitted them to see inside the eye and detect tumors before they invaded into surrounding tissues, thereby allowing the tumors to be removed in their early stages.

Paradoxically, this impressive medical advance had an unexpected consequence: A new form of retinoblastoma emerged that had not been seen before. Before the ophthalmoscope was invented, few children with retinoblastoma survived to adulthood. Now almost 90% are cured and go on to live normal lives. When these retinoblastoma survivors grow up and have families of their own, their children may also be at high risk for retinoblastoma. Normally a child has less than a 1-in-20,000 chance of developing retinoblastoma, but in the families of some retinoblastoma survivors, roughly 50% of the children develop the disease (Figure 8-2). This particular form of retinoblastoma, which runs in family clusters, is called *familial retinoblastoma* to distinguish it from the *sporadic retinoblastomas* seen in children with no family history of the disease. About 40% of retinoblastoma cases are familial and the rest are sporadic. Children with the familial form of the disease often develop multiple tumors in both eyes, whereas sporadic retinoblastoma is almost always a single tumor.

The Two-Hit Model Predicts That Two Mutations Are Needed to Trigger the Development of Retinoblastoma

The pattern of inheritance seen in familial retinoblastoma is consistent with the transmission of a single mutant gene from parent to offspring. But what about the sporadic form of the disease? Is the same gene involved, or do the two forms of eye cancer arise by different mechanisms? In 1971, Alfred Knudson proposed the **two-hit model** to address this issue. According to his theory, the presence of two mutations is required to create a retinoblastoma. In the sporadic form of the disease, where no mutations are inherited, the two mutations must arise spontaneously within the same cell. Because typical mutation rates in human cells are about one in a million per gene per cell division, the chance that the two required mutations will occur in the same cell is extremely remote.

In familial retinoblastoma, one mutation is already inherited from a parent and is therefore present in all body cells, including those of the retina. An individual retinal cell therefore needs to sustain only one additional mutation to create the two mutant genes required to produce a cancer. Since there are more than ten million cells in the retina and since the mutation rate is about one in a million per gene per cell division, it is likely that one or more retinal cells will spontaneously incur the second mutation required for cancer to arise. That would explain why retinoblastoma rates are so high in families with the hereditary form of the disease and why children in such families often develop multiple tumors.

However, the preceding model leaves an important question unanswered: Do the two mutations involve different genes, or are we dealing with two defective alleles of the same gene, one residing on each member of a chromosome pair? Microscopic examination of the chromosomes of retinoblastoma cells provided an early clue. Cells from individuals with familial retinoblastoma were sometimes found to exhibit a deleted segment in a particular region of chromosome 13. Moreover, the deletion was observed not just in tumor cells but in all cells of the body, suggesting that the inherited defect leading to retinoblastoma involved a gene located in the deleted region of chromosome 13. A similar deletion is occasionally observed in sporadic retinoblastomas as well (although in this case only in tumor cells and not elsewhere in the body). Taken together, such observations suggest that chromosome 13 contains a gene whose loss is associated with both the hereditary and nonhereditary forms of retinoblastoma.

By itself, inheriting a single deletion in this region of chromosome 13 does not cause cancer. For cancer to develop, a mutation involving the same region of the second copy of chromosome 13 must arise during the many rounds of cell division that occur as the retina is formed. This second mutation was eventually localized to a small region of chromosome 13 containing a gene that was given the name ***RB*** (for "retinoblastoma").

So Knudson's two-hit theory for retinoblastoma could finally be restated in more precise terms (Figure 8-3). The "two hits" required for the development of retinoblastoma involve the two copies of the *RB* gene that we each inherit, one from our father and one from our mother. Each "hit" is a mutation (or deletion) that disrupts the function of one copy of the *RB* gene. According to this model, the *RB* mutation is *recessive* because both alleles must be mutant (or deleted) for cancer to arise. However, the *cancer risk syndrome* in which children inherit a greatly elevated risk

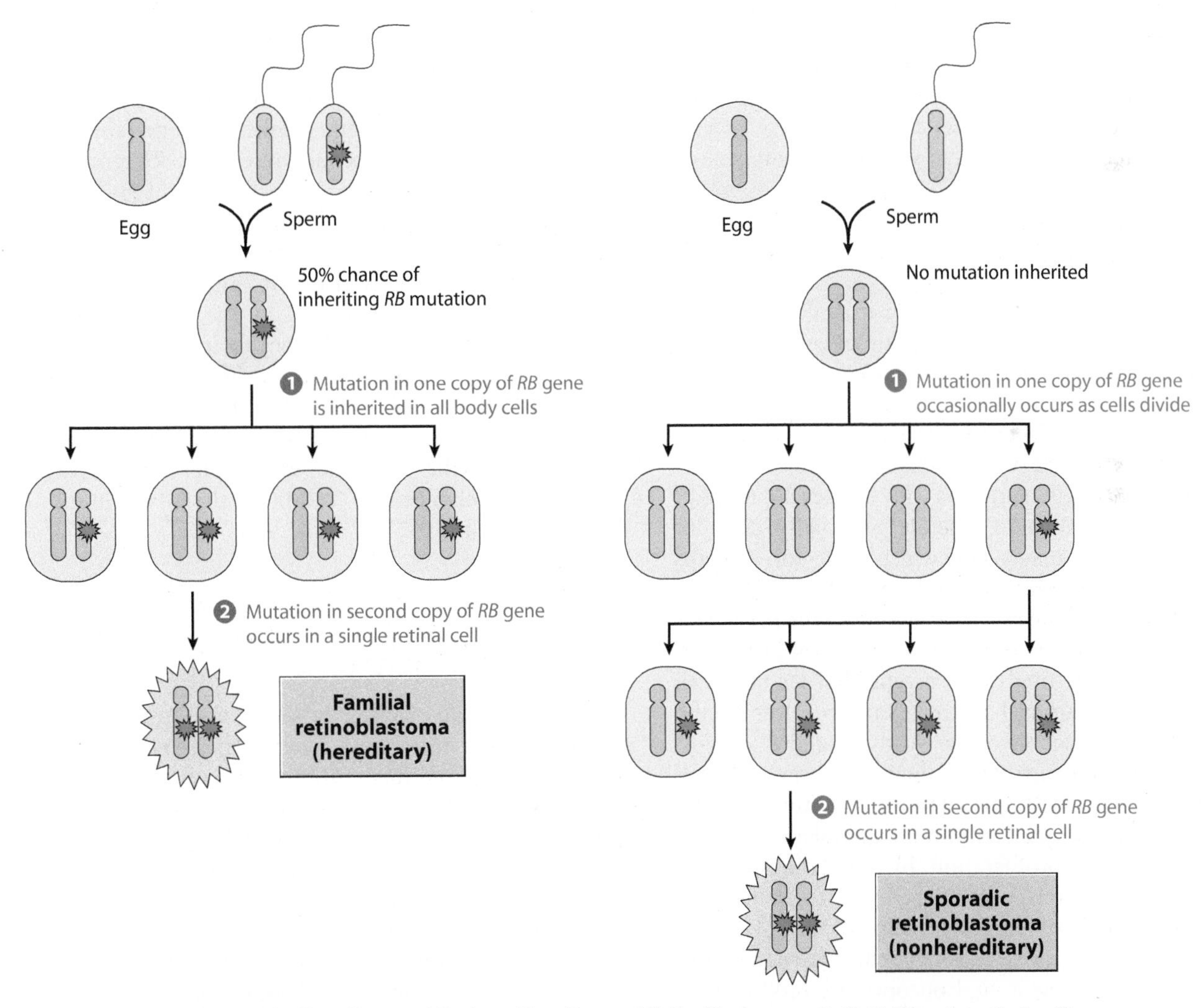

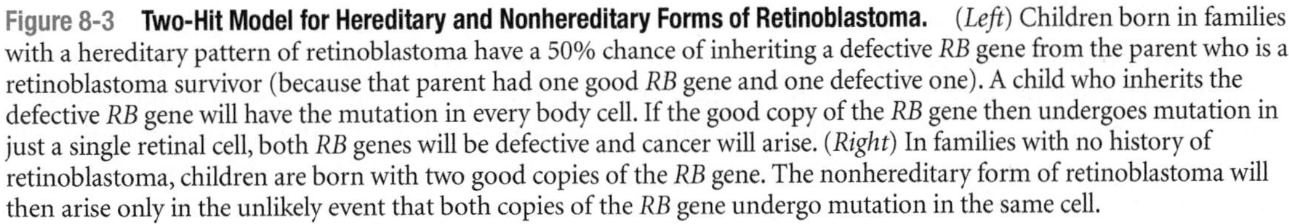
Figure 8-3 Two-Hit Model for Hereditary and Nonhereditary Forms of Retinoblastoma. (*Left*) Children born in families with a hereditary pattern of retinoblastoma have a 50% chance of inheriting a defective *RB* gene from the parent who is a retinoblastoma survivor (because that parent had one good *RB* gene and one defective one). A child who inherits the defective *RB* gene will have the mutation in every body cell. If the good copy of the *RB* gene then undergoes mutation in just a single retinal cell, both *RB* genes will be defective and cancer will arise. (*Right*) In families with no history of retinoblastoma, children are born with two good copies of the *RB* gene. The nonhereditary form of retinoblastoma will then arise only in the unlikely event that both copies of the *RB* gene undergo mutation in the same cell.

of developing cancer is a *dominant* trait because only a single *RB* mutation needs to be inherited to create a very high cancer risk. If a child inherits such a defective *RB* gene, mutation of the good copy of the *RB* gene in a single retinal cell is all that is needed for cancer to develop. Since there are millions of cells in the retina and since normal mutation rates are about one in a million per gene per cell division, the chance that such an inactivating mutation will occur in at least one retinal cell is quite high. Roughly 90% of the children who inherit a mutant *RB* gene from one parent eventually acquire this second mutation and develop cancer. In other words, the penetrance of retinoblastoma in children who inherit a single defective *RB* gene is 90%.

In families with no history of retinoblastoma, children are born with two good copies of the *RB* gene. As a result, retinoblastoma only arises in the very unlikely event that both copies of the *RB* gene happen to undergo mutation in the same cell, leading to the same genetic outcome as in familial retinoblastoma. Retinoblastoma is one of a very few cancers in which inactivation of both copies of a single gene appears to be so crucial in leading directly to cancer. As we will see in Chapter 10, it is more common for cancers to arise as a multistep process involving sequential mutations in several different genes.

The *RB* Gene Is a Suppressor of Cell Proliferation

What kind of gene is *RB* that makes it so central to the development of cancer? The only group of cancer-causing genes we have discussed so far are *oncogenes*, which were introduced and defined in Chapter 7 as a class of genes whose presence can trigger the development of cancer. In contrast, *RB* is a gene whose *absence* (or inactivation) rather than *presence* is associated with the development of cancer. Such a gene must operate on totally different principles than do oncogenes, which code for proteins that promote cell proliferation and survival.

It turns out that the normal function of the *RB* gene is not to promote cell proliferation and survival but rather to *restrain* them. Getting rid of a gene that restrains cell proliferation and survival is analogous to releasing the brake pedal on a car; in other words, it allows cells to proliferate in an uncontrolled fashion. The way in which the *RB* gene performs its normal restraining function is by coding for the Rb protein, a molecule that plays a key role in controlling cell proliferation by regulating passage through the restriction point of the cell cycle. Chapter 2 introduced the role of the Rb protein in preventing cell proliferation when growth factors are absent (p. 26), and Chapter 7 showed how a viral oncoprotein produced by the human papillomavirus inactivates the Rb protein, thereby contributing to uncontrolled cell proliferation and cervical cancer (p. 136). The role of the Rb protein in controlling progression through the cell cycle helps explain why retinoblastoma is only observed in young children. By the time a child reaches 5 or 6 years of age, the retina is fully formed and most of the retinal cells exit from the cell cycle and permanently stop dividing. Once that happens, these cells are no longer susceptible to uncontrolled proliferation because the cell cycle has been permanently stopped.

Genes like *RB*—whose loss opens the door to excessive cell proliferation and cancer—are referred to as **tumor suppressor genes**. This name is slightly misleading, however, because the normal function of tumor suppressor genes is to restrain cell proliferation in general, not just the formation of tumors. But if the *RB* gene is involved in the general restraint of cell proliferation, shouldn't its loss lead to cancers in tissues other than the retina? In fact, retinoblastoma is not the only cancer to appear in children who inherit *RB* mutations. Such individuals often develop a second type of cancer as well, usually osteosarcoma but sometimes leukemia, melanoma, lung cancer, or bladder cancer. The tumor cells in these cancers exhibit mutations in the second copy of the *RB* gene, just as occurs with retinoblastoma.

RB mutations have also been detected in several kinds of cancer that do not involve a hereditary predisposition. For example, small cell lung carcinomas—a type of cancer that arises mainly in cigarette smokers with no family predisposition to cancer—frequently exhibit *RB* mutations that are created by the carcinogens present in tobacco smoke rather than being inherited. Noninherited *RB* mutations have likewise been implicated in the development of some bladder and breast cancers. The *RB* gene therefore plays a broader role in the development of cancer than its name would seem to imply. The gene is called "*RB*" because it was initially discovered in families at high risk for retinoblastoma, but its role in the control of cell proliferation and cancer is much broader. We will return to the *RB* gene in Chapter 10, where its molecular mechanism of action is described in more detail (see Figure 10-3).

Familial Adenomatosis Polyposis Is a Colon Cancer Syndrome Caused by Inherited Mutations in the *APC* Gene

The *RB* gene is an example of a gene that imparts a high risk for a rare childhood cancer when inherited in a mutant (or deleted) form. Another group of genes create hereditary predispositions for some of the common adult cancers. For example, about 5% of all colon cancer cases are directly linked to inherited mutations in a small group of genes that increase colon cancer risk. Although 5% is not a very high proportion, it represents a large number of people because colon cancer is among the most frequently encountered cancers in adults. About 150,000 new cases of colon cancer are diagnosed in the United States each year, so 5% of 150,000 equals 7500 new colon cancers caused by high-risk hereditary factors. In the case of retinoblastoma, a much higher proportion of the cases are hereditary (about 40%), but retinoblastoma itself is very rare. Roughly 200 new cases of retinoblastoma are detected

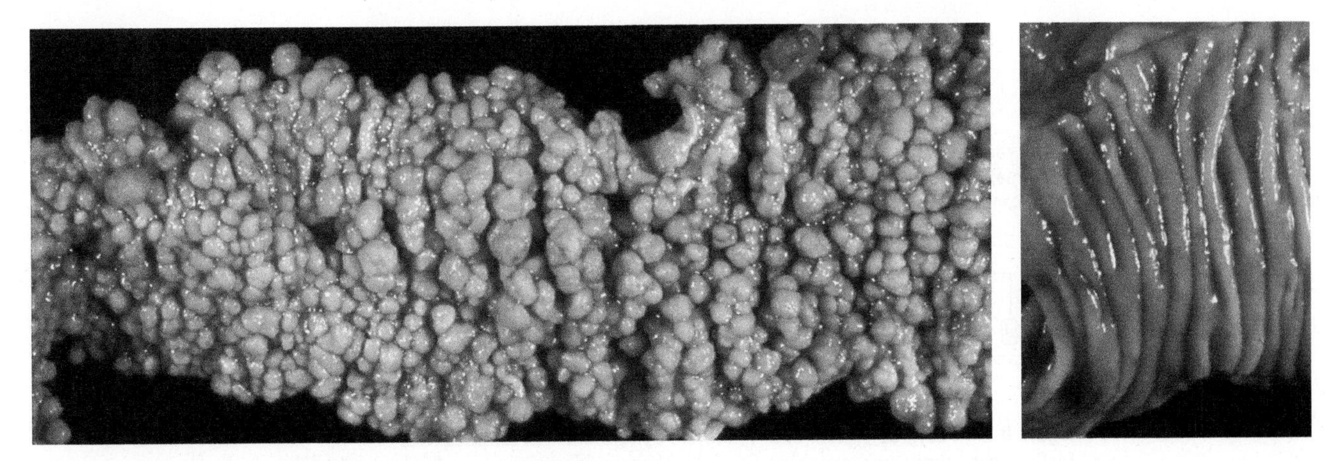

Figure 8-4 Colon Polyps in Familial Adenomatous Polyposis. The photograph on the left shows the inner surface of the colon in a person with familial adenomatous polyposis. Hundreds of polyps can be seen in this short stretch of colon. For comparison, the smooth surface of a normal colon is shown on the right. [Courtesy of Gerald Abrams.]

each year, so a 40% rate corresponds to 80 new cases of hereditary retinoblastoma compared with 7500 new cases of hereditary colon cancer.

Mutations in several different genes have been implicated in hereditary colon cancers. One is the gene responsible for **familial adenomatous polyposis**, an inherited condition in which numerous polyps develop in the colon. (Recall from Chapter 1 that the term *polyp* refers to a tiny mass of tissue that arises from the inner lining of a hollow organ and protrudes into the lumen.) Most colon polyps are tiny *adenomas*—that is, benign tumors composed of glandular cells. The presence of polyps that are adenomas explains why the disease is called familial *adenoma*tous *polyp*osis. In people who inherit this condition, the inner surface of the colon becomes carpeted with hundreds or even thousands of benign polyps (Figure 8-4), and at least one of the polyps is likely to turn malignant by the time the person is 40 years old. Unless the entire colon is removed, virtually everyone with familial adenomatous polyposis will eventually develop colon cancer. Such individuals are also at increased risk for cancers of the bile duct, small intestine, and stomach. In Japan, where stomach cancer is more common than in the United States, the risk of stomach cancer may even be greater than that for colon cancer.

The inherited mutation responsible for familial adenomatous polyposis resides in a gene called ***APC*** (for Adenomatous Polyposis Coli). As in the case of the *RB* gene, the two-hit model applies to the behavior of the *APC* gene. Individuals with familial adenomatous polyposis inherit a single defective or missing copy of *APC* from one parent, and an inactivating mutation then occurs in the second copy of the gene in a colon epithelial cell. The resulting loss of *APC* function leads to uncontrolled cell proliferation that can set the stage for cancer development. The *APC* gene, like the *RB* gene, is therefore an example of a tumor suppressor gene—that is, a gene whose loss or inactivation is associated with the development of cancer.

The *APC* gene normally exerts its restraining effects on cell proliferation by producing a protein, also called APC, that inhibits the *Wnt signaling pathway*. ("Wnt" is an abbreviation for "wingless-type," which refers to the fate of fruit flies with mutations in this pathway.) The Wnt pathway plays a prominent role in regulating cell proliferation and differentiation, especially during embryonic development. When *APC* mutations arise that cause a loss of functional APC protein, the resulting uncontrolled activity of the Wnt pathway leads to enhanced cell proliferation. Epithelial cells lining the colon are especially sensitive to such effects. In people who inherit an *APC* mutation, excessive proliferation of the colon epithelium leads to the formation of numerous polyps and, sooner or later, colon cancer. This connection between *APC* mutations and colon cancer is not restricted to hereditary colon cancers; mutations in *APC* also occur in about two-thirds of the more common forms of colon cancer that arise in people with no family history of the disease. We will return to a discussion of the *APC* gene and its mechanism of action in Chapter 10, where the Wnt pathway is described in detail (see Figure 10-8).

The Li-Fraumeni Syndrome Is Caused by Inherited Defects in the *p53* Gene

The two hereditary syndromes described thus far each involve an inherited risk for one main type of cancer, either retinoblastoma for individuals inheriting a mutant *RB* gene or colon cancer for those inheriting a mutant *APC* gene. In the hereditary condition to be described next, called the **Li-Fraumeni syndrome**, susceptibility to cancer in general rather than to a specific type of cancer is transmitted from parent to offspring. The family pedigree shown in Figure 8-5 illustrates such a case. The woman marked by the arrow is typical of someone who might seek genetic counseling because of a fear that "cancer runs in the family." Her father had colon cancer when he was 40, a sarcoma at age 45, and lung cancer

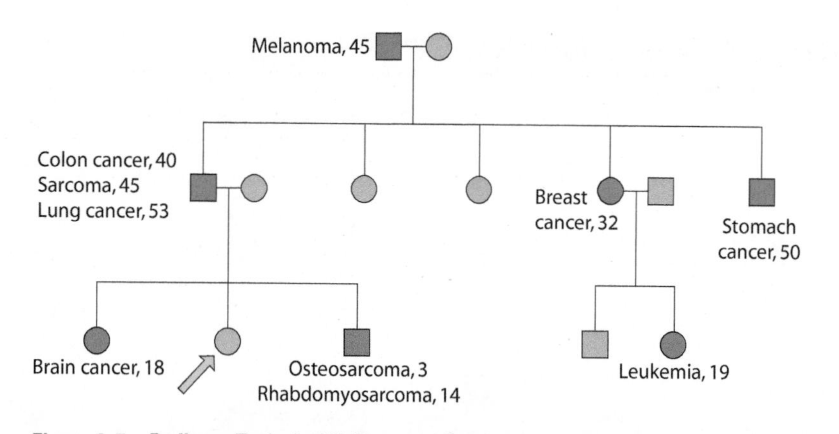

Figure 8-5 Pedigree Typical of Li-Fraumeni Syndrome. In families with Li-Fraumeni syndrome, roughly half of the offspring develop cancer (*red symbols*). The cancers are of a wide variety of types, and a single person may develop several different types of cancer in succession. The arrow points to the person described in the text who sought genetic counseling because of a suspicion that "cancer runs in the family." The numbers represent the ages at which cancer developed in each individual.

when he was 53; her sister had brain cancer as a teenager; her brother developed an osteosarcoma at age 3 and rhabdomyosarcoma (skeletal muscle cancer) at age 14; a cousin had leukemia as a teenager; an aunt died of breast cancer; an uncle had stomach cancer; and her grandfather died of melanoma.

What clearly distinguishes this scenario from other hereditary syndromes is the variety of cancers involved. Individuals inheriting a mutant *RB* or *APC* gene have roughly a 90% chance of developing a single type of cancer, either retinoblastoma in the case of *RB* or colon cancer in the case of *APC*. The mutation responsible for Li-Fraumeni syndrome also confers about a 90% risk of developing cancer, but no single cancer predominates. Cancers commonly associated with Li-Fraumeni syndrome include osteosarcomas, breast cancers, leukemias, adrenal carcinomas, brain tumors, soft tissue sarcomas, melanomas, and cancers of the stomach, colon, and lung. These cancers arise in adults as well as children, but the age of onset is usually earlier than would be typical for the particular type of cancer involved. A single individual may develop several different kinds of cancer in succession, which rarely happens with nonhereditary cancers.

As you might expect from the diversity of the cancers involved, Li-Fraumeni syndrome arises from defects in a gene that is critical for the control of cell proliferation and survival in many tissues. The gene in question is the ***p53* gene** (also called ***TP53*** in humans). We have already encountered the *p53* gene, and the p53 protein it produces, in several different contexts. Chapter 2 introduced the role of the p53 protein in the pathway that causes cells with damaged DNA to self-destruct by apoptosis (p. 27); Chapter 6 described how sunlight-induced mutations in the *p53* gene trigger the development of skin cancer (p. 106); and Chapter 7 showed how a viral oncoprotein produced by the human papillomavirus binds to and promotes destruction of the p53 protein, thereby contributing to the development of cervical cancer (p. 135). The diversity of these examples suggests that the p53 protein is broadly important in protecting against the development of cancer.

It is therefore not surprising that individuals with Li-Fraumeni syndrome, who inherit a mutation in one copy of the *p53* gene, are at great risk of developing cancer. The chances are about one in a million per dividing cell that a mutation in the second copy of the *p53* gene will arise. Since billions of cells divide throughout the body during a person's lifetime, the probability is extremely high that a mutation disrupting the second, good copy of the *p53* gene will appear in at least one cell somewhere in the body. Since cancers develop when *p53* function is lost, the *p53* gene is another example (like *RB* and *APC*) of a tumor suppressor gene.

Li-Fraumeni syndrome is an extremely rare condition; only a few hundred families have been diagnosed with the condition worldwide. That does not mean, however, that *p53* mutations are rare in human cancers. To the contrary, *p53* mutations are detected in almost half of all nonhereditary cancers, making it the most commonly mutated gene in human cancer. In some cases, a linkage between *p53* mutations and specific environmental carcinogens has been clearly established. For example, the role of sunlight-induced *p53* mutations in nonmelanoma skin cancers was already discussed in Chapter 6. In addition, about half of all lung cancers have been found to exhibit *p53* mutations in DNA bases that are attacked by the polycyclic hydrocarbons present in tobacco smoke. Carcinogen-induced mutations in *p53* are also common in many other types of cancer, including cancers of the colon, pancreas, ovary, bladder, liver, stomach, and breast. The role played by the *p53* gene in cancer development will be discussed further in Chapter 10, which describes the mechanism of action of the p53 protein in detail (see Figure 10-5).

Table 8-1 Some Examples of Hereditary Cancer Syndromes

Gene Class/ Syndrome Name	Mutant Genes	Inheritance Pattern of Cancer Risk Syndrome*	Main Types of Cancer
Tumor Suppressor Genes (loss-of-function mutations in *gatekeeper genes* that restrain cell proliferation)			
Cowden syndrome	*PTEN*	Dominant	Breast, thyroid
Familial adenomatous polyposis	*APC*	Dominant	Colon
Familial retinoblastoma	*RB*	Dominant	Retinoblastoma, osteosarcoma
Li-Fraumeni syndrome	*p53*	Dominant	Sarcomas, breast, leukemias, adrenal, brain, others
Neurofibromatosis type 1	*NF1*	Dominant	Neurofibrosarcoma, glioma
Neurofibromatosis type 2	*NF2*	Dominant	Meningioma (benign)
Von Hippel-Lindau	*VHL*	Dominant	Kidney
Wilms tumor	*WT*	Dominant	Kidney
Tumor Suppressor Genes (loss-of-function mutations in *caretaker genes* involved in DNA repair)			
Ataxia telangiectasia	*ATM*	Recessive	Lymphoma
Bloom syndrome	*BLM*	Recessive	Multiple
Familial breast cancer	*BRCA1, BRCA2*	Dominant	Breast, ovary
Fanconi anemia	*FANCA, FANCB, FANCC, FANCD2, FANCE, FANCF, FANCG, FANCI, FANCJ, FANCL, BRCA2*	Recessive	Leukemia
Hereditary nonpolyposis colon cancer	*MSH2, MSH3, MSH4, MSH5, MSH6, PMS1, PMS2, MLH1*	Dominant	Colon
Xeroderma pigmentosum	*XPA, XPB, XPC, XPD, XPE, XPF, XPG, XPV (POLη)*	Recessive	Skin
Proto-oncogene (gain-of function mutations)			
Multiple endocrine neoplasia type II	*RET*	Dominant	Thyroid

*In this table, the terms *dominant* and *recessive* refer to whether one or both copies of the mutant gene must be inherited to create the high-risk *cancer susceptibility syndrome*, not cancer itself.

Hereditary Cancer Syndromes Arising from Defects in Genes That Restrain Cell Proliferation Share Several Features in Common

The three inherited cancer syndromes described thus far—familial retinoblastoma, familial adenomatous polyposis, and Li-Fraumeni syndrome—share several features in common. (1) Each is caused by a mutation in a single tumor suppressor gene (*RB*, *APC*, or *p53*) inherited from one parent. (2) The elevated risk of developing cancer is a *dominant* trait because only a single mutation needs to be inherited to impart a very high risk of developing cancer, usually in the range of 90% to 100%. (3) For cancer to actually arise, the second copy of the gene must also become inactivated (the "second hit" of the two-hit model). The likelihood of that happening in at least one dividing cell is very high. (4) If a person with one of these hereditary syndromes has children, each child will have a 50:50 chance of inheriting the gene defect and its associated cancer risk (see Figure 8-3, *left*). (5) Children who are among the 50% that do not receive the genetic defect will not be at increased risk for cancer and will not pass the risk on to their children. (6) The gene mutations responsible for these hereditary cancer syndromes involve the inactivation or loss of a tumor suppressor gene whose normal role is to restrain cell proliferation. (7) Carcinogen-induced mutations in the same tumor suppressor genes can cause nonhereditary cancers.

In addition to *RB*, *APC*, and *p53*, several other tumor suppressor genes behave according to the same principles (Table 8-1, *top*). However, even though these genes all share the property of restraining cell proliferation and survival, they differ significantly in the molecular mechanisms involved. A detailed description of some of these mechanisms, and the signaling pathways that are affected, is provided in Chapter 10.

HEREDITARY RISK: GENES AFFECTING DNA REPAIR AND GENETIC STABILITY

The hereditary cancer syndromes discussed thus far are caused by loss-of-function mutations in tumor suppressor genes whose normal role is to restrain cell proliferation and survival. A second group of tumor suppressor genes act through a fundamentally different set of mechanisms involving DNA repair and chromosome stability. Whereas the first group of tumor suppressors produce proteins involved in the control of cell proliferation, and whose loss therefore leads *directly* to tumor formation, the loss of genes involved in repairing DNA or maintaining chromosome stability acts *indirectly* by permitting an increased mutation rate for all genes. This increased mutation rate increases the likelihood that random mutations will disrupt genes that do impact cell proliferation directly.

The terms **gatekeepers** and **caretakers** are used to distinguish between the two classes of tumor suppressors. Genes like *RB*, *APC*, and *p53*, whose normal role is to restrain cell proliferation, are considered to be "gatekeepers" because their loss opens the gates to tumor formation directly. Genes involved in DNA maintenance and repair, on the other hand, are viewed as "caretakers" that preserve the integrity of a cell's **genome** (its total genetic information). The loss of a caretaker gene does not directly affect cell proliferation. Instead, it leads to disruptions in DNA maintenance and repair that cause an increased mutation rate for all genes (including gatekeepers), and it is only through subsequent mutation of these additional genes that cancer arises. We will now examine some of the hereditary cancer syndromes that are caused by mutations in caretaker genes.

Figure 8-6 A Child with Xeroderma Pigmentosum Playing Outdoors. Exposure to daylight is dangerous for individuals with xeroderma pigmentosum, so NASA has designed a special space suit to protect them from the ultraviolet radiation emitted by the sun. [Courtesy of the *Orlando Sentinel*.]

Xeroderma Pigmentosum Is an Inherited Sensitivity to Sunlight-Induced Skin Cancer

The first reports of a connection between DNA repair and cancer susceptibility came from the study of a rare hereditary disease known as **xeroderma pigmentosum**. Individuals with this condition are so sensitive to ultraviolet radiation that sunlight can be fatal. Exposure to even a few minutes of daylight is enough to cause severe burning and various skin cancers, including basal cell carcinoma, squamous cell carcinoma, and melanoma. Because it is only safe for them to go outside at night, children with xeroderma pigmentosum are sometimes referred to as "children of the moon." There is even a special camp in upstate New York, called *Camp Sundown*, that allows affected children to participate in normal recreational activities but on a different time clock: Outdoor activities begin after sundown and take place during the night, allowing the children to be back indoors and safely behind drawn curtains by sunrise. Scientists from NASA have taken an interest in the problem, working on the design of a special space suit that provides enough sunlight protection to allow children with xeroderma pigmentosum to play outdoors occasionally (Figure 8-6).

Xeroderma pigmentosum does not exhibit the same inheritance pattern as the previously discussed cancer syndromes, which involved a single mutation inherited from a person's mother or father. Xeroderma pigmentosum requires two mutant copies of the same gene to be inherited, one from each parent. Because it is unlikely that both parents will carry a mutant version of the same gene, xeroderma pigmentosum is a very rare condition. However, if each parent does happen to carry a mutation in the same responsible gene (along with a second, normal copy), their children will have a 50% chance of inheriting the mutant version from their father and a 50% chance of inheriting a mutant version from their mother. The overall probability that any given child will inherit both mutations and have xeroderma pigmentosum is therefore 50% × 50% = 25%.

Since two mutant copies of the same gene must be inherited to cause the disease, xeroderma pigmentosum exhibits a recessive pattern of inheritance. Family pedigrees for xeroderma pigmentosum therefore look different from the pedigrees for familial retinoblastoma and Li-Fraumeni syndrome shown earlier in the chapter. In familial retinoblastoma and Li-Fraumeni syndrome, which exhibit a dominant pattern of inheritance, parents who transmit the high-risk trait to their children are at high risk for cancer themselves because the trait is dominant, and 50% of their children will acquire the high-risk condition (see Figures 8-2 and 8-5). With cancer syndromes that exhibit a recessive pattern of inheritance, such as xeroderma pigmentosum, the parents do not have a high cancer risk because they each carry only a single recessive gene; the high-risk condition will, however, appear in 25% of their offspring (Figure 8-7).

Despite their differing patterns of inheritance, the underlying behavior of the tumor suppressor genes involved in recessive and dominant cancer syndromes is not as different as these distinctions seem to imply. With recessive cancer syndromes, a high risk for cancer arises when two defective copies of a gene are inherited and normal gene function is therefore lost (Figure 8-8a). With dominant cancer syndromes (Figure 8-8b), inheriting one defective copy imparts a high risk of developing cancer, but cancer only arises after the second copy is mutated or lost (the "second hit" of the two-hit model). The final

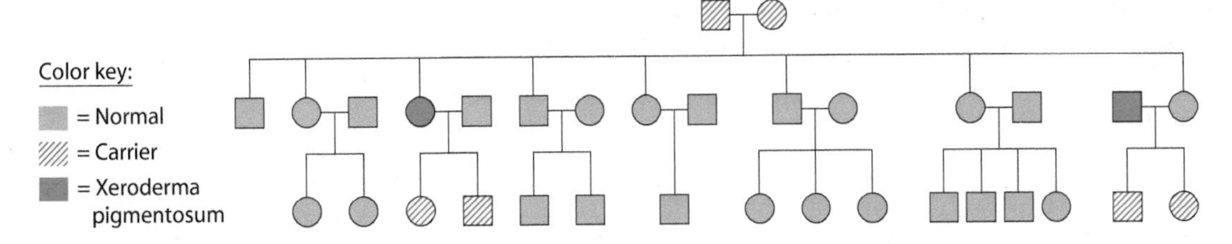

Figure 8-7 Pedigree Typical of Xeroderma Pigmentosum. Xeroderma pigmentosum requires two mutant copies of the same gene to be inherited, one from each parent. If both parents carry such a single mutation, they will exhibit no disease symptoms but will function as "carriers" who can pass the gene along to children. Each child will have a 50% chance of inheriting the mutant version from each parent, so the overall probability of inheriting both mutations and developing xeroderma pigmentosum is 50% × 50% = 25%. In the first generation of offspring shown in this pedigree, two of eight children develop xeroderma pigmentosum (*red symbols*). Because both copies of the responsible gene are defective in these two individuals, all their children will receive a defective copy and will therefore become carriers (four *striped symbols* in the second generation of offspring). Some of the offspring in the first generation may have received a single mutant gene and may therefore be carriers, but that cannot be determined from such a pedigree.

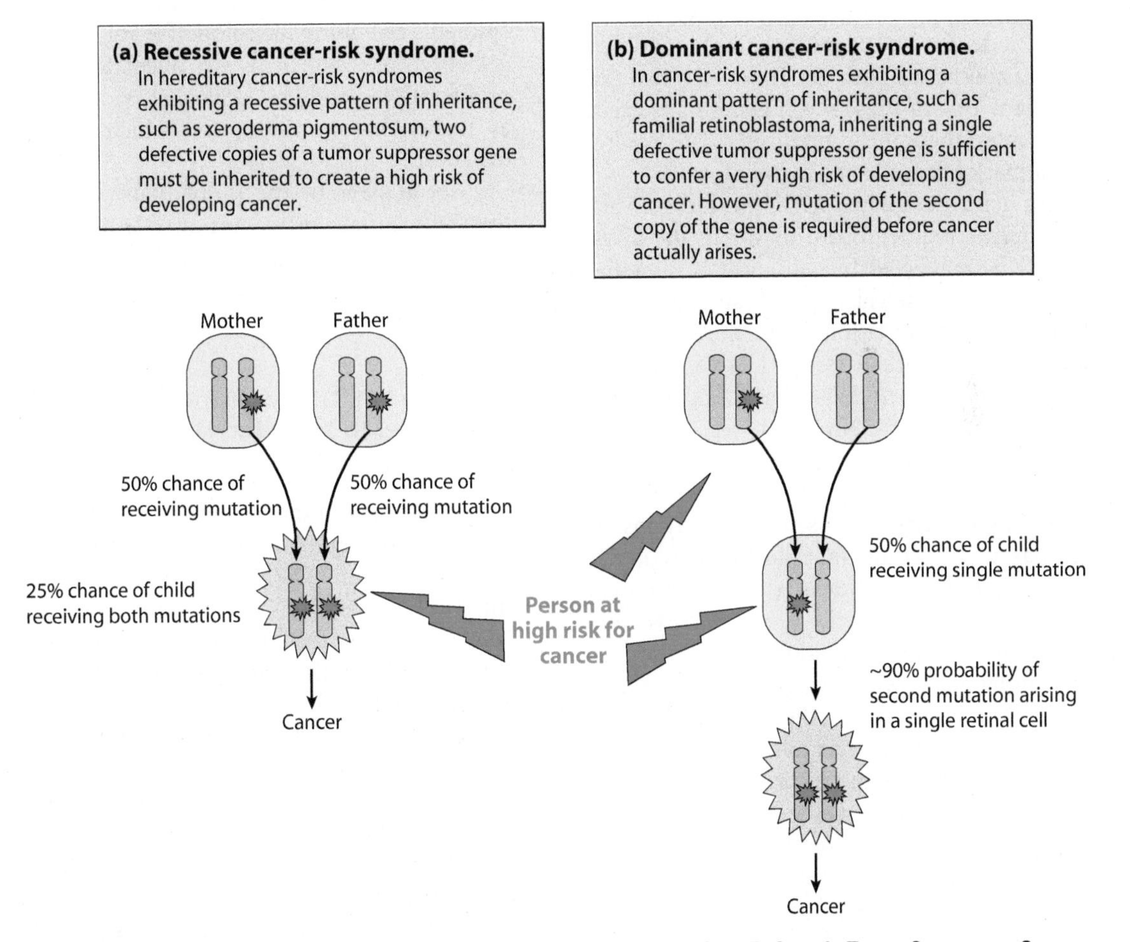

Figure 8-8 Inheritance Patterns of High-Risk Cancer Syndromes Arising from Defects in Tumor Suppressor Genes. Recessive cancer-risk syndromes require inheritance of two defective copies of a tumor suppressor gene to create a high cancer risk, whereas dominant cancer-risk syndromes require inheritance of only a single defective tumor suppressor gene to create a high cancer risk. Despite these differing patterns of inheritance, the behavior of the underlying tumor suppressor genes is comparable. In each case, cancer only arises after both copies of the gene become defective. So in both situations, the behavior of the tumor suppressor gene itself is recessive; that is, both copies must be disrupted before cancer arises.

result is therefore similar in both cases; that is, both copies of a tumor suppressor gene lose their function.

Xeroderma Pigmentosum Is Caused by Inherited Defects in Excision Repair

The susceptibility to skin cancer that is the hallmark of xeroderma pigmentosum has been traced to inherited defects in DNA repair. In Chapter 6, we saw that ultraviolet radiation absorbed by skin cells leads to DNA mutations—especially pyrimidine dimers—that can cause cancer if the mutations are not repaired. One mechanism for repairing pyrimidine dimers is **excision repair**, a pathway that recognizes major distortions in the DNA double helix and uses a series of enzymes to excise the damaged region and fill the resulting gap with the correct sequence of nucleotides (see Figure 2-16).

It was first reported in the late 1960s that cells from individuals with xeroderma pigmentosum are unable to perform excision repair. DNA mutations therefore accumulate and cancer eventually arises. Subsequent studies have revealed that mutations in seven different genes can cause xeroderma pigmentosum through their effects on excision repair. Each of these seven genes, designated *XPA* through *XPG*, codes for an enzyme involved in a different step of the excision repair pathway. Inheriting two defective copies of any one of these seven genes halts excision repair and creates the cancer predisposition syndrome that is the hallmark of xeroderma pigmentosum.

An eighth gene, designated *XPV*, produces a variant form of xeroderma pigmentosum in which the excision repair pathway is unaffected but individuals nonetheless inherit an increased sensitivity to sunlight-induced cancers. This particular mutation affects the enzyme DNA polymerase η (eta), a special form of DNA polymerase that catalyzes *translesion synthesis*—that is, the synthesis of new error-free stretches of DNA across regions in which the template strand is damaged (p. 32). DNA polymerase η is capable of accurately replicating DNA in regions where pyrimidine dimers are present, correctly inserting the proper bases. Therefore, inherited defects in DNA polymerase η, like inherited defects in excision repair, interfere with the ability of cells to correct pyrimidine dimers created by ultraviolet radiation.

Because individuals with xeroderma pigmentosum cannot repair pyrimidine dimers, they exhibit skin cancer rates that are 2000-fold higher than normal. Moreover, their average age for developing skin cancer is 8 years old, compared with age 60 for the general population. The defects in DNA repair associated with xeroderma pigmentosum also cause a 20-fold increase in cancers of the brain, lung, stomach, breast, uterus, and testes, as well as leukemias.

Hereditary Nonpolyposis Colon Cancer Is Caused by Inherited Defects in Mismatch Repair

Earlier in the chapter we discussed familial adenomatous polyposis, a colon cancer syndrome caused by inheriting a mutant copy of the *APC* gene. A second hereditary syndrome, called **hereditary nonpolyposis colon cancer (HNPCC)**, also increases a person's risk of developing colon cancer, in this case because of an inherited defect in DNA repair. Cancer again tends to arise from colon polyps, although people with HNPCC typically have only a few polyps rather than the hundreds or thousands of polyps observed with familial adenomatous polyposis. (*Polyposis* literally means "numerous polyps", so the presence of only a few polyps in HNPCC explains why the term *nonpolyposis* is included as part of its name.)

The inherited mutations responsible for HNPCC disrupt the **mismatch repair** pathway (p. 33), which is normally responsible for correcting errors involving bases that are incorrectly paired between the two strands of the DNA double helix. Mismatch repair requires the participation of many different proteins, and defects in genes coding for at least eight of these proteins have been implicated in HNPCC. The disease exhibits a dominant pattern of inheritance, which means that inheriting a single mutant copy of any of the eight genes is sufficient to create an increased susceptibility to developing colon cancer. However, cancer only arises if the second, functional copy of the affected gene becomes mutated in a proliferating epithelial cell lining the colon. If that happens, the affected cell becomes deficient in mismatch repair and its proliferation produces cells that accumulate mutations at higher-than-normal rates, which can in turn lead to cancer.

About 75% of the individuals who inherit one of the eight mutant genes responsible for HNPCC eventually develop colon cancer. Smaller increases in risk are also observed for cancers of the uterus, ovary, stomach, and kidney.

Mutations in the *BRCA1* and *BRCA2* Genes Are Linked to Inherited Risk for Breast and Ovarian Cancer

Among women in the Western world, breast cancer is the second most frequent type of cancer (next to skin cancer) and the second most common cause of cancer deaths (next to lung cancer). Current statistics suggest that about 1 in every 8 women in the United States will develop breast cancer within her lifetime. Breast cancer risk is roughly doubled in women who have had a close blood relative with the disease (mother, sister, or daughter), and two close relatives increases the risk about fivefold. Risk is also increased by a history of breast cancer in more distant family members (e.g., aunt or grandmother), which may be on either the mother's or father's side of the family. However, having a close relative with breast cancer does not always mean that the disease runs in the family. Breast cancer is a common disease, and many women will have a relative with breast cancer solely by chance.

About 10% of all breast cancer cases can be traced to a high-risk hereditary predisposition, usually created by an inherited mutation in either the ***BRCA1*** or ***BRCA2*** **gene** (BRCA is an abbreviation for BReast CAncer). Inherited

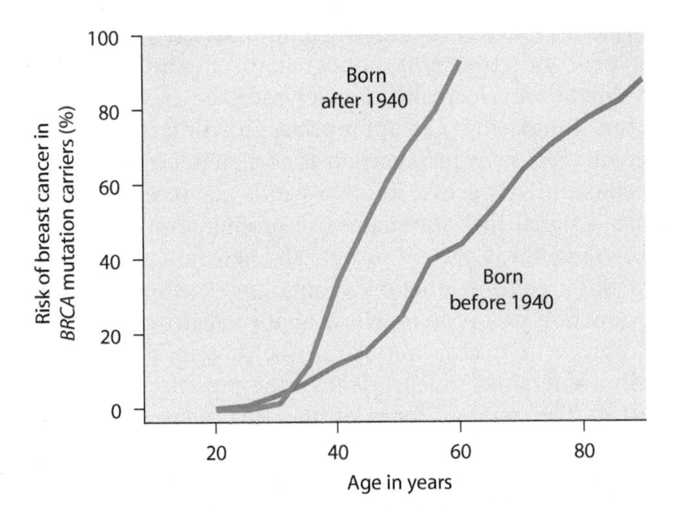

Figure 8-9 Effect of Birth Year on Breast Cancer Risk from *BRCA* Mutations. Women born after 1940 who carry mutations in the *BRCA1* or *BRCA2* genes have higher breast cancer rates (at any given age) than women born before 1940 who carry the same mutations. This difference indicates that nongenetic factors influence breast cancer risk, even in individuals who inherit high-risk mutations. [Data from M.-C. King et al., *Science* 302 (2003): 643 (Figure 1D).]

defects in these genes are responsible for *familial breast cancer*, which is characterized by the onset of breast cancer at an early age and, in some cases, cancer in both breasts or breast and ovarian cancer in the same individual. Depending on the exact nature of the particular *BRCA1* or *BRCA2* mutation that is inherited, risk typically falls in the range of 40% to 80% for breast cancer and 15% to 65% for ovarian cancer. Interestingly, women born after 1940 who carry mutations in the *BRCA1* or *BRCA2* genes have a higher risk of developing breast (but not ovarian) cancer on an age-adjusted basis than do women with the same mutations who were born before 1940 (Figure 8-9). Such findings indicate that nongenetic factors significantly affect the risk of developing breast cancer, even in individuals who inherit a high-risk predisposition.

When the *BRCA1* and *BRCA2* genes were first discovered, they were thought to play a role in the control of cell proliferation. Later studies revealed, however, that the proteins produced by these genes are involved in repairing DNA damage, especially double-strand breaks. It has therefore been concluded that *BRCA1* and *BRCA2* mutations do not directly open the gates to excessive cell proliferation but instead hamper the process of DNA repair, thereby increasing the rate of subsequent mutations that can lead to cancer. This conclusion is consistent with the finding that mutations involving the classic "gatekeeper" tumor suppressors, such as *RB*, *APC*, and *p53*, are observed in both hereditary and nonhereditary cancers because they act directly on cell proliferation and directly open the gates to uncontrolled proliferation and cancer. Mutations involving *BRCA1* and *BRCA2*, on the other hand, are rarely seen in nonhereditary cancers, presumably because they are "caretaker" genes that do not directly control cell proliferation and so are only indirectly related to cancer development. We will return to the *BRCA* genes in Chapter 10, where their role in DNA repair will be described in more detail (see Figure 10-13).

Inherited Defects in DNA Repair Underlie Ataxia Telangiectasia, Bloom Syndrome, and Fanconi Anemia

Inherited defects in DNA repair are responsible for several additional high-risk cancer syndromes in addition to the ones described thus far. Although they all arise from inherited defects in DNA repair, these syndromes exhibit a remarkable diversity of symptoms. One striking example is an inherited condition known as **ataxia telangiectasia** (*ataxia* = "lack of coordination," *telangiectasia* = "dilation of capillaries"). The first abnormality seen in children with ataxia telangiectasia, usually arising between 1 and 3 years of age, is an inability to walk steadily caused by degeneration of the cerebellum, which is the part of the brain that governs muscle coordination and balance. Other symptoms include abnormal eye movements, slurred speech, retarded growth, recurrent infections arising from a deficient immune system, and red marks on the skin and inner surface of the eyelids caused by abnormal dilation of small blood vessels. The preceding symptoms are accompanied by a roughly 40% risk of developing cancer, mostly lymphomas and leukemias, but also cancers of the skin, breast, stomach, pancreas, ovary, and brain.

The gene responsible for ataxia telangiectasia, called the ***ATM* gene** (for Ataxia Telangiectasia Mutated), exhibits a recessive pattern of inheritance in which two mutant *ATM* genes must be inherited, one from each parent, to create the syndrome. Neither parent is likely to exhibit any disease symptoms because they each carry a single mutant *ATM* gene whose effects are recessive and thus are not expressed. The *ATM* gene codes for a protein involved in the **DNA damage response**, an intricate network of cellular pathways invoked as a protective response to assaults on DNA integrity. The ATM protein plays a central role both in detecting the occurrence of DNA damage—especially double-strand breaks—and in activating a cascade of appropriate responses. In Chapter 10, this role of the ATM protein in responding to DNA damage will be described in detail (see Figures 10-5 and 10-13).

Deficiencies in the DNA damage response also occur in several other hereditary syndromes, each of which exhibits a distinctive pattern of symptoms and cancer risks. For example, individuals with **Bloom syndrome** are characterized by short stature, sun-induced facial rashes, immunodeficiency, decreased fertility, and an elevated risk of developing cancer before age 20, most commonly lymphomas, leukemias, and cancers of the mouth, stomach, larynx, lung, esophagus, colon, skin, breast, and cervix. The *BLM* gene, whose mutation causes Bloom syndrome, exhibits a recessive pattern of inheritance and codes for a *DNA helicase*, which is a protein that unwinds the DNA double helix during the repair of double-strand breaks and other types of DNA damage.

In **Fanconi anemia**, the main symptom is the inability of the bone marrow to produce a sufficient number of blood cells, accompanied by skeletal malformations, organ deformities, reduced fertility, and marked predisposition to developing leukemias and squamous cell carcinomas. Fanconi anemia exhibits a recessive pattern of inheritance and is caused by inactivating mutations in any of at least 11 different genes. The proteins produced by these genes interact with one another and with various components of DNA damage response pathways, including the proteins produced by the *ATM* and *BRCA1* genes. One of the 11 genes that can cause Fanconi anemia is *BRCA2*, the same gene that causes familial breast cancer. *BRCA2* exhibits a dominant pattern of inheritance for familial breast cancer and a recessive pattern of inheritance for Fanconi anemia. In other words, the familial breast cancer syndrome arises from the inheritance of a single mutant allele of *BRCA2*, and Fanconi anemia arises from the inheritance of two mutant alleles.

HEREDITARY RISK: OTHER GENES AND ISSUES

The high-risk cancer syndromes discussed thus far all stem from mutations in tumor suppressor genes, either gatekeeper genes involved in controlling cell proliferation and survival or caretaker genes involved in DNA maintenance and repair. Tumor suppressors, however, are not the only genes that affect hereditary cancer risk. The remainder of the chapter will briefly consider several other types of genes that influence cancer susceptibility and will then conclude with a brief discussion of genetic testing.

Multiple Endocrine Neoplasia Type II Is Caused by an Inherited Mutation in a Proto-oncogene

The hereditary cancer syndromes we have been considering so far all involve *loss-of-function* mutations—that is, on mutations that inactivate or delete genes, or cause them to produce nonfunctional products. By definition, the affected genes are tumor suppressor genes because a tumor suppressor gene is defined as a gene whose loss of function leads to cancer. *Gain-of-function* mutations, which cause a gene to produce a protein exhibiting new or excessive activity, can also influence hereditary cancer risk. Perhaps the best characterized syndrome that works in this way is **multiple endocrine neoplasia type II**, an inherited condition associated with the development of both benign and malignant tumors of endocrine glands. The disease usually starts in childhood and is characterized by developmental abnormalities stemming from the overproduction of specific hormones. About 70% of affected individuals develop cancer by age 70, usually thyroid cancer.

Multiple endocrine neoplasia type II is caused by inheriting a single mutant ***RET* gene**, which codes for the Ret receptor protein. The Ret receptor is located on the surface of endocrine cells, where it binds external growth factors and transmits a signal that stimulates cell proliferation. Normally the Ret receptor is only active when stimulated by an appropriate growth factor, but the mutant *RET* gene produces an abnormal Ret receptor that is constitutively active; in other words, the receptor transmits a signal that stimulates cell proliferation whether a growth factor is present or not. The new function created in this "gain-of-function" mutation is therefore the production of a protein whose abnormal structure allows it, unlike the normal version of the protein, to stimulate cell proliferation independent of the presence of growth factor. The mutant form of the *RET* gene is thus an **oncogene** because its presence can lead to cancer, and the normal *RET* gene is a **proto-oncogene** because it is closely related to, and can be converted into, an oncogene.

Figure 8-10 summarizes the fundamental difference in the behavior of mutations involving proto-oncogenes and tumor suppressor genes. In essence, tumor suppressor genes are genes in which loss-of-function mutations lead to cancer, and proto-oncogenes are genes in which gain-of-function mutations lead to cancer. To lose the function of a tumor suppressor gene, both copies must usually undergo an inactivating mutation (or deletion). In contrast, mutation of one copy of a proto-oncogene can be sufficient to create an oncogene whose presence contributes to the development of cancer.

Inherited Variations in Immune Function and Metabolic Enzymes Can Influence Cancer Risk

The role played by the immune system in helping protect against cancer has been mentioned in several earlier contexts. For example, we have seen that cancer risk increases when immune function is disrupted either through HIV infection or in organ transplant recipients who use immunosuppressive drugs (see Figures 7-5 and 2-22). It is therefore not surprising that inherited deficiencies in immune function can increase cancer risk as well. In some cases, disruptions in immune function are an indirect effect of inherited mutations whose primary effect is not on the immune system itself. Examples include ataxia telangiectasia and Bloom syndrome, which arise from inherited defects in DNA repair that trigger increased mutation rates but also impair the ability of lymphocytes to perform their normal immune functions. The resulting loss in immune activity may contribute to the elevated cancer rates seen in such syndromes, along with the primary role played by defective DNA repair in increasing mutation rates in the cells that are actually destined to become cancerous.

Primary immunodeficiency diseases, caused by inherited mutations that directly incapacitate the immune system, are likewise associated with modest increases in cancer risk, although an increased susceptibility to infections is usually the main symptom. The most frequently encountered cancers are lymphomas, often triggered by EBV infections that proceed unrestrained in

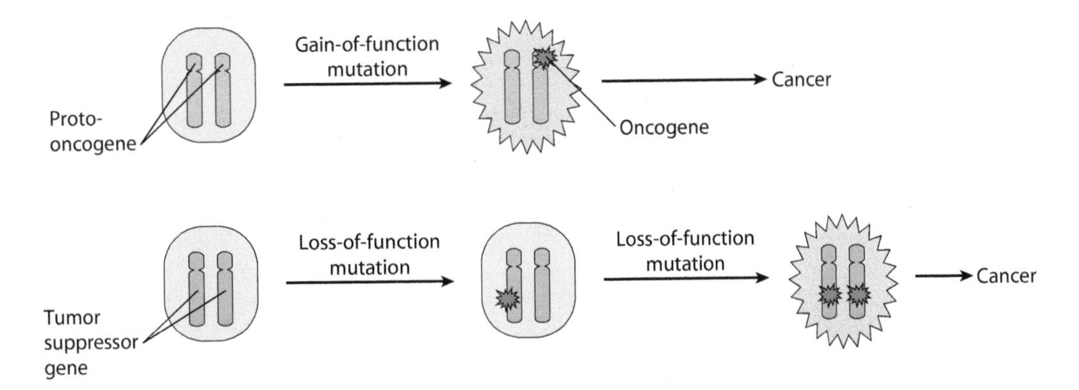

Figure 8-10 Behavior of Proto-oncogenes Compared with Tumor Suppressor Genes. (*Top*) A proto-oncogene is a gene whose mutation converts it into an oncogene that can trigger the development of cancer. Because only a single oncogene is required to produce this effect, oncogenes are said to exhibit dominant behavior. (*Bottom*) A tumor suppressor gene is a gene whose loss or inactivation is associated with the development of cancer. To lose the function of a tumor suppressor, both copies of the gene must become nonfunctional. Tumor suppressor genes are therefore recessive. Recall, however, that cancer-risk syndromes involving tumor suppressor genes can exhibit either a dominant or recessive pattern of inheritance (see Figure 8-8). The explanation for this apparent paradox is that inheriting a single defective tumor suppressor gene can be sufficient to create a high-risk predisposition to developing cancer, even though the second gene must still undergo mutation for cancer to actually arise.

the absence of an effective immune response. (Recall from Chapter 7 that immunodeficiencies caused by malaria or HIV are similarly associated with an increased incidence of EBV-induced lymphomas.)

Inherited deficiencies in immune function have an indirect effect on cancer risk rather than directly targeting the cells destined to form tumors. Another type of indirect effect stems from hereditary differences in the liver enzymes that metabolize foreign chemicals. As we saw in Chapter 5, some carcinogens must be metabolically activated by cytochrome P450 enzymes in the liver before they can trigger mutations and cancer. Because of this requirement, inherited differences in liver enzymes may influence a person's susceptibility to developing cancer. For example, cigarette smokers who inherit certain forms of cytochrome P450 have been found to exhibit higher lung cancer rates than those exhibited by other smokers (p. 92).

Cancer Susceptibility Is Influenced by Inheritance of Small-Risk as Well as High-Risk Genes

High-risk mutations—that is, mutations responsible for the conspicuous clustering of cancer within families—account for no more than 5% of cancer cases overall. However, the role of heredity in determining cancer risk is not limited to high-risk mutations; it also includes the smaller contributions of many other genes. The individual effects of these small-risk genes are not strong enough to create obvious patterns of cancer inheritance, but as a group, their effects on cancer susceptibility can be significant.

Small-risk genes are difficult to identify because they do not create obvious family patterns of cancer inheritance. Identifying such genes is also complicated by the fact that cancer arises through an interaction between heredity and environment. With high-risk inherited mutations, the impact on cancer rates is so great that it overshadows environmental factors. The effects of small-risk genes, on the other hand, are easily obscured by environmental or lifestyle conditions. For example, suppose a variant form of a gene were to exist that doubles a person's risk of developing lung cancer. In practice, it would be difficult to detect the existence of such an increase because differences in smoking behavior (or radon or asbestos exposure) have a much larger influence on lung cancer rates and would therefore tend to obscure the effects of the smaller-risk genetic factor.

Differences that have been observed in the cancer rates of various racial and ethnic groups illustrate how hard it can be to determine the exact role played by heredity. Consider, for example, the black and white populations of the United States, which exhibit several distinctive differences in cancer patterns. One pattern that is easy to explain involves melanoma, a form of skin cancer whose incidence is about 15 times higher among whites than blacks. The explanation for this difference was discussed in Chapter 6 and is quite straightforward: People with darkly pigmented skin produce large amounts of *melanin*, a pigment that absorbs ultraviolet radiation and thereby lowers sunlight-induced mutation rates in the skin.

Other racial differences in cancer rates are not as easily explained. For example, blacks living in the United States have higher cancer rates than whites for most common cancers, including colon, lung, and prostate cancers, although not breast cancer (Figure 8-11, *left*). In theory, such differences might be caused by subtle variations in small-risk cancer genes, but nonhereditary

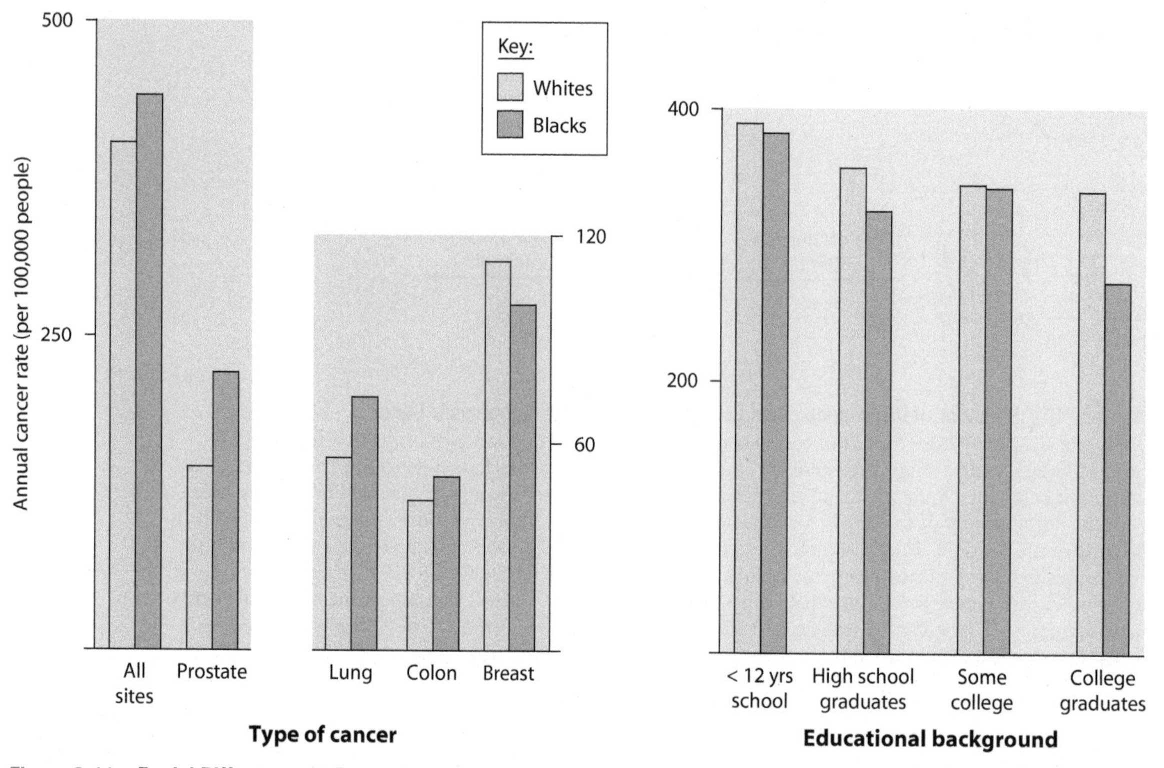

Figure 8-11 Racial Differences in Cancer Incidence. (*Left*) Blacks living in the United States have higher cancer rates than whites for most common cancers, including prostate, lung, and colon cancers, although not breast cancer. (*Right*) When cancer rates are compared among individuals with comparable levels of education, blacks have similar (or even lower) cancer rates than whites. Thus many of the disparities in cancer incidence observed among blacks and whites are probably linked to socioeconomic variables rather than to genetic differences. [Based on data from M. J. Thun and P. A. Wingo in *Holland-Frei Cancer Medicine*, 5th ed. (R. C. Bast et al., eds., Lewiston, NY: Decker, 2000), Chapter 23 (Tables 23.8 and 23.9).]

mechanisms are equally plausible. One possibility that has received serious consideration is socioeconomic status. Figure 8-11 (*right*) illustrates the findings of a study that compared cancer rates by race for individuals of varying educational levels, which is an indicator of a person's overall socioeconomic environment. The data show that when cancer rates are compared among individuals with equivalent levels of education, blacks have similar (or even lower) cancer rates than whites. So rather than being related to genetic differences between racial groups, many of the disparities in cancer incidence observed among blacks and whites may be linked to socioeconomic variables. The risk factors associated with lower socioeconomic status that are most likely to affect cancer rates include exposure to tobacco and alcohol, poor nutrition, lack of physical activity, and obesity.

Genetic Testing for Cancer Predisposition Has Benefits as Well as Risks

The main focus of this chapter has been on cancer tendencies that run in families. Of course, if several people in the same family develop cancer, it does not necessarily mean that heredity is responsible. As mentioned in Chapter 1, cancer is a common disease, and the presence of several cases of cancer in a single family may be a coincidence. Shared lifestyle or environmental factors may also be responsible for the clustering of cancer cases within a family. Nonetheless, when people suspect that a genetic predisposition may be involved because cancer rates in their family are unusually high, the answers to several questions can provide clues as to whether the condition is likely to be hereditary: (1) Have several family members developed the *same type of cancer* without an obvious environmental or lifestyle explanation, such as smoking cigarettes? (2) Are family members developing cancer during childhood or earlier in adulthood than is typical? (3) Have individual family members developed multiple primary cancers of the same type or different types of cancer in succession? If the preceding patterns exist, a variety of laboratory tests can be performed to analyze a person's DNA for inherited mutations. Such **genetic testing** is currently available for almost all the cancer-susceptibility genes listed in Table 8-1, and additional tests are being developed at a rapid pace.

Before deciding to pursue genetic testing, it is important to understand the potential risks as well as benefits of testing (Table 8-2). To begin with, the failure to find a cancer-risk mutation only tells you about the gene for which the test is designed, and hence a different mutation could be present that is not detected by the procedures being used. Or, the test might detect a type of mutation that has not been seen before and is thus of uncertain significance. It is therefore important to be prepared for the possibility of ambiguous results.

Table 8-2 Some Possible Risks and Benefits of Cancer Predisposition Testing

Potential Risks	Potential Benefits
Ambiguous results	Ends uncertainty
Results cause anxiety	Possible absence of mutation
Lack of prevention strategies	Improved medical care
Discrimination	Clarifies risk for relatives
Strain on family relationships	Informed reproductive choices

If genetic testing does indicate the presence of a high-risk mutation, this discovery might produce significant fear and anxiety, and may also raise the possibility of discrimination by insurance companies or employers. A strain on family relationships is possible because other family members might not welcome the news that a high-risk cancer gene runs in the family. Moreover, the ability to detect high-risk mutations is not always accompanied by effective medical strategies for treatment and prevention. Some people therefore risk being told that they have a high probability of developing cancer and that little can be done about it.

On the positive side, a benefit of genetic testing is that it ends uncertainty about a person's status, especially if the individual is already worried because cancer seems to run in the family. A test that indicates that high-risk mutations are not present will be especially good news. And for people who discover that they do carry a high-risk mutation, frequent and rigorous medical screening for early diagnosis and, where relevant, preventive therapies or behavioral changes may improve their chances of survival. The results of a genetic test may also help clarify the risks for other family members and can reveal whether the possibility exists of passing on a high-risk cancer gene to one's children.

In the end, deciding whether or not to pursue genetic testing is a personal choice that requires careful homework and a good understanding of your own psychological nature. If two people are confronted with the same exact set of circumstances, the use of genetic testing to peek into the crystal ball and predict the future may be an appropriate choice for one person and not for the other.

Summary of Main Concepts

- The overall impact of heredity on cancer risk is small compared with that of environmental and lifestyle variables, but it can be a major factor for individuals who inherit high-risk mutations.
- Familial retinoblastoma is caused by inheritance of a defective or missing *RB* gene, which creates a roughly 90% risk of developing retinoblastoma. For retinoblastoma to actually arise, the second copy of the *RB* gene must undergo mutation during development of the retina (the "second hit" of the two-hit model). The normal function of the *RB* gene is to restrain cell proliferation by producing the Rb protein, which halts progression through the cell cycle when growth factors are absent. The loss of Rb function allows cell proliferation to proceed unrestrained by the need for growth factors.
- Inheritance of a defective or missing *APC* gene causes familial adenomatous polyposis, which is characterized by a high-risk predisposition for colon cancer. In people with this inherited condition, the inner surface of the colon becomes carpeted with hundreds or thousands of polyps, at least one of which is likely to become malignant by age 40. The *APC* gene codes for a protein that restrains cell proliferation by inhibiting the Wnt signaling pathway; loss of *APC* function allows the Wnt pathway to stimulate cell proliferation without restraint.
- Inheritance of a defective or missing *p53* gene causes Li-Fraumeni syndrome, which is characterized by a high-risk susceptibility to a broad spectrum of cancers. The *p53* gene codes for the p53 protein, a central molecule in the pathway that prevents cells with damaged DNA from proliferating. In addition to its role in Li-Fraumeni syndrome, the *p53* gene is mutated in almost half of all nonhereditary tumors, making it the most frequently mutated gene in human cancer.
- The *RB*, *APC*, and *p53* genes are examples of tumor suppressor genes that produce proteins whose normal role is to restrain cell proliferation; such genes are called "gatekeepers" because their loss directly opens the gates to excessive cell proliferation and tumor formation. A second group of tumor suppressor genes are involved in DNA repair and the maintenance of chromosome stability. Such genes are viewed as "caretakers" that maintain genetic stability and whose loss does not directly influence cell proliferation; instead, defects in caretaker genes lead to the accumulation of other mutations that can in turn trigger excessive cell proliferation and the development of cancer.
- Xeroderma pigmentosum is caused by an inherited defect in DNA repair that makes individuals so sensitive to ultraviolet radiation that exposure to even

a few minutes of daylight can cause severe burning of the skin and skin cancer. Individuals with xeroderma pigmentosum inherit two defective copies of a single gene, one from each parent. The affected gene may be any of seven genes involved in excision repair or the gene coding for DNA polymerase η, an enzyme that can synthesize new stretches of error-free DNA in regions where the template strand is damaged.

- Breast cancer risk is doubled in women who have had a close relative with breast cancer, and two close relatives increases the risk about fivefold. Roughly 10% of all breast cancer cases can be traced to hereditary predisposition, mostly to mutations in the *BRCA1* and *BRCA2* tumor suppressor genes. The proteins produced by these genes are involved in repairing DNA damage, especially double-strand breaks.
- Hereditary nonpolyposis colon cancer is caused by defects in DNA mismatch repair. Other hereditary syndromes stemming from defects in DNA repair include ataxia telangiectasia, Bloom syndrome, and Fanconi anemia.
- Inherited mutations in tumor suppressor genes (either gatekeepers or caretakers) are loss-of-function mutations that inactivate a gene's function. Hereditary cancer risk can also be created by gain-of-function mutations. For example, multiple endocrine neoplasia type II is caused by an inherited gain-of-function mutation in the *RET* proto-oncogene, which codes for a growth factor receptor. The mutation converts the *RET* proto-oncogene into an oncogene that produces an abnormal receptor that is constitutively active, thereby leading to excessive cell proliferation.
- The role of heredity in determining cancer risk includes high-risk mutations that cause cancers to cluster in particular families along with the smaller contributions of other genes whose effects are not strong enough to create obvious patterns of cancer inheritance. People who suspect that high-risk mutations may run in their family may choose to undergo genetic testing, but they should first make sure that they understand the potential risks as well as the expected benefits of such tests.

Key Terms for Self-Testing

Hereditary Risk: Genes Involved in Restraining Cell Proliferation

familial (hereditary) cancer (p. 141)
sporadic (nonhereditary) cancer (p. 141)
diploid (p. 141)
homologous chromosomes (p. 141)
alleles (p. 141)
homozygous (p. 141)
heterozygous (p. 141)
dominant (p. 141)
recessive (p. 141)
penetrance (p. 141)
retinoblastoma (p. 142)
two-hit model (p. 142)
RB gene (p. 143)
tumor suppressor gene (p. 144)
familial adenomatous polyposis (p. 145)
APC gene (p. 145)
Li-Fraumeni syndrome (p. 145)
p53 gene (*TP53* gene) (p. 146)

Hereditary Risk: Genes Affecting DNA Repair and Genetic Stability

gatekeeper (p. 148)
caretaker (p. 148)
genome (p. 148)
xeroderma pigmentosum (p. 148)
excision repair (p. 150)
hereditary nonpolyposis colon cancer (HNPCC) (p. 150)
mismatch repair (p. 150)
BRCA1 and *BRCA2* genes (p. 150)
ataxia telangiectasia (p. 151)
ATM gene (p. 151)
DNA damage response (p. 151)
Bloom syndrome (p. 151)
Fanconi anemia (p. 152)

Hereditary Risk: Other Genes and Issues

multiple endocrine neoplasia type II (p. 152)
RET gene (p. 152)
oncogene (p. 152)
proto-oncogene (p. 152)
genetic testing (p. 154)

Suggested Reading

Hereditary Risk: Genes Involved in Restraining Cell Proliferation

Evans, G. Inherited Predispositions to Cancer. In *The Cancer Handbook* (M. Alison, ed.). London: Nature Publishing Group, 2002, Chapter 4.

Hernandez-Boussard, T. M., and P. Hainaut. A specific spectrum of *p53* mutations in lung cancer from smokers: Review of mutations compiled in the IARC *p53* database. *Environ. Health Perspect.* 106 (1998): 385.

Li, F. P., and J. F. Fraumeni Jr. Prospective study of a family cancer syndrome. *JAMA* 247 (1982): 2692.

Knudson, A. G., Jr. Mutation and cancer: statistical study of retinoblastoma. *Proc. Natl. Acad. Sci. USA* 68 (1971): 820.

Weinberg, R. A. Finding the anti-oncogene. *Sci. Amer.* 259 (September 1988): 44.

Hereditary Risk: Genes Affecting DNA Repair and Genetic Stability

Cleaver, J. E. Xeroderma pigmentosum: the first of the cellular caretakers. *Trends Biochem. Sci.* 26 (2001): 398.

Cleaver, J. E. Cancer in xeroderma pigmentosum and related disorders of DNA repair. *Nature Reviews Cancer* 5 (2005): 564.

King, M.-C., J. H. Marks, and J. B. Mandel. Breast and ovarian cancer risks due to inherited mutations in *BRCA1* and *BRCA2*. *Science* 302 (2003): 643.

Kinzler, K. W., and B. Vogelstein. Gatekeepers and caretakers. *Nature* 386 (1997): 761.

Moses, R. E. DNA damage processing defects and disease. *Annu. Rev. Genomics Hum. Genet.* 2 (2001): 41.

Nathanson, K. L., R. Wooster, and B. L. Weber. Breast cancer genetics: What we know and what we need. *Nature Medicine* 7 (2001): 552.

Shiloh, Y. ATM and related protein kinases: Safeguarding genome integrity. *Nature Reviews Cancer* 3 (2003): 155.

Hereditary Risk: Other Genes and Issues

de la Chapelle, A. Genetic predisposition to colorectal cancer. *Nature Reviews Cancer* 4 (2004): 769.

Frank, S. A. Genetic predisposition to cancer—insights from population genetics. *Nature Reviews Genetics* 5 (2004): 764.

Kolonel, L. N., D. Altshuler, and B. E. Henderson. The multiethnic cohort study: exploring genes, lifestyle and cancer risk. *Nature Reviews Cancer* 4 (2004): 519.

Lerman, C., and A. E. Shields. Genetic testing for cancer susceptibility: the promise and the pitfalls. *Nature Reviews Cancer* 4 (2004): 235.

Marx, S. J. Molecular genetics of multiple endocrine neoplasia types 1 and 2. *Nature Reviews Cancer* 5 (2005): 367.

Miller, B. A., et al. *Racial/Ethnic Patterns of Cancer in the United States 1988–1992.* Bethesda, MD: National Cancer Institute, 1996.

Ponder, B. Genetic testing for cancer risk. *Science* 278 (1997): 1050.

Schneider, K. *Counseling About Cancer*, 2d ed. New York: Wiley-Liss, 2002.

Willett, W. C. Balancing life-style and genomics research for disease prevention. *Science* 296 (2002): 695.

9 Oncogenes

A large body of evidence points to the role played by gene mutations in cancer. As we saw in Chapters 4 through 8, some cancer-causing mutations are triggered by chemicals or radiation, some are caused by infectious agents, and some are inherited. Spontaneous mutations and errors in DNA replication are additional sources of mutation. Yet despite the various ways in which mutations arise in cancer cells, the final result is always the same: Genes involved in controlling cell proliferation and survival are altered in ways that change either the behavior or the concentration of the proteins they produce.

The affected genes fall into two main categories. The first, called **oncogenes**, are genes whose *presence* can contribute to uncontrolled cell proliferation and cancer. The second, known as **tumor suppressor genes**, are genes whose *absence* (or inactivation) can likewise contribute to uncontrolled cell proliferation and cancer. In this chapter and the next, the properties of oncogenes and tumor suppressor genes will be examined in depth to see how they exert their effects at the molecular level. This chapter, devoted to oncogenes, begins with a discussion of how oncogenes arise and then addresses the question of how the proteins produced by oncogenes contribute to the development of cancer.

HOW CELLULAR ONCOGENES ARISE

Oncogenes can arise inside cells in two fundamentally different ways. One mechanism, already covered in Chapter 7, involves the participation of cancer viruses that introduce oncogenes into the cells they infect. The alternative, to be described in the first half of this chapter, is based on a series of mechanisms that convert normal cellular genes into oncogenes, often as a result of exposure to carcinogenic agents.

Cellular Oncogenes Arise from Proto-oncogenes

Those normal cellular genes that can be converted into oncogenes are referred to as **proto-oncogenes**. Despite their harmful-sounding name, proto-oncogenes are not bad genes simply lying in wait for an opportunity to foster the development of cancer. Rather they are *normal genes that make essential contributions to the regulation of cell proliferation and survival.* The term *proto-oncogene* simply implies that if and when the structure or activity of a proto-oncogene is disrupted by certain kinds of mutations, the altered form of the gene can cause cancer. In genetic terms, such mutations are considered to be "gain-of-function" mutations because they create a new function, namely the ability to induce tumor formation, that is not originally present in proto-oncogenes. Thus proto-oncogenes are normal genes that contribute to normal cell function, but they can also be converted into oncogenes, which are dysfunctional genes that produce proteins that perform a decidedly abnormal function—that is, inducing the development of cancer.

Cellular Oncogenes Were Initially Detected in Gene Transfer Experiments

Oncogenes are not present in normal cells, so how do they arise in cancer cells? In some cancers, an infecting virus simply brings one or more **viral oncogenes** into the cells it infects. But what about cancers that are not caused by viruses, which is the situation with most human cancers? Do they also possess oncogenes, and if so, where do they come from?

Evidence that oncogenes are indeed present in non-viral cancers first emerged from *gene transfer* experiments in which DNA isolated from tumor cells was introduced into normal cells and tested for its ability to transform them into cancer cells. The initial studies of this type, carried out in the laboratories of Robert Weinberg and Geoffrey Cooper in the early 1980s, focused on the behavior of DNA isolated from human bladder cancers. DNA was extracted from bladder cancer cells and applied to cultures of normal mouse cells under experimental conditions that stimulate DNA **transfection**—that is, uptake of foreign DNA into the cells and its incorporation into their chromosomes. Upon being transfected with tumor cell DNA, some of the mouse cells started to proliferate, and when the proliferating cells were injected back into mice, the animals developed cancer (see Figure 2-20). In contrast, transfection with DNA extracted from normal human cells did not transform mouse cells into cancer cells. It was therefore concluded that the DNA of bladder cancer cells contains genetic information, not present in normal DNA, that is capable of causing cancer.

Because these bladder cancer transfection experiments employed total cellular DNA samples containing thousands of different genes, further work was needed to identify the cancer-causing gene(s) contained in such complex DNA mixtures. This goal was pursued using **gene cloning** techniques, a set of procedures that allow DNA molecules to be broken into gene-sized fragments that can be reproduced in large amounts. When gene cloning was applied to bladder cancer DNA, it led to the isolation and identification of a single mutant gene that is responsible for the ability of the transfected DNA to cause cancer.

This mutant gene, which was the first human oncogene to be identified, is called a **cellular oncogene** to distinguish it from the *viral oncogenes* that are brought into cells by cancer viruses. Although it is not a viral gene, the oncogene isolated from bladder cancer DNA did turn out to be related in base sequence to a viral oncogene, namely the v-*ras* oncogene found in retroviruses that cause sarcomas in rats. (The name *ras* was originally derived from "ra̲t s̲arcoma.") By convention, human versions of the v-*ras* oncogene are named using capital letters and italics—that is, they are called ***RAS* oncogenes.** *RAS* oncogenes, like retroviral oncogenes, are altered forms of normal cellular proto-oncogenes. In the case of *RAS* oncogenes, we will see shortly that the corresponding ***RAS* proto-oncogene** is a normal gene that produces a protein involved in a normal pathway for stimulating cell proliferation. So genes that produce this protein are linked to cancer in two different ways: (1) v-*ras* oncogenes are present in certain retroviruses that cause sarcomas in rats, and (2) closely related *RAS* oncogenes appear in some human cancers that are not caused by viruses.

Following the discovery of the *RAS* oncogene—the first example of a human cellular oncogene—subsequent transfection experiments using DNA from other human tumor types revealed the existence of more than a dozen additional oncogenes. However, only about 20% of human cancers turn out to have cellular oncogenes when tested in this way. Such results do not mean that other human cancers lack oncogenes, but simply that oncogenes cannot always be detected using DNA transfection techniques. As we will see shortly, the reason is related to differences in the mechanisms by which cellular oncogenes arise. These mechanisms can be grouped into five basic categories: (1) *point mutation*, (2) *gene amplification*, (3) *chromosomal translocation*, (4) *DNA rearrangement*, and (5) *insertional mutagenesis*. In the following sections, each mechanism is described in detail.

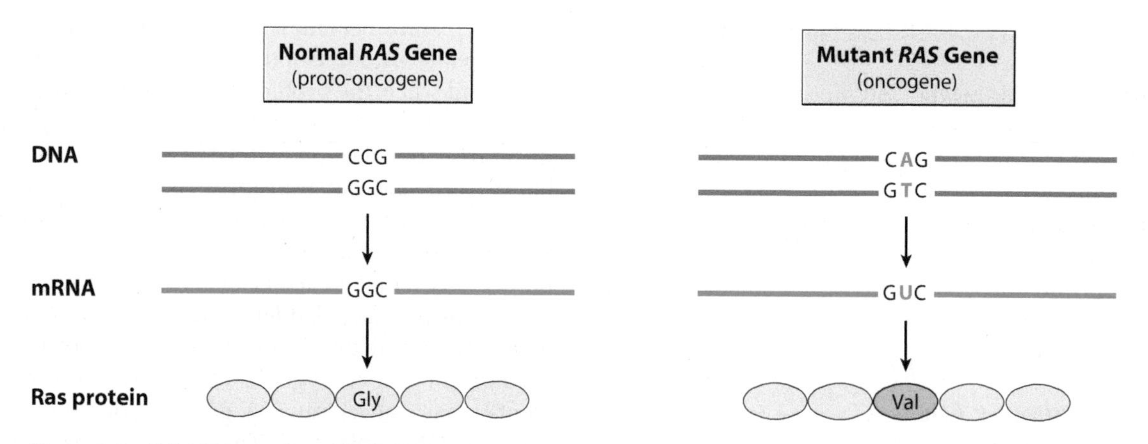

Figure 9-1 Point Mutation in a *RAS* Oncogene. *RAS* oncogenes typically differ from normal *RAS* proto-oncogenes in only a single nucleotide base. In this example, a single nucleotide mutation converts a normal *RAS* proto-oncogene into an oncogene that codes for an abnormal Ras protein in which a single amino acid is converted from glycine (Gly) to valine (Val).

Mechanism 1: Point Mutations Can Convert Proto-oncogenes into Oncogenes

When the *RAS* oncogene was first isolated from human bladder cancer cells, the question immediately arose as to how this abnormal *RAS* gene (an oncogene) differs from the normal *RAS* gene (a proto-oncogene). Analyzing the base sequences of the two genes provided the answer: *RAS* oncogenes typically differ from the normal *RAS* proto-oncogene in a single nucleotide base. In other words, changing just a single base in the nucleotide sequence of a normal gene is sufficient to convert it into a gene that can cause cancer. Figure 9-1 illustrates an example of such a mutation that is frequently encountered in *RAS*. In this particular case, changing a single base in the *RAS* proto-oncogene causes the resulting *RAS* oncogene to produce an abnormal Ras protein in which a single amino acid is converted from glycine to valine. [Recall from Table 7-3 that it is common practice to name human oncogenes using italicized capital letters (e.g., *RAS* gene) and to designate the protein encoded by the same gene without italics and often with only the initial letter capitalized (e.g., Ras protein).]

The preceding example represents the simplest mechanism for creating an oncogene from a proto-oncogene: A single nucleotide in a proto-oncogene simply undergoes mutation, thereby creating an oncogene that codes for a protein that differs in a single amino acid from the normal protein encoded by the gene. Such mutations in a single nucleotide base are called **point mutations**. *RAS* oncogenes produced by point mutation have been detected in many types of cancer, including cancers of the bladder (where they were originally discovered), lung, colon, pancreas, and thyroid. A point mutation can occur at any of several different sites within a *RAS* oncogene, and the particular site involved appears to be influenced by the carcinogen that caused it. For example, asbestos, vinyl chloride, and dimethylbenzanthracene each trigger mutations at different locations within the *RAS* gene. The ability of these carcinogens to cause cancer may therefore stem from their ability to mutate different nucleotides in the same gene.

Point mutations that convert proto-oncogenes into oncogenes are not restricted to genes coding for Ras proteins. The *RET* gene (whose role in *multiple endocrine neoplasia type II* was discussed in Chapter 8) can be converted by various point mutations into *RET* oncogenes that code for altered Ret receptor proteins. Individuals inheriting a single copy of such a mutant gene exhibit a roughly 70% risk of developing cancer within their lifetime, usually thyroid cancer. In addition to playing a role in hereditary cancers, *RET* oncogenes containing point mutations also appear in some sporadic thyroid cancers, where they are thought to be caused by exposure to environmental carcinogens.

Mechanism 2: Gene Amplification Can Convert Proto-oncogenes into Oncogenes

A second, fundamentally different mechanism for creating cellular oncogenes uses the process of **gene amplification** to create multiple, duplicate copies of the same gene. Gene amplification is accomplished by replicating the DNA located in a specific chromosomal region numerous times in succession, thereby creating dozens, hundreds, or even thousands of copies of the same stretch of DNA. Chromosome regions containing amplified genes often exhibit a distinctive, abnormal appearance that can be recognized when chromosomes are examined by light microscopy (Figure 9-2). The main types of abnormalities are **homogeneously staining regions** (**HSRs**) and **double minutes** (**DMs**). HSRs are chromosome regions that stain homogeneously rather than exhibiting the alternating pattern of light and dark bands that is typical of normal chromosomes. In contrast, DMs are independent, chromosome-like bodies that are much smaller than typical chromosomes, often appearing as spherical, paired structures. Both types of chromosomal abnormalities represent regions of amplified DNA containing from several dozen to several hundred copies of one or more genes.

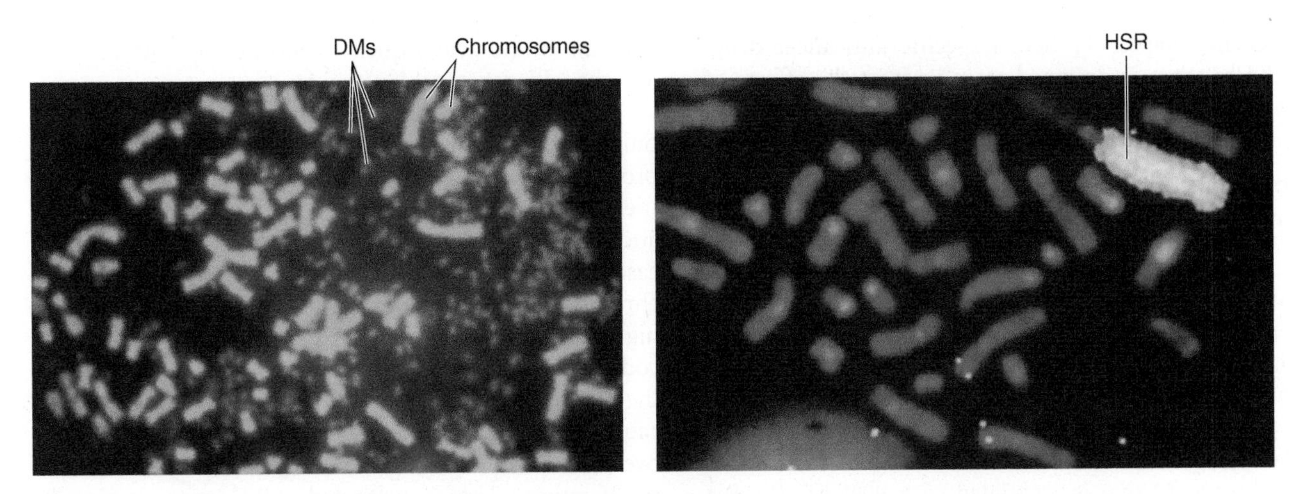

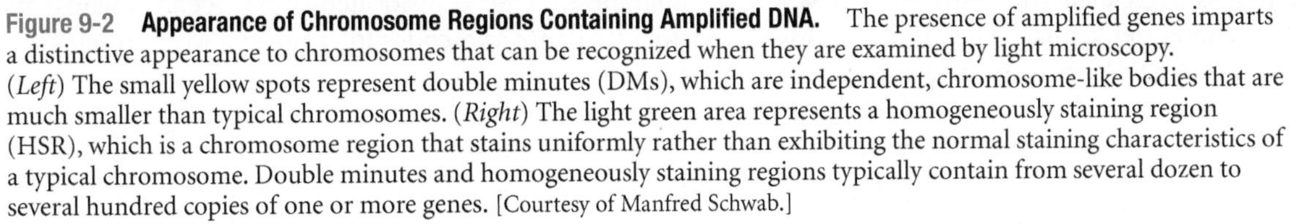

Figure 9-2 Appearance of Chromosome Regions Containing Amplified DNA. The presence of amplified genes imparts a distinctive appearance to chromosomes that can be recognized when they are examined by light microscopy. (*Left*) The small yellow spots represent double minutes (DMs), which are independent, chromosome-like bodies that are much smaller than typical chromosomes. (*Right*) The light green area represents a homogeneously staining region (HSR), which is a chromosome region that stains uniformly rather than exhibiting the normal staining characteristics of a typical chromosome. Double minutes and homogeneously staining regions typically contain from several dozen to several hundred copies of one or more genes. [Courtesy of Manfred Schwab.]

When a gene is amplified, most (if not all) of the multiple copies are actively expressed. The total amount of protein produced by an amplified gene is therefore much greater than when the gene is present in only a single copy per chromosome. However, the base sequence of each copy of an amplified gene is usually normal. So unlike point mutations, which create oncogenes that produce an abnormal protein, gene amplification typically yields oncogenes that produce normal proteins but in excessive quantities.

Members of the ***MYC* gene** family—individually called *MYC*, *MYCL*, and *MYCN*—are among the most commonly encountered oncogenes to arise by gene amplification in human cancers. (Recall from Chapter 7 that some cancer viruses also introduce abnormal versions or trigger abnormal behavior of *MYC* genes.) Amplified *MYC* genes have been detected in a diverse array of tumors, including cancers of the breast, ovary, uterine cervix, lung, and esophagus. These amplified genes produce excessive amounts of normal Myc protein, whose role in the control of cell proliferation will be discussed later in the chapter. The extent of *MYC* gene amplification—that is, the number of gene copies per cell—varies among people with the same type of cancer, but tends to remain constant for a given person's tumor. In some cases, a connection has been detected between the number of gene copies and tumor behavior. For example, neuroblastomas with extensively amplified *MYCN* genes are more likely to invade and metastasize, and have lower survival rates, than neuroblastomas in which the *MYCN* gene is less amplified (Figure 9-3).

In addition to *MYC* genes, more than a dozen other examples of amplified oncogenes have been detected in human cancers. Depending on the particular type of cancer involved, amplified oncogenes are present in anywhere from 5% to 50% of cancer cases and exhibit repetition frequencies as high as 500 copies or more. An amplified oncogene of significant clinical importance is the ***ERBB2* gene**, which appears in amplified form in about 25% of all breast and ovarian cancers. As was the case for *MYCN* gene amplification in neuroblastomas, individuals whose tumors exhibit higher copy numbers of the *ERBB2* gene tend to exhibit poorer survival rates. We

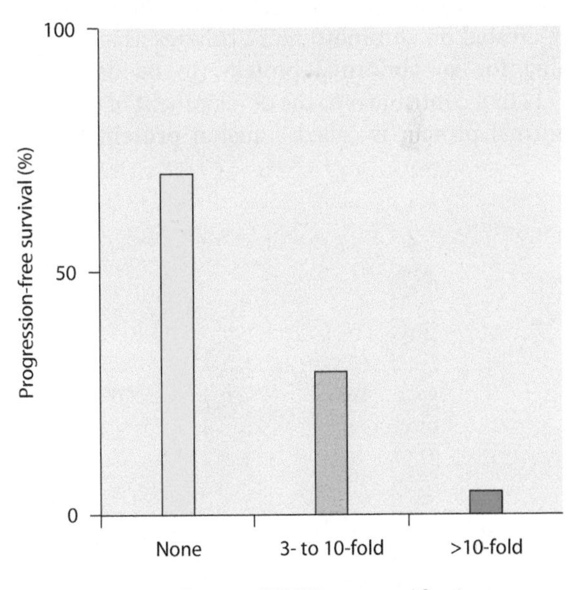

Figure 9-3 Relationship Between *MYCN* Gene Amplification and Neuroblastoma Survival Rates. The percentage of neuroblastoma patients who survive 18 months after diagnosis without disease progression is indicated for groups exhibiting different degrees of *MYCN* gene amplification. The data reveal that neuroblastomas with extensively amplified *MYCN* genes have lower survival rates than neuroblastomas in which the *MYCN* gene is less amplified. [Based on data from R. C. Seeger et al., *New England J. Med.* 313 (1985): 1111.]

will see in Chapter 11 that a specific anticancer drug, called *Herceptin*, is designed to counteract the effects of the overactive *ERBB2* gene in such individuals.

Mechanism 3: Chromosomal Translocations Can Convert Proto-oncogenes into Oncogenes

The third mechanism for creating oncogenes is by **chromosomal translocation**, a process in which a piece of one chromosome is broken off and moved to another chromosome. Several examples of chromosomal translocations have already been mentioned in previous chapters. For example, in Chapter 2 we briefly encountered the **Philadelphia chromosome**, an abnormal version of chromosome 22 that is associated with 90% of all cases of chronic myelogenous leukemia. The Philadelphia chromosome is created by DNA breakage near the ends of chromosomes 9 and 22, followed by reciprocal exchange of DNA between the two chromosomes. This translocation creates a new oncogene comprised of portions of two normal genes that were initially located on chromosome 9 and chromosome 22, respectively.

One of the two genes, called *ABL*, resides near the end of chromosome 9; the other gene, called *BCR*, resides near the end of chromosome 22. During chromosomal translocation, a break occurs within the *ABL* gene on chromosome 9 and within the *BCR* gene on chromosome 22, and the two broken ends of the chromosomes are then switched (Figure 9-4). As a result, the translocated versions of chromosomes 9 and 22 each acquire a new **fusion gene**—that is, a gene containing sequences derived from two different genes spliced together. The ***BCR-ABL*** fusion gene created on chromosome 22 behaves as an oncogene, coding for an abnormal protein (to be described on p. 171) that contributes to the development of cancer. This abnormal protein is called a **fusion protein** because it contains amino acid sequences encoded by both the *BCR* and the *ABL* genes.

Creating genes that code for fusion proteins is not the only way in which chromosomal translocations can produce oncogenes. Another mechanism is illustrated by the chromosomal translocation involving chromosomes 8 and 14 that occurs in Burkitt's lymphoma (p. 135). In this case, the entire *MYC* proto-oncogene is moved from chromosome 8 to 14, where it becomes situated next to a highly active region of chromosome 14 that contains genes coding for antibody molecules (see Figure 7-12). Moving the *MYC* gene so close to the highly active antibody genes causes *MYC* to become activated also, thereby leading to an overproduction of Myc protein that in turn stimulates cell proliferation. The translocated *MYC* gene retains its normal structure and codes for a normal Myc protein, but it still behaves as an oncogene because its new location on chromosome 14 causes the gene to be overexpressed.

Thus chromosomal translocations contribute to cancer development through two distinct mechanisms, either by fusing two genes together to form an oncogene coding for a fusion protein or by activating the expression of a proto-oncogene by placing it near a highly active gene. Chromosomal translocations acting in these ways have been detected in a variety of human cancers, especially leukemias and lymphomas. Besides the two examples described thus far (chronic myelogenous leukemia and Burkitt's lymphoma), acute myelogenous leukemias have been found to exhibit a diverse array of chromosomal translocations, including translocations between chromosomes 3 and 5, chromosomes 6 and 9, chromosomes 7 and 11, chromosomes 8 and 16, chromosomes 9 and 12, chromosomes 12 and 22, and chromosomes 16 and 21. In each case, researchers have identified the two genes whose juxtaposition or fusion contributes to the development of cancer.

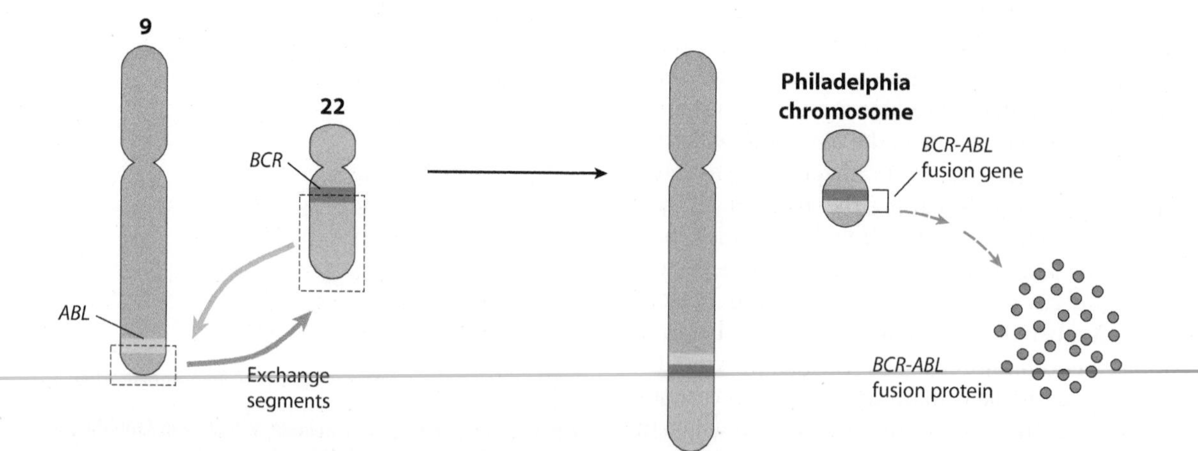

Figure 9-4 Chromosomal Translocation That Creates the Philadelphia Chromosome and Its *BCR-ABL* Oncogene. The Philadelphia chromosome is created by DNA breakage near the end of chromosome 9 (within the *ABL* gene) and near the end of chromosome 22 (within the *BCR* gene), followed by reciprocal exchange of DNA between the two chromosomes. This translocation creates a shortened version of chromosome 22, called the Philadelphia chromosome, which contains a *BCR-ABL* fusion gene whose protein product contributes to the development of cancer. (See Figure 2-19 for a photograph of the Philadelphia chromosome.)

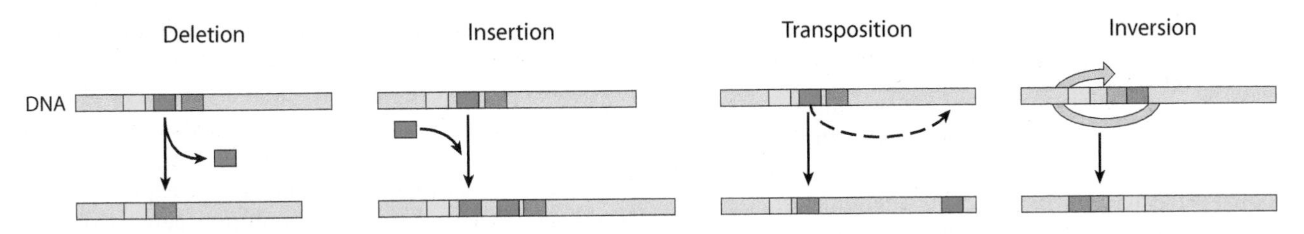

Figure 9-5 **Types of DNA Rearrangements.** Deletions and insertions refer to DNA segments that are either lost or added, respectively. Transpositions involve the movement of a DNA segment from one location to another, and inversions are DNA segments that have been excised and then reinserted backwards in the same location.

Mechanism 4: Local DNA Rearrangements Can Convert Proto-oncogenes into Oncogenes

The fourth mechanism for creating oncogenes involves localized **DNA rearrangements** in which the movement of DNA base sequences in a particular region of a chromosome disrupts the expression or structure of a proto-oncogene located in that region. As shown in Figure 9-5, DNA rearrangements can be grouped into four basic categories. The simplest types involve short stretches of DNA that are either lost or added during normal DNA replication, thereby causing mutations known as **deletions** or **insertions**, respectively. Rearrangements in which DNA segments are moved from one location to another are called **transpositions**, and DNA segments that have been excised and then reinserted backwards in the same location are termed **inversions**.

DNA rearrangements are frequently detected in human tumor cells, especially in certain types of cancer. In thyroid cancers, for example, rearrangements are present in nearly 50% of the tumors examined. Some of these rearrangements illustrate how a simple chromosomal inversion can create an oncogene from two perfectly normal genes. One well-studied example involves two genes, named *NTRK1* and *TPM3*, that both reside on the same chromosome. *NTRK1* codes for a growth factor receptor and *TPM3* codes for *tropomyosin*, a totally unrelated protein involved in muscle contraction and cell motility. In some thyroid cancers, a DNA inversion occurs that causes one end of the *TPM3* gene to become fused to the opposite end of the *NTRK1* gene, thereby creating a fusion gene called the ***TRK* oncogene** (Figure 9-6). The *TRK* oncogene produces a Trk fusion protein whose amino acid sequence is determined partly by the *NTRK1* receptor gene and partly by the *TPM3* gene.

The Trk fusion protein contains the part of the normal receptor molecule that exhibits *tyrosine kinase* activity, which means that it catalyzes the phosphorylation of the amino acid tyrosine in target proteins. As we will see later in this chapter, the tyrosine kinase site of a normal receptor only becomes active after a growth factor binds to the receptor and causes adjacent receptor molecules to cluster together as *dimers* (two molecules joined together). The Trk fusion protein, however, contains a portion of the tropomyosin molecule that forms *coiled coils*, which are structures that cause protein chains to join together as dimers. As a result, the fusion protein forms a permanent dimer and its tyrosine kinase is permanently activated (see Figure 9-6, step ④). The Trk fusion protein therefore contributes to cancer development by acting as a permanently activated receptor that continually stimulates cell proliferation, regardless of whether its growth factor is present.

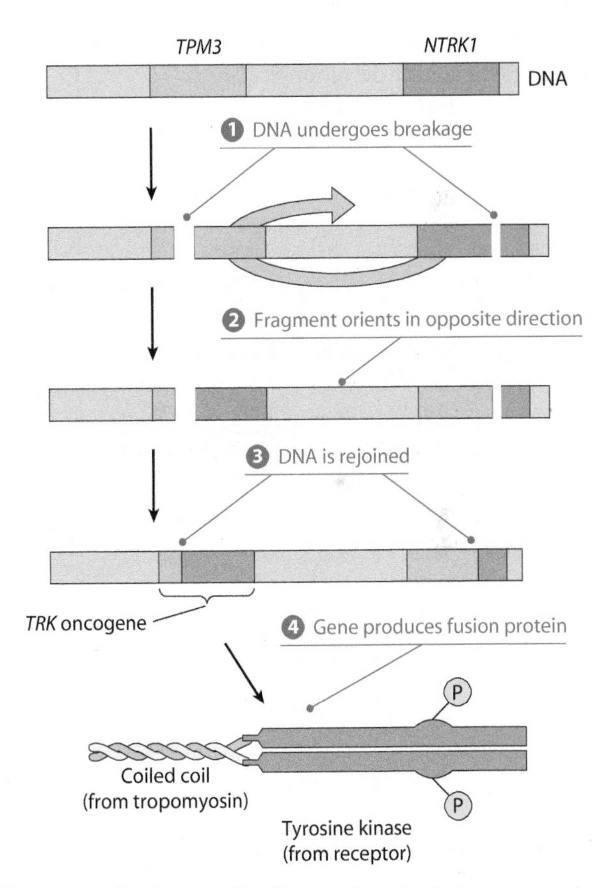

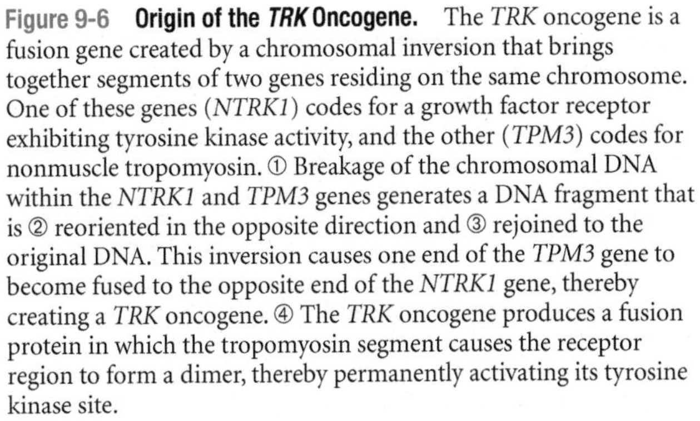

Figure 9-6 **Origin of the *TRK* Oncogene.** The *TRK* oncogene is a fusion gene created by a chromosomal inversion that brings together segments of two genes residing on the same chromosome. One of these genes (*NTRK1*) codes for a growth factor receptor exhibiting tyrosine kinase activity, and the other (*TPM3*) codes for nonmuscle tropomyosin. ① Breakage of the chromosomal DNA within the *NTRK1* and *TPM3* genes generates a DNA fragment that is ② reoriented in the opposite direction and ③ rejoined to the original DNA. This inversion causes one end of the *TPM3* gene to become fused to the opposite end of the *NTRK1* gene, thereby creating a *TRK* oncogene. ④ The *TRK* oncogene produces a fusion protein in which the tropomyosin segment causes the receptor region to form a dimer, thereby permanently activating its tyrosine kinase site.

Mechanism 5: Insertional Mutagenesis Can Convert Proto-oncogenes into Oncogenes

Elucidation of the final mechanism for creating oncogenes emerged from the discovery that some cancer viruses possess no oncogenes of their own, but instead convert a cell's own genes into oncogenes. This phenomenon, called **insertional mutagenesis**, was introduced in Chapter 7 when we described how the *avian leukosis virus* causes cancer by inserting itself next to the *MYC* proto-oncogene, thereby triggering overproduction of the Myc protein (see Figure 7-11, *top*). As we saw, the underlying mechanism involves the presence of special viral sequences called *long terminal repeats* (*LTRs*), which are located at both ends of the genome of retroviruses. LTRs contain sequences that promote gene transcription so efficiently that they activate the transcription of cellular genes that lie near the inserted viral genes. Therefore, if a retrovirus happens to randomly insert its genes near a proto-oncogene, the LTRs may stimulate transcription of the proto-oncogene and trigger overproduction of a normal cellular protein that can contribute to cancer development.

Because virally induced cancers are more common in animals than they are in humans, the best understood examples of insertional mutagenesis have come from the study of animal viruses. Nonetheless, there are reasons for believing that a similar phenomenon applies to humans as well. We saw in Chapter 7, for example, that insertional mutagenesis can inadvertently take place when retroviruses are used in *gene therapy* experiments to ferry healthy genes into cells exhibiting genetic defects. At least one retrovirus employed for such purposes occasionally integrates itself next to a proto-oncogene whose abnormal expression has been found to trigger leukemia (p. 137).

Summing Up: Cellular Oncogenes Arise from Proto-oncogenes by Mechanisms That Alter Gene Structure or Expression

The general features of the five preceding mechanisms for converting proto-oncogenes into cellular oncogenes are summarized in Figure 9-7. In some cases, the structure of a proto-oncogene is altered in a way that causes it to produce an abnormal protein. In other cases, the expression of a proto-oncogene is enhanced, thereby leading to excessive production of a normal protein. The existence of

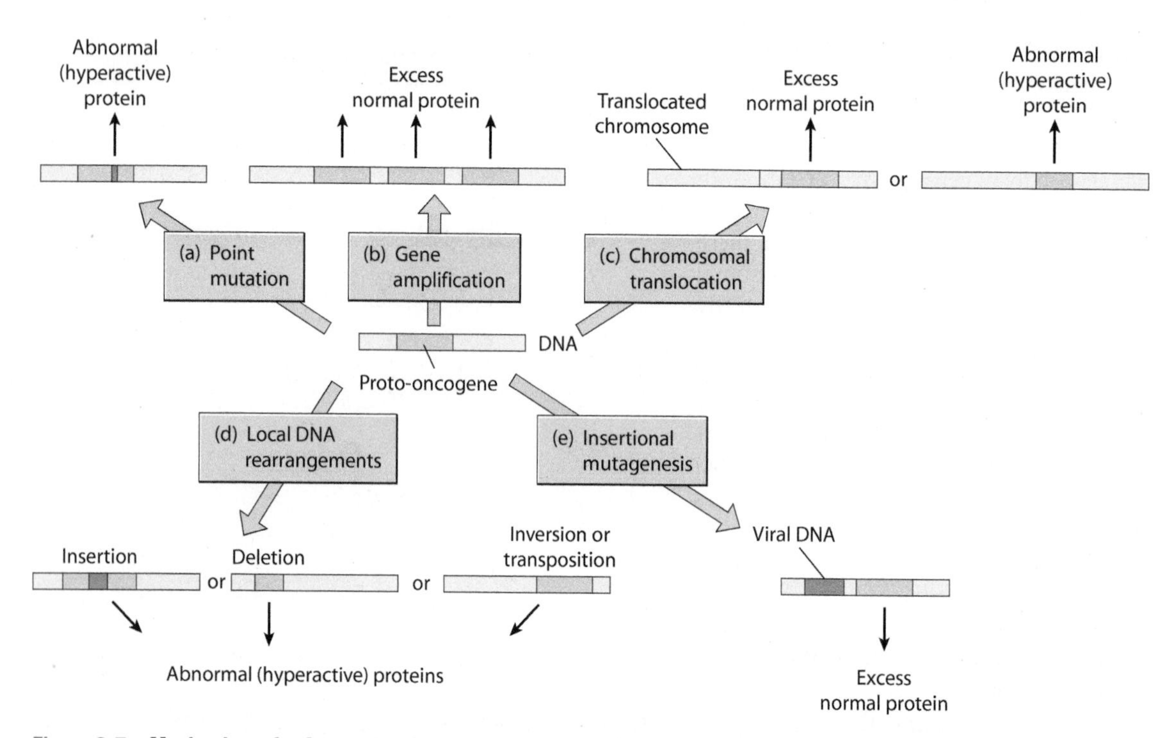

Figure 9-7 Mechanisms for Converting Proto-oncogenes into Oncogenes. In some cases, the structure of a proto-oncogene is altered in a way that causes it to produce an abnormal protein. In other cases, the expression of a proto-oncogene is enhanced, thereby leading to excessive production of a normal protein. (**a**) Point mutation involves a single nucleotide substitution that creates an oncogene coding for an abnormal protein differing in a single amino acid from the normal protein produced by the proto-oncogene. (**b**) Gene amplification involves the production of multiple gene copies that are actively expressed, thereby producing excessive amounts of a normal protein. (**c**) Chromosomal translocations involve the exchange of chromosome segments from one chromosome to another. This exchange may either fuse two genes together to form an oncogene coding for an abnormal protein or it may place a proto-oncogene next to a highly active gene, thereby inducing the translocated proto-oncogene to become more active than normal. (**d**) Local DNA rearrangements such as insertions, deletions, inversions, and transpositions can disrupt the structure of proto-oncogenes and cause them to produce abnormal proteins. (**e**) Insertional mutagenesis occurs when viral DNA is integrated into a host chromosome near a cellular proto-oncogene. The inserted DNA may stimulate the expression of the proto-oncogene and cause it to produce too much protein.

these alternative strategies helps explain why the oncogenes present in human cancers cannot always be detected using the DNA transfection technique described earlier in the chapter. (Recall from the discussion on p. 159 that only about 20% of human cancers have oncogenes that can be detected by DNA transfection.) The DNA transfection approach—in which DNA isolated from cancer cells is introduced into normal cells and tested for its ability to transform them into cancer cells—is best suited for detecting oncogenes containing mutations that cause them to produce an abnormal protein. In the case of oncogenes that simply produce too much of a normal protein, this altered expression is not likely to persist after chromosome structure has been disrupted and DNA fragments are isolated, which is how DNA is prepared for use in transfection studies.

PROTEINS PRODUCED BY ONCOGENES

Thus far, we have seen that five distinct mechanisms exist for converting a proto-oncogene into an oncogene, which then produces either a structurally abnormal protein or a normal protein in excessive amounts. In either case, the question arises as to how these proteins cause cancer. Addressing such a question is a complex task because more than 100 oncogenes have been identified to date and the proteins they produce fall into a variety of categories, including growth factors, receptors, enzymes that catalyze protein phosphorylation, and proteins that bind to and regulate the activity of DNA or other proteins (Table 9-1). Yet despite this diversity, a unifying theme can be found. As we will see in the following pages, *most of the proteins produced by oncogenes are components of signaling pathways that promote cell proliferation and survival.* By producing abnormal versions or excessive quantities of proteins involved in these pathways, oncogenes disrupt normal signaling mechanisms and foster the excessive proliferation and inappropriate survival of cancer cells.

Oncogenes Typically Code for Components of Signaling Pathways That Activate Cell Proliferation

Normally a cell will not grow and divide unless it is stimulated by an appropriate growth factor, which triggers proliferation by activating signaling pathways involving dozens of molecules within the targeted cell. Oncogenes code for proteins that are participants in such signaling pathways, but rather than producing the proper amount of a correct protein, an oncogene produces either an abnormal protein or an excessive amount of a normal one. In either case, the net result is the unregulated activation of a signaling pathway and hence uncontrolled cell proliferation, even in the absence of growth factors.

In most cells, proliferation is controlled by multiple signaling pathways that function in overlapping networks. Yet despite the overall complexity, these pathways tend to share some features in common. In general, the binding of a growth factor to a receptor located on the outer surface of a cell leads to receptor activation; the activated receptor then stimulates a series of molecules that relay information to various compartments within the cell, including the nucleus; and some of the relay molecules reaching the nucleus trigger changes in gene expression that stimulate cell proliferation or promote cell survival.

A good example of these principles is provided by the **Ras-MAPK pathway**, which plays a central role in stimulating normal cell proliferation (and which often behaves abnormally in cancer cells). As shown in Figure 9-8, the Ras-MAPK pathway involves six main steps. (1) A growth factor binds to a cell-surface receptor. (2) The activated receptor becomes phosphorylated. (3) The phosphorylated receptor binds to adaptor proteins that relay the signal to Ras proteins associated with the inner surface of the plasma membrane. (4) Activated Ras triggers a cascade of intracellular protein phosphorylation reactions that lead to activation of a protein kinase called *MAPK*. (5) Activated MAPK enters the nucleus and phosphorylates proteins called *transcription factors*, which bind to DNA and activate the transcription of specific genes. (6) The activated genes produce proteins that stimulate cell proliferation. Among these proteins are cell-cycle regulators such as *Cdk* and *cyclin*, which stimulate progression through the restriction point and into S phase.

Although the Ras-MAPK pathway is only one of several signaling mechanisms used by cells for controlling cell proliferation, it is a good starting point for discussing how oncogenes work because it illustrates the main types of proteins produced by oncogenes, and the pathway often functions abnormally in human cancers. In the following six sections we will look at each step of the Ras-MAPK pathway in more depth, examining how it operates normally and providing examples of oncogenes that cause it to act in a hyperactive or uncontrolled fashion. Other signaling pathways that tend to behave abnormally in cancer cells will also be described later in this (and the next) chapter as we discuss the roles played by oncogenes (and tumor suppressor genes) in the development of cancer.

Some Oncogenes Produce Growth Factors

The first step in the Ras-MAPK pathway involves the binding of a **growth factor** to a target cell (see Figure 9-8, step ①). That normal cell proliferation requires the presence of an appropriate growth factor can be demonstrated by placing cells in a culture medium containing nutrients and vitamins but no growth factors. Under these conditions, progression through the cell cycle is halted during G1 (at the restriction point) and cell proliferation ceases. Progression through the cell cycle can be restarted by adding small amounts of blood serum, which contains several growth factors. One is **platelet-derived growth factor** (**PDGF**), a protein produced by blood platelets that stimulates the proliferation of connective tissue cells. Another is **epidermal growth factor** (**EGF**), a protein

Table 9-1 Selected Oncogenes Grouped Together by Protein Function

Oncogene Name	Protein Produced	Oncogene Origin	Common Cancer Type*
1. Growth factors			
v-sis	PDGF	Simian sarcoma virus	Sarcomas (monkeys)
COL1A1-PDGFB	PDGF	Chromosomal translocation	Fibrosarcoma
2. Receptors			
v-erb-b	EGF receptor	Avian erythroblastosis virus	Leukemia (chickens)
RET	Ret receptor	Point mutation, chromosomal translocation	Thyroid
TRK	Nerve growth factor receptor	DNA rearrangement (inversion)	Thyroid
ERBB2	ErbB2 receptor	Amplification	Breast, ovary
v-mpl	Thrombopoietin receptor	Myeloproliferative leukemia virus	Leukemia (mice)
3. Plasma membrane G proteins			
v-K-ras	Ras	Harvey sarcoma virus	Sarcomas (rats)
v-H-ras	Ras	Kirsten sarcoma virus	Sarcomas (rats)
KRAS	Ras	Point mutation	Pancreas, colon, lung, others
HRAS	Ras	Point mutation	Bladder
NRAS	Ras	Point mutation	Leukemias
4. Intracellular protein kinases			
BRAF	Raf kinase	Point mutation	Melanoma
v-src	Src kinase	Rous sarcoma virus	Sarcomas (chickens)
SRC	Src kinase	DNA rearrangement	Colon
TEL-JAK2	Jak kinase	Chromosomal translocation	Leukemias
BCR-ABL	Abl kinase	Chromosomal translocation	Chronic myelogenous leukemia
5. Transcription factors			
MYC	Myc	Chromosomal translocation	Burkitt's lymphoma
MYCN	Myc	Amplification	Neuroblastoma
MYCL	Myc	Amplification	Small cell lung cancer
v-myc	Myc	Avian myelocytomatosis virus	Leukemia (chickens)
c-myc	Myc	Insertional mutagenesis (triggered by avian leukosis virus)	Leukemia (chickens)
v-ets	Ets	Avian erythroblastosis virus	Leukemia (chickens)
v-jun	Jun	Avian sarcoma virus	Sarcomas (chickens)
v-fos	Fos	Murine osteosarcoma virus	Bone (mice)
6. Cell cycle or cell death regulators			
CYCD1	Cyclin	Amplification, chromosomal translocation	Breast Lymphoma
CDK4	Cdk	Amplification	Sarcoma, glioblastoma
BCL2	Bcl2	Chromosomal translocation	Non-Hodgkin's lymphoma
MDM2	Mdm2	Amplification	Sarcomas

*Cancers are in humans unless otherwise specified. Only the most frequent cancer types are listed.

widely distributed in normal tissues that acts on a variety of cell types, mainly (but not exclusively) of epithelial origin. The normal function of growth factors is to stimulate the cell proliferation that is required during events such as embryonic development, tissue regeneration, and wound repair. For example, blood platelets accumulate at sites of tissue injury and release the growth factor PDGF, which is instrumental in stimulating the cell proliferation needed for wound healing.

Proliferation of a target cell population normally depends on growth factors produced by other cells that detect the need for tissue growth and produce the appropriate growth factors. But what would happen if a cell were to produce a growth factor that stimulates its own proliferation? This scenario would create an uncontrolled situation in which cells proliferate to create more cells of the same type, which then produce more of the growth factor that continues to stimulate proliferation of the same cells. One of the first oncogenes found to trigger such a scenario was v-*sis*, a gene from the *simian sarcoma virus* that causes sarcomas in monkeys. The v-*sis* oncogene codes for a mutant form of the growth factor PDGF. When the virus infects a monkey cell whose proliferation is normally stimulated by PDGF, the mutant PDGF produced by the v-*sis* oncogene continually stimulates the cell's own proliferation (in contrast to the normal situation in which cells are only exposed to PDGF when it is released from surrounding blood platelets). The net result is that PDGF produced by the v-*sis* oncogene causes the infected cells to constantly stimulate their own uncontrolled proliferation.

This type of mechanism, originally discovered in an animal retrovirus, is now known to apply to human cancers as well. In certain human sarcomas, a chromosomal translocation creates a fusion gene in which part of the PDGF gene is joined to part of an unrelated gene that codes for *collagen*, a protein component of the extracellular matrix. The fused gene (*COL1A1-PDGFB*) behaves as an oncogene because it produces PDGF in an uncontrolled fashion, thereby causing cells containing the gene to continually stimulate their own proliferation.

Some Oncogenes Produce Receptor Proteins

The next step in the Ras-MAPK pathway involves the plasma membrane **receptors** that transmit signals from growth factors to the cell interior (see Figure 9-8, step ②). Growth factor receptors are typically *transmembrane* proteins, which means that one end of the receptor is exposed outside the cell and the other end is exposed inside the cell. The exterior portion of the receptor contains a binding site for its corresponding growth factor, and the end protruding inside the cell transmits signals to the cell interior, usually by acting as a **protein kinase**. As mentioned in Chapter 2, a protein kinase is an enzyme that catalyzes *protein phosphorylation* (the attachment of phosphate groups to protein molecules). Receptors involved in the Ras-MAPK pathway specifically phosphorylate the amino acid tyrosine in target proteins, so such receptors are called **receptor tyrosine kinases**.

Binding of a growth factor to its receptor site exposed at the outer cell surface leads to activation of the receptor's tyrosine kinase site protruding inside the cell. This process of *receptor activation* is accomplished through the ability

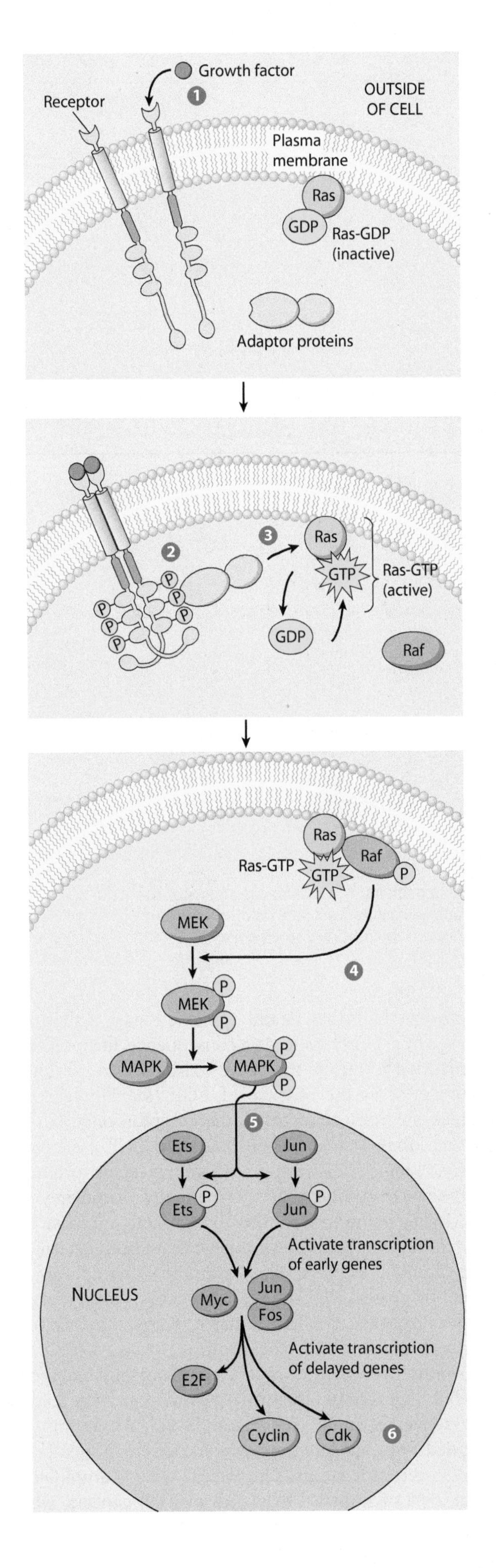

Figure 9-8 The Ras-MAPK Pathway. Signaling molecules that use the Ras-MAPK pathway trigger a chain of events involving six main steps: ① binding of a growth factor (such as PDGF or EGF) to its receptor; ② clustering and autophosphorylation of the receptor; ③ activation of Ras; ④ activation of a cascade of intracellular protein kinases (Raf, MEK, and MAPK); ⑤ activation or production of nuclear transcription factors (such as Ets, Jun, Fos, Myc, and E2F); and ⑥ synthesis of cyclin and Cdk molecules. The resulting Cdk-cyclin complexes catalyze the phosphorylation of Rb and hence trigger passage through the G1 restriction point and into S phase. MAPK = mitogen-activated protein kinase.

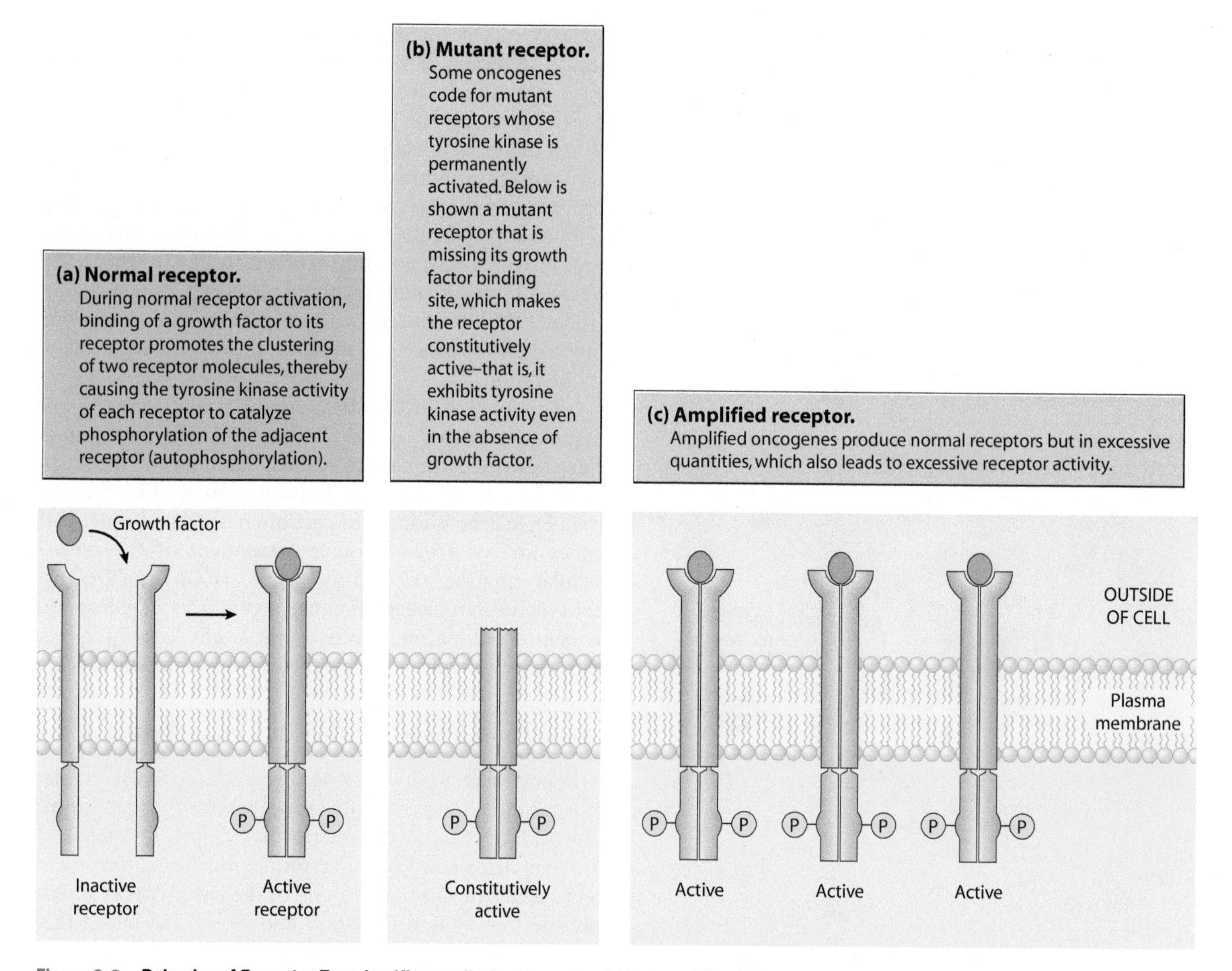

Figure 9-9 Behavior of Receptor Tyrosine Kinases Under Normal and Abnormal Conditions. (**a**) Normal receptors only exhibit tyrosine kinase activity after a growth factor has bound to them. (**b**) Some oncogenes code for mutant receptors whose tyrosine kinase is permanently activated. (**c**) Other oncogenes produce normal receptors but in excessive quantities, which also leads to excessive receptor activity.

of growth factors to promote or stabilize the clustering of two or more receptor molecules into configurations that activate their tyrosine kinase sites. For example, the growth factor EGF is a single protein chain that binds to two receptor molecules simultaneously, thereby joining two receptors together to form a *dimer* (Figure 9-9a). After receptor molecules have become clustered and activated by the binding of growth factor, the tyrosine kinase activity of each receptor catalyzes the phosphorylation of the adjacent receptor at multiple sites. Since receptors are phosphorylating other receptor molecules of the same type, the process is referred to as **autophosphorylation.**

Several dozen oncogenes are known to code for receptor tyrosine kinases. Many of these oncogenes produce mutant receptors whose tyrosine kinase activity is permanently activated rather than being dependent on a growth factor to trigger activation. An example is the v-*erb-b* oncogene, found in the *avian erythroblastic leukemia virus* that causes a cancer of red blood cells in chickens. The v-*erb-b* oncogene produces an altered version of the receptor for EGF that retains tyrosine kinase activity but lacks the EGF binding site. As a consequence, the receptor is **constitutively active**—that is, it exhibits tyrosine kinase activity even in the absence of EGF (see Figure 9-9b), whereas the normal form of the receptor only exhibits tyrosine kinase activity when bound to EGF. Because its tyrosine kinase is always active, the receptor produced by the v-*erb-b* oncogene permanently stimulates the Ras-MAPK pathway and thereby triggers excessive cell proliferation. Comparable oncogenes have been detected in some human cancers. For example, thyroid cancers frequently possess *RET* or *TRK* oncogenes, which code for mutant receptor tyrosine kinases whose uncontrolled activity stimulates cell proliferation.

Another group of oncogenes produces normal receptors but in excessive quantities, which can also lead to hyperactive signaling (see Figure 9-9c). An example is provided by the *ERBB2* gene, which codes for a member of the EGF receptor family. The *ERBB2* gene is amplified in about 25% of human breast and ovarian cancers, where the multiple copies of the *ERBB2* gene produce excessive amounts of a normal receptor protein. The presence of so many receptor molecules at the cell surface leads to a magnified response to growth factor binding and hence excessive cell proliferation.

Receptors do not always possess their own tyrosine kinase activity. In the case of some receptors, binding of growth factor instead causes the activated receptor to stimulate the activity of an independent tyrosine kinase molecule. One such tyrosine kinase is Jak, a central component of a signaling mechanism called the **Jak-STAT pathway**. As shown in Figure 9-10, binding of growth factors to receptors involved in this pathway causes the receptors to activate Jak molecules, which in turn catalyze the phosphorylation of cytoplasmic proteins called **STATs** (an abbreviation for Signal Transducers and Activators of Transcription). The phosphorylated STAT molecules then join together and move from the cytoplasm to the nucleus, where they trigger changes in gene expression that stimulate cell proliferation. One oncogene that codes for a receptor involved in the Jak-STAT pathway has been detected in the *myeloproliferative leukemia virus*, which causes leukemia in mice. The oncogene, called v-*mpl*, codes for a mutant version of the receptor for *thrombopoietin*, which is a growth factor that uses the Jak-STAT pathway to stimulate the production of blood platelets.

Some Oncogenes Produce Plasma Membrane G Proteins

After a growth factor has bound to and activated its receptor, a number of different signaling pathways can be triggered by the activated receptor. In the case of Ras-MAPK signaling, the phosphorylated tyrosines created on the receptor by autophosphorylation serve as binding sites for *adaptor proteins* that relay the signal to **Ras**, a protein associated with the inner surface of the plasma membrane. The Ras protein is a member of a class of molecules called **G proteins** because their activity is regulated by the two small nucleotides *GTP* (guanosine

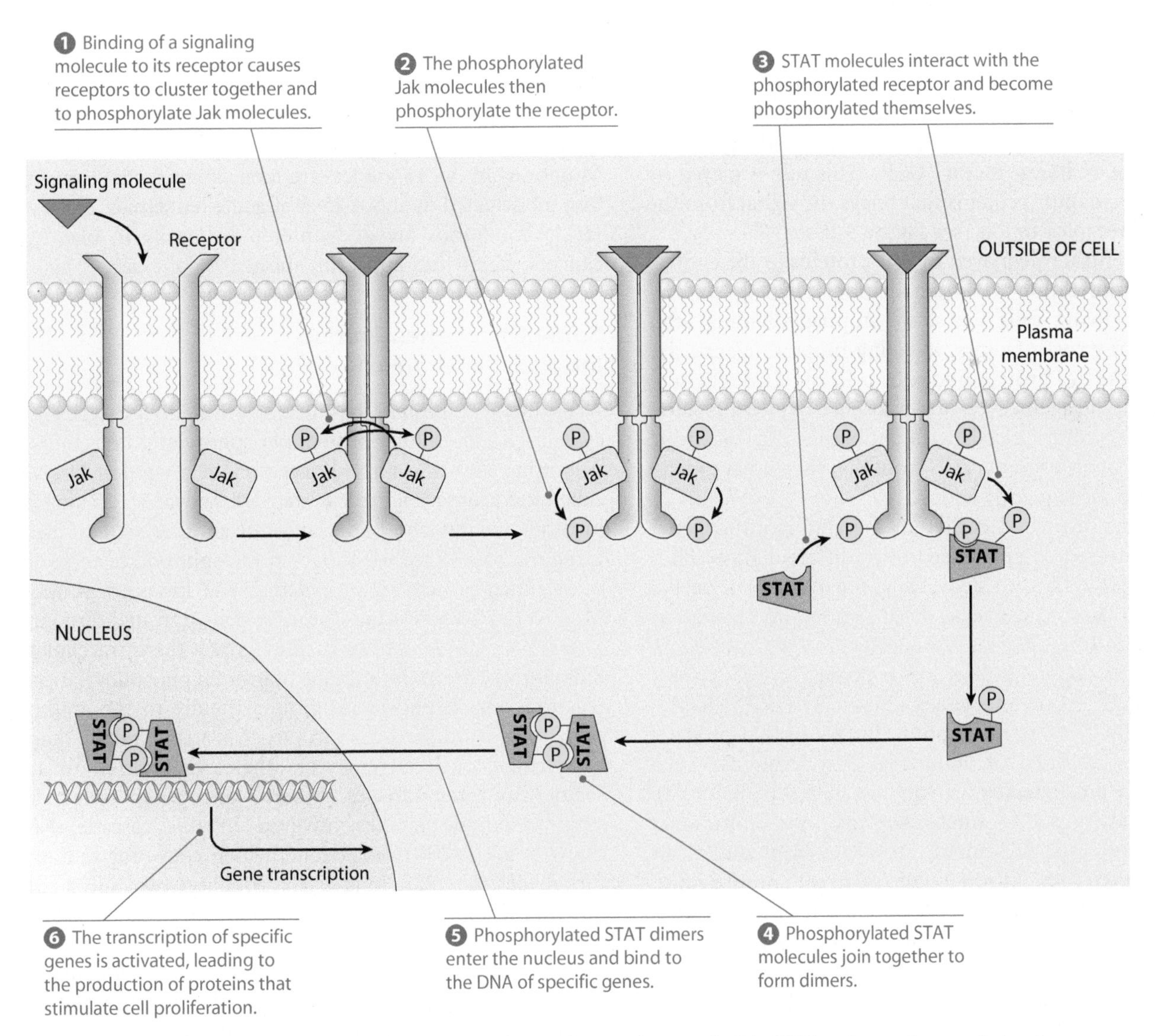

Figure 9-10 The Jak-STAT Signaling Pathway. Signaling molecules that use the Jak-STAT pathway trigger a chain of events involving six main steps that leads to the production of proteins that stimulate cell proliferation.

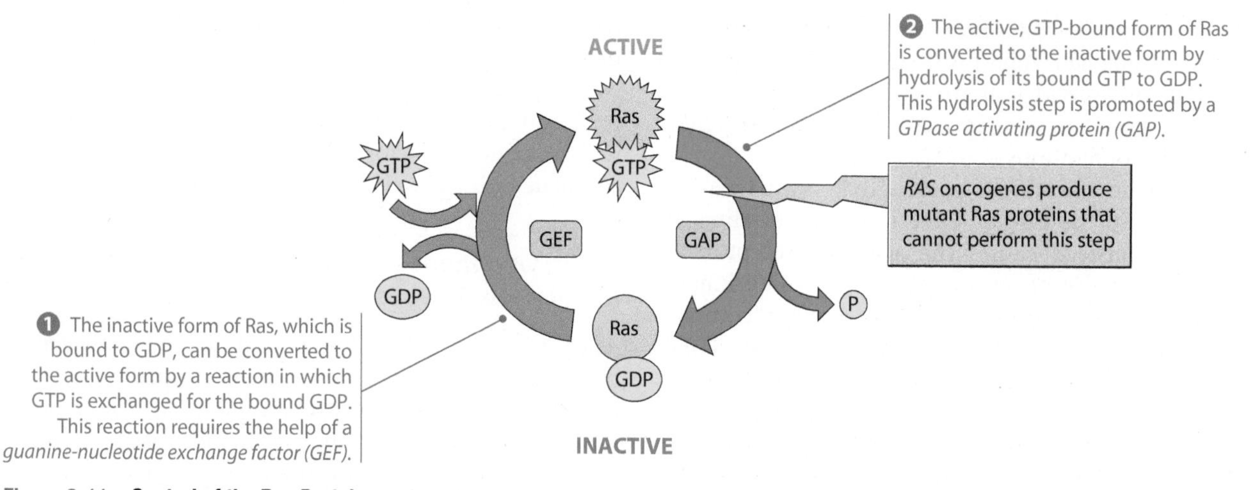

Figure 9-11 Control of the Ras Protein. The activity of the Ras protein is determined by whether it is bound to GTP or GDP. Some oncogenes produce Ras proteins that do not carry out the GTP hydrolysis step (②), thereby maintaining Ras in a permanently activated state.

triphosphate) and *GDP* (guanosine diphosphate). G proteins are molecular switches whose "on" or "off" state depends on whether they are bound to GTP or to GDP (Figure 9-11). In the absence of receptor stimulation, Ras is normally bound to GDP and is inactive. To become active, it must release GDP and acquire GTP in a reaction that requires the help of another protein called a **guanine-nucleotide exchange factor** (**GEF**). This role is played by one of the adaptor proteins that relays the signal from the activated receptor to Ras (see Figure 9-8, step ③).

The central role played by Ras proteins in the control of cell proliferation has been demonstrated by experiments involving cells that have stopped dividing after removal of growth factors. Injecting mutant, hyperactive forms of the Ras protein into such cells can cause them to begin dividing again, even in the absence of growth factor. Conversely, injecting cells with antibodies that inactivate the Ras protein prevents cells from dividing when growth factors are subsequently added.

As was described earlier in the chapter, oncogenes coding for Ras proteins arise in two different ways: They may be brought into cells by a retrovirus (as occurs mainly in animal cancers), or they may be created by point mutations in proto-oncogenes (as is common in human cancers). Human cells possess three closely related *RAS* proto-oncogenes known as *HRAS*, *KRAS*, and *NRAS*; each can incur point mutations that produce oncogenes coding for abnormal Ras proteins. Such mutations are detected in roughly 30% of all human cancers, making *RAS* oncogenes the most commonly encountered type of human oncogene. Point mutations in *RAS* oncogenes cause a single incorrect amino acid to be inserted at one of several possible locations within the Ras protein. The net result is often a hyperactive Ras protein that retains bound GTP instead of degrading it to GDP, thereby maintaining the protein in a permanently activated state. In this hyperactive state, the Ras protein continually sends a stimulatory signal to the rest of the Ras-MAPK pathway, regardless of whether an appropriate growth factor is bound to the cell's growth factor receptors.

Of the three types of *RAS* genes, *KRAS* is the most frequently mutated in human cancers. Point mutations in *KRAS* are present in about 30% of lung cancers, 50% of colon cancers, and up to 90% of pancreatic cancers. Mutations in *NRAS* are less frequent in epithelial cancers but are detected in about 25% of acute leukemias. Finally, *HRAS* mutations are encountered primarily in bladder cancers, where they appear in about 10% of cases.

Some Oncogenes Produce Intracellular Protein Kinases

After the Ras protein has been activated, it triggers a cascade of intracellular protein phosphorylation reactions, beginning with the phosphorylation of a protein kinase called **Raf kinase** (Figure 9-8, step ④). Activated Raf kinase in turn catalyzes phosphorylation of another intracellular protein kinase called MEK, which phosphorylates another intracellular protein kinase called a **MAP kinase** or simply **MAPK** (an abbreviation for Mitogen-Activated Protein Kinase). Unlike receptor tyrosine kinases, the intracellular kinases involved in this cascade of protein phosphorylation reactions attach phosphate groups mainly to the amino acids serine and threonine in target proteins, rather than to tyrosine. Such enzymes are therefore referred to as **serine/threonine kinases**. Several oncogenes code for serine/threonine kinases involved in this cascade. An example is the ***BRAF*** oncogene, which codes for mutant forms of the Raf kinase in roughly two-thirds of human melanomas and at a lower frequency in a variety of other cancers.

In addition to the preceding types of *serine/threonine* kinases, several *tyrosine* kinases are likewise involved in intracellular signaling pathways. Unlike the receptor tyrosine kinases described earlier, which span the plasma membrane, these *intracellular* tyrosine kinases may be nuclear, cytoplasmic, or associated with the inner surface of the plasma membrane. Intracellular tyrosine kinases do not possess receptor sites and are therefore referred to as **nonreceptor tyrosine kinases**. Three examples are briefly described below.

1. **Src kinase.** The Src kinase is an intracellular tyrosine kinase found in normal cells that can induce cancer when present in an abnormal form. For example, we have already seen that the v-*src* oncogene of the Rous sarcoma virus produces an abnormal Src kinase that triggers sarcoma development in chickens (p. 133). In addition, mutations in the human *SRC* gene are associated with certain forms of colon cancer. Src kinase and related members of the Src kinase family interact with a broad range of growth factor receptors. In some cases, these intracellular tyrosine kinases transmit signals from activated receptors that do not possess their own tyrosine kinase activity. In other situations, they augment the activity of receptors that do possess tyrosine kinase activity. For example, Src kinase associates with receptors activated by the growth factors EGF and PDGF, catalyzing the phosphorylation of these receptors in a way that enhances the ability of the activated receptor to trigger the next steps in the signaling pathway.

2. **Jak kinase.** In the Jak-STAT pathway, receptors transmit signals to the cell interior by activating intracellular tyrosine kinases that are members of the Jak family (see Figure 9-10). In certain leukemias, a translocation between chromosomes 9 and 12 has been identified that creates a fusion gene called *TEL-JAK2*. This oncogene codes for a fusion protein in which the catalytic region of a Jak kinase is fused to a segment of an unrelated protein that causes the tyrosine kinase activity of Jak to become permanently activated. The resulting stimulation of the Jak-STAT signaling pathway leads to excessive cell proliferation.

3. **Abl kinase.** The Abl tyrosine kinase, which is produced by the *ABL* proto-oncogene, functions in the cell nucleus as part of a normal signaling pathway that causes cells with damaged DNA to self-destruct by apoptosis. In chronic myelogenous leukemia, a chromosomal translocation event involving chromosomes 9 and 22 creates a *Philadelphia chromosome* in which segments of the *BCR* and *ABL* genes are fused together to form a *BCR-ABL* oncogene (see Figure 9-4). The *BCR-ABL* oncogene codes for an abnormal version of the Abl tyrosine kinase that remains in the cytoplasm and therefore cannot trigger apoptosis. So in this particular case, an oncogene fosters an excessive accumulation of proliferating cells by enhancing their survival rather than by stimulating their proliferation.

The preceding examples represent just a few of the many intracellular protein kinases whose uncontrolled activity can contribute to cancer development by stimulating pathways that activate cell proliferation, promote cell survival, or both.

Some Oncogenes Produce Transcription Factors

Some of the intracellular protein kinases that are activated in cells stimulated by growth factors will in turn trigger changes in **transcription factors**, which are proteins that bind to DNA and alter the expression of specific genes. Activation of transcription factors is a common feature of the signaling pathways that control cell proliferation and survival. It occurs in both the Ras-MAPK and Jak-STAT pathways described in this chapter and in several additional pathways to be covered in the next chapter. In the case of the Ras-MAPK pathway, activated MAP kinases enter the nucleus and phosphorylate several different transcription factors, including Jun and members of the Ets family of proteins (see Figure 9-8, step ⑤). These activated transcription factors then stimulate the transcription of "early genes" that code for the production of other transcription factors, including the proteins Myc, Fos, and Jun, which then activate the transcription of a family of "delayed genes." One of these latter genes codes for the E2F transcription factor, whose role in controlling progression through the restriction point of the cell cycle will be described in Chapter 10 (see Figure 10-3).

Oncogenes that produce altered forms or excessive quantities of specific transcription factors have been detected in a broad range of human and animal cancers. Among the most common are oncogenes coding for Myc transcription factors, which control the expression of numerous genes involved in cell proliferation and survival. Chapter 7 described three examples of virally induced cancers in which an oncogene coding for a Myc protein plays a central role. The first example involved the *avian leukosis virus*, a retrovirus that causes leukemia in chickens by an insertional mutagenesis event in which the proviral DNA is integrated near the normal cellular gene coding for Myc. Insertion of the proviral DNA enhances the rate at which the nearby Myc gene is transcribed, thereby leading to overproduction of the normal Myc protein. The second example involved the *avian myelocytomatosis virus*, a retrovirus possessing a v-*myc* oncogene that induces cancer by producing an abnormal version of the Myc protein. The final example was Burkitt's lymphoma, a human cancer triggered by the Epstein-Barr virus. Burkitt's lymphoma is associated with a chromosomal translocation in which the *MYC* gene is translocated to chromosome 14, bringing it into close proximity to genes coding for antibody molecules (see Figure 7-12). This event leads to excessive production of the Myc protein in cells where antibody genes are active—that is, in lymphocytes.

Burkitt's lymphoma is only one of several human cancers in which the Myc protein is overproduced. In these other cancers, gene amplification rather than chromosomal translocation is usually responsible for the excessive production of Myc. For example, amplification of the *MYC* gene is frequently observed in small cell lung cancers and, to a lesser extent, in a wide range of other carcinomas, including 20% to 30% of breast and ovarian cancers. Two other members of the *MYC* gene family, which code for slightly different versions of the Myc protein, have also been implicated in cancer development. One is called *MYCN* (because it was first discovered in neuroblastomas), and the other is *MYCL* (first discovered in lung cancers). About 30% to 40% of small cell lung cancers exhibit amplification of the *MYC, MYCN,* or *MYCL* gene. *MYCN* is also amplified in other tumor types, including neuroblastomas and glioblastomas. In neuroblastomas, patients whose tumor cells exhibit *MYCN* gene amplification have poorer survival rates than patients whose tumors do not (see Figure 9-3).

Myc is just one of several transcription factors known to be produced by oncogenes. In animal retroviruses, for example, the oncogenes v-*fos*, v-*jun*, v-*myb*, v-*ets*, and v-*erb-a* each codes for a different transcription factor. Oncogenes coding for a variety of different transcription factors have also been reported in human cancers, although *MYC* family members are still the most prevalent.

Some Oncogenes Produce Cell Cycle or Cell Death Regulators

In the final step of growth factor signaling pathways, transcription factors activate the expression of genes coding for proteins involved in cell proliferation (see step ⑥ in Figures 9-8 and 9-10). The activated genes include those coding for *cyclin-dependent kinases* (*Cdks*) and *cyclins*, whose roles in controlling passage through key points in the cell cycle were described in Chapter 2. Several human oncogenes produce proteins in this category. For example, a cyclin-dependent kinase gene called *CDK4* is amplified in certain sarcomas and glioblastomas, and the cyclin gene *CYCD1* is often amplified in breast cancers and is altered by chromosomal translocation in some lymphomas. The presence of such oncogenes causes the production of excessive amounts or hyperactive versions of Cdk-cyclin complexes, which then stimulate progression through the cell cycle.

Stimulating progression through the cell cycle is not the only way of increasing the number of proliferating cells. The number of cells in a growing tumor is also influenced by the rate at which cells die. Normal tissues maintain a carefully regulated balance between cell proliferation and cell death, but in tumors this balance is disrupted in ways that lead to a progressive increase in dividing cells. This increase can arise from enhanced cell proliferation, decreased cell death, or some combination of the two. Most of the oncogenes discussed in this chapter act mainly by stimulating cell proliferation, but a few oncogenes act primarily or solely by inhibiting cell death. One example is the *BCL2* gene, which codes for a protein called **Bcl2**. The Bcl2 protein resides on the outer surface of mitochondria and acts as a retraining influence on the pathway by which cells are destroyed by apoptosis (see Figure 2-13, *bottom*). In non-Hodgkin's lymphomas, a common chromosomal translocation causes the *BCL2* gene to produce too much Bcl2. The excessive amounts of Bcl2 block the pathway for apoptosis, thereby leading to a progressive accumulation of cells that would otherwise have been destroyed. Another gene that affects cell death, called *MDM2*, is amplified in some human sarcomas and produces excessive amounts of a protein (Mdm2) that inhibits the ability of cells to self-destruct by apoptosis. (The mechanism by which the Mdm2 protein exerts its influence on apoptosis will be covered in the next chapter.) Oncogenes such as *BCL2* and *MDM2* help cancer cells evade the apoptotic pathways that would otherwise trigger their destruction.

Summing Up: Oncogene-Induced Disruptions in Signaling Pathways Exhibit Some Common Themes

In discussing how oncogenes work, this chapter has focused on two signaling pathways: the Ras-MAPK pathway and the Jak-STAT pathway. In reality, these two signaling mechanisms are components of a larger network of pathways, involving numerous branches and shared components, that work together to determine whether cells will proliferate, stop proliferating, or die. Yet despite the complexity of the branched interconnections, the various pathways involved in controlling cell proliferation and survival share some common features. First, binding of a growth factor to its receptor leads to receptor activation. Next, the activated receptor triggers a complex chain of events that includes a series of protein phosphorylation reactions. These protein phosphorylations then trigger changes in transcription factors that alter the expression of specific genes. Finally, the activated or inhibited genes produce proteins that influence cell proliferation and cell death.

Oncogenes exert their harmful effects by producing excessive quantities or hyperactive versions of proteins involved in these steps. The net result is that the *presence* of an oncogene leads to excessive cell proliferation and, in some cases, diminished cell death. In the next chapter, we will see how the *absence* (or loss of function) of a tumor suppressor gene can likewise lead to excessive cell proliferation and diminished cell death.

Summary of Main Concepts

- Proto-oncogenes are normal genes that play essential roles in the control of cell proliferation and survival. The term *proto-oncogene* simply implies that if and when the structure or activity of a proto-oncogene is altered, the resulting mutant form of the gene, called an oncogene, can cause cancer. Proto-oncogenes are converted into oncogenes by five different mechanisms: point mutation, gene amplification, chromosomal translocation, local DNA rearrangements, and insertional mutagenesis. In some cases, the structure of a proto-oncogene is altered in a way that causes it to produce an abnormal protein. In other cases, the expression of a proto-oncogene is enhanced, thereby leading to excessive production of a normal protein. In either case, the net result is the unregulated activation of a signaling pathway that leads to excessive cell proliferation or survival.
- Cellular oncogenes were initially detected by gene transfer experiments in which DNA isolated from tumor cells was introduced into normal cells and tested for its ability to transform them into cancer cells. This approach is useful mainly for detecting oncogenes that code for abnormal proteins.
- Most oncogenes code for proteins that are components of pathways that stimulate cell proliferation, such as the Ras-MAPK and Jak-STAT pathways. Instead of producing the proper amount of a correct protein, however, an oncogene produces either an abnormal protein or excessive amounts of a normal one, thereby leading to excessive activity of the signaling pathway.
- The proteins involved in growth factor signaling pathways fall into six main categories: growth factors, receptors, plasma membrane G proteins, intracellular protein kinases, transcription factors, and cell cycle or cell death regulators. Oncogenes can code for proteins in any of these categories. For example, oncogenes have been identified that produce (1) growth factors such as PDGF molecules in an uncontrolled fashion; (2) receptor tyrosine kinases in which the growth factor binding site is disrupted, thereby leading to unregulated tyrosine kinase activity; (3) hyperactive forms of the G protein Ras; (4) abnormal forms of intracellular protein kinases such as Raf, Src, Jak, and Abl; (5) excessive quantities or hyperactive forms of transcription factors such as Myc; and (6) abnormal forms or excessive quantities of cell cycle or cell death regulators such as cyclins, cyclin-dependent kinases, Bcl2, and Mdm2.
- The Ras-MAPK and Jak-STAT signaling mechanisms are components of a larger interconnected network of pathways, involving numerous branches and shared components, that work together to determine whether cells will proliferate, stop proliferating, or die. Oncogenes exert their harmful effects by producing excessive quantities or hyperactive versions of proteins involved in these signaling networks. The net result is an unregulated activation of signaling pathways that leads to uncontrolled cell proliferation, diminished cell death, or both.

Key Terms for Self-Testing

oncogene (p. 158)
tumor suppressor gene (p. 158)

How Cellular Oncogenes Arise

proto-oncogene (p. 159)
viral oncogene (p. 159)
transfection (p. 159)
gene cloning (p. 159)
cellular oncogene (p. 159)
RAS oncogene (p. 159)
RAS proto-oncogene (p. 159)
point mutation (p. 160)
gene amplification (p. 160)
homogeneously staining region (HSR) (p. 160)
double minute (DM) (p. 160)
MYC gene (p. 161)
ERBB2 gene (p. 161)
chromosomal translocation (p. 162)
Philadelphia chromosome (p. 162)
fusion gene (p. 162)
BCR-ABL (p. 162)
fusion protein (p. 162)
DNA rearrangement (p. 163)
deletion (p. 163)
insertion (p. 163)
transposition (p. 163)
inversion (p. 163)
TRK oncogene (p. 163)
insertional mutagenesis (p. 164)

Proteins Produced by Oncogenes

Ras-MAPK pathway (p. 165)
growth factor (p. 165)
platelet-derived growth factor (PDGF) (p. 165)
epidermal growth factor (EGF) (p. 165)
receptor (p. 167)
protein kinase (p. 167)
receptor tyrosine kinase (p. 167)
autophosphorylation (p. 168)
constitutively active (p. 168)
Jak-STAT pathway (p. 169)
STATs (p. 169)
Ras (p. 169)
G protein (p. 169)
guanine-nucleotide exchange factor (GEF) (p. 170)
Raf kinase (p. 170)
MAP kinase (MAPK) (p. 170)
serine/threonine kinase (p. 170)
BRAF (p. 170)
nonreceptor tyrosine kinase (p. 171)
Src kinase (p. 171)
Jak kinase (p. 171)
Abl kinase (p. 171)
transcription factor (p. 171)
Bcl2 (p. 172)

Suggested Reading

How Cellular Oncogenes Arise

Butti, M. G., et al. A sequence analysis of the genomic regions involved in the rearrangements between TPM3 and NTRK1 genes producing TRK oncogenes in papillary thyroid carcinomas. *Genomics* 28 (1995): 15.

Olopade, I. O., O. M. Sobulo, and J. D. Rowley. Recurring Chromosome Rearrangements in Human Cancer. In *Holland-Frei Cancer Medicine*, 5th ed. (R. C. Bast et al., eds.). Lewiston, NY: Decker, 2000, Chapter 6.

Pierotti, M. A., et al. Oncogenes. In *Holland-Frei Cancer Medicine*, 5th ed. (R. C. Bast et al., eds.). Lewiston, NY: Decker, 2000, Chapter 4.

Rowley, J. D. Chromosome translocations: dangerous liaisons revisited. *Nature Reviews Cancer* 1 (2001): 245.

Schwab, M. Amplification of oncogenes in human cancer cells. *BioEssays* 20 (1998): 473.

Seeger, R. C., et al. Association of multiple copies of the N-myc oncogene with rapid progression of neuroblastoma. *New England J. Med.* 313 (1985): 1111.

Weinberg, R. A. A molecular basis of cancer. *Sci. Amer.* 249 (November 1983): 126.

Proteins Produced by Oncogenes

Blume-Jensen, P., and T. Hunter. Oncogenic kinase signalling. *Nature* 411 (2001): 355.

Cory, S., and J. M. Adams. The BCL2 family: regulators of the cellular life-or-death switch. *Nature Reviews Cancer* 2 (2002): 647.

Fearon, E. R. Oncogenes and Tumor Suppressor Genes. In *Clinical Oncology* (M. D. Abeloff, ed.). New York: Churchill Livingstone, 2000, Chapter 5.

Futreal, P. A., et al. A census of human cancer genes. *Nature Reviews Cancer* 4 (2004): 177.

Malumbres, M., and M. Barbacid. *RAS* oncogenes: the first 30 years. *Nature Reviews Cancer* 3 (2003): 7.

Martin, G. S. The hunting of the Src. *Nature Reviews Molecular Cell Biol.* 2 (2001): 467.

Pelengaris, S., M. Khan, and G. Evan. c-MYC: more than just a matter of life and death. *Nature Reviews Cancer* 2 (2002): 764.

Yeatman, T. J. A renaissance for SRC. *Nature Reviews Cancer* 4 (2004): 470.

Tumor Suppressor Genes and Cancer Overview

10

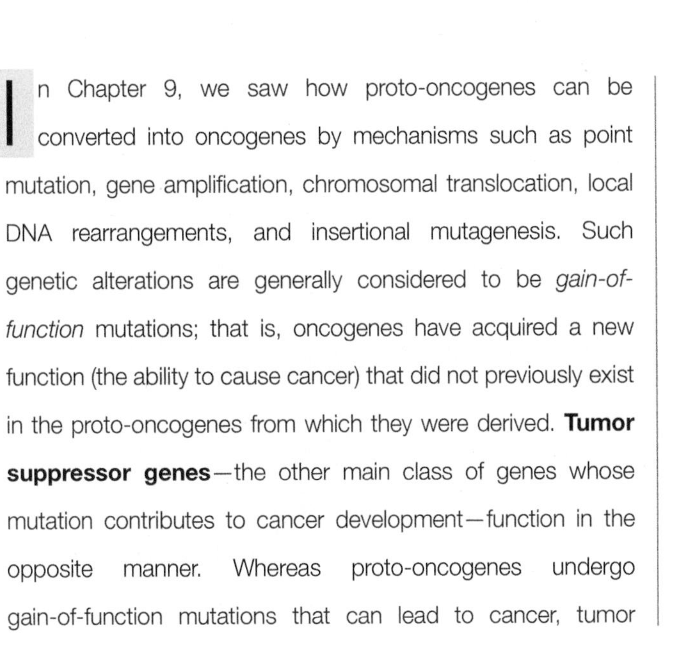

In Chapter 9, we saw how proto-oncogenes can be converted into oncogenes by mechanisms such as point mutation, gene amplification, chromosomal translocation, local DNA rearrangements, and insertional mutagenesis. Such genetic alterations are generally considered to be *gain-of-function* mutations; that is, oncogenes have acquired a new function (the ability to cause cancer) that did not previously exist in the proto-oncogenes from which they were derived. **Tumor suppressor genes**—the other main class of genes whose mutation contributes to cancer development—function in the opposite manner. Whereas proto-oncogenes undergo gain-of-function mutations that can lead to cancer, tumor suppressor genes undergo *loss-of-function* mutations that can likewise lead to cancer.

Of the roughly 25,000 genes present in human cells, only a few dozen exhibit the properties of tumor suppressors. Since losing the function of just one of these genes can be the critical event that causes cancer to arise, each tumor suppressor must perform an extremely important task. In this chapter, we will examine the nature of tumor suppressor genes and the ways in which their loss can lead to cancer. After describing the functions of tumor suppressors, we will conclude with an overview of the roles played by all types of gene mutations, along with nonmutational changes, in converting normal cells into cancer cells.

TUMOR SUPPRESSOR GENES: ROLES IN CELL PROLIFERATION AND CELL DEATH

By definition, tumor suppressors are genes whose loss or inactivation can lead to cancer, a condition characterized by increased cell proliferation and decreased cell death. It is therefore logical to suspect that the normal function of a tumor suppressor gene would be the opposite—namely, to *inhibit* cell proliferation or *promote* cell death—and so the loss of such functions would cause increased cell proliferation or decreased cell death. We will see in the first part of the chapter that many tumor suppressor genes behave in precisely this way.

Cell Fusion Experiments Provided the First Evidence for the Existence of Tumor Suppressor Genes

The first indication that cells might contain genes whose loss is associated with the development of cancer came from experiments using a technique called **cell fusion**. In 1960, a research team in Paris headed by Georges Barski discovered that cells of two different types grown in culture will occasionally fuse together to form hybrid cells containing the chromosomes of both original cell types. Shortly thereafter Henry Harris reported that cell fusion can be artificially induced by treating cells with inactivated forms of a particular type of virus called *Sendai virus*. Treatment with the virus causes the plasma membranes of two cells to fuse with each other, creating a combined cell in which the nuclei of the two original cells share the same cytoplasm. When the cell subsequently divides, the two separate nuclei break down and a single new nucleus is formed that contains chromosomes derived from both of the original cells. Such a cell, containing a nucleus with chromosomes derived from two different cells, is called a **hybrid cell**.

Experiments in which cancer cells were fused with normal cells provided some important early insights into the genetic basis for the abnormal behavior of cancer cells. Based on our current understanding of oncogenes, you might expect that the hybrid cells created by fusing cancer cells with normal cells would have acquired oncogenes from the original cancer cell and would therefore exhibit uncontrolled proliferation, just like a cancer cell. In fact, that is not what usually happens; the fusion of cancer cells with normal cells almost always yields hybrid cells that initially behave like the normal parent and do not form tumors (Figure 10-1). Such results, first reported in the late 1960s, provided the earliest evidence that normal cells contain genes that can suppress tumor growth and reestablish normal controls on cell proliferation.

Although fusing cancer cells with normal cells generally yields hybrid cells that lack the ability to form tumors, it does not mean that these cells are normal. When they are allowed to grow for extended periods in culture, the hybrid cells often revert back to the malignant, uncontrolled behavior of the original cancer cells. Reversion to malignant behavior is associated with the loss of certain chromosomes, suggesting that these particular chromosomes contain genes that had been suppressing the ability to form tumors. Such observations eventually led to the naming of the lost genes as "tumor suppressor genes."

As long as hybrid cells retain both sets of original chromosomes—that is, chromosomes derived from both the cancer cells and the normal cells—the ability to form tumors is suppressed. Tumor suppression is even observed when the original cancer cells possess an oncogene, such as a mutant *RAS* gene, that is actively expressed in the hybrid cells. This means that tumor suppressor genes located in the chromosomes of normal cells are able to overcome the effects of a *RAS* oncogene present in a cancer cell chromosome. The ability to form tumors only reappears after the hybrid cell loses a chromosome containing a critical tumor suppressor gene.

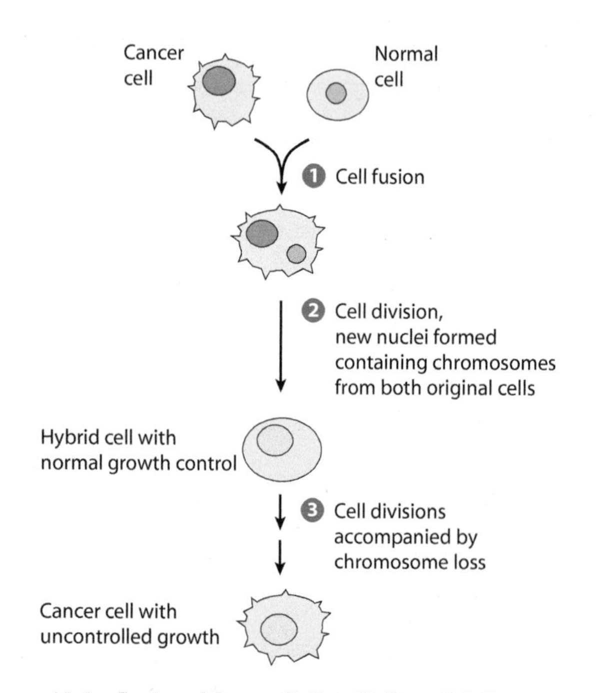

Figure 10-1 Fusion of Cancer Cells with Normal Cells. Cells can be artificially induced to fuse together by exposing them to inactivated Sendai virus (or to several other inactivated viruses or chemical treatments). ① When a cancer cell is fused to a normal cell, the initial result is a cell in which the nuclei of the two original cells share the same cytoplasm. ② During the next cell division, the separate nuclei break down and a single new nucleus is formed containing chromosomes derived from the two original cells. These initial hybrid cells usually exhibit normal growth control and do not form tumors. ③ After dividing for extended periods in culture, the hybrid cells often revert back to the uncontrolled proliferation of the original cancer cells and acquire the ability to form tumors. This reversion is accompanied by the loss of chromosomes containing tumor suppressor genes.

Studies of Inherited Chromosomal Defects and Loss of Heterozygosity Have Led to the Identification of Several Dozen Tumor Suppressor Genes

Although cell fusion experiments provided early evidence for the existence of tumor suppressor genes, identifying these genes did not turn out to be a simple task. By definition, the existence of a tumor suppressor gene only becomes evident after its function has been lost. How do scientists go about identifying something whose very existence is unknown until it disappears?

One approach is based on the fact that defects in tumor suppressors are responsible for several hereditary cancer syndromes. Members of cancer-prone families often inherit a defective tumor suppressor gene from one parent, thereby elevating their cancer risk because a single mutation in the other copy of that tumor suppressor gene can then lead to cancer. Microscopic examination of cells obtained from individuals in such families sometimes reveals the existence of gross chromosomal defects. For example, certain individuals with *familial retinoblastoma* exhibit a deleted segment in a specific region of one copy of chromosome 13, not just in cancer cells but in all cells of the body. To determine whether a tumor suppressor is located in the region that has undergone deletion, scientists have simply examined retinoblastoma cells to see which gene has become mutated in the comparable region of the second copy of chromosome 13. As described in Chapter 8, this approach first led to the discovery of the *RB* tumor suppressor gene (p. 143).

The loss of tumor suppressor genes is not restricted to hereditary cancers. These genes may also be lost or inactivated through random mutations that strike a particular target tissue, leading to the mutation or loss of both copies of the same gene. You might think that the most straightforward way for that to happen would be through two independent mutations randomly occurring in sequence. However, the mutation rate for any given gene is about one in a million per cell division, so the chance of two independent mutations affecting two copies of the same gene is extremely remote.

After a single copy of a tumor suppressor gene has undergone mutation, a more efficient approach for disrupting the remaining normal copy is through a phenomenon known as **loss of heterozygosity**, so named because the initial state, in which one abnormal and one normal gene copy are present, is called the *heterozygous* state. Getting rid of the remaining normal copy therefore causes the heterozygous state to be lost. Loss of heterozygosity is more common than you might expect; whereas individual gene mutations arise at a rate of one in a million per gene per cell division, loss of heterozygosity is as frequent as once in a thousand cell divisions and tends to affect large regions of DNA encompassing hundreds of different genes.

Figure 10-2 illustrates several ways in which loss of heterozygosity may arise. In one mechanism, called *mitotic nondisjunction*, the two duplicated copies of a given chromosome fail to separate (disjoin) at the time of mitosis, so both copies go to one daughter cell and the other daughter cell receives no copies. As seen in Figure 10-2b, the latter cell will no longer be heterozygous for any genes contained on the missing chromosome. A second mechanism involves *mitotic recombination*, in which homologous chromosomes exchange DNA sequences when they line up during the process of mitosis. Figure 10-2c shows how such an exchange could lead to loss of heterozygosity. A third mechanism, called *gene conversion*, occurs when the DNA molecules from two homologous chromosomes line up next to each other and copy base sequence information from one to the other. In this way, a DNA region that was originally present in two different versions in the two members of a homologous pair of chromosomes can be made identical by copying DNA sequence information from one chromosome to the other chromosome (Figure 10-2d).

The existence of the preceding mechanisms means that if a cell happens to acquire a random mutation that inactivates one copy of a tumor suppressor gene, loss of heterozygosity might either replace the normal copy with the defective version or remove it entirely. Loss of heterozygosity usually affects hundreds of neighboring genes simultaneously, making it relatively easy to detect. You simply analyze a large number of known genes, searching for those that are present in two different versions in the normal cells of a cancer patient but are present in only one version in the same person's cancer cells. When genes exhibiting this behavior are detected, it is likely that they lie near a tumor suppressor gene whose loss of heterozygosity is actually responsible for the cancerous growth.

Geneticists have performed thousands of searches looking for chromosomal regions that exhibit loss of heterozygosity in cancer cells. This approach, along with the study of chromosomal defects associated with hereditary cancer syndromes, has led to the identification of several dozen tumor suppressor genes. Once they had been identified, two obvious questions arose: What are the normal functions of these genes, and how does disruption of their functions lead to cancer? Answers will be provided in the next several sections, as we describe the properties of the proteins produced by several prominent tumor suppressor genes.

The *RB* Tumor Suppressor Gene Produces a Protein That Restrains Passage Through the Restriction Point

The first tumor suppressor gene to be isolated and characterized was the ***RB* gene**, a gene whose loss in hereditary retinoblastoma was described in Chapter 8.

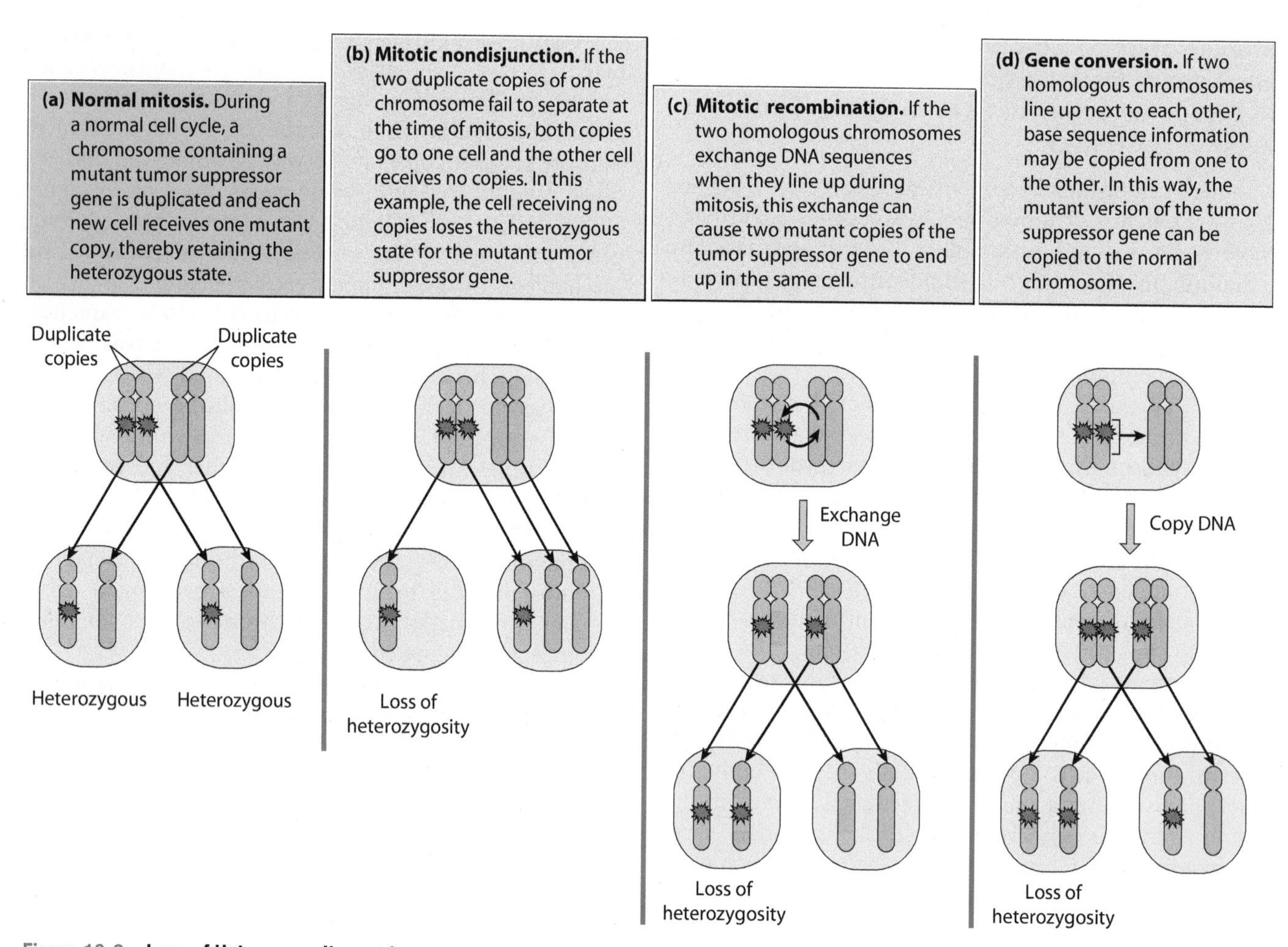

Figure 10-2 Loss of Heterozygosity. After a tumor suppressor gene has undergone mutation in one chromosome, the normal copy present in the other homologous chromosome may be disrupted through a phenomenon called loss of heterozygosity. Three mechanisms for loss of heterozygosity are illustrated here: (**b**) mitotic nondisjunction, (**c**) mitotic recombination, and (**d**) gene conversion. A chromosome region that has undergone loss of heterozygosity in cancer cells but not in normal cells is likely to contain a tumor suppressor gene within it.

The protein produced by the *RB* gene, called the **Rb protein** (or simply Rb), restrains cell proliferation in the absence of growth factors. As was described in Chapter 2, the Rb protein normally exerts this action by halting the cell cycle at the restriction point. In cells that have been exposed to an appropriate growth factor, however, signaling pathways trigger the production of Cdk-cyclin complexes that catalyze the phosphorylation of Rb. Phosphorylated Rb can no longer exert its inhibitory effects and so the cells are free to pass through the restriction point and into S phase.

The molecular mechanism by which Rb exerts this control over the restriction point is summarized in Figure 10-3. Prior to phosphorylation, Rb binds to the **E2F transcription factor**, a protein that (in the absence of bound Rb) activates the transcription of genes coding for enzymes and other proteins required for initiating DNA replication. As long as the Rb protein remains bound to E2F, the E2F molecule is inactive and these genes stay silent, thereby preventing cells from entering into S phase. However, in a cell that has been stimulated to divide (e.g., by the addition of growth factors), the activation of growth signaling pathways leads to the production of Cdk-cyclin complexes that catalyze the phosphorylation of Rb. Phosphorylation abolishes the ability of Rb to bind to E2F, thus allowing E2F to activate the transcription of genes whose products are required for entry into S phase.

Because the normal purpose of Rb is to halt the cell cycle in the absence of growth factors, *RB* mutations that lead to the loss or inactivation of the Rb protein remove this restraining influence on the cell cycle and lead to excessive proliferation. Such mutations leading to a loss of Rb function are observed in some hereditary as well as environmentally caused forms of cancer. Certain cancer viruses also disrupt Rb function. For example, the human papillomavirus (HPV), whose role in cervical cancer was described in Chapter 7, has an oncogene that codes for the **E7 oncoprotein**, which binds to Rb (see Figure 7-13, *bottom*). When bound to E7, the Rb protein cannot perform its normal function of restraining passage through the restriction point and cell proliferation therefore proceeds unchecked, even in the absence of growth factors. Cancers triggered by a loss of Rb function can thus arise in two fundamentally

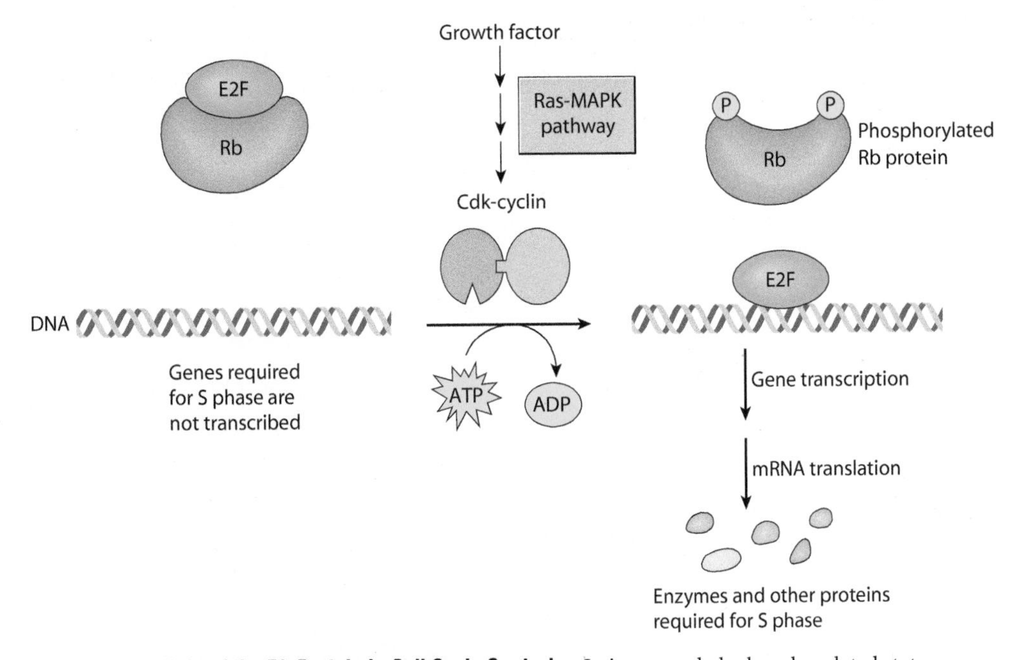

Figure 10-3 Role of the Rb Protein in Cell Cycle Control. In its normal, dephosphorylated state, the Rb protein binds to the E2F transcription factor. This binding prevents E2F from activating the transcription of genes coding for proteins required for DNA replication, which are needed before the cell can pass through the restriction point and into S phase. In cells that have been stimulated by growth factors, signaling pathways such as the Ras-MAPK pathway trigger the production of Cdk-cyclin complexes that catalyze Rb phosphorylation. The phosphorylated Rb can no longer bind to E2F, which allows E2F to activate gene transcription and trigger the onset of S phase. At the time of the subsequent mitosis (not shown), the phosphate groups are removed from Rb so that it can once again inhibit E2F.

different ways: through mutations that delete or disrupt both copies of the *RB* gene and through the action of viral oncoproteins that bind to and inactivate the Rb protein.

Table 10-1 Some Examples of Tumor Suppressor Genes

Gene	Pathway Affected
Gatekeeper Genes	
APC	Wnt signaling
CDKN2A	Rb and p53 signaling
PTEN	PI3K-Akt signaling
RB	Restriction point control
SMAD4	TGFβ-Smad signaling
TGFβ receptor	TGFβ-Smad signaling
p53	DNA damage response
Caretaker Genes	
ATM	DNA damage response
BRCA1, BRCA2	Double-strand break repair
MSH2, MSH3, MSH4, MSH5, MSH6, PMS1, PMS2, MLH1	DNA mismatch repair
XPA, XPB, XPC, XPD, XPE, XPF, XPG,	DNA excision repair
XPV (POLη)	Translesion synthesis

The *p53* Tumor Suppressor Gene Produces a Protein That Prevents Cells with Damaged DNA from Proliferating

Since the discovery of the *RB* gene in the mid-1980s, dozens of additional tumor suppressor genes have been identified (Table 10-1). One of the most important is the ***p53* gene** (also called *TP53* in humans), which produces the **p53 protein**. The *p53* gene is mutated in a broad spectrum of different tumor types, and almost half of the close to the ten million people diagnosed worldwide with cancer each year will have *p53* mutations, making it the most commonly mutated gene in human cancers (Figure 10-4).

The p53 protein is sometimes called the "guardian of the genome" because of the central role that it plays in protecting cells from the effects of DNA damage. Figure 10-5 illustrates how this function is performed. When cells are exposed to DNA-damaging agents, such as ionizing radiation or toxic chemicals, the damaged DNA triggers the activation of an enzyme called **ATM kinase**, which catalyzes the phosphorylation of p53 and several other target proteins. (The ATM kinase is produced by the *ATM* tumor suppressor gene, whose role in ataxia telangiectasia was described on p. 151 in Chapter 8.) Phosphorylation of p53 by the ATM kinase prevents it from interacting with **Mdm2**, a protein that would otherwise mark p53 for destruction by linking it to a small protein called **ubiquitin**. Mdm2 is one of numerous

Figure 10-4 Prevalence of *p53* Mutations in Human Cancers. The percentage of tumors exhibiting mutations in the *p53* gene is illustrated for various types of human cancer. The *p53* gene is the most commonly mutated gene in human cancers. [Based on data from *IARC TP53 Mutation Database* (Lyon, France: International Agency for Research on Cancer, 2004).]

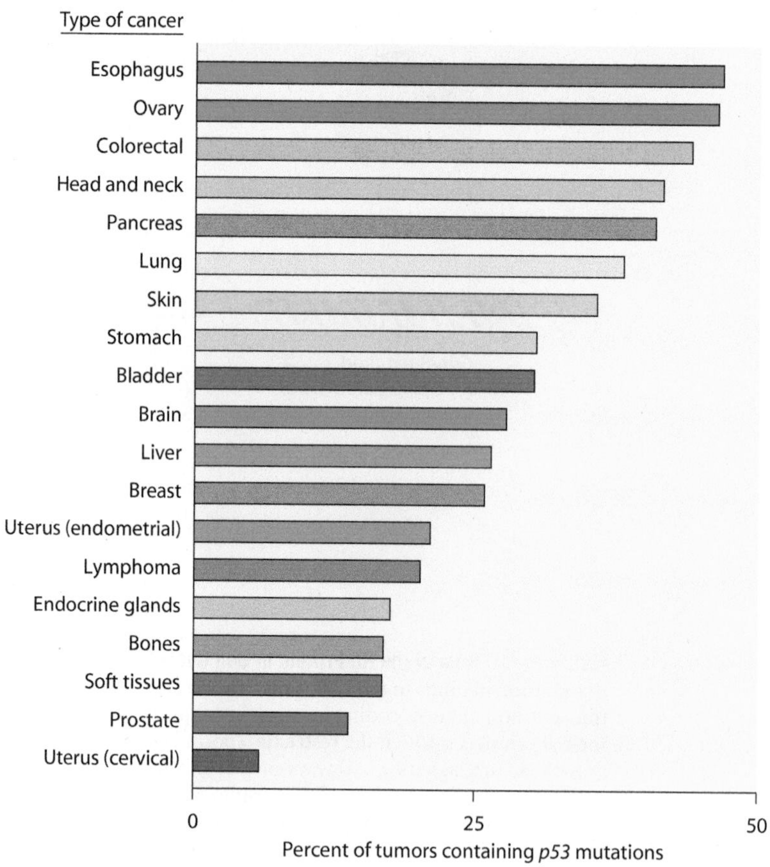

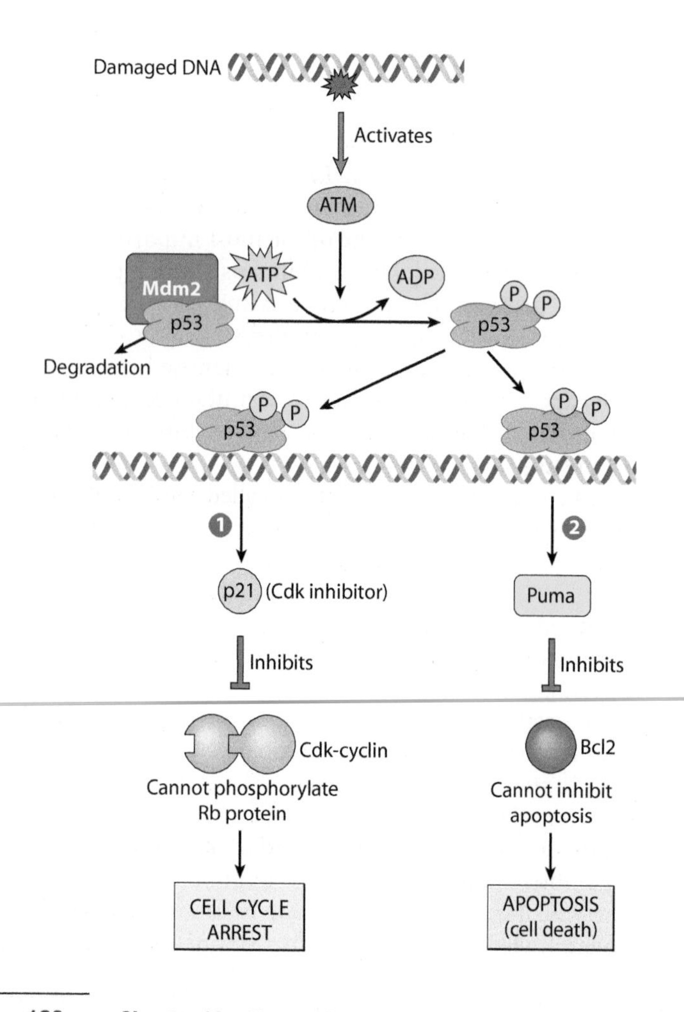

Figure 10-5 Role of the p53 Protein in Responding to DNA Damage. Damaged DNA activates the ATM protein kinase, leading to phosphorylation of the p53 protein. Phosphorylation stabilizes p53 by blocking its interaction with Mdm2, a protein that would otherwise mark p53 for degradation by attaching it to ubiquitin (see Figure 10-6). When the interaction between p53 and Mdm2 is blocked by p53 phosphorylation, the phosphorylated p53 protein accumulates and triggers two events. ① The p53 protein binds to DNA and activates transcription of the gene coding for the p21 protein, a Cdk inhibitor. The resulting inhibition of Cdk-cyclin prevents phosphorylation of the Rb protein, leading to cell cycle arrest at the restriction point. ② When the DNA damage cannot be repaired, p53 activates genes coding for a group of proteins involved in triggering cell death by apoptosis. A key protein is Puma, which promotes apoptosis by binding to, and blocking the action of, the apoptosis inhibitor Bcl2.

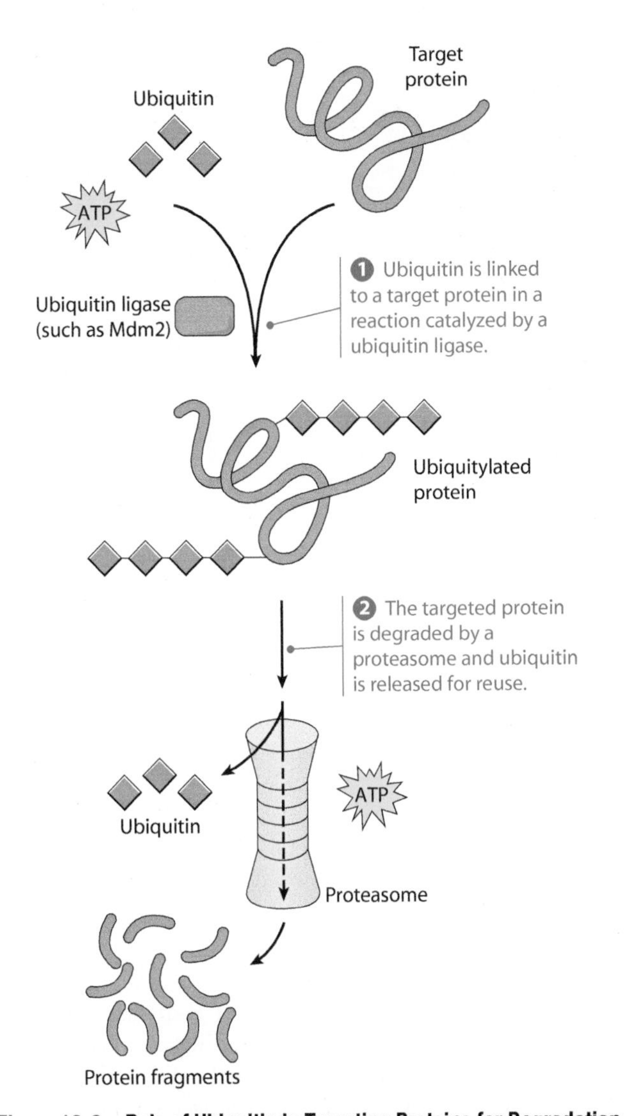

Figure 10-6 Role of Ubiquitin in Targeting Proteins for Degradation. A common mechanism for targeting proteins for destruction involves tagging them with a small protein called ubiquitin. Cells possess a variety of enzymes called ubiquitin ligases, each of which attaches ubiquitin molecules to a specific set of target proteins. Mdm2 is one example of such a ubiquitin ligase. Proteins that have been linked to ubiquitin are degraded by proteasomes, the cell's main protein-degrading apparatus. ATP provides energy required by this pathway at two different points.

proteins in the cell, called *ubiquitin ligases*, that attach ubiquitin molecules to a specific set of proteins. As shown in Figure 10-6, the normal function of ubiquitin is to direct molecules to the **proteasome**, the cell's main protein destruction machine. After p53 has been phosphorylated by ATM in response to DNA damage, the Mdm2 ubiquitin ligase can no longer attach ubiquitin chains to p53. As a result, the p53 protein accumulates in cells containing damaged DNA rather than being degraded by the ubiquitin-mediated proteasome pathway.

The accumulating p53 in turn activates two types of events: *cell cycle arrest* and *cell death*. Both responses are based on the ability of p53 to act as a transcription factor that binds to DNA and activates specific genes. Among the targeted genes is the gene coding for the **p21 protein**, a member of a class of molecules called **Cdk inhibitors** because they block the activity of Cdk-cyclin complexes. The p21 protein inhibits the Cdk-cyclin complex that would normally phosphorylate Rb, thereby halting the cell cycle at the restriction point and providing time for the DNA damage to be repaired. At the same time, p53 also activates the production of DNA repair enzymes. If the damage cannot be successfully corrected, p53 then activates genes that produce proteins involved in triggering cell death by apoptosis. A key protein in this pathway, called **Puma** ("p53 upregulated modulator of apoptosis"), promotes apoptosis by binding to and inactivating the Bcl2 protein, a normally occurring inhibitor of apoptosis (see Figure 2-13).

By triggering cell cycle arrest or cell death in response to DNA damage, the p53 protein prevents genetically altered cells from proliferating and passing the damage on to future cell generations. Mutations that disrupt p53 function therefore increase cancer risk because they permit cells with damaged DNA to survive and reproduce. For example, individuals who inherit a mutant *p53* gene from one parent have an elevated risk of developing cancer because they only require one additional mutation to inactivate the second copy of the gene. This high-risk hereditary condition, called the *Li-Fraumeni syndrome*, was described in detail in Chapter 8. Most *p53* mutations, however, are not inherited; they are caused by exposure to DNA-damaging chemicals and radiation. To cite but two examples, carcinogenic chemicals in tobacco smoke have been found to trigger point mutations in the *p53* gene of lung cells, and the ultraviolet radiation in sunlight has been shown to cause *p53* mutations in skin cells (p. 106).

When exposure to carcinogenic chemicals or radiation creates mutations in the *p53* gene, you might expect that both copies of the gene would need to be inactivated before functional p53 protein would be lost. In some cases, however, mutation of one copy of the *p53* gene may be sufficient to disrupt the p53 protein, even when the other copy of the gene is normal. The apparent explanation is that the p53 molecule is constructed from four protein chains bound together to form a *tetramer*. As shown in Figure 10-7, the presence of even one mutant chain in such a tetramer can be enough to prevent the p53 protein from functioning normally. When a mutation in one copy of the *p53* gene causes the p53 protein to be inactivated in this way, even in the presence of a normal copy of the gene, it is called a **dominant negative mutation**.

Mutating the *p53* gene is not the only mechanism for disrupting p53 function; the p53 protein can also be targeted directly by certain viruses. For example, human papillomavirus—whose E7 oncoprotein inactivates the Rb protein—produces another molecule, called the **E6 oncoprotein**, which binds to and targets the p53 protein for destruction (see Figure 7-13, *top*). The ability of human papillomavirus to cause cancer is therefore linked

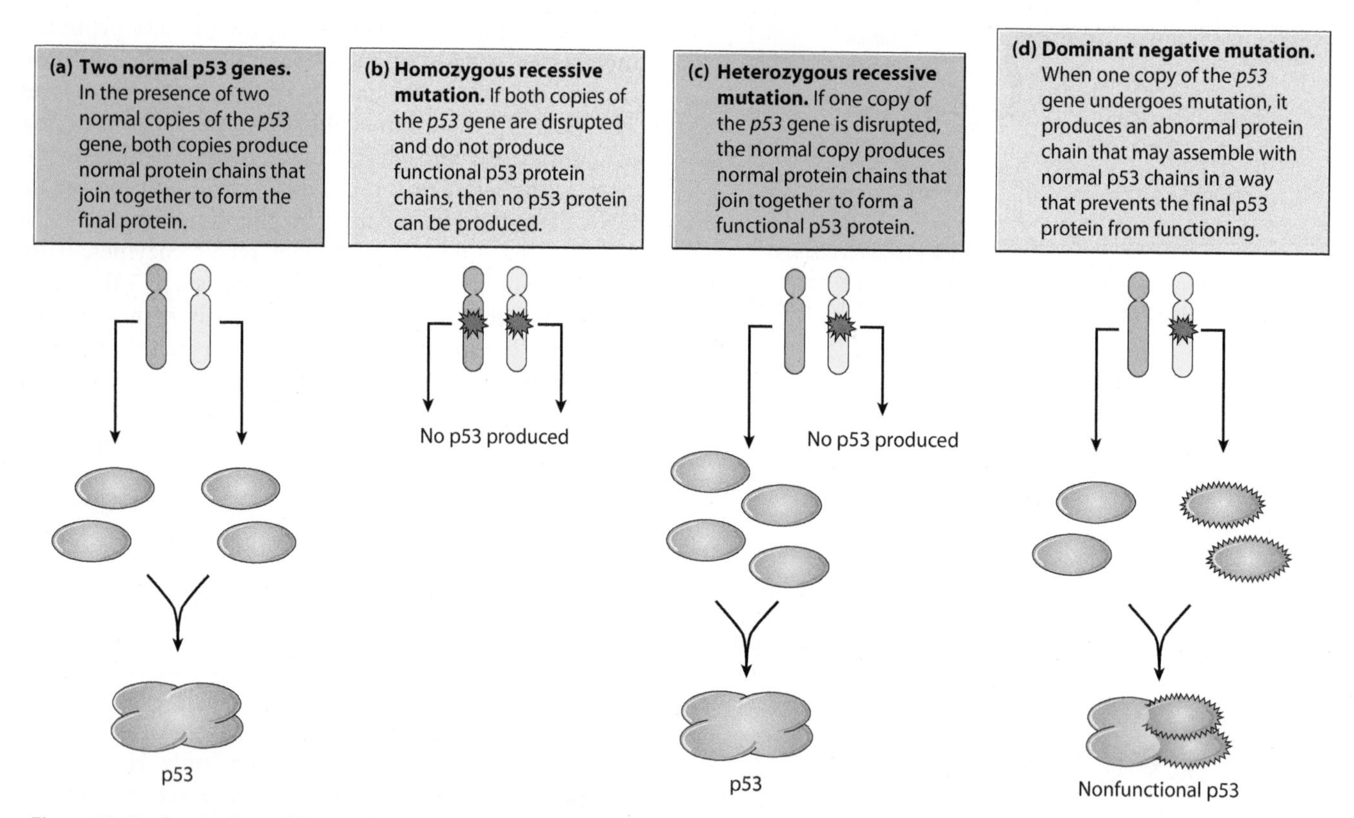

Figure 10-7 Comparison of Recessive and Dominant Negative *p53* Mutations. Loss of p53 protein function can arise either through (**b**) homozygous recessive mutations that disrupt both copies of the *p53* gene or through (**d**) a dominant negative mutation in which the abnormal protein chain produced by one mutant *p53* gene causes the p53 protein to be inactivated, even in the presence of a normal copy of the *p53* gene.

to its capacity to block the action of proteins produced by both the *RB* and *p53* tumor suppressor genes.

The *APC* Tumor Suppressor Gene Codes for a Protein That Inhibits the Wnt Signaling Pathway

The next tumor suppressor to be discussed is, like the *p53* gene, a frequent target for cancer-causing mutations; in this case, however, cancers arise mainly in one organ, namely the colon. The gene in question, called the ***APC* gene**, is the tumor suppressor whose association with *familial adenomatous polyposis* was introduced in Chapter 8. Individuals with this condition inherit a defective *APC* gene that causes thousands of polyps to grow in the colon and imparts a nearly 100% risk of developing colon cancer for individuals who live to the age of 60. Although familial adenomatous polyposis is quite rare, accounting for less than 1% of all colon cancers, *APC* mutations are also associated with the more common forms of colon cancer that arise in people with no family history of the disease. In fact, recent studies suggest that roughly two-thirds of all colon cancers involve *APC* mutations.

The *APC* gene codes for a protein involved in the **Wnt pathway**, a signaling mechanism that plays a prominent role in activating cell proliferation during embryonic development. As shown in Figure 10-8, the central component of the Wnt pathway is a protein called *β-catenin*. Normally, β-catenin is prevented from functioning by a multiprotein *destruction complex* that consists of the APC protein combined with the proteins *axin* and *glycogen synthase kinase 3* (*GSK3*). When assembled in such an APC-axin-GSK3 complex, GSK3 catalyzes the phosphorylation of β-catenin. The phosphorylated β-catenin then becomes a target for a ubiquitin ligase that attaches it to ubiquitin, thereby marking the phosphorylated β-catenin for degradation by proteasomes (see Figure 10-6). The net result is a low concentration of β-catenin, which makes the Wnt pathway inactive.

The Wnt pathway is turned on by signaling molecules called *Wnt proteins*, which bind to and activate cell surface *Wnt receptors*. The activated receptors stimulate a group of proteins that inhibit the axin-APC-GSK3 destruction complex and thereby prevent the degradation of β-catenin. The accumulating β-catenin then enters the nucleus and interacts with transcription factors that activate a variety of genes, including some that stimulate cell proliferation.

Mutations causing abnormal activation of the Wnt pathway have been detected in numerous cancers. Most of them are loss-of-function mutations in the *APC* gene that are either inherited or, more commonly, triggered by environmental carcinogens. The resulting absence of functional APC protein prevents the axin-APC-GSK3 complex from assembling and β-catenin therefore accumulates,

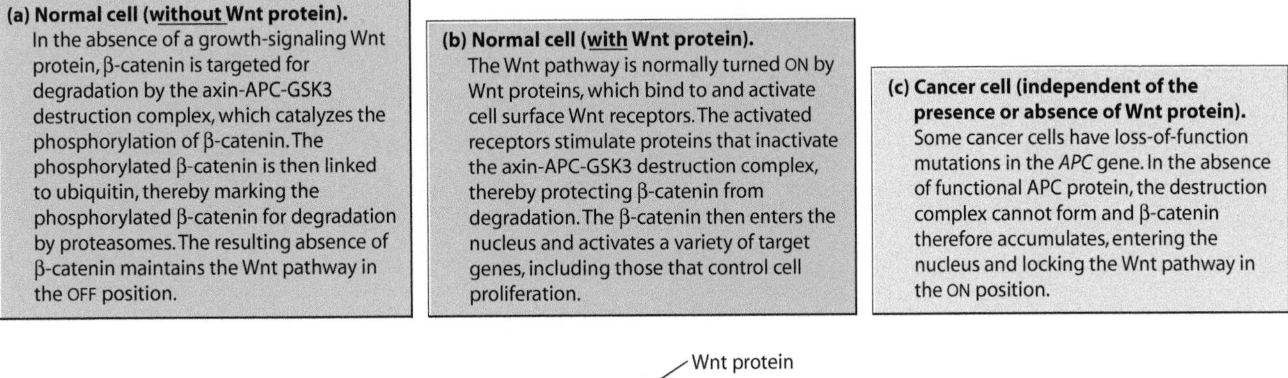

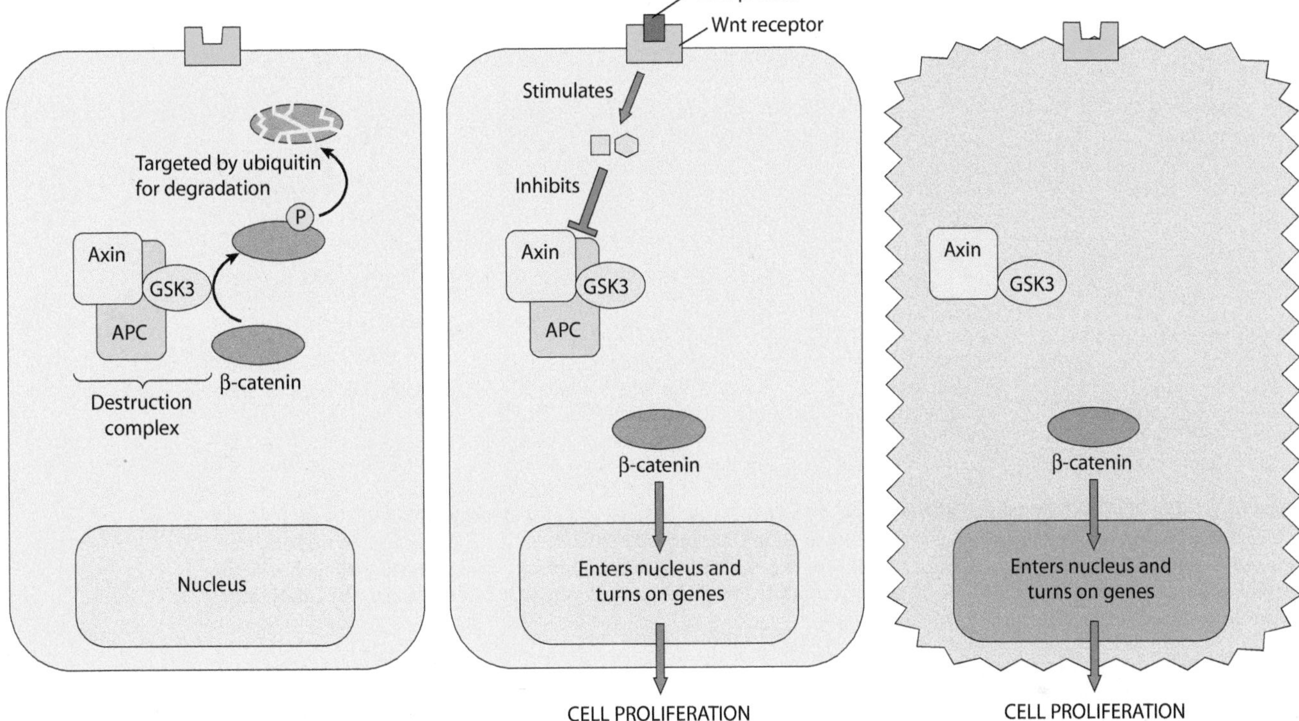

Figure 10-8 The Wnt Signaling Pathway. In normal cells, shown in (**a**) and (**b**), the Wnt pathway is only active in the presence of an external Wnt protein. In cancer cells (**c**), the Wnt pathway is active regardless of the presence or absence of Wnt protein.

locking the Wnt pathway in the ON position and sending the cell a persistent signal to divide.

The *PTEN* Tumor Suppressor Gene Codes for a Protein That Inhibits the PI3K-Akt Signaling Pathway

Cell proliferation is controlled through an interconnected network of pathways with numerous branches and shared components. A good example is provided by growth factors that activate the Ras-MAPK pathway, whose central role in activating cell proliferation was a focus of Chapter 9. When a growth factor binds to a receptor that activates Ras-MAPK signaling, the receptor usually activates several other pathways at the same time.

One of these additional pathways, called the **PI3K-Akt pathway**, involves an enzyme called **phosphatidylinositol 3-kinase** (abbreviated as **PI 3-kinase** or **PI3K**). As shown in Figure 10-9, PI 3-kinase undergoes activation when it binds to phosphorylated tyrosines found in receptors that have been stimulated by growth factor binding. (Recall from Figure 9-8 that a similar mechanism is involved in triggering the Ras-MAPK pathway.) PI 3-kinase then catalyzes the addition of a phosphate group to a plasma membrane lipid called **PIP_2** (phosphatidylinositol-4,5-bisphosphate), which converts PIP_2 into **PIP_3** (phosphatidylinositol-3,4,5-trisphosphate). PIP_3 in turn recruits protein kinases to the inner surface of the plasma membrane, leading to phosphorylation and activation of a protein kinase called **Akt**. Through its ability to catalyze the phosphorylation of several key target proteins, Akt suppresses apoptosis and inhibits cell cycle arrest. The net effect of the PI3K-Akt signaling pathway is therefore to promote cell survival and proliferation.

Dysfunctions in PI3K-Akt signaling have been detected in a number of different cancers. For example,

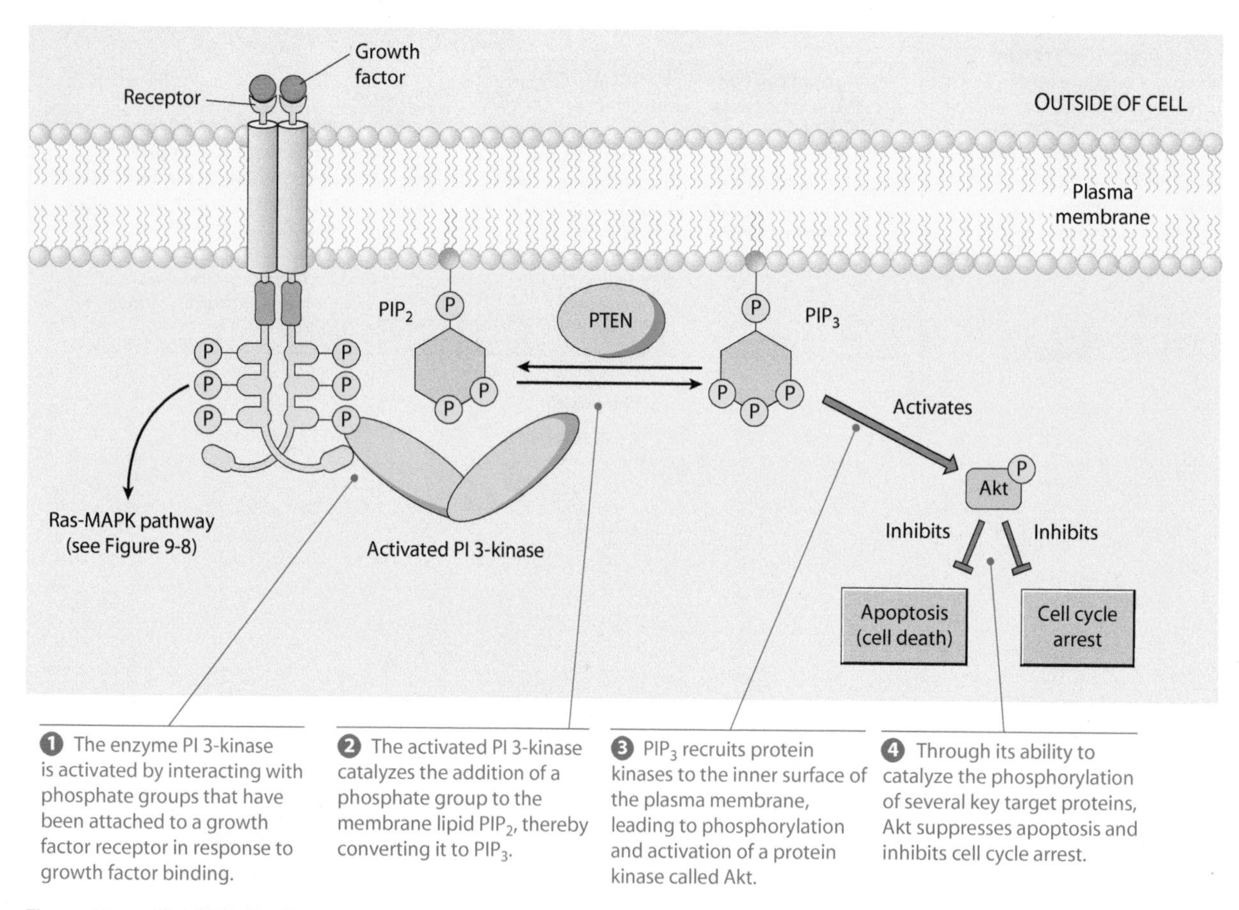

Figure 10-9 The PI3K-Akt Signaling Pathway. Growth factors that bind to receptor tyrosine kinases activate several pathways in addition to the Ras-MAPK pathway described in Chapter 9. The PI3K-Akt pathway shown in this diagram leads to the activation of Akt, a protein kinase that suppresses apoptosis and inhibits cell cycle arrest by phosphorylating several key target proteins. The PI3K-Akt pathway is inhibited by PTEN, an enzyme that reverses step ② by catalyzing the breakdown of PIP_3 to PIP_2.

AKT gene amplification occurs in some ovarian and pancreatic cancers, and a *v-akt* oncogene coding for a mutant Akt protein is present in an animal retrovirus that causes thymus cancers in mice. In such cases, excessive production or activity of the Akt protein leads to hyperactivity of the PI3K-Akt pathway and hence an enhancement of cell proliferation and survival.

Conversely, inhibitors of PI3K-Akt signaling can function as tumor suppressors. A prominent example is **PTEN**, an enzyme that removes a phosphate group from PIP_3 and thus abolishes its ability to activate Akt. In cells that are not being stimulated by growth factors, the intracellular concentration of PIP_3 is kept low by the action of PTEN and the PI3K-Akt pathway is therefore inactive. When loss-of-function mutations disrupt the ability to produce PTEN, the cell cannot degrade PIP_3 efficiently and its concentration rises. The accumulating PIP_3 in turn activates Akt, thereby leading to enhanced cell proliferation and survival (even in the absence of growth factors). Mutations that reduce PTEN activity are found in up to 50% of prostate cancers and glioblastomas, 35% of uterine endometrial cancers, and to varying extents in ovarian, breast, liver, lung, kidney, thyroid, and lymphoid cancers.

Some Tumor Suppressor Genes Code for Components of the TGFβ-Smad Signaling Pathway

Growth factors are usually thought of as being molecules that *stimulate* cell proliferation, but some growth factors have the opposite effect: They inhibit cell proliferation. An example is **transforming growth factor β(TGFβ)**, a protein that may either stimulate or inhibit cell proliferation, depending on the cell type and context. TGFβ is especially relevant for tumor development because it is a potent inhibitor of epithelial cell proliferation, and roughly 90% of human cancers are carcinomas—that is, cancers of epithelial origin.

TGFβ exerts its inhibitory effects on cell proliferation through the **TGFβ-Smad pathway** illustrated in Figure 10-10. The first step in this pathway is the binding of TGFβ to a cell surface receptor. Like many other growth factor receptors, the receptors for TGFβ catalyze protein phosphorylation reactions, although in this case the amino acids *serine* and *threonine* rather than tyrosine are phosphorylated. TGFβ binds to two types of receptors, called type I and type II receptors, located on the surface of its target cells. Upon binding

❶ In the absence of TGFβ, the type I and type II receptors for TGFβ are not clustered or phosphorylated.

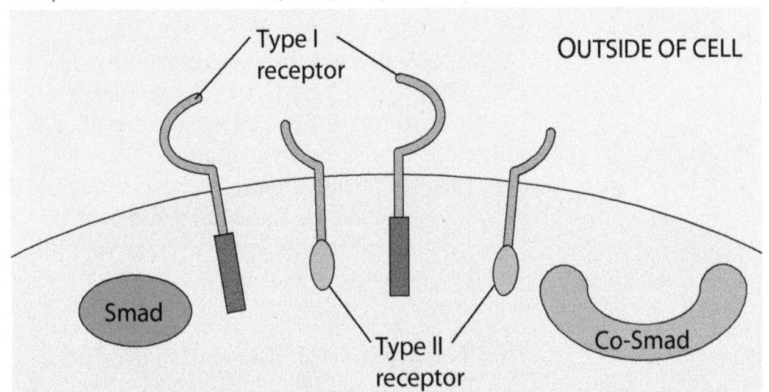

❷ Binding of TGFβ results in clustering of type I and type II receptors, followed by phosphorylation of type I receptors by type II receptors.

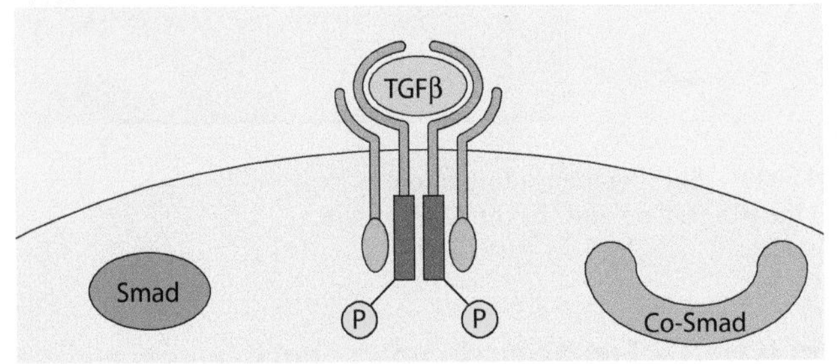

❸ The activated type I receptors phosphorylate Smad proteins.

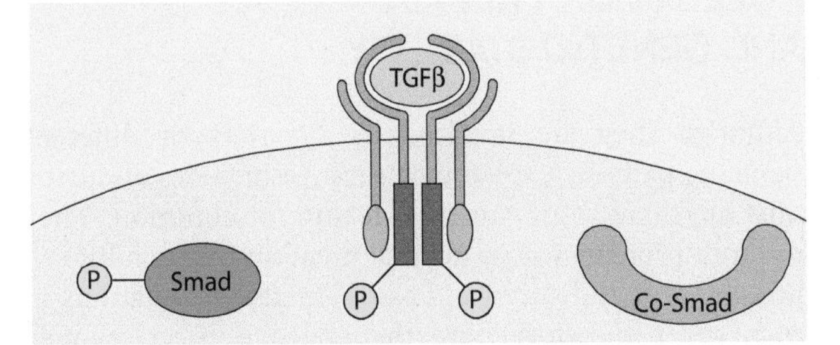

❹ The phosphorylated Smads bind to other Smads (co-Smads), and together they enter the nucleus to activate the transcription of genes that inhibit cell proliferation. Two such genes produce the Cdk inhibitors p15 and p21, which halt the cell cycle by inhibiting the Cdk-cyclin complexes whose actions are required for progression through key transition points in the cell cycle.

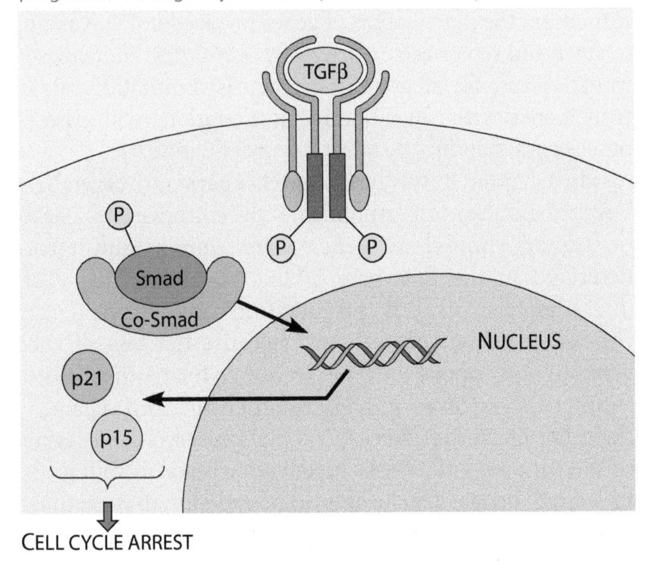

Figure 10-10 The TGFβ-Smad Signaling Pathway. Transforming growth factor beta (TGFβ) is a potent inhibitor of epithelial cell proliferation that exerts its effects through Smad proteins, which activate genes coding for proteins that halt the cell cycle. Loss-of-function mutations affecting the TGFβ-Smad pathway are frequently encountered in human cancers.

of TGFβ, type II receptors phosphorylate type I receptors. The type I receptors then phosphorylate a class of proteins known as **Smads**, which bind to an additional protein (a "co-Smad") and move into the nucleus.

Once inside the nucleus, the Smad complex activates the expression of genes that inhibit cell proliferation. Two key genes produce the **p15 protein** and the p21 protein, which both function as Cdk inhibitors (the p21 protein was already mentioned earlier in the chapter when we covered the mechanism of p53-mediated cell cycle arrest). The p15 and p21 proteins halt progression through the cell cycle by inhibiting the Cdk-cyclin complexes whose actions are required for passing through key transition points in the cycle.

Components of the TGFβ-Smad signaling pathway are frequently inactivated in human cancers. For example, loss-of-function mutations in the TGFβ receptor are common in colon and stomach cancers, and occur in some cancers of the breast, ovary, and pancreas as well. Loss-of-function mutations in Smad proteins are likewise observed in a variety of cancers, including 50% of all pancreatic cancers and about 30% of colon cancers. Such evidence indicates that the genes coding for TGFβ receptors and Smads both qualify as tumor suppressors.

One Gene Produces Two Tumor Suppressor Proteins: p16 and ARF

Thus far, this chapter has described the relationship between tumor suppressor genes and several signaling pathways for inhibiting cell proliferation or promoting cell death. The next tumor suppressor to be covered, known as the ***CDKN2A* gene**, exhibits the rather unusual property of coding for two different proteins that act independently on two of these pathways, the Rb pathway and the p53 pathway.

How does the *CDKN2A* gene produce two entirely different tumor suppressor proteins? Because the genetic code is read three bases at a time, changing the start point by one or two nucleotides will completely change the message contained in a base sequence. For example, the sequence AAAGGGCCC can be read in three different **reading frames** starting from the first, second, or third base—that is, starting as AAA-GGG . . ., AAG-GGC . . ., or AGG-GCC . . ., respectively. A shift in the normal reading frame usually creates a garbled message that does not code for a functional protein. In the case of the *CDKN2A* gene, however, a shift in the reading frame

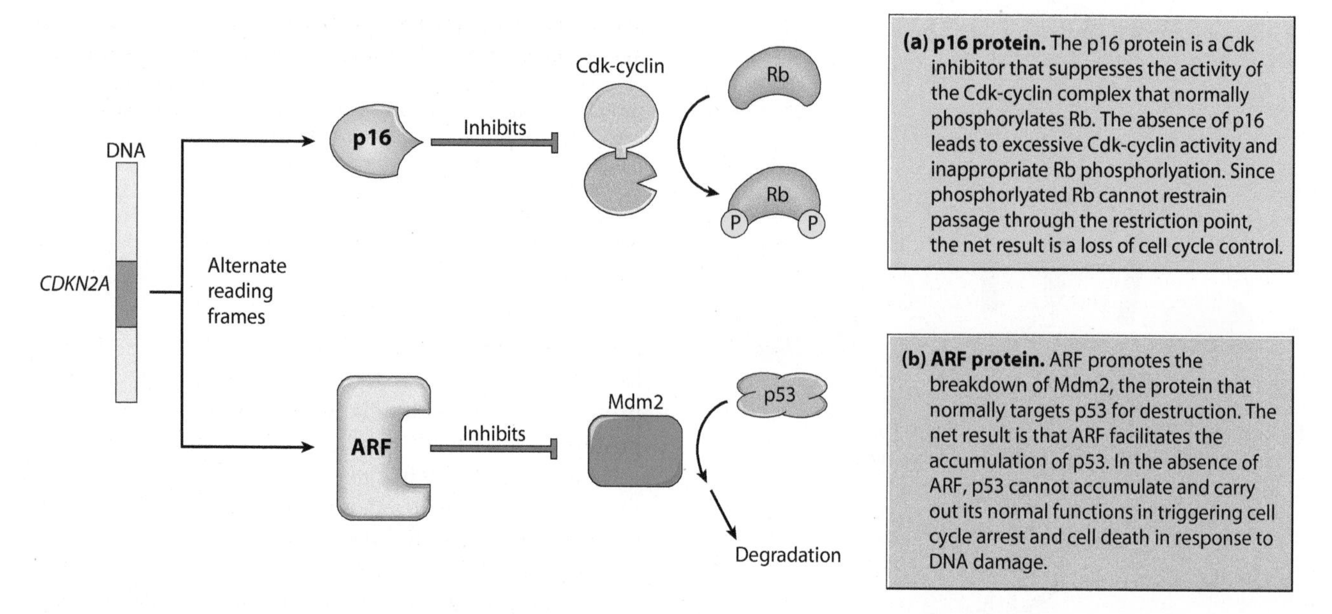

Figure 10-11 Two Tumor Suppressor Proteins Produced by the *CDKN2A* Gene. Shifting the reading frame of the *CDKN2A* gene allows it to produce two different tumor suppressors: (**a**) the p16 protein and (**b**) the ARF protein.

leads to the production of an alternative protein that is fully functional.

The first of the two proteins produced by the *CDKN2A* gene is the **p16 protein** (also called INK4a), a Cdk inhibitor that suppresses the activity of the Cdk-cyclin complex that normally phosphorylates the Rb protein. Loss-of-function mutations affecting p16 lead to excessive Cdk-cyclin activity and inappropriate Rb phosphorylation. Since the phosphorylated form of Rb cannot restrain the cell cycle at the restriction point, the net result is a loss of cell cycle control.

The second protein produced by the *CDKN2A* gene is called the **ARF** (for Alternative Reading Frame) **protein.** Although they are produced by the same gene, p16 and ARF are completely different proteins exhibiting no sequence similarity. Whereas p16 is a Cdk inhibitor, ARF binds to and promotes the degradation of Mdm2, the ubiquitin ligase that normally targets p53 for destruction by tagging it with ubiquitin (see Figures 10-5 and 10-6). By promoting the degradation of Mdm2, ARF facilitates the stabilization and accumulation of p53. Conversely, loss-of-function mutations affecting ARF interfere with the ability of p53 to accumulate and perform its function in triggering cell cycle arrest and cell death.

The *CDKN2A* gene therefore influences cell proliferation and survival through two independent proteins: the p16 protein, which is required for proper Rb signaling, and the ARF protein, which is required for proper p53 signaling (Figure 10-11). Loss-of-function mutations in *CDKN2A* have been observed in numerous human cancers, including 15% to 30% of all cancers originating in the breast, lung, pancreas, and bladder. Deletion of both copies of the *CDKN2A* gene, which leads to complete absence of both the p16 and ARF proteins, is common in such cases.

TUMOR SUPPRESSOR GENES: ROLES IN DNA REPAIR AND GENETIC STABILITY

Although they are involved in a variety of different signaling pathways, the tumor suppressor genes discussed thus far share a fundamental feature in common: They produce proteins whose normal function is to inhibit cell proliferation and survival. Loss-of-function mutations in such genes therefore have the opposite effect, namely increased cell proliferation and survival.

A second group of tumor suppressors act through their effects on DNA repair and the maintenance of chromosome integrity. Unlike genes that exert direct effects on cell proliferation and whose inactivation can lead directly to tumor formation, the inactivation of genes involved in DNA maintenance and repair acts *indirectly* by permitting an increased mutation rate for all genes. This increased mutation rate in turn increases the likelihood that alterations will arise in other genes that directly affect cell proliferation.

In Chapter 8, the terms **gatekeepers** and **caretakers** were introduced to distinguish between these two classes of tumor suppressor genes. The tumor suppressors described in the first part of this chapter, which exert direct effects on cell proliferation and survival, are considered to be "gatekeepers" because the loss of such genes directly opens the gates to tumor formation. Tumor suppressors involved in DNA maintenance and repair, on the other hand, are "caretakers" that preserve the integrity of the genome and whose inactivation leads to mutations in other genes (including gatekeepers) that actually trigger the development of cancer. In the following sections, we will examine the functions of some of these caretaker genes.

Genes Involved in Excision and Mismatch Repair Help Prevent the Accumulation of Localized DNA Errors

Cancer cells accumulate mutations at rates that can be hundreds or even thousands of times higher than normal. This condition, called **genetic instability**, does not by itself disrupt the normal controls on cell proliferation. In fact, most of the mutations that arise in genetically unstable cells are likely to be harmful mutations that hinder cell survival. But elevated mutation rates also increase the probability that occasional mutations will arise that allow cells to escape from the normal constraints on cell proliferation and survival. Cells that randomly incur such mutations will tend to outgrow their neighbors, an important first step in the development of cancer. Increased mutation rates also facilitate tumor progression in which cells acquire additional traits—for example, faster growth rate, increased invasiveness, ability to survive in the bloodstream, resistance to immune attack, ability to grow in other organs, resistance to drugs, and evasion of death-triggering mechanisms—that allow cancers to become increasingly more aggressive.

Genetic instability occurs in several different forms that differ in their underlying mechanisms. The simplest type is caused by defects in the DNA repair mechanisms that cells use for correcting localized errors involving one or a few nucleotides. These localized errors typically arise either from exposure to DNA-damaging agents or from base-pairing mistakes that take place during DNA replication. Chapter 2 described the two types of repair mechanisms employed for correcting such errors. *Excision repair*, described on page 31, is capable of repairing abnormal bases created by exposure to DNA-damaging agents, and *mismatch repair*, described on page 33, is used for correcting inappropriately paired bases that arise spontaneously during DNA replication.

Individuals who inherit loss-of-function mutations involving genes required for either of these repair mechanisms exhibit an increased cancer risk. For example, we saw in Chapter 8 that inherited mutations in excision repair genes cause *xeroderma pigmentosum*, a hereditary cancer syndrome involving an extremely high risk for skin cancer (p. 150). In a similar fashion, inherited mutations in genes coding for proteins involved in mismatch repair are responsible for *hereditary nonpolyposis colon cancer* (*HNPCC*), a hereditary syndrome associated with a high risk for colon cancer (p. 150).

Although both of these hereditary syndromes involve a striking increase in cancer risk, xeroderma pigmentosum exhibits a recessive pattern of inheritance and HNPCC exhibits a dominant pattern of inheritance. In other words, inheriting an elevated cancer risk requires two defective copies of an excision repair gene but only one defective copy of a mismatch repair gene. The reason for this difference appears to be related to how many steps are required to create genetic instability in the two cases (Figure 10-12). In a person who inherits a single defective mismatch repair gene, all that is required to start accumulating DNA errors at a high rate is for the second copy of the gene to undergo mutation. This second "hit" will immediately permit uncorrected errors to accumulate during normal DNA replication because of the absence of mismatch repair. In contrast, if

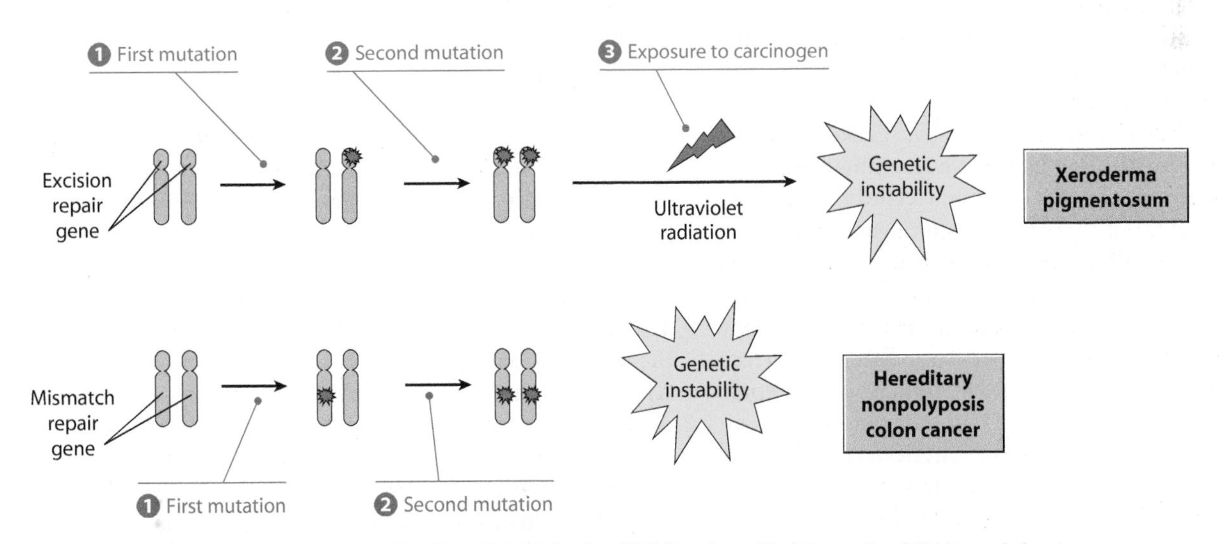

Figure 10-12 Routes to Genetic Instability Based on Defective DNA Repair. (*Top*) In a cell exhibiting a defect in a single excision repair gene, an inactivating mutation in the second copy of the gene does not immediately cause a large number of DNA errors to arise. The cell must first be exposed to a DNA-damaging agent, such as ultraviolet light, before large numbers of mutations will appear. Such a scenario is observed in xeroderma pigmentosum. (*Bottom*) In a cell exhibiting a defect in a single mismatch repair gene, all that is required to start accumulating DNA errors at a high rate is for the second copy of the gene to undergo mutation. This second mutation will disable the mismatch repair pathway, which immediately allows uncorrected errors to accumulate during normal DNA replication. Such a scenario is observed in hereditary nonpolyposis colon cancer.

a person were to inherit a single defective excision repair gene, subsequent mutation of the second copy of the gene would debilitate excision repair but would not immediately lead to the accumulation of mutations. A third step, namely exposure to a DNA-damaging agent such as ultraviolet light, is needed to actually create the mutations. Thus more steps are needed to create genetic instability involving excision repair than is the case for mismatch repair.

Inherited mutations in genes required for excision or mismatch repair create a dramatic increase in the risk for certain hereditary cancers, but mutations in these two classes of genes are less important for most nonhereditary forms of cancer. Nonetheless, mutations in excision or mismatch repair have been detected in about 15% of colon cancers and in several other kinds of cancer as well, suggesting that deficiencies in DNA repair occasionally contribute to the genetic instabilities observed in nonhereditary cancers.

Proteins Produced by the *BRCA1* and *BRCA2* Genes Assist in the Repair of Double-Strand DNA Breaks

Another type of genetic instability exhibited by cancer cells involves their tendency to acquire gross abnormalities in chromosome structure and number. Such *chromosomal instabilities* can be caused by defects in a variety of different tumor suppressors, including the ***BRCA1*** and ***BRCA2*** **genes** introduced in Chapter 8. Women who inherit a mutation in one of the *BRCA* genes typically exhibit a lifetime cancer risk of 40% to 80% for breast cancer and 15% to 65% for ovarian cancer. The *BRCA1* and *BRCA2* genes were initially thought to exert their effects directly on cell proliferation, but later studies revealed that they produce proteins involved in pathways for sensing DNA damage and performing the necessary repairs.

The two *BRCA* tumor suppressor genes code for large nuclear proteins that bear little resemblance to one another. An early clue regarding their cellular role came from the observation that cells deficient in either of the BRCA proteins exhibit large numbers of chromosomal abnormalities, including broken chromosomes and chromosomal translocations. The apparent reason for these abnormalities is that the two BRCA proteins are involved in the process by which cells repair double-strand breaks in DNA. Double-strand breaks are more difficult to repair than single-strand breaks because with single-strand breaks, the remaining strand of the DNA double helix remains intact and can serve as a template for aligning and repairing the defective strand. In contrast, double-strand breaks completely cleave the DNA double helix into two separate fragments and the repair machinery is therefore confronted with the problem of identifying the correct two fragments and rejoining their broken ends without losing any nucleotides.

As we saw in Chapter 2, the two main ways of repairing double-strand breaks are *nonhomologous end-joining* and *homologous recombination* (see Figure 2-18). Of the two mechanisms, **homologous recombination** is less prone to error because it uses the DNA present in the unbroken homologous chromosome to serve as a template for guiding the repair of the DNA from the broken chromosome. Repairing double-strand breaks by homologous recombination is a complex process that requires the participation of a large number of different proteins, including BRCA1 and BRCA2. The pathway is activated by the same ATM kinase whose role in detecting and responding to DNA damage was introduced earlier in this chapter (see Figure 10-5). We have already seen that in response to DNA damage, the ATM kinase catalyzes the phosphorylation of the p53 protein, which then halts the cell cycle to permit time for repair to occur. The ATM kinase also phosphorylates and activates more than a dozen additional proteins involved in cell cycle control and DNA repair, including BRCA1 and other molecules required for repairing double-strand breaks.

Figure 10-13 shows that the mechanism for repairing double-strand breaks by homologous recombination involves two main phases. First, a group of proteins called the **Rad50 exonuclease complex** removes nucleotides from one strand of the broken end of a DNA double helix to expose a single-stranded segment on the opposite strand. In the second phase, a multiprotein assembly called the **Rad51 repair complex** carries out a "strand invasion" reaction in which the exposed single-stranded DNA segment at the end of the broken DNA molecule displaces one of the two strands of the intact DNA molecule being used as a template. In this step, Rad51 first coats the single-stranded DNA; the coated strand then invades and moves along the target DNA double helix until it reaches a complementary sequence. Once it has been located, the complementary sequence is used as a template for guiding repair of the broken DNA.

Although their roles are not completely understood, the BRCA1 and BRCA2 proteins are both required for efficient repair of double-strand breaks. BRCA2 binds tightly to and controls the activity of Rad51, the central protein responsible for carrying out strand invasion during repair by homologous recombination. BRCA1 is associated with both the Rad50 exonuclease complex and the Rad51 repair complex. Moreover, it is known that ATM phosphorylates BRCA1 in response to DNA damage, suggesting that BRCA1 plays an early role in activating the pathway for repairing double-strand breaks. Cells deficient in either BRCA1 or BRCA2 are extremely sensitive to carcinogenic agents that produce double-strand DNA breaks. In such cells, double-strand breaks can only be repaired by error-prone mechanisms, such as nonhomologous end-joining, that lead to broken, rearranged, and translocated chromosomes. The resulting chromosomal instability is thought to play a large role in the cancer risks exhibited by women who inherit *BRCA1* or *BRCA2* mutations.

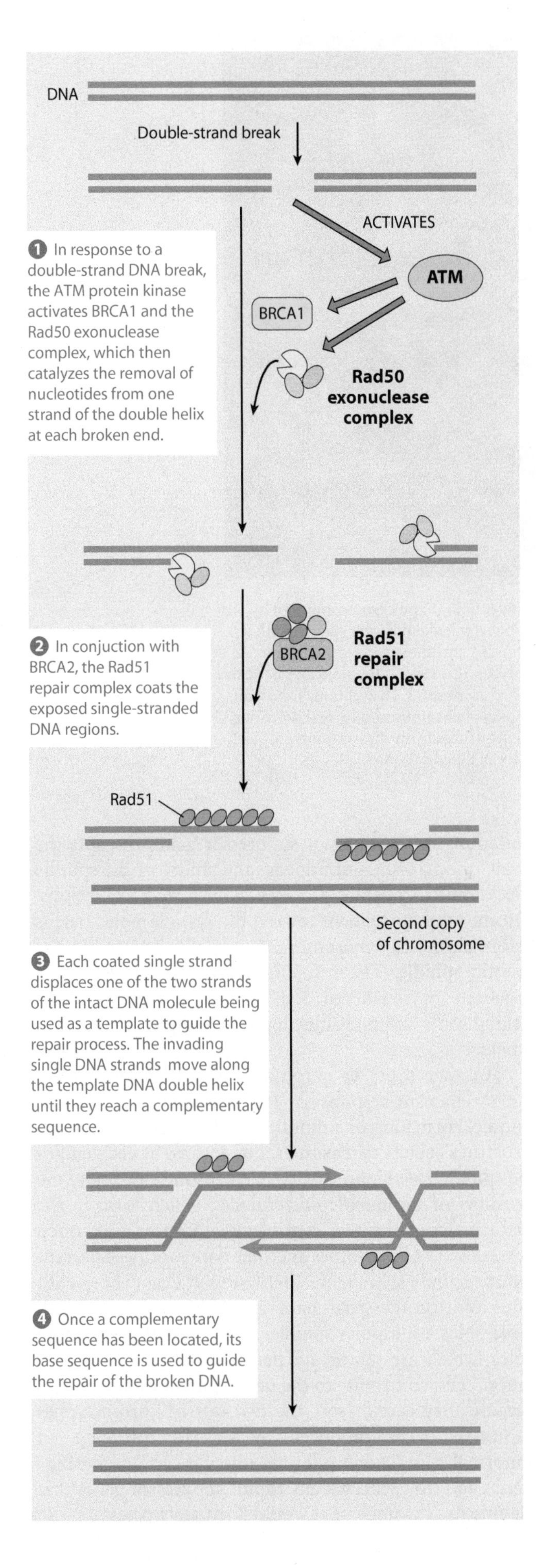

Mutations in Genes That Influence Mitotic Spindle Behavior Can Lead to Chromosomal Instabilities

We have just seen how broken and translocated chromosomes arise in cancer cells as a result of mutations that disrupt tumor suppressor genes needed for repairing double-strand DNA breaks. Another chromosomal abnormality frequently observed in cancer cells is the tendency for whole chromosomes to be lost or gained, thereby leading to **aneuploid** cells that possess an abnormal number of chromosomes (Figure 10-14). The various mechanisms that underlie the development of aneuploidy are just beginning to be unraveled, but evidence already points to the existence of tumor suppressor genes whose loss contributes to this type of chromosomal instability.

To explain how these tumor suppressors work, we first need to review the normal mechanisms used by cells for sorting and parceling out chromosomes during cell division. In a normal cell cycle, chromosomal DNA is first replicated during S phase to create duplicate copies of each chromosome, and the duplicate copies are then separated into the two new cells formed by the subsequent mitotic cell division. Accurate separation of the duplicated chromosomes is accomplished by attaching the chromosomes to the *mitotic spindle*, which separates and moves the chromosomes in a way that ensures that each new cell receives a complete set of chromosomes (Figure 10-15).

A critical moment occurs at the end of **metaphase**, when the chromosomes line up at the center of the mitotic spindle just before being parceled out to the two new cells. If chromosome movement toward opposite spindle poles were to begin before the chromosomes are all attached to the spindle, a newly forming cell might receive extra copies of some chromosomes and no copies of others. To protect against this possible danger, cells possess a control mechanism called the **spindle checkpoint** that monitors chromosome attachment to the spindle and prevents chromosome movement from beginning until all chromosomes are properly attached. In the absence of such a mechanism, there would be no guarantee that each newly forming cell would receive a complete set of chromosomes (see Figure 10-15, *bottom right*).

The key to the spindle checkpoint is the **anaphase-promoting complex**, a multiprotein complex that triggers the onset of **anaphase**—the stage of mitosis when the chromosomes move toward opposite poles of the mitotic

Figure 10-13 Pathway for Repairing Double-Strand DNA Breaks by Homologous Recombination. In response to a double-strand DNA break, the Rad50 exonuclease and Rad51 repair complexes use the DNA present in the unbroken homologous chromosome to serve as a template for guiding the repair of the DNA from the broken chromosome.

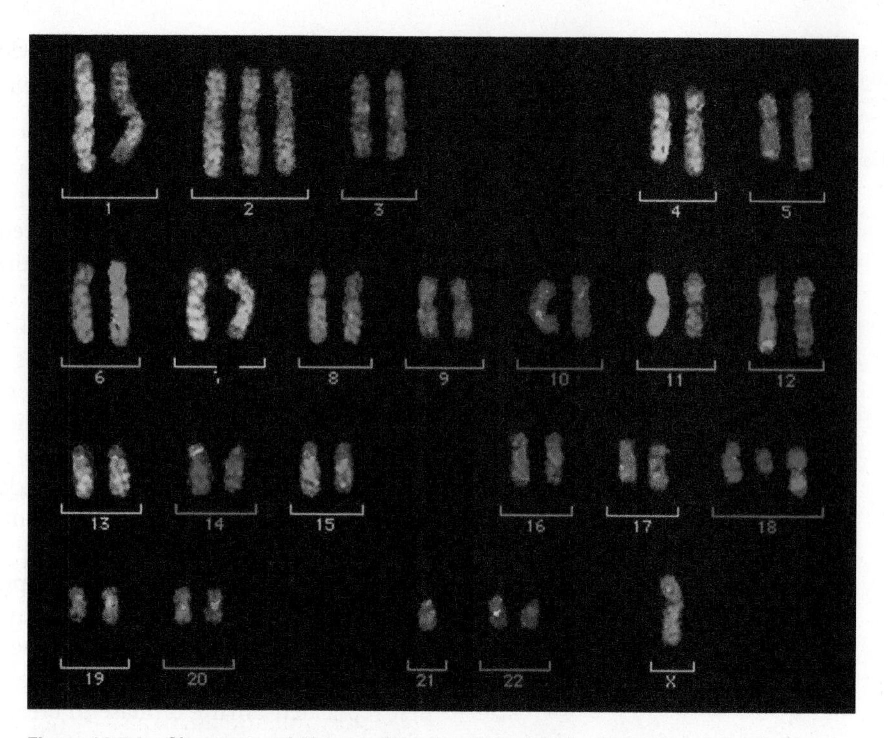

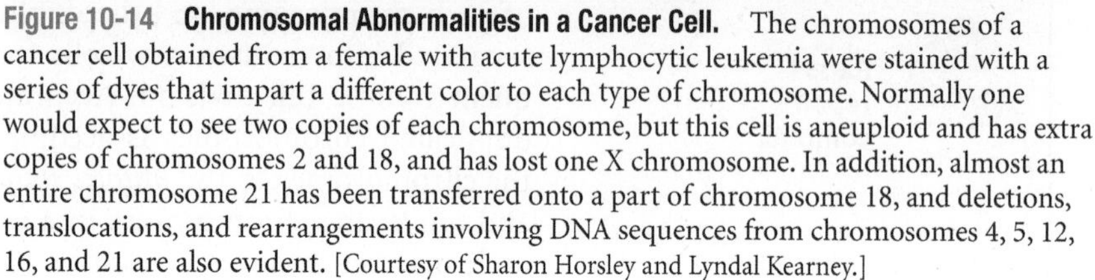

Figure 10-14 Chromosomal Abnormalities in a Cancer Cell. The chromosomes of a cancer cell obtained from a female with acute lymphocytic leukemia were stained with a series of dyes that impart a different color to each type of chromosome. Normally one would expect to see two copies of each chromosome, but this cell is aneuploid and has extra copies of chromosomes 2 and 18, and has lost one X chromosome. In addition, almost an entire chromosome 21 has been transferred onto a part of chromosome 18, and deletions, translocations, and rearrangements involving DNA sequences from chromosomes 4, 5, 12, 16, and 21 are also evident. [Courtesy of Sharon Horsley and Lyndal Kearney.]

spindle. As shown in Figure 10-16a, the anaphase-promoting complex initiates chromosome movement by activating *separase*, an enzyme that breaks down proteins called *cohesins* that hold the duplicated chromosomes together. As long as they are joined together by cohesins, the duplicated chromosomes cannot separate from each other and move toward opposite spindle poles.

To prevent premature separation, chromosomes that are not yet attached to the mitotic spindle send a "wait" signal that inhibits the anaphase-promoting complex, thereby blocking the activation of separase. The "wait" signal is transmitted by proteins that are members of the **Mad** and **Bub** families. The Mad and Bub proteins bind to chromosomes that are unattached to the mitotic spindle and are converted into a Mad-Bub multiprotein complex, which inhibits the anaphase-promoting complex by blocking the action of one of its essential activators, the Cdc20 protein (see Figure 10-16b). After the chromosomes have all become attached to the spindle, the Mad and Bub proteins are no longer converted into this inhibitory complex and the anaphase-promoting complex is free to initiate the onset of anaphase.

Mutations that cause the loss or inactivation of Mad or Bub proteins have been linked to certain types of cancer, which indicates that genes coding for some of the Mad and Bub proteins behave as tumor suppressor genes. A lack of Mad or Bub proteins caused by loss-of-function mutations in these tumor suppressor genes disrupts the "wait" mechanism and impedes the ability of the spindle checkpoint to operate properly. Under such conditions, chromosome movement toward the spindle poles begins before all the chromosomes are properly attached to the mitotic spindle. The result is a state of chromosomal instability in which cell division creates aneuploid cells lacking some chromosomes and possessing extra copies of others.

Another route to chromosomal instability involves the mechanism responsible for assembling the mitotic spindle. Formation of a mitotic spindle requires two small structures called **centrosomes**, one located at each end of the spindle (see Figure 10-15). Centrosomes promote the assembly of the *spindle microtubules*, which form in the space between the two centrosomes. Cancer cells often possess extra centrosomes and therefore produce aberrant mitotic spindles. In Figure 10-17, we see a cancer cell with three centrosomes that have assembled a spindle with three poles. Multipolar spindles containing three or more poles, which are rare in normal tissues but common in cancer cells, contribute to the development of aneuploidy because they cannot sort the two sets of chromosomes accurately. Cells produced by mitosis involving an abnormal spindle will often be missing certain chromosomes and thus will lack any tumor suppressor genes that the missing chromosomes would normally possess.

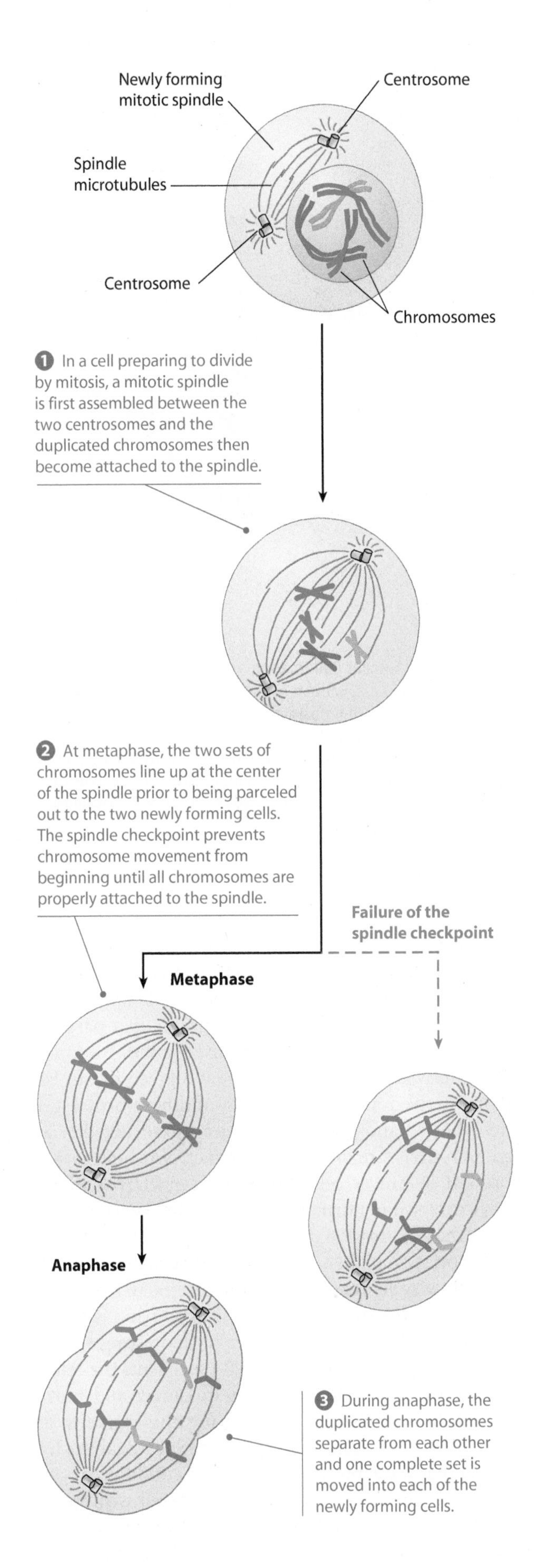

Figure 10-15 Distributing Chromosomes During Mitosis. In a cell undergoing mitosis, the spindle checkpoint prevents chromosome movement from beginning until all chromosomes are properly attached to the spindle. If the spindle checkpoint fails to operate properly, chromosome movement can begin prematurely and the newly forming cells are in danger of receiving extra copies of some chromosomes and no copies of others.

OVERVIEW OF CARCINOGENESIS

Over the course of the past several chapters, we have encountered several categories of cancer-related genes. Mutations in these genes, and the genetic instability that facilitates the accumulation of such mutations, are centrally involved in the mechanisms by which cancers arise. Yet one cannot explain the behavior of a malignant tumor by pointing solely to gene mutations. The final part of this chapter will provide a broad overview of the role played not just by mutations but also by nonmutational changes in converting normal cells into cancer cells.

Cancers Vary in Their Gene Expression Profiles

Mutations that create oncogenes or disrupt the function of tumor suppressor genes are central to the development of cancer, but they do not explain all the cellular changes that accompany the conversion of normal cells into cancer cells. Many of the properties exhibited by cancer cells are triggered not by gene mutations, but by switching on (or off) the expression of normal genes, thereby leading to increases (or decreases) in the production of hundreds of different proteins. The term **epigenetic change** is employed when referring to such alterations that are based on changing the expression of a gene rather than mutating it.

Measuring epigenetic changes requires techniques that can monitor the expression of thousands of genes simultaneously. One very powerful tool is the **DNA microarray**, a fingernail-sized, thin chip of glass or plastic that has been spotted at fixed locations with thousands of DNA fragments corresponding to various genes of interest. A single microarray may contain 10,000 or more spots, each representing a different gene. To determine which genes are being expressed in any given cell population, one begins by extracting molecules of messenger RNA (mRNA), which represent the products of gene transcription. The mRNA is then copied with *reverse transcriptase*, an enzyme that makes single-stranded DNA copies that are complementary in sequence to each mRNA (p. 129). The resulting single-stranded DNA (called *cDNA* for *complementary DNA*) is then attached to a fluorescent dye. When the microarray is bathed with the fluorescent cDNA, each cDNA molecule binds or *hybridizes* by complementary base-pairing to the spot containing the specific gene to which it corresponds.

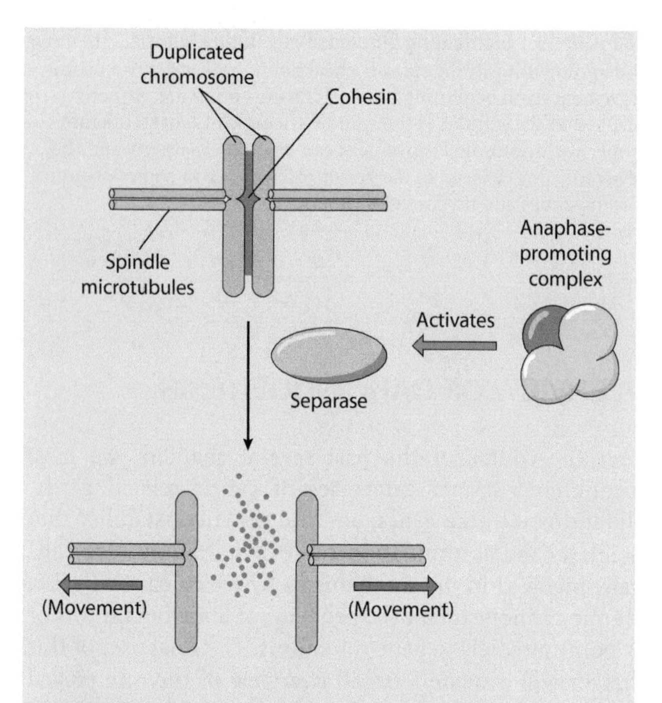

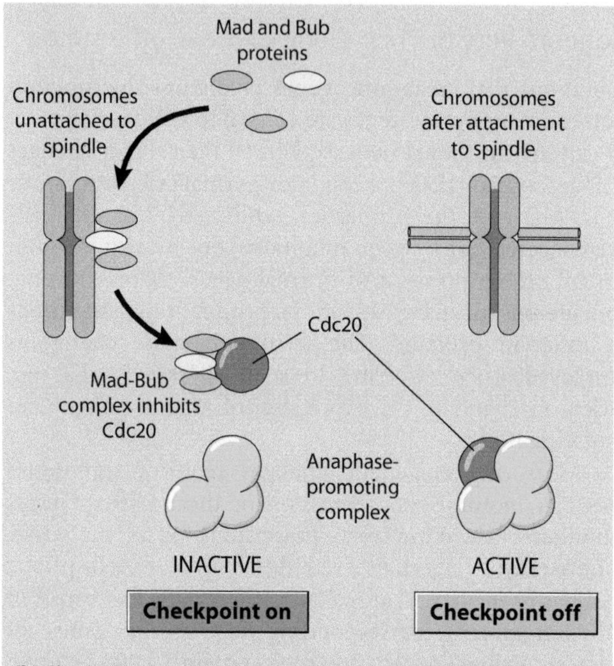

Figure 10-16 The Anaphase-Promoting Complex and the Spindle Checkpoint. (**a**) The anaphase-promoting complex triggers the onset of anaphase by activating separase, an enzyme that degrades the cohesin proteins that hold duplicated chromosomes together. After cohesin breakdown, the duplicated chromosomes are free to move toward opposite spindle poles. (**b**) The spindle checkpoint ensures that anaphase does not begin until all chromosomes are attached to the mitotic spindle. Chromosomes that are not yet attached to the mitotic spindle send a "wait" signal ("checkpoint on") by converting Mad and Bub proteins into a Mad-Bub complex, which inhibits the anaphase-promoting complex by blocking one of its essential activators, the Cdc20 protein. After the chromosomes have become attached to the spindle, this "wait" signal ceases ("checkpoint off") and the anaphase-promoting complex becomes active.

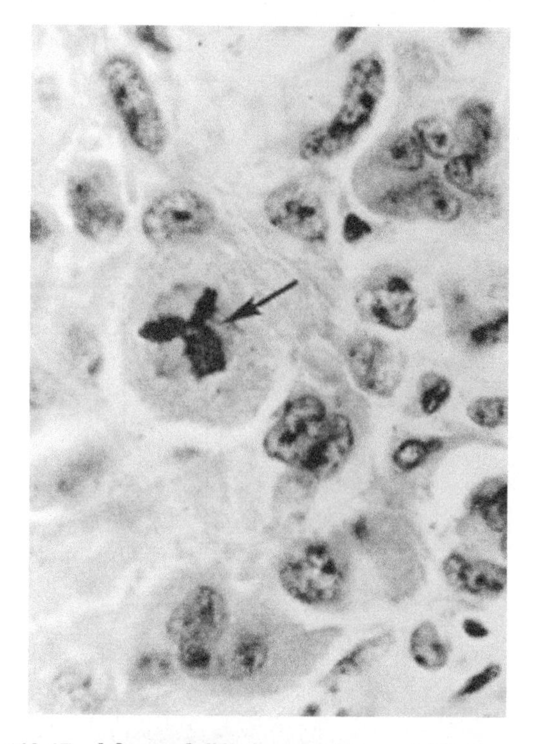

Figure 10-17 A Cancer Cell Undergoing Abnormal Mitosis. This photograph of a specimen of cancer tissue viewed with a light microscope shows a cell with an abnormal mitotic spindle possessing three spindle poles. Such spindles, which cannot separate chromosomes properly into the two newly forming cells, are created by the presence of three centrosomes rather than the normal two. [From S. L. Robbins, *Textbook of Pathology* (Philadelphia, PA: W. B. Saunders, 1957).]

Figure 10-18 illustrates how DNA microarrays can be used to create a *gene expression profile* that compares the patterns of gene expression in cancer cells and a corresponding population of normal cells. In this particular example, two fluorescent dyes are used: a red dye to label cDNAs derived from cancer cells and a green dye to label cDNAs derived from the corresponding normal cells. When the red and green cDNAs are mixed together and placed on a DNA microarray, the red cDNAs bind to genes expressed in cancer cells and the green cDNAs bind to genes expressed in normal cells. Red spots therefore represent higher expression of a gene in cancer cells, green spots represent higher expression of a gene in normal cells, yellow spots (caused by a mixture of red and green fluorescence) represent genes whose expression is roughly the same, and black spots (absence of fluorescence) represent genes expressed in neither cell type. Thus the relative expression of thousands of genes in cancer and normal cells can be compared by measuring the intensity and color of the fluorescence of each spot. Such analyses have revealed that the expression of hundreds of different genes is typically altered in cancer cells compared with normal cells of the same tissue. Moreover, significant variations in gene expression are often detected when the same type of cancer is examined in different patients, a phenomenon

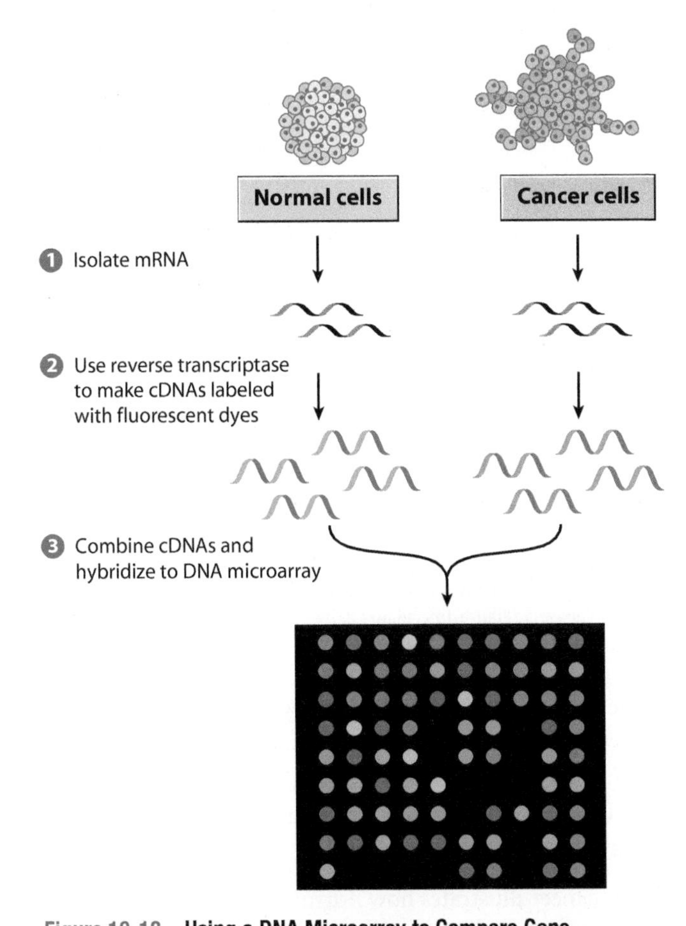

Figure 10-18 Using a DNA Microarray to Compare Gene Expression Profiles in Normal and Cancer Cells. In this example, gene expression in cancer cells and normal cells is compared by isolating mRNA from each cell population, using reverse transcriptase to make cDNA copies of the mRNA, and attaching a green fluorescent dye to the normal cell cDNAs and a red fluorescent dye to the cancer cell cDNAs. A DNA microarray containing thousands of DNA fragments representing different genes is then bathed with a mixture of the two cDNA populations (only a small section of the microarray is illustrated). Each cDNA binds (hybridizes) by complementary base-pairing to the spot containing the specific gene to which it corresponds. Red spots therefore represent genes expressed preferentially in cancer cells, green spots represent genes expressed preferentially in normal cells, yellow spots represent genes whose expression is similar in the two cell populations, and dark regions (missing spots) represent genes that are not expressed in either cell type.

whose usefulness in cancer diagnosis and treatment will be discussed in Chapter 11.

The changes in gene expression commonly exhibited by cancer cells arise in several difference ways. One well-documented mechanism involves *epigenetic silencing* by **DNA methylation**, a process in which methyl groups are attached to the base C in DNA at sites where it is located adjacent to the base G. In vertebrate DNA, these –CG– sequences are preferentially located near the beginning of genes (about half of all human genes are associated with –CG– sites). When –CG– sequences undergo methylation, the transcription of adjacent genes is inhibited or "silenced." Most –CG– sites are unmethylated in normal cells, but extensive methylation is often seen in cancer cells, where it leads to the inappropriate silencing of a variety of different genes. Tumor suppressor genes are frequently among the genes to be silenced by this mechanism. In fact, the tumor suppressor genes of cancer cells are inactivated by epigenetic silencing at least as often as they are inactivated by DNA mutation. Loss of gene function through inappropriate methylation may therefore be as important to cancer cells as mutation-induced loss of function.

Colon Cancer Illustrates How a Stepwise Series of Mutations Can Lead to Malignancy

Chapter 5 introduced the idea that cancer arises via a multistep process in which cellular properties gradually change over time as mutations confer new traits that impart selective advantages to the cells in which they arise (see Figure 5-17). Now that we have described the main classes of cancer-related genes and the molecular pathways in which they participate, it is appropriate to return to the concept of multistep carcinogenesis to see how a specific sequence of gene mutations can lead to cancer.

Current estimates indicate that there are more than 100 different oncogenes and several dozen tumor suppressor genes. For cancer to arise, it is rarely sufficient to have a defect in just one of these genes, nor is it necessary for a large number to be involved. Instead, each type of cancer tends to be characterized by a small handful of mutations involving the inactivation of tumor suppressor genes as well as the conversion of proto-oncogenes into oncogenes. In other words, creating a cancer cell usually requires that the brakes on cell growth (tumor suppressor genes) be released and the accelerators for cell growth (oncogenes) be activated.

This principle is nicely illustrated by the stepwise progression toward malignancy observed in colon cancer. Scientists have isolated DNA from a large number of colon cancer patients and examined it for the presence of mutations. The most common pattern to be detected is the presence of a *KRAS* oncogene (a member of the *RAS* gene family) accompanied by loss-of-function mutations in the tumor suppressor genes *APC*, *p53*, and *SMAD4*. Rapidly growing colon cancers tend to exhibit all four genetic alterations, whereas benign tumors have only one or two, suggesting that mutations in the four genes occur in a stepwise fashion that correlates with increasingly aggressive behavior.

As shown in Figure 10-19, the earliest mutation to be routinely detected is loss of function of the *APC* gene, which frequently occurs in small polyps before cancer has even arisen. Mutations in *KRAS* tend to be seen when the polyps get larger, and mutations in *SMAD4* and *p53* usually appear as cancer finally begins to develop. These mutations, however, do not always occur in the same sequence or with the same exact set of genes. For example, *APC* mutations are found in about two-thirds of all colon cancers, which means that the *APC* gene is normal in one out of every three cases. Analysis of tumors containing

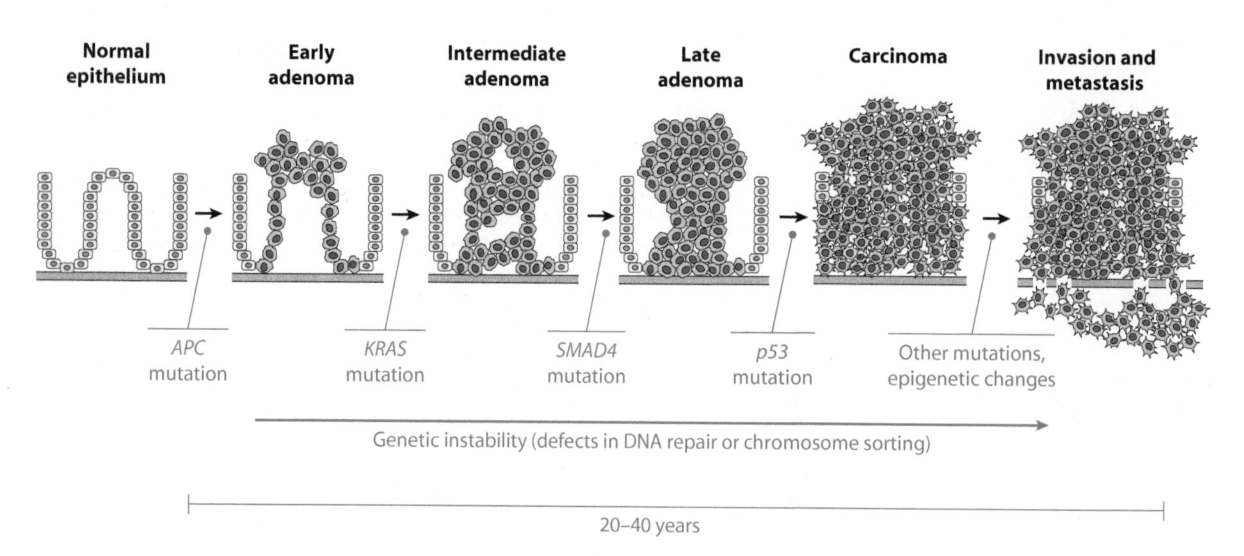

Figure 10-19 Stepwise Model for the Development of Colon Cancer. Colon cancer often arises through a stepwise series of mutations involving the *APC*, *KRAS*, *SMAD4*, and *p53* genes. Each successive mutation is associated with increasingly abnormal cell behavior. [Modified from R. J. Davies et al., *Nature Reviews Cancer* 5 (2005): 199 (Figure 2).]

normal APC genes has revealed that many of them possess oncogenes that produce an abnormal, hyperactive form of *β-catenin*, a protein that—like the APC protein—is involved in Wnt signaling (see Figure 10-8). Because APC inhibits the Wnt pathway and β-catenin stimulates it, mutations leading to the loss of APC and mutations that create hyperactive forms of β-catenin have the same basic effect: Both enhance cell proliferation by increasing the activity of the Wnt pathway.

Another pathway frequently disrupted in colon cancer is the *TGFβ-Smad pathway* (p. 184), which *inhibits* rather than stimulates epithelial cell proliferation. Loss-of-function mutations in genes coding for components of this pathway, such as the TGFβ receptor or Smad4, are commonly detected in colon cancers. Such mutations disrupt the growth-inhibiting activity of the TGFβ-Smad pathway and thereby contribute to enhanced cell proliferation.

Overall, the general principle illustrated by the various colon cancer mutations is that different tumor suppressor genes and oncogenes can affect the same pathway, and it is the disruption of particular signaling pathways that is important in cancer cells rather than the particular gene mutations through which the disruption is achieved (Table 10-2).

The Various Causes of Cancer Can Be Brought Together into a Single Model

Colon cancer illustrates how normal cells can be converted into cancer cells by a small number of genetic changes, each affecting a particular pathway and conferring some type of selective advantage. Of course, colon cancer is just one among dozens of different human cancers, and the few genes commonly mutated in colon cancer are only a tiny fraction of the more than 100 different oncogenes and tumor suppressor genes. When various kinds of tumors are compared, it is found that different combinations of gene mutations can lead to cancer and that each type of cancer tends to exhibit its own characteristic mutation patterns.

Despite this variability, a number of shared principles are apparent in the various routes to cancer. An overview is provided by the model illustrated in Figure 10-20, which begins with the four main causes of cancer: chemicals, radiation, infectious agents, and heredity.

Table 10-2 Some Common Mutations in Human Colon Cancer

Gene	Protein	Category	Pathway	Effect of Mutation
APC	APC	Tumor suppressor	Wnt pathway	Loss of function increases Wnt signaling
CTNNB1	β-catenin	Oncogene	Wnt pathway	Gain of function increases Wnt signaling
KRAS	Ras	Oncogene	Ras-MAPK pathway	Gain of function increases Ras-MAPK signaling
SMAD4	Smad4	Tumor suppressor	TGFβ-Smad pathway	Loss of function decreases TGFβ-Smad signaling*
TGFBR2	TGFβ receptor II	Tumor suppressor	TGFβ-Smad pathway	Loss of function decreases TGFβ-Smad signaling*
p53	p53	Tumor suppressor	DNA-damage response	Loss of function promotes genetic instability
Mismatch repair genes	Various enzymes	Tumor suppressors	DNA repair	Loss of function promotes genetic instability

Note: Mutations that have similar effects are grouped together.
*Because the TGFβ-Smad signaling pathway *inhibits* epithelial cell proliferation, decreased activity in this pathway will lead to enhanced cell proliferation.

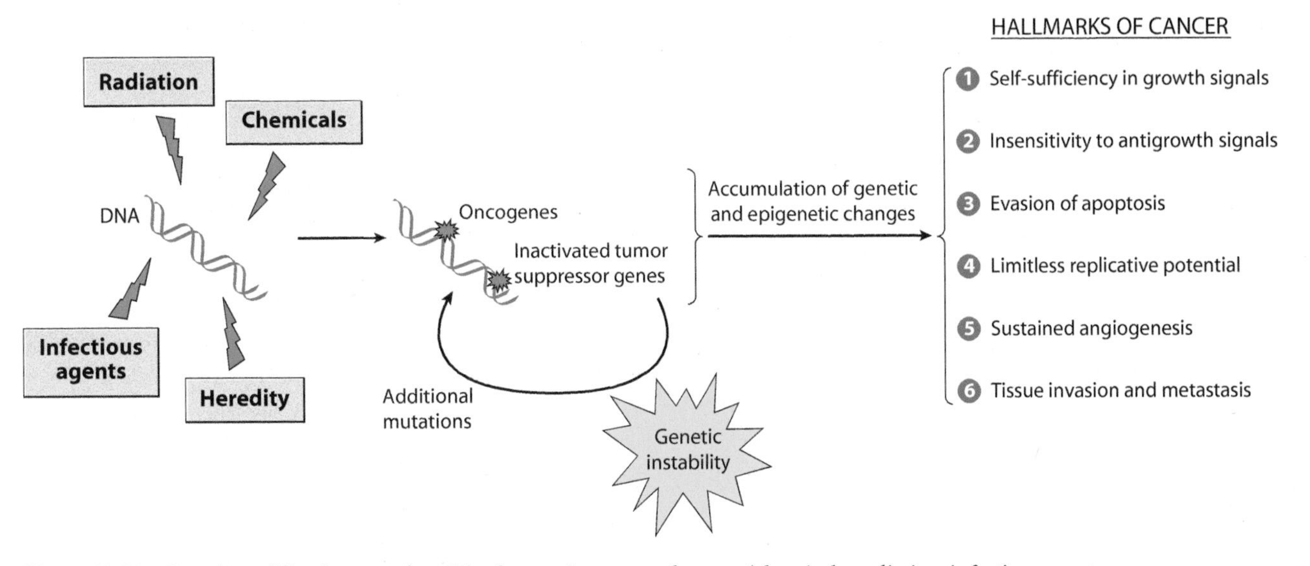

Figure 10-20 Overview of Carcinogenesis. The four main causes of cancer (chemicals, radiation, infectious agents, and heredity) trigger DNA alterations that create oncogenes and disrupt tumor suppressor genes. Genetic instability facilitates the acquisition of additional mutations that, along with accompanying epigenetic changes, eventually lead to the six "hallmark" traits: self-sufficiency in growth signals, insensitivity to antigrowth signals, evasion of apoptosis, limitless replicative potential, sustained angiogenesis, and tissue invasion and metastasis.

Chapters 5 through 8 described how each of these four factors contributes to the development of malignancy. While the details may differ, the bottom line is that one way or another, each of the four causes of cancer leads to DNA alterations.

In the case of either viruses that introduce specific oncogenes into cells or cancer syndromes that arise from inherited gene defects, the DNA alterations involve a specific gene. Most of the DNA mutations induced by carcinogens, on the other hand, are random. The higher the dose and potency of the carcinogen, the greater the DNA damage and therefore the greater the probability that a random mutation will disrupt a critical gene. But critical genes (proto-oncogenes and tumor suppressor genes) represent only a tiny fraction of the chromosomal DNA, so the random nature of mutation means that luck plays a significant role; if two people are exposed to the same dose of a carcinogen, one may develop cancer while the other does not simply because random mutations happen to damage a critical proto-oncogene or tumor suppressor gene in the unlucky individual.

The random nature of mutation contributes to the long period of time that is usually required for cancer to develop. Moreover, when DNA repair mechanisms and DNA damage checkpoints are operating properly, many mutations are either repaired or the cells containing them are destroyed by apoptosis. Taken together, such considerations may help explain why cancer is largely a disease of older age. For cancer to develop, cells need to gradually accumulate a stepwise series of appropriate mutations involving the inactivation of tumor suppressor genes as well as the conversion of proto-oncogenes into oncogenes. These mutations, along with accompanying epigenetic changes, eventually create a group of traits referred to as the "hallmarks" of cancer.

Summing Up: The Hallmarks of Cancer

Since various combinations of mutations involving tumor suppressor genes and oncogenes can lead to cancer, the question arises as to whether there are any common principles that would help simplify the picture. As a unifying concept, Douglas Hanahan and Robert Weinberg have proposed that a series of six acquired traits are common to, and essential for, the development of cancer, but each can be acquired through a variety of different genetic (and epigenetic) mechanisms. According to such a view, the variety of genetic changes observed in cancer cells simply reflects a variety of routes to these six essential alterations, which are referred to as the "hallmarks of cancer." Each of the six hallmark traits has already been described elsewhere in this book, but they will now be summarized as a final overview of the road map to cancer.

1. Self-Sufficiency in Growth Signals. Cells do not normally proliferate unless they are stimulated to do so by growth factors, which bind to receptors that trigger various pathways for stimulating cell division. Cancer cells escape from such a requirement by producing molecules that allow a cancer cell to stimulate its own proliferation. This is accomplished mainly by oncogenes that produce excessive quantities or hyperactive versions of proteins involved in growth signaling pathways. Oncogenes in this category may code for growth factors, receptors, G proteins, protein kinases, transcription factors, or cell cycle regulators (see Table 9-1). Often the net result is uncontrolled activation of the Ras-MAPK pathway. About 30% of all human cancers produce mutant Ras proteins that provide an ongoing stimulus for the cell to proliferate independent of growth factors, and mutations affecting other components of the Ras-MAPK pathway are common as well.

2. Insensitivity to Antigrowth Signals. Normal tissues are protected from excessive cell proliferation by a variety of inhibitory mechanisms. Cancer cells must evade these antigrowth signals if they are to continue proliferating. Most antigrowth signals act during late G1 and exert their effects through the Rb protein, whose phosphorylation regulates passage through the restriction point and into S phase. One of the best-documented antigrowth factors is TGFβ, a potent inhibitor of epithelial cell proliferation discussed earlier in the chapter. TGFβ exerts its antigrowth effects through the TGFβ-Smad pathway (see Figure 10-10), which produces Cdk inhibitors that block Rb phosphorylation and thereby prevent passage through the restriction point. The TGFβ-Smad pathway of cancer cells can be disrupted by a variety of different mechanisms, including mutations, epigenetic changes, and interactions with viral oncoproteins. For example, loss-of-function mutations in tumor suppressor genes coding for TGFβ receptors and Smad proteins are common in several human cancers. Mutations in the *RB* gene can also make cells insensitive to the antigrowth effects of TGFβ or other antigrowth factors that exert their effects through the Rb protein. And in cancers caused by certain viruses, the Rb protein is inactivated by viral proteins, such as the E7 oncoprotein of HPV. The bottom line is that antigrowth pathways converging on the Rb protein are disabled in a majority of human cancers.

3. Evasion of Apoptosis. Tumor growth is affected not just by the rate of cell proliferation but also by the rate of cell death. The primary source of regulated cell death is *apoptosis*, the process described in Chapter 2 in which cells commit suicide by activating protein-degrading enzymes called *caspases*. Apoptosis is normally triggered by cellular abnormalities such as DNA damage, overexpression of oncogenes, or oxygen deficiency, any (or all) of which are likely to occur in tumors. Mechanisms for evading apoptosis are therefore crucial to the survival of cancer cells. The most common mechanism involves mutations in the *p53* gene, which occur in almost half of all human cancers. The resulting loss or inactivation of the p53 protein removes a central molecule in the pathway by which DNA damage or other abnormalities trigger apoptotic cell death. Several oncogenes also interfere with this pathway. An example is the *BCL2* gene, which codes for an inhibitor of apoptosis. By producing excessive amounts of the Bcl2 protein, *BCL2* oncogenes block the pathway that would normally trigger apoptosis, thereby permitting abnormal cells to continue proliferating. Another oncogene that inhibits cell death is *MDM2*, which produces the Mdm2 protein that promotes the degradation of p53 (see Figure 10-5). Through such mechanisms, cancer cells manage to evade the apoptotic pathways that would otherwise trigger their destruction.

4. Limitless Replicative Potential. The net result of the preceding three traits is to uncouple cancer cells from the signals that normally balance cell proliferation with the need for new cells. The resulting lack of growth control is not sufficient to ensure unlimited proliferation, however. As indicated in Chapter 2, most cells cannot divide indefinitely because they lose a small amount of DNA at the ends of their chromosomal DNA molecules each time they divide. If the telomeric DNA sequences located at the end of each chromosome become worn away, the chromosome ends become unstable. Such unprotected chromosomes tend to fuse with each other, creating joined chromosomes that separate improperly at the time of cell division and become fragmented. To circumvent such chromosomal disarray, which would otherwise lead to cell death, cancer cells activate mechanisms for replenishing telomere sequences. In most cancer cells, telomere maintenance is accomplished by activating the gene coding for the enzyme *telomerase*, which adds new copies of the telomeric repeat sequence to the ends of existing DNA molecules (p. 22). A few cancer cells activate an alternative mechanism for maintaining telomeres that involves the exchange of sequence information between chromosomes. By one mechanism or the other, cancer cells maintain telomere length above a critical threshold and thereby attain limitless replicative potential—that is, the ability to divide indefinitely and thereby achieve immortality.

5. Sustained Angiogenesis. An essential step in converting a tiny localized mass of cancer cells into an invasive and metastatic tumor is the growth of a network of blood vessels that penetrate into the tumor, supplying nutrients and oxygen and removing waste products. Without this process of *angiogenesis*, tumors will not grow beyond a few millimeters in size. As described in Chapter 3, the ability to trigger blood vessel growth is determined by the balance between molecules that inhibit angiogenesis and molecules that stimulate angiogenesis. Cancer cells initially lack the ability to trigger angiogenesis, but at some point during early tumor development they acquire the ability to do so. A common strategy is for cancer cells to activate the transcription of genes that produce angiogenesis stimulators and to depress the transcription of genes that produce angiogenesis inhibitors. The mechanisms that underlie such changes in gene expression are not well understood, but in some cases they have been linked to the activities of known tumor suppressor genes or oncogenes. For example, the p53 protein activates transcription of the gene coding for thrombospondin, an inhibitor of angiogenesis; hence, loss of p53 function, which occurs in many human cancers, can cause thrombospondin concentrations to fall. Likewise, excessive activity of the Ras-MAPK pathway caused by *RAS* oncogenes leads to decreased expression of the thrombospondin gene as well as an increased expression of the gene that produces VEGF (an activator of angiogenesis).

6. Tissue Invasion and Metastasis. The ability to invade into surrounding tissues and metastasize to distant sites is the defining characteristic that distinguishes a cancerous growth from a benign tumor. Although most human cancers eventually invade and metastasize, these traits are often the last ones to emerge. As we saw in Chapter 3, invasion and metastasis are complex events involving a variety of genetic and biochemical mechanisms. Three acquired properties appear to be especially important: decreased cell-cell adhesion, increased motility, and the production of proteases that degrade the extracellular matrix and basal lamina. Decreased adhesiveness frequently stems from deficits in *E-cadherin*, a cell-cell adhesion protein that binds epithelial cells to one another. E-cadherin is lost in the majority of epithelial cancers by either mutation, decreased gene expression, or destruction of the E-cadherin protein. Inherited mutations in the *CDH1* gene coding for E-cadherin have been linked to a hereditary form of stomach cancer, indicating that E-cadherin qualifies as a tumor suppressor whose loss is sufficient to predispose individuals to developing cancer. Alterations in other molecules involved in adhesion, motility, and the production of proteases also play roles in invasion and metastasis, but the genetic and biochemical mechanisms that govern these shifts in cancer cells are incompletely understood and seem to differ among tumor types and tissue environments.

An Enabling Trait: Genetic Instability. To acquire the preceding six traits, cancer cells need to accumulate more mutations than would normally be expected. Mutation rates are usually quite low because cells possess DNA repair pathways for correcting DNA errors and checkpoint mechanisms that monitor for DNA and chromosomal integrity before allowing the cell cycle to proceed. Taken together, these mechanisms ensure that mutations are so rare that the multiple mutations found in cancer cells would be unlikely to occur within a person's lifetime. Cells must therefore acquire defects in their DNA repair or checkpoint mechanisms and become genetically unstable before they can accumulate enough mutations for cancer to arise. The most common route to genetic instability involves defects in the p53 DNA-damage response pathway. Mutations in the *p53* gene are detected in almost half of all human cancers; when mutations affecting other components of the p53 pathway are included as well, the p53-mediated damage response is defective for one reason or another in most, if not all, human cancers. Defects in genes coding for proteins involved in DNA repair or chromosome maintenance also contribute to genetic instability. Genetic instability is placed in a separate category from the previously discussed "hallmark" traits, which are directly involved in the proliferation and spread of cancer cells, because genetic instability simply represents the means that enables evolving populations of cancer cells to acquire the six hallmark traits.

Summary of Main Concepts

- Tumor suppressors are genes whose loss or inactivation can lead to cancer.
- The first group of tumor suppressors to be discovered consists of genes that suppress cell proliferation and survival. Early evidence for the existence of such genes came from cell fusion experiments in which cancer cells were fused with normal cells. The resulting hybrid cells almost always fail to form tumors, suggesting that normal cells contain genes that suppress tumor growth.
- Several dozen tumor suppressor genes have subsequently been identified using two main approaches. One tactic involves the study of hereditary cancer syndromes in which individuals inherit a defective tumor suppressor gene from one parent. The other strategy is based on the discovery that cancer cells that have incurred a mutation in one copy of a tumor suppressor gene often undergo loss of heterozygosity in which the second, normal copy of the gene is lost or replaced with the defective version. Because loss of heterozygosity usually affects hundreds of neighboring genes simultaneously, such events are easy to detect and often point to the location of a tumor suppressor gene.
- Genes coding for Rb, p53, APC, PTEN, TGFβ receptors, Smads, p16, and ARF are examples of "gatekeeper" tumor suppressor genes that produce proteins involved in restraining cell proliferation and survival. The Rb protein restrains cell proliferation by halting progression through the restriction point of the cell cycle when growth factors are not present. The p53 protein triggers cell cycle arrest and apoptosis in cells that have sustained DNA damage, thereby preventing genetically damaged cells from proliferating. The APC protein is a part of a multiprotein complex that degrades β-catenin and thereby restrains the Wnt signaling pathway. The PTEN protein degrades PIP_3 and thereby acts as a retraining influence on the PI3K-Akt growth signaling pathway. TGFβ receptors and Smads are components of the TGFβ-Smad signaling pathway, which inhibits epithelial cell proliferation. Finally, the p16 and ARF proteins, which are produced by the same gene, are required for proper Rb and p53 signaling, respectively. Because the preceding proteins are all involved in restraining cell proliferation or survival, loss-of-function mutations in the genes coding for any of these proteins can lead to an excessive accumulation of dividing cells.
- A second group of tumor suppressors, known as "caretakers," are involved in DNA repair and chromosome maintenance. Loss-of-function mutations in caretaker genes act indirectly by permitting an increased mutation rate for all genes. This

condition, known as genetic instability, increases the likelihood that mutations will arise in genes that directly affect cell proliferation.

- The simplest type of genetic instability arises from loss-of-function mutations in genes that produce proteins required for fixing localized DNA errors by excision or mismatch repair. An additional type of genetic instability involves the tendency of cancer cells to acquire gross abnormalities in chromosome structure and number. Such chromosomal instabilities may arise from mutations that interfere either with chromosome sorting mechanisms or with mechanisms for repairing double-strand DNA breaks.

- Many of the properties that are exhibited by cancer cells stem from epigenetic changes in which the expression of normal genes is switched on or off. Measuring epigenetic changes is facilitated by the use of DNA microarrays, which allow the activity of thousands of genes to be measured simultaneously. DNA methylation, a common epigenetic mechanism for silencing gene activity, is excessive in many human cancers and can cause the inappropriate silencing of numerous genes, including tumor suppressor genes.

- Each type of human cancer is characterized by a relatively small handful of gene mutations involving both oncogenes and tumor suppressor genes. Although mutation patterns differ among cancers, this variation simply reflects a diversity of routes to the same set of hallmark traits: self-sufficiency in growth signals, insensitivity to antigrowth signals, ability to evade apoptosis, limitless replicative potential, sustained angiogenesis, and tissue invasion and metastasis.

Key Terms for Self-Testing

tumor suppressor gene (p. 175)

Tumor Suppressor Genes: Roles in Cell Proliferation and Cell Death

cell fusion (p. 176)
hybrid cell (p. 176)
loss of heterozygosity (p. 177)
RB gene (p. 177)
Rb protein (p. 178)
E2F transcription factor (p. 178)
E7 oncoprotein (p. 178)
p53 gene (*TP53* gene) (p. 179)
p53 protein (p. 179)
ATM kinase (p. 179)
Mdm2 (p. 179)
ubiquitin (p. 179)
proteasome (p. 181)
p21 protein (p. 181)
Cdk inhibitor (p. 181)
Puma (p. 181)
dominant negative mutation (p. 181)
E6 oncoprotein (p. 181)
APC gene (p. 182)
Wnt pathway (p. 182)
PI3K-Akt pathway (p. 183)
phosphatidylinositol 3-kinase (PI 3-kinase or PI3K) (p. 183)
PIP_2 (p. 183)
PIP_3 (p. 183)
Akt (p. 183)
PTEN (p. 184)
transforming growth factor β (TGFβ) (p. 184)
TGFβ-Smad pathway (p. 184)
Smad (p. 185)
p15 protein (p. 185)
CDKN2A gene (p. 185)
reading frame (p. 185)
p16 protein (p. 186)
ARF protein (p. 186)

Tumor Suppressor Genes: Roles in DNA Repair and Genetic Stability

gatekeeper (p. 186)
caretaker (p. 186)
genetic instability (p. 187)
BRCA1 and *BRCA2* genes (p. 188)
homologous recombination (p. 188)
Rad50 exonuclease complex (p. 188)
Rad51 repair complex (p. 188)
aneuploid (p. 189)
metaphase (p. 189)
spindle checkpoint (p. 189)
anaphase-promoting complex (p. 189)
anaphase (p. 189)
Mad proteins (p. 190)
Bub proteins (p. 190)
centrosome (p. 190)

Overview of Carcinogenesis

epigenetic change (p. 191)
DNA microarray (p. 191)
DNA methylation (p. 193)

Suggested Reading

Tumor Suppressor Genes: Roles in Cell Proliferation and Cell Death

Bienz, M. The subcellular destinations of APC proteins. *Nature Reviews Molec. Cell Bio.* 3 (2002): 328.

Bullock, A. N., and A. R. Fersht. Rescuing the function of mutant p53. *Nature Reviews Cancer* 1 (2001): 68.

De Caestecker, M. P., E. Piek, and A. B. Roberts. Role of transforming growth factor-β signaling in cancer. *J. Natl. Cancer Inst.* 92 (2000): 1388.

Harris, H., G. Klein, P. Worst, and T. Tachibana. Suppression of malignancy by cell fusion. *Nature* 223 (1969): 363.

Levine, A. J., C. A. Finlay, and P. W. Hinds. P53 is a tumor suppressor gene. *Cell* S116 (2004): S67.

Massague, J., S. W. Blain, and R. S. Lo. TGFβ signaling in growth control, cancer, and heritable disorders. *Cell* 103 (2000): 295.

Mayo, L. D., and D. B. Donner. The PTEN, Mdm2, p53 tumor suppressor-oncoprotein network. *Trends Biochem. Sci.* 27 (2002): 462.

Miller, J. R. Wnt Signal Transduction. In *The Cancer Handbook* (M. Alison, ed.). London: Nature Publishing Group, 2002, Chapter 14.

Polakis, P. Wnt signaling and cancer. *Genes & Develop.* 14 (2000): 1837.

Qing, J., and R. Derynck. Signalling by TGF-β. In *The Cancer Handbook* (M. Alison, ed.). London: Nature Publishing Group, 2002, Chapter 13.

Reya, T., and H. Clevers. Wnt signalling in stem cells and cancer. *Nature* 434 (2005): 843.

Sherr, C. J., and F. McCormick. The RB and p53 pathways in cancer. *Cancer Cell* 2 (2002): 103.

Soussi, T., and C. Béroud. Assessing TP53 status in human tumours to evaluate clinical outcome. *Nature Reviews Cancer* 1 (2001): 233.

Sulis, M. L., and R. Parsons. PTEN: from pathology to biology. *Trends Cell Biol.* 13 (2003): 478.

Vivanco, I., and C. L. Sawyers. The phosphatidylinositol 3-kinase-Akt pathway in human cancer. *Nature Reviews Cancer* 2 (2002): 489.

Vogelstein, B., D. Lane, and A. J. Levine. Surfing the p53 network. *Nature* 408 (2000): 307.

Vousden, K. H., and X. Lu. Live or lie die: the cell's response to p53. *Nature Reviews Cancer* 2 (2002): 594.

Yu, J., and L. Zhang. No PUMA, no death: Implications for p53-dependent apoptosis. *Cancer Cell* 4 (2003): 248.

Tumor Suppressor Genes: Roles in DNA Repair and Genetic Stability

D'Andrea, A. D., and M. Grompe. The Fanconi anemia/BRCA pathway. *Nature Reviews Cancer* 3 (2003): 23.

Hoeijmakers, J. H. J. Genome maintenance mechanisms for preventing cancer. *Nature* 411 (2001): 366.

Jallepalli, P. V., and C. Lengauer. Chromosome segregation and cancer: cutting through the mystery. *Nature Reviews Cancer* 1 (2001): 109.

Kohn, K. W., and V. A. Bohr. Genomic Instability and DNA Repair. In *The Cancer Handbook* (M. Alison, ed.). London: Nature Publishing Group, 2002, Chapter 7.

Lengauer, C., K. W. Kinzler, and B. Vogelstein. Genetic instabilities in human cancers. *Nature* 396 (1998): 643.

Nigg, E. A. Centrosome aberrations: cause or consequence of cancer progression? *Nature Reviews Cancer* 2 (2002): 815.

Scully, R., and D. M. Livingston. In search of the tumour-suppressor functions of BRCA1 and BRCA2. *Nature* 408 (2000): 429.

Shiloh, Y. ATM and related protein kinases: safeguarding genome integrity. *Nature Reviews Cancer* 3 (2003): 155.

Sieber, O. M., K. Heinimann, and I. P. M. Tomlinson. Genomic instability—the engine of tumorigenesis. *Nature Reviews Cancer* 3 (2003): 701.

Venkitaraman, A. R. Cancer susceptibility and the functions of BRCA1 and BRCA2. *Cell* 108 (2002): 171.

Overview of Carcinogenesis

Feinberg, A. P., and B. Tycko. The history of cancer epigenetics. *Nature Reviews Cancer* 4 (2004): 143.

Hahn, W. C., and R. A. Weinberg. Modelling the molecular circuitry of cancer. *Nature Reviews Cancer* 2 (2002): 331.

Hanahan, D., and R. A. Weinberg. The hallmarks of cancer. *Cell* 100 (2000): 57.

Issa, J.-P. CpG island methylator phenotype in cancer. *Nature Reviews Cancer* 4 (2004): 988.

Jones, P. A., and S. B. Baylin. The fundamental role of epigenetic events in cancer. *Nature Reviews Genetics* 3 (2002): 415.

Sherr, C. J. Principles of tumor suppression. *Cell* 116 (2004): 235.

Ushijima, T. Detection and interpretation of altered methylation patterns in cancer cells. *Nature Reviews Cancer* 5 (2005): 223.

11 Cancer Screening, Diagnosis, and Treatment

Great strides have been made over the past several decades in elucidating the genetic and biochemical changes that underlie the development of cancer. In fact, most of the information described in the preceding two chapters about cancer-related genes and their effects on cell signaling pathways has emerged since 1980. One of the hopes for such research is that the growing understanding of these underlying molecular abnormalities will lead to improved strategies for cancer diagnosis and treatment.

The bottom line, of course, is that people want a "cure for cancer." Hardly a week goes by without another newspaper or television story reporting the latest research developments that might lead to the long-awaited cure. However, this highly publicized search for a cancer cure seems to imply that we lack effective approaches for dealing with the disease. In fact, nothing could be further from the truth. While better treatments are certainly needed and are being vigorously pursued, we will see in these final two chapters that many people who develop cancer can already be cured or could have prevented their cancer from arising in the first place.

CANCER SCREENING AND DIAGNOSIS

The success rates for current cancer treatments are strongly influenced by the stage at which the disease is diagnosed. When cancer is detected early and tumor cells are still localized to their initial site of origin, cure rates tend to be very high, even for cancers that would otherwise have a poor prognosis (Figure11-1). Unfortunately, many cancers are difficult to detect in their early stages and by the time they are diagnosed, metastasis may have already occurred. If cancers were routinely detected at an earlier stage, many cancer deaths could be prevented.

Early detection is a feasible goal because despite the common perception that cancer arises rapidly and with little warning, most cancers develop slowly and only become aggressive and invasive after the gradual passage of time (usually measured in years rather than weeks or months). A prolonged window of opportunity therefore exists for detecting the disease in its earlier stages when treatments are more likely to be effective.

Cancer Has Few Symptoms That Arise Early or Are Specific for the Disease

The first thing to signal the presence of a disease is usually some type of physical symptom that prompts a visit to a doctor and helps guide the diagnosis. Few generalizations are possible about the symptoms of cancer because it can arise almost anywhere in the body. When a cancer grows beyond a tiny localized clump of cells into a larger mass that invades surrounding tissues, symptoms may begin to be triggered as the tumor impinges on surrounding structures and organs. For example, if a tumor presses on a nerve it may cause pain, or it might disrupt blood vessels and cause bleeding. The location of symptoms varies widely, depending on the type of cancer involved. After a cancer has metastasized, symptoms may appear in other parts of the body and, in some cases, may represent the first signs of disease.

Tumors tend to produce few or no symptoms when they are small and localized, so it is difficult to come up with reliable guidelines to help people detect cancer early. For many years, the American Cancer Society publicized a list of seven warning signs that are possible indicators of the presence of cancer:

1. Change in bowel or bladder habits.
2. A sore that does not heal.
3. Unusual bleeding or discharge.
4. Thickening or lump in the breast or elsewhere.
5. Indigestion or difficulty swallowing.
6. Obvious change in a wart or mole.
7. Nagging cough or hoarseness.

Any of the preceding symptoms might be a sign of cancer, but the list has two shortcomings that limit its usefulness. First, when one of these symptoms does arise because of cancer, it may not appear until the disease has advanced to a relatively late stage. Second, none of the listed symptoms

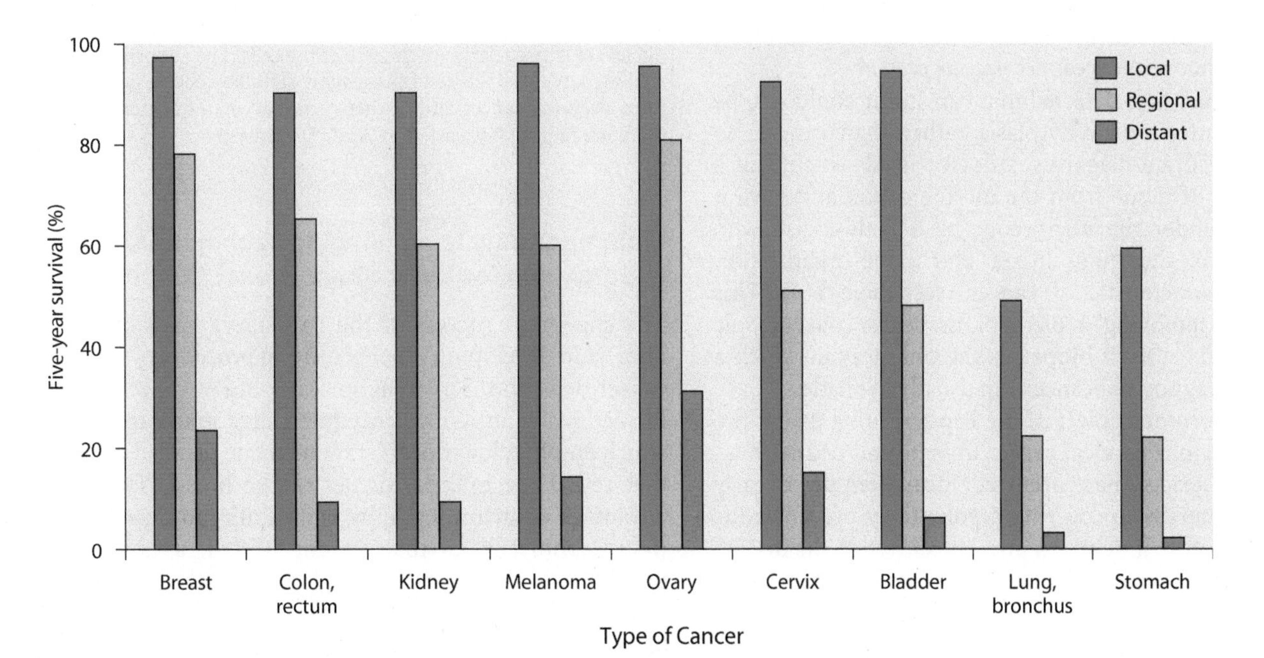

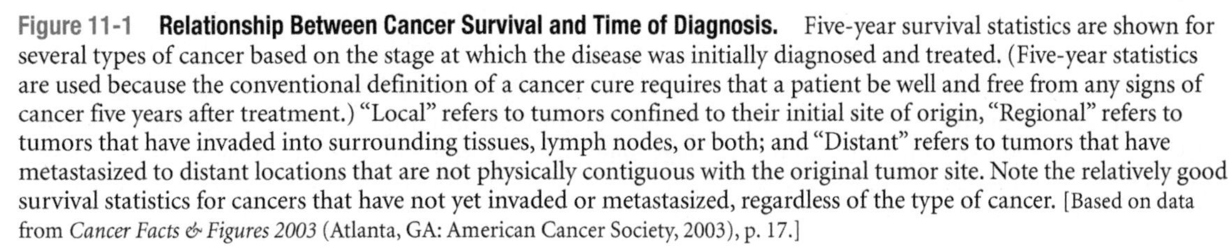
Figure 11-1 Relationship Between Cancer Survival and Time of Diagnosis. Five-year survival statistics are shown for several types of cancer based on the stage at which the disease was initially diagnosed and treated. (Five-year statistics are used because the conventional definition of a cancer cure requires that a patient be well and free from any signs of cancer five years after treatment.) "Local" refers to tumors confined to their initial site of origin, "Regional" refers to tumors that have invaded into surrounding tissues, lymph nodes, or both; and "Distant" refers to tumors that have metastasized to distant locations that are not physically contiguous with the original tumor site. Note the relatively good survival statistics for cancers that have not yet invaded or metastasized, regardless of the type of cancer. [Based on data from *Cancer Facts & Figures 2003* (Atlanta, GA: American Cancer Society, 2003), p. 17.]

is specific for cancer, and in most cases a person exhibiting one or more of the symptoms will not actually have cancer. Yet despite these shortcomings, each warning sign in the list indicates a condition that should be assessed by a doctor because if it does signal the presence of cancer, the outcome will be much better if treatment is started early.

The Pap Smear Illustrates That Early Detection Can Prevent Cancer Deaths

Because cancers do not produce many symptoms in their early stages, there is an urgent need for better screening procedures that can detect the presence of cancer before any symptoms appear. The most successful screening technique thus far has been the **Pap smear**, a procedure for the early detection of cancer of the uterine cervix that was first introduced in the early 1930s by George Papanicolaou (for whom it is named). The rationale underlying the Papanicolaou procedure is that the microscopic appearance of cancer cells is so distinctive compared to normal cells that it is possible to detect the likely presence of a malignancy by simply examining a few isolated cells under a microscope.

To obtain cells for examination, a doctor inserts a small brush and spatula into the vagina and uses them to scrape cells from the surface of the uterine cervix. The cells are then smeared across a glass slide—hence the name Pap "smear"—and the slide is stained and examined with a microscope. If the cells in the specimen are found to exhibit abnormal features, such as numerous cells undergoing mitosis, cells with large irregular nuclei, and prominent variations in cell size and shape (Figure 11-2), it is a sign that cervical cancer *may be present*.

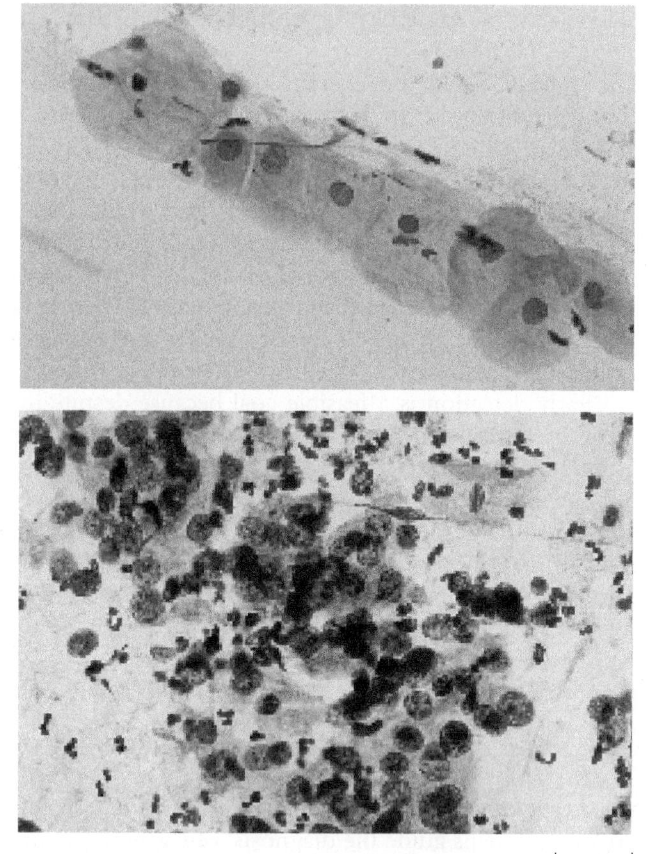

Figure 11-2 **Normal and Abnormal Pap Smears.** (*Top*) In a normal Pap smear, the cells are relatively uniform in size and contain small spherical nuclei. (*Bottom*) In this abnormal Pap smear, marked variations in cell size and shape are evident, and the nuclei are larger relative to the size of the cells. The abnormalities exhibited by these isolated cells suggest that they may be derived from a cervical cancer, and further examination of the uterus is therefore required. [Courtesy of Gerald D. Abrams.]

Abnormal cells detected in a Pap smear could also be a sign of infection or dysplasia rather than cancer. To make an accurate diagnosis, a doctor needs to cut out a small piece of tissue from the uterine cervix and have it examined under the microscope by a pathologist, who looks for the alterations in cell and tissue organization that are characteristic of cancer (see Table 1-4). This process of removing a piece of tissue for microscopic examination, called a **biopsy**, is the only way in which a definitive diagnosis of cancer can usually be made.

The enormous power of the Pap screening procedure is that it permits cervical cancer to be found before invasion and metastasis have occurred; it can even detect early dysplasia, thereby uncovering a potentially precancerous condition before it develops into cancer. When the disease is caught in these early stages and treated appropriately, the survival rate is virtually 100%. Widespread use of the Pap smear has lowered the death rate from cervical cancer in the United States more than 75% since the procedure was first introduced in the 1930s, thereby preventing hundreds of thousands of cancer deaths. Quite recently the usefulness of the Pap smear has been further improved by combining it with a DNA test for the human papillomavirus (HPV), which is linked to the development of cervical cancer (p. 122).

Mammography Is an Imaging Technique Used in Screening for Early Stage Breast Cancers

The enormous success of the Pap smear has stimulated interest in developing other screening procedures for early cancer detection. Screening for early breast cancer is performed using an X-ray procedure called **mammography**, which employs low-dose X-rays to create detailed pictures that reveal the internal tissues of the breast. The main advantage of mammography is its ability to detect small tissue abnormalities that cannot be felt, although the procedure is not accurate enough to prove the presence or absence of cancer. If mammography reveals the presence of an abnormal tissue mass, a biopsy must be performed to determine whether it is actually a cancer.

Some disagreement has been voiced in the medical and scientific literature as to the exact age when women should begin mammography. At the time when this book was being written, the American Cancer Society and governmental agencies were recommending mammography every one to two years for women aged 40 and older. Some scientists, however, have advocated caution in recommending mammography for women in their 40s because radiation itself can cause cancer. Although the level of radiation used in modern mammography is quite low, the younger a woman is when she begins annual mammography, the greater the lifetime dose of radiation she will eventually receive. The benefits derived from starting mammography at a younger age therefore need to be balanced against the risks accumulated through radiation exposure. The only way to resolve this issue is through careful statistical analysis of the actual, real-life experiences of women who have used mammography.

Such analyses suggest that mammography is most beneficial for two groups of individuals: (1) women with an unusually high risk of developing breast cancer (mainly those with a mother or sister who have already developed the disease) and (2) women over 50, who are at higher risk of developing breast cancer because they are older. The risk of dying of breast cancer for these older women appears to be reduced up to 30% through the routine use of mammography. The data suggest that younger women, in the 40 to 50 age group, may also derive some benefit from mammography, although the apparent reduction in mortality is much smaller than for older women and the statistics are not as compelling. It is therefore possible that future recommendations concerning the exact age when women should begin routine mammography may change slightly as we acquire more experience and data concerning the use of this screening tool.

Colonoscopy, X-ray Procedures, and the Fecal Occult Blood Test Are Used in Screening for Early Stage Colorectal Cancers

Cancers of the colon and rectum cause more deaths in the United States than any type of cancer other than lung cancer, so there is a great need for effective screening procedures that can reduce the death rate through early detection. A variety of screening options are currently available, including direct visual examination of the colon and rectum, X-ray imaging, and biochemical testing of feces.

Procedures for direct examination of the colon and rectum utilize a slender, flexible, fiber-optic tube that is inserted into the anus, through the rectum (the last 6 to 9 inches of the large intestine), and into the colon, thereby allowing doctors to see the inner surface of the colon and rectum. **Colonoscopy** uses a version of this instrument called a *colonoscope*, which is long enough to visualize the entire length of the colon (a shorter version is used for *sigmoidoscopy*, a similar procedure that allows only the left part of the colon to be seen). If any abnormal growths are found during colonoscopy, they can be biopsied or, in the case of small polyps, removed entirely with a tiny instrument that is inserted through the colonoscope.

X-ray imaging techniques are also employed in screening for colorectal cancer. In one procedure known as a *barium enema*, a physician administers a liquid containing barium through the anus and into the rectum and colon. The barium compound helps improve the contrast of the image when X-ray pictures are taken. However, the resulting X-ray images are still not as accurate as colonoscopy in detecting early cancers, and use of this technique has been dropping over the past decade. A newer X-ray procedure, called *virtual colonoscopy*, employs an X-ray scanner to take multiple pictures of the colon at various angles; the resulting data are then assembled by a computer to produce a three-dimensional reconstruction of the inner surface of the colon. Virtual colonoscopy is not quite as sensitive as an actual colonoscopy for detecting the smallest polyps, but it is effective in detecting larger polyps and is currently being evaluated as a noninvasive alternative to conventional colonoscopy.

Another screening strategy involves biochemical testing of fecal samples for indications that cancer might be growing in the colon or rectum. Because colorectal cancers tend to bleed intermittently, one useful indicator is the presence of blood in the feces. A screening test called the **fecal occult blood test** (**FOBT**) is designed to detect such blood, even when it is present in amounts that are too small to be seen (hence the name "occult," which means "hidden"). With an application stick, a dab of a stool specimen is simply smeared on a chemically treated card that is then tested in a laboratory for evidence of blood. If blood is detected, colonoscopy may be necessary to find the source of the bleeding. FOBT is not as sensitive or specific as colonoscopy, but it is less expensive, noninvasive, and more practical for routine testing, and its use has been shown to reduce death rates from colorectal cancer.

Other substances present in fecal samples might also indicate the presence of cancer. For example, we saw in Chapter 10 that colon cancer cells often have mutations in the *APC* gene. Since colorectal cancers continually shed cells from the inner surface of the colon or rectum, small amounts of DNA from these cells end up in the feces. Researchers have shown that techniques capable of detecting tiny amounts of DNA can identify the presence of DNA containing *APC* mutations in fecal samples obtained from individuals with colorectal cancer. Although the procedure is not yet in routine clinical use, such approaches are currently being evaluated, and it is possible that screening stool samples for DNA mutations may improve our ability to detect early colorectal cancer in the not-too-distant future.

Blood Tests for Cancer Screening Include the PSA Test for Prostate Cancer as Well as Experimental New Proteomic Techniques

The ideal screening test would allow doctors to detect early stage cancers anywhere in the body with one simple procedure, such as a blood test. Prostate cancer is an example of a cancer that can sometimes be detected in this way. Men over the age of 50 are advised by many doctors to get a **PSA test**, which measures how much *prostate-specific antigen*

(*PSA*) is present in the bloodstream. PSA is a protein produced by the prostate gland that normally appears in only tiny concentrations in the blood. If a PSA blood test reveals a high concentration of PSA, it indicates the existence of a prostate problem that might be an infection, hyperplasia, or cancer. A biopsy therefore needs to be performed to determine whether cancer is actually present. If cancer is diagnosed, blood tests for PSA may be used again later to determine how many cancer cells remain after treatments are administered to remove or destroy the cancer.

A number of other cancers also produce proteins that are released into the bloodstream in elevated amounts. Two of these proteins were mentioned in Chapter 2: *alpha-fetoprotein*, which is produced by some liver cancers, and *carcinoembryonic antigen* (*CEA*), which is produced by some colon, stomach, pancreatic, and lung cancers. Another example is *CA125*, a protein released into the bloodstream by many ovarian cancers. The presence of such proteins in the blood is not reliable enough to be used for routine cancer screening, but they are measured in some cancer patients to monitor the course of a person's disease and its responsiveness to treatment.

Researchers are currently trying to identify other proteins in the blood that might be better indicators of the presence of specific cancers. One technique being applied to the problem is known as **proteomic analysis** (the term *proteome* refers to all the proteins produced by a cell or organism). The key to proteomic analysis is *mass spectrometry*, a high-speed, extremely sensitive method for identifying proteins based on differences in mass and electrical charge. Mass spectrometry permits rapid examination of the protein makeup of a tiny sample, such as a drop of blood, creating a snapshot of thousands of proteins at once (Figure 11-3). To overcome the difficulty in trying to look at patterns involving thousands of proteins to see which ones might indicate the presence of cancer, computer scientists have developed software programs to tackle the job. These computer programs can compare the complex proteomic patterns seen in blood samples from individuals with or without cancer and identify small changes that are associated with certain kinds of cancer.

One of the first cancers to be investigated in this way was ovarian cancer. When ovarian cancer is detected before it spreads beyond the ovaries, the five-year survival rate is better than 95%. Early disease has few symptoms, however, and relatively few cases are detected early, so less than 50% of women with the disease end up surviving more than five years. Hence better detection techniques are urgently needed. Using proteomic analysis, scientists have uncovered a diagnostic proteomic pattern in the blood of women with ovarian cancer that does not appear in the blood of other women. Initial studies indicated that the test had a specificity of about 95%, which means that 95 out of 100 women exhibiting the abnormal pattern of blood proteins will have ovarian cancer. Subsequent reports suggest that it may be possible to improve the sensitivity even further, and a commercial version of such a test is under development. Questions have been raised, however, concerning the reliability of this approach, and the validity of the test needs to be independently confirmed. The development of proteomic screening is thus still in its infancy, and much work remains to be done before we will know whether cancers can be reliably detected in their early stages using proteomic analysis to identify small changes in blood protein composition.

False Negatives, False Positives, and Overdiagnosis Are Some of the Problems Encountered with Cancer Screening Tests

Screening tests are intended to reduce cancer deaths through early detection, but two criteria must be met before we can be confident that a given test is actually

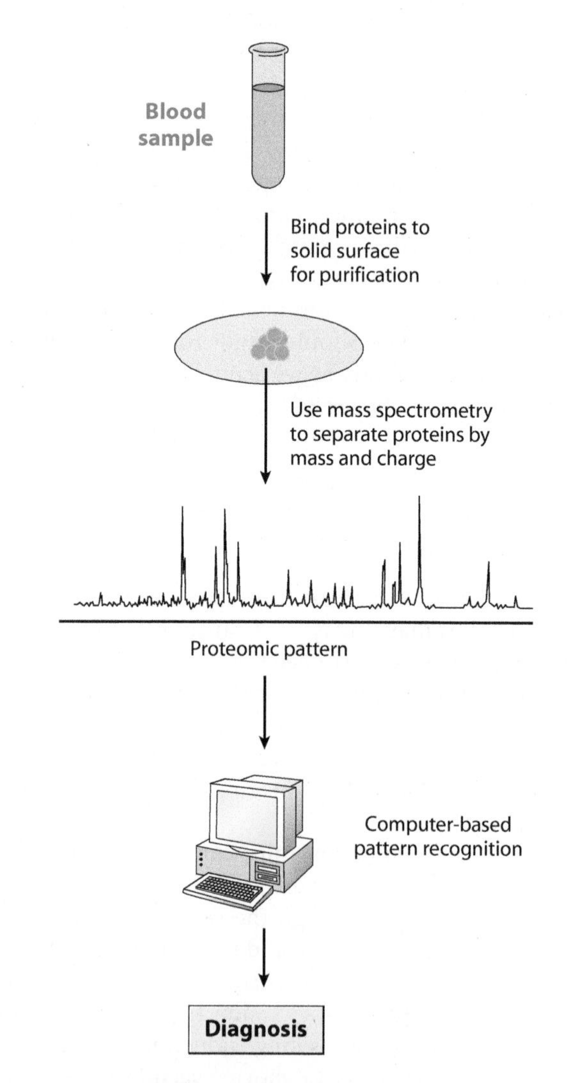

Figure 11-3 **Cancer Screening Using Proteomic Analysis.** Recently developed procedures allow rapid analysis of the protein makeup of a tiny sample of blood, creating a snapshot of thousands of proteins at once. The key to this approach is mass spectrometry, a sensitive technique for separating proteins based on differences in their mass and electrical charge. Computer-based pattern recognition programs compare the proteomic patterns seen in blood samples from individuals with or without cancer and learn to identify small changes that are associated with certain kinds of cancer.

accomplishing this objective. First, effective methods must be available for treating any cancers that are detected through early screening. Second, screening tests must exhibit sufficient *sensitivity* and *specificity* to minimize the potential problems that would be created by incorrect test results.

The *sensitivity* of a cancer screening test addresses the question: What percentage of the people with a given type of cancer will have their cancer detected when a screening test is used? If a test is not very sensitive, there will be many **false negative** results in which a person who has cancer obtains a negative test result—that is, the cancer is not detected. A test that yields many false negatives will not be very effective at reducing cancer deaths and may even be counterproductive, giving a false sense of security to people who do have cancer.

In contrast, the *specificity* of a screening test addresses the question: What percentage of the people who do not have cancer are correctly identified as being free of the disease? If a screening test is not very specific, it will yield many **false positive** results; that is, people will test positive even though they do not have cancer. A false positive can lead to unnecessary and costly follow-up procedures as well as anxiety to the person involved. Moreover, the number of false positives can easily exceed the number of correct positives if the overall cancer rate is relatively low in the population being screened, which is the case for many cancers (Figure 11-4).

To illustrate the relevance of these concepts to a real-life situation, let us briefly consider the FOBT screening test for the early detection of colorectal cancer. The FOBT test has a reported specificity as high as 98% (2% false positives), which at first impression sounds like a high level of specificity. However, the annual incidence of colorectal cancer in the United States for the overall population is currently about 55 cases per 100,000. So, if 100,000 people were randomly tested once per year using the FOBT test, the result would be 2% $\times$ 100,000 = 2000 false positives in a population that has only 55 new cases of colorectal cancer. Thus a 2% false positive rate yields so many incorrect results relative to the number of real cancer cases as to make the procedure almost useless for random testing. On the other hand, colorectal cancer rates increase dramatically as people get older, rising roughly tenfold between 30 and 50 years of age and another five-fold between 50 and 70 years of age. If FOBT testing is restricted to older populations, the ratio of real cancer cases to false positives increases dramatically and the test becomes more useful.

It may seem counterintuitive, but another problem that can arise with cancer screening techniques is the detection of cancers that would not otherwise have been a health hazard. This situation, known as **overdiagnosis**, has been documented for prostate cancer, a common disease in older men that often arises in slow-growing or dormant forms. Current research suggests that at least 30% of the men whose prostate cancers are detected through PSA screening have tumors that would not otherwise have been detected or created any health problems during their lifetime. Diagnosing cancer in such individuals is likely to

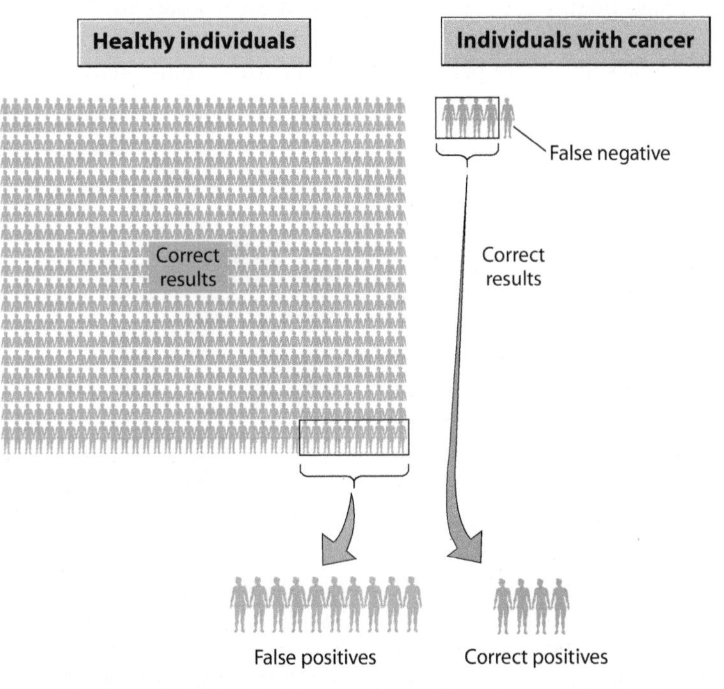

Figure 11-4 The Impact of False Positives on Cancer Screening. Color shading is used to indicate people who do (red) or do not (green) have cancer. This diagram shows that when cancer rates in a population are relatively low (which is the case for most cancers), the number of false positives given by a cancer screening test can easily exceed the number of correct positives, thereby limiting the usefulness of the test.

cause anxiety and lead to unneeded treatments with potentially adverse side effects. On the other hand, men who avoid the PSA test to avert the possibility of overdiagnosis face the risk of missing the detection of an aggressive prostate cancer whose early detection could have saved their lives. What we clearly need is better screening tests that can distinguish between aggressive cancers that are potentially life threatening and those cancers that are not serious and would be better left alone.

A Biopsy Can Diagnose the Presence of Cancer Before Invasion and Metastasis Have Begun

When the results of a cancer screening test are "positive" (i.e., abnormal), it does not mean that a person necessarily has cancer, but it does indicate the need for follow-up evaluation to determine the exact nature of the underlying problem. A definitive diagnosis generally requires that a biopsy specimen be taken from the suspected tumor site and examined with a microscope, looking for the changes in cell structure, mitotic rate, and tissue organization that signal the presence of malignancy (see Table 1-4).

Although a biopsy is always desirable, it may be difficult to obtain the required tissue specimen when the tumor is located in a relatively inaccessible site, such as the brain or in an organ deep within the body. For example, headaches and other neurological symptoms occasionally indicate the presence of a brain tumor, but a brain biopsy would not be attempted to obtain an initial diagnosis. In situations in which performing a biopsy would be difficult, or when the precise location of a possible tumor needs to be determined before it can be biopsied, diagnostic **imaging techniques** are used to take pictures of the inside of the body.

Imaging techniques utilize X-rays, magnetic fields, or ultrasound to form images. In the case of X-ray imaging, a variety of approaches are available. Conventional X-ray procedures (for example, a chest X-ray) may be useful for the initial identification of tissue abnormalities. In some organs, administering an electron opaque liquid containing barium can help increase the contrast of the X-ray pictures. The highest resolution images are produced by a **computed tomography scan (CT scan)**, a technique in which an X-ray scanner moves around the body taking multiple pictures that are then assembled by a computer into a series of detailed cross-sectional images. **Magnetic resonance imaging (MRI)** involves a similar approach using strong magnets and radio waves instead of X-rays, thereby generating a more accurate image. **Ultrasound imaging** uses sound waves and their echoes to produce a picture of internal body structures. High-frequency sound waves are transmitted into the region of the body being studied, the echoes are picked up by a receiving instrument, and a computer converts the data into a visible image.

If one or more imaging techniques reveals an abnormal tissue mass within the body, a biopsy is usually performed. Microscopic examination of a biopsy specimen permits a definitive diagnosis of cancer to be made, even if the tumor has not yet invaded or metastasized. In other words, even though cancer is defined as a growth that can spread by invasion and metastasis, microscopic examination permits a tumor harboring this potential to be identified before it has actually invaded surrounding tissues or metastasized to other areas of the body. In essence, doctors are able to see into the future, predicting that a tumor will invade and metastasize at some future time even though it has not yet done so.

A cancer that has been diagnosed before invasion has taken place is said to be *in situ* ("in place"). This term is most commonly applied to cancers of epithelial origin—that is, carcinomas—because epithelial cell layers are separated from underlying tissues by a distinct boundary structure that makes it relatively easy to determine whether a cancer has begun to spread. This boundary structure, called the **basal lamina**, is a thin, dense layer of protein-containing material that forms a barrier between an epithelial cell layer and the underlying tissue (Figure 11-5). A cancer that has not yet invaded through the basal lamina is called a **carcinoma *in situ***, meaning that it is still in its preinvasive stage.

Cancer Diagnosis Includes Information Regarding the Stage of the Disease

The verdict that cancer is present is only the beginning of a complete cancer diagnosis. One of the next issues that needs to be addressed is the question of how far a person's cancer has progressed. In Chapter 1, we learned that **tumor staging** uses three main criteria to establish a stage number that reflects how early a cancer has been detected: (1) the size of the primary tumor and the extent of its spread into nearby tissues, (2) the extent to which cancer cells have spread to regional lymph nodes, and (3) the extent to

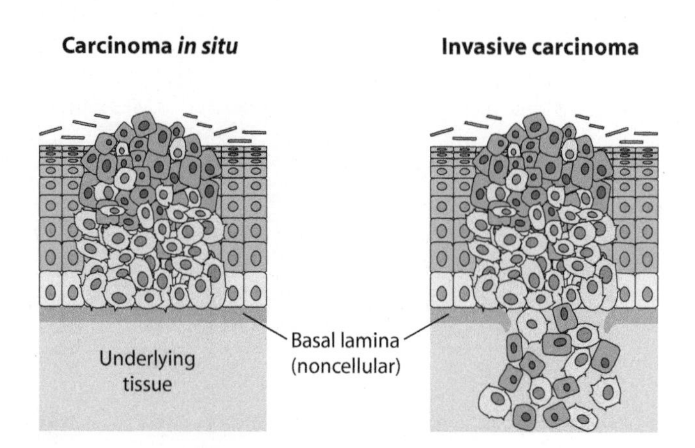

Figure 11-5 Distinction Between Carcinoma *in Situ* and an Invasive Cancer. Epithelial cell layers are separated from underlying tissue by a dense layer of protein-containing material known as the basal lamina. A cancer that has not invaded through the basal lamina is called a carcinoma *in situ* because it is still "in place"—that is, restricted to its original location. After cancer cells have penetrated through the basal lamina, they can invade into surrounding tissues, enter the bloodstream, and metastasize to distant sites.

which distant metastases are evident. A low stage number means that a cancer has been caught earlier and that treatment is more likely to be successful (see Figure 1-7, *right*).

Sometimes a biopsy reveals tissue abnormalities that occur even earlier, prior to the formation of an actual tumor. For example, the development of cancer in some tissues is preceded by a distinct period of *dysplasia* (abnormal cell proliferation accompanied by the loss of normal tissue organization). Areas of dysplasia may revert back to normal behavior, or the abnormalities can become more severe and gradually develop into cancer. Dysplasia is a common condition in the uterus, where it is often discovered when the uterine cervix is biopsied after an abnormal Pap smear. The conversion of dysplasia into cancer typically requires several years, providing a significant "window of opportunity" during which the dysplastic region can be removed or destroyed to prevent cancer from arising. Even when the condition is not diagnosed until carcinoma *in situ* has developed, these preinvasive tumors are relatively easy to treat because they have not yet begun to invade and spread.

A slightly different sequence of precancerous stages has been uncovered by screening tests for colon cancer. Colonoscopy sometimes reveals the presence of colon polyps that, upon microscopic examination, turn out to be adenomas (benign tumors of gland cells). As we saw in Chapter 10, the cells in a polyp can acquire subsequent mutations, usually over a period of several years, that convert the polyp first into a localized, preinvasive adenocarcinoma (carcinoma *in situ*) and then into an invasive adenocarcinoma. In other words, a benign tumor is often the first step on the road to malignancy in the colon. Removal of polyps while they are still benign is therefore an effective way of decreasing a person's risk of developing colon cancer.

The preceding examples reinforce a point that has been made several times in earlier chapters: Cancer arises through a multistep process that begins with early cellular abnormalities, such as dysplasia or benign neoplasia, followed by conversion into a preinvasive cancer that progresses into an invasive cancer that eventually metastasizes. The initial stages of this process usually take a significant amount of time, providing ample opportunity for early diagnosis and treatment. When caught at any stage prior to the onset of invasion and metastasis, the disease can almost always be treated successfully.

Cancer Diagnosis Includes Information Regarding the Microscopic Appearance and Molecular Properties of the Tumor Cells

The diagnosis that a person has cancer is generally accompanied by information concerning the cancer's site of origin and the cell type involved. The site of origin may or may not be the site in which the cancer was initially detected. For example, cancer discovered in the bones of the spinal column might turn out to be lung cancer that has metastasized to bone, or cancer discovered in the liver might turn out to be stomach cancer that has metastasized to the liver. The type of cancer in such situations is always defined by the location of the primary tumor. In other words, lung cancer that has metastasized to bone is still lung cancer, not bone cancer, and stomach cancer that has metastasized to the liver is stomach cancer, not liver cancer. Knowing the site of origin is important because it determines the type of cancer and provides information as to how the cancer is likely to behave and how it should be treated.

After a tumor's site of origin has been determined, additional information can be provided by microscopic examination of the biopsy specimen to determine the exact type of cells involved. For example, there are different types of lung cancer, different types of stomach cancer, and different types of skin cancer, each determined by the identity of the cell type that has become malignant. Knowing the cell type, like knowing the site of origin, provides information regarding likely tumor behavior and guidance as to the most appropriate treatment.

For cancers arising in the same site and involving the same cell type, further distinctions can be made based on the severity of the abnormalities that are observed during microscopic examination of the biopsy specimen. As discussed in Chapter 1, pathologists have devised systems for **tumor grading** in which cancers of the same type are assigned different numerical grades based on the extent of the cellular and tissue disruptions that are seen with a microscope (p. 11). Lower-grade cancers have a more normal appearance and often a better prognosis for long-term survival than do higher-grade cancers.

Biochemical tests for molecular components can further refine the picture regarding likely tumor behavior and appropriate treatment strategies. For example, breast cancer specimens are often tested for the presence of estrogen receptors, which are protein molecules involved in the mechanism by which estrogen stimulates the proliferation of breast cells. Breast cancers that possess estrogen receptors tend to have a better prognosis than cancers without estrogen receptors and are more likely to respond to hormone therapies.

Cancer cells also exhibit numerous changes in gene expression that can provide information about how tumors are likely to behave. One widely used approach for measuring gene expression is DNA microarray analysis, a technique whose ability to monitor the activity of thousands of genes simultaneously was described in Chapter 10 (see Figure 10-18). Experiments involving DNA microarrays have led to the identification of differing patterns of gene expression among tumors of the same type that allow predictions to be made regarding tumor behavior. In the case of breast cancer, for example, the expression of 21 key genes turns out to be a good indicator of whether a given tumor is likely to metastasize. Based on this discovery, a test called *Oncotype DX* has been devised that measures the activity of these 21 genes and converts the data into a single number called a *recurrence score*. As shown in Figure 11-6, women with breast

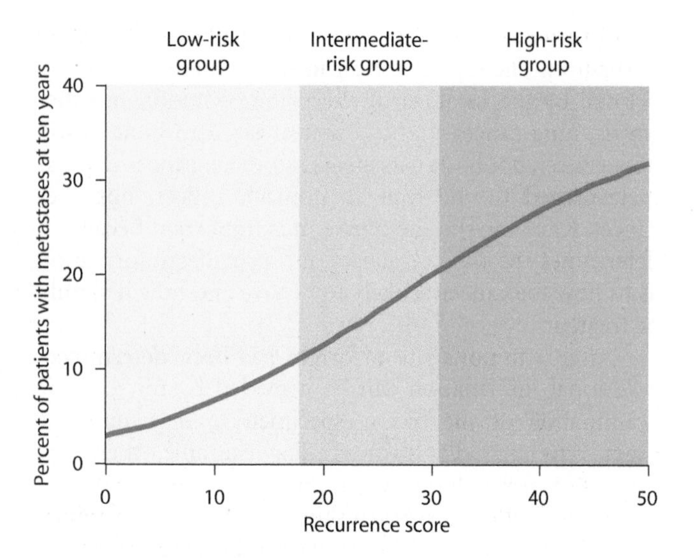

Figure 11-6 Ability of a Gene Expression Test to Predict Future Cancer Metastases. The data presented here are for the Oncotype DX gene expression test, which measures the expression of 21 key genes in breast cancer tissue and converts the data into a recurrence score. Higher recurrence scores indicate a higher likelihood of future metastases after the original breast cancer has been removed surgically. The breast cancers in this study had estrogen receptors, had not spread to lymph nodes at the time of initial diagnosis, and were treated with tamoxifen after surgery. [Data from S. Paik et al., *New England J. Med.* 351 (2004): 2817 (Figure 4).]

cancer whose tumors have a high recurrence score are more likely to develop metastases than are women whose tumors exhibit a low recurrence score. Such information is useful in guiding treatment strategies because patients with higher recurrence scores derive more benefit from subsequent chemotherapy.

SURGERY, RADIATION, AND CHEMOTHERAPY

People diagnosed with cancer have a variety of treatment options available that depend both on the type of cancer they have and how far it has spread. The ultimate goal of traditional cancer treatments is the complete removal or destruction of cancer cells accompanied by minimal damage to normal tissues. This goal is usually pursued through a combination of *surgery* (when possible) to remove the primary tumor, followed (if necessary) by *radiation*, *chemotherapy*, or both to destroy any remaining cancer cells.

Surgery Can Cure Cancers When They Have Not Yet Metastasized

Surgical techniques for removing tumors were first described more than three thousand years ago, making surgery the oldest approach for treating cancer. Its early use, however, was severely limited by the excruciating pain caused in the absence of anesthetics and by the extremely high death rate from infections. The modern era of surgery was ushered in by the discovery of ether anesthesia in the 1840s and by the introduction of carbolic acid to inhibit bacterial infections in the 1860s. By 1890, these innovations had made it possible to perform the first **mastectomy**—that is, complete removal of the breast in women with breast cancer. This milestone was followed in the early 1900s by the development of surgical techniques for removing tumors from virtually every organ of the body.

When people think of cancer surgery, they usually picture a doctor using a scalpel to cut out the tumor and perhaps surrounding tissues. Although that is certainly the most common surgical technique, a variety of newer procedures using different types of instruments have broadened the concept of what surgery is. For example, *laser surgery* utilizes a highly focused beam of laser light to cut through tissue or to vaporize certain cancers, such as those occurring in the cervix, larynx (voice box), liver, rectum, or skin. *Electrosurgery*, which involves high-frequency electrical current, is sometimes used to destroy cancer cells in the skin and mouth. *Cryosurgery* involves the use of a liquid nitrogen spray or a very cold probe to freeze and kill cancer cells. This technique is utilized for the treatment of certain prostate cancers and for precancerous conditions of the cervix such as dysplasia. Finally, *high-intensity focused ultrasound* (*HIFU*) is a technique that focuses acoustic energy at a selected location within the body, where the absorbed energy heats and destroys cancer cells with minimum damage to surrounding tissues.

When cancer is diagnosed before a primary tumor has spread to other sites, surgical removal of a tumor can usually cure the disease. In fact, most cancer cures are achieved in this way. But cancers arising in internal organs are difficult to detect in their early stages and have often metastasized by the time they are diagnosed. Sometimes the metastatic tumors formed at distant sites are large enough to also be detected and surgically removed; in other cases, the body has simply been seeded with tiny clumps of cancer cells, known as *micrometastases*, that are too small to be detected. Because roughly half of all cancers (excluding skin cancers) have started to metastasize by the time they are diagnosed, surgical removal of the primary tumor is frequently followed by radiation, chemotherapy, or both to attack any disseminated cells that were not removed during surgery.

The growing use of follow-up radiation and chemotherapy has allowed surgeons to decrease the amount of surgery they need to perform on the average cancer patient. For example, the standard treatment for breast cancer between 1900 and 1970 was the *radical mastectomy*, a drastic and disfiguring operation that involves complete surgical removal of the breast along with the underlying chest muscles and lymph nodes of the armpit. However, radical mastectomies are rarely performed today because such extensive tissue removal has not been found to improve survival compared to less drastic procedures. From 1970 to 1990 the most common procedure was the *modified radical mastectomy*, which involves removal of the breast and lymph nodes but not the chest muscles. Today more than half of all breast cancer patients are treated by *partial mastectomy* (*lumpectomy*), which removes just the tumor and a small amount of surrounding normal tissue. Surgery is usually followed by radiation therapy to the breast to destroy any cancer cells that may remain in the area.

Radiation Therapy Kills Cancer Cells by Triggering Apoptosis or Mitotic Death

If a tumor has invaded into surrounding tissues and possibly metastasized to distant sites, surgery may not be able to remove all cancer cells from the body. In some cases, surgery is not even practical. For example, the location of a brain tumor may make it impossible to remove the tumor without causing unacceptable brain damage, and leukemias cannot be treated surgically because the cancer cells reside mainly in the bloodstream. When surgery is insufficient by itself or impractical, other treatments are used (often after surgery) to destroy any cancer cells that may still reside in the body.

One type of treatment is **radiation therapy**, which uses high-energy X-rays or other forms of *ionizing radiation* to kill cancer cells. Ionizing radiation removes electrons from water and other intracellular molecules, thereby generating highly reactive *free radicals* that attack DNA. In Chapter 6 we saw that the resulting DNA damage can actually cause cancer to arise. Ironically, the same type of radiation is used in higher doses to kill cancer cells in people who already have the disease. Radiation treatments do create a small risk that a second cancer will develop in the future, but the risk is far outweighed by the potential benefit of curing a cancer that already exists.

High doses of radiation kill cancer cells in two different ways. First, DNA damage caused by the radiation treatment activates the p53 signaling pathway, which triggers cell death by apoptosis. Lymphomas and cancers arising in reproductive tissues are particularly sensitive to this type of radiation-induced apoptosis. However, more than half of all human cancers have mutations that disable the p53 protein or other components of the p53 signaling pathway. As a consequence, p53-induced apoptosis plays only a modest role in the response of most cancers to radiation treatment.

Radiation also kills cells by causing chromosomal damage that is so severe that it prevents cells from progressing through mitosis, and the cells die while trying to divide. Because this process of **mitotic death** only occurs at the time of cell division, cells that divide more frequently are more susceptible to mitotic death than cells that divide less frequently (or are not dividing at all). This difference in susceptibility makes rapidly growing cancers more sensitive to the killing effects of radiation than slower-growing cancers and also helps protect nondividing or slowly dividing normal cells in the surrounding tissue from being killed by the radiation.

Radiation Treatments Are Designed to Minimize Damage to Normal Tissues

To minimize damage to normal tissues, radiation treatments must be accurately focused on those regions of the body that contain tumor cells. This goal, called *radiation planning*, is accomplished by taking X-ray pictures that define the three-dimensional boundaries of the tumor and then using that information to guide a moving beam of high-energy radiation that is directed toward the target region from a number of different angles. Such an approach allows maximum radiation to be directed at the tumor area with minimal exposure to surrounding tissues.

The effectiveness of radiation therapy is determined to a large extent by differences in the survival rates of normal versus cancer cells after irradiation. If the difference in survival rates is small and the entire radiation dose is administered as a single treatment, the survival curves will closely track one another and there will be little difference in the numbers of cancer cells and normal cells killed (Figure 11-7, *left*). It might be possible to destroy a tumor this way, but it would be at the expense of a large amount of damage to normal tissue. If the same total amount of radiation is administered as a series of lower doses, however, small differences in the survival rates of normal and cancer cells after each treatment become magnified as the treatments are repeated multiple times (see Figure 11-7, *right*). By the end of the series of treatments, all cancer cells could be destroyed while maintaining enough normal cells

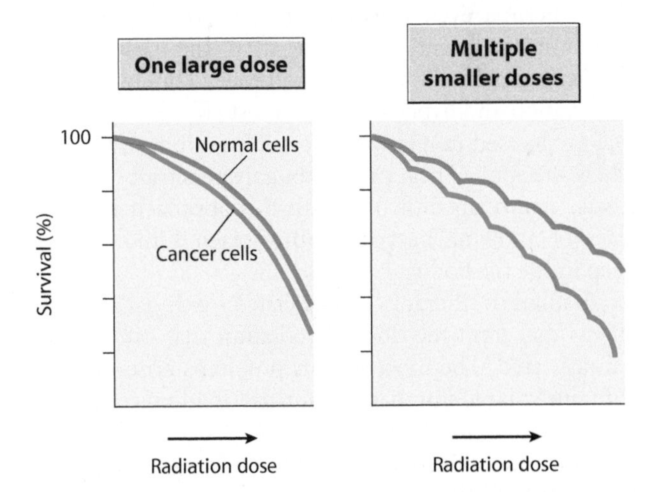

Figure 11-7 Effectiveness of Single Versus Multiple Radiation Doses in Cancer Treatment. (*Left*) If the difference in survival rates of normal and cancer cells after intense radiation exposure is small, there will be little difference in the percentage of cancer cells killed compared to normal cells following a single dose of radiation. (*Right*) If the radiation is administered as a series of smaller doses, thereby providing time for cells to repair radiation damage between exposures, small differences in the survival rates after each treatment become magnified as treatments are repeated multiple times. [Adapted from A. S. Lichter in *Clinical Oncology* (M. D. Abeloff, ed., New York: Churchill Livingstone, 2000), Chapter 18 (Figure 18-29).]

to avoid serious tissue damage. For this reason, radiation therapy is usually divided into multiple treatments administered over several weeks or months.

An alternative approach for minimizing damage to normal tissues, called **brachytherapy**, uses a radiation source that can be inserted directly within (or close to) the tumor. For example, early stage prostate cancer is sometimes treated by implanting small radioactive pellets, about the size of a grain of rice, directly into the prostate gland. The pellets emit low doses of radiation for weeks or months and are simply left in place after the radiation has all been emitted. The advantage of this approach is that most of the radiation is concentrated in the prostate gland itself, sparing surrounding tissues such as the bladder and rectum.

Another technique for improving the effectiveness of radiation therapy involves agents that sensitize tumor cells to the killing effects of radiation. One group of drugs, known as *hypoxic radiosensitizers*, mimic oxygen and are taken up by cancer cells, which frequently tend to be hypoxic (deficient in oxygen). Radiation creates more cellular damage in the presence of adequate oxygen, so the uptake of these drugs by cancer cells increases the effectiveness of radiation therapy. Combining radiation treatments with certain anticancer drugs, such as *fluorouracil* and *platinum compounds*, can likewise enhance the effectiveness of radiation treatments. The properties of these and related anticancer drugs will be described shortly, when we discuss the topic of cancer chemotherapy.

Raising the temperature of tumor tissue by a few degrees—a technique known as **hyperthermia**—also sensitizes cells to the killing effects of radiation. Hyperthermia even works when it is administered *after* radiation treatment, suggesting that the heat may be interfering with cellular repair pathways. The combination of radiation and hyperthermia is most effective for tumors that are located in relatively accessible regions of the body, where the applied heat can thoroughly penetrate the tumor tissue. The main difficulty with this approach is finding ways of applying heat to hard-to-reach tumors located deep inside the body.

Radiation therapy is associated with various side effects that limit the dose of radiation that can be safely administered. The most serious problems arise in tissues containing large numbers of normal dividing cells, which are also susceptible to radiation-induced killing. For example, radiation damage to the dividing cells that line the gastrointestinal tract causes nausea, vomiting, and diarrhea. And damage to dividing cells in the bone marrow reduces the production of one or more types of blood cells, which can lead to anemia, defective blood clotting, and immune deficiencies that increase the susceptibility to infections. The likelihood that such side effects will be severe depends to a great extent on the location of the tumor and its sensitivity to radiation-induced killing. Some cancers are very sensitive to radiation and can be destroyed with modest doses that elicit minimal side effects, whereas other cancers require high radiation doses and are more difficult to control using radiation (Table 11-1).

Table 11-1 Radiation Sensitivity of Selected Cancers

Sensitivity to Radiation Treatment	Type of Cancer
Very responsive to radiation	Hodgkin's disease
	Non-Hodgkin's lymphomas
	Seminoma (testicles)
	Neuroblastoma
	Retinoblastoma
Moderately responsive to radiation	Head and neck cancer
	Breast cancer
	Prostate cancer
	Cervical cancer
	Esophageal cancer
	Rectal cancer
	Lung cancer
Poorly responsive to radiation	Melanoma
	Glioblastoma
	Kidney cancer
	Pancreatic cancer
	Sarcomas

Data from A. S. Lichter in *Clinical Oncology* (M. D. Abeloff, ed., New York: Churchill Livingstone, 2000, Chapter 18).

Chemotherapy Involves the Use of Drugs That Circulate in the Bloodstream to Reach Cancer Cells Wherever They May Reside

The third main approach for treating cancer (in addition to surgery and radiation) is **chemotherapy**, which involves the use of drugs that either kill cancer cells or interfere with the ability of cancer cells to proliferate. Chemotherapy is especially well suited for treating cancers that have already metastasized because drugs circulate through the bloodstream to reach cancer cells wherever they may have spread, even if the metastasizing cells have not yet formed visible tumors. This also means, however, that the toxic side effects commonly associated with chemotherapy can occur anywhere in the body because most anticancer drugs, like radiation, are toxic to dividing cells in general.

Despite its various side effects, chemotherapy has been successfully applied to a wide range of cancers. In some cases, as with certain forms of leukemia, chemotherapy may cure cancer by itself. More commonly, chemotherapy is employed in conjunction with surgery, radiation, or both. Dozens of anticancer drugs are currently available and the best choice will vary, depending on the type and stage of the cancer being treated. Based on differences in the way they work, the various drugs can be grouped into several distinct categories (Table 11-2). In the following sections, each category will be discussed in turn.

Table 11-2 Examples of Some Drugs Used in Cancer Chemotherapy

Class	Examples	Mechanism of Action
1. Antimetabolites	Methotrexate	Folic acid antagonist
	Fluorouracil	Pyrimidine analog
	Cytarabine	Pyrimidine analog
	Capecitabine	Pyrimidine analog
	Gemcitabine	Pyrimidine analog
	Mercaptopurine	Purine analog
	Thioguanine	Purine analog
2. Alkylating and platinating agents	Mechlorethamine (nitrogen mustard)	DNA crosslinking agent
	Cyclophosphamide	DNA crosslinking agent
	Chlorambucil	DNA crosslinking agent
	Melphalan	DNA crosslinking agent
	BCNU (bischloroethyl nitrosourea)	DNA crosslinking agent
	Cisplatin (Platinol)	DNA crosslinking agent
3. Antibiotics	Doxorubicin	Topoisomerase II inhibitor*
	Daunorubicin	Topoisomerase II inhibitor
	Mitomycin	DNA crosslinking agent
	Bleomycin	DNA strand breaks
4. Plant-derived drugs	Etoposide	Topoisomerase II inhibitor
	Teniposide	Topoisomerase II inhibitor
	Topotecan	Topoisomerase I inhibitor
	Irinotecan	Topoisomerase I inhibitor
	Vinblastine	Antimicrotubule agent
	Vincristine	Antimicrotubule agent
	Taxol	Antimicrotubule agent
5. Hormone therapy	Tamoxifen	Blocks estrogen receptors (in breast)
	Arimidex	Aromatase inhibitor
	Leuprolide	Inhibitor of androgen production
	Flutamide	Blocks androgen receptors
	Prednisone	Glucocorticoid

*There are two forms of topoisomerase, called topoisomerase I and topoisomerase II.

Antimetabolites Disrupt DNA Synthesis by Substituting for Molecules Involved in Normal Metabolic Pathways

Antimetabolites, the first group of chemotherapeutic drugs that we will consider, are molecules that resemble substances involved in normal cellular metabolism. This resemblance causes enzymes to bind to antimetabolites in place of the normal molecules, thereby disrupting essential metabolic pathways and poisoning the cell. Most of the antimetabolites used in cancer chemotherapy disrupt pathways required for normal DNA synthesis and repair.

The use of this approach for treating cancer was pioneered in the 1940s by Sidney Farber, who had been studying the nutritional needs of children with leukemia. Farber initially believed that vitamin therapy might help children fight off the disease, so he provided them with supplements of various vitamins, including the B vitamin, *folic acid*. Unexpectedly, the added folic acid made the leukemias grow even faster. While that was certainly not the desired result, it raised an intriguing possibility: If cancer growth is stimulated by excess folic acid, blocking the action of folic acid might have the opposite effect and restrain the disease.

Farber therefore decided to treat some of his patients with folic acid **analogs**, which are chemical derivatives of folic acid that can substitute for the natural molecule and thereby disrupt any pathways in which folic acid is normally involved. When one analog, called *aminopterin*, was given to several children who were very sick with leukemia, the children quickly regained their health and returned to virtually normal lives. Unfortunately, the improvement turned out to be only temporary, but these transient remissions caused a stir of excitement and stimulated the hunt for other antimetabolites whose effects might be more permanent than those of aminopterin.

The resulting search led to the discovery of *methotrexate*, a derivative of folic acid that efficiently binds to and inhibits the enzyme *dihydrofolate reductase* (Figure 11-8). Dihydrofolate reductase catalyzes the

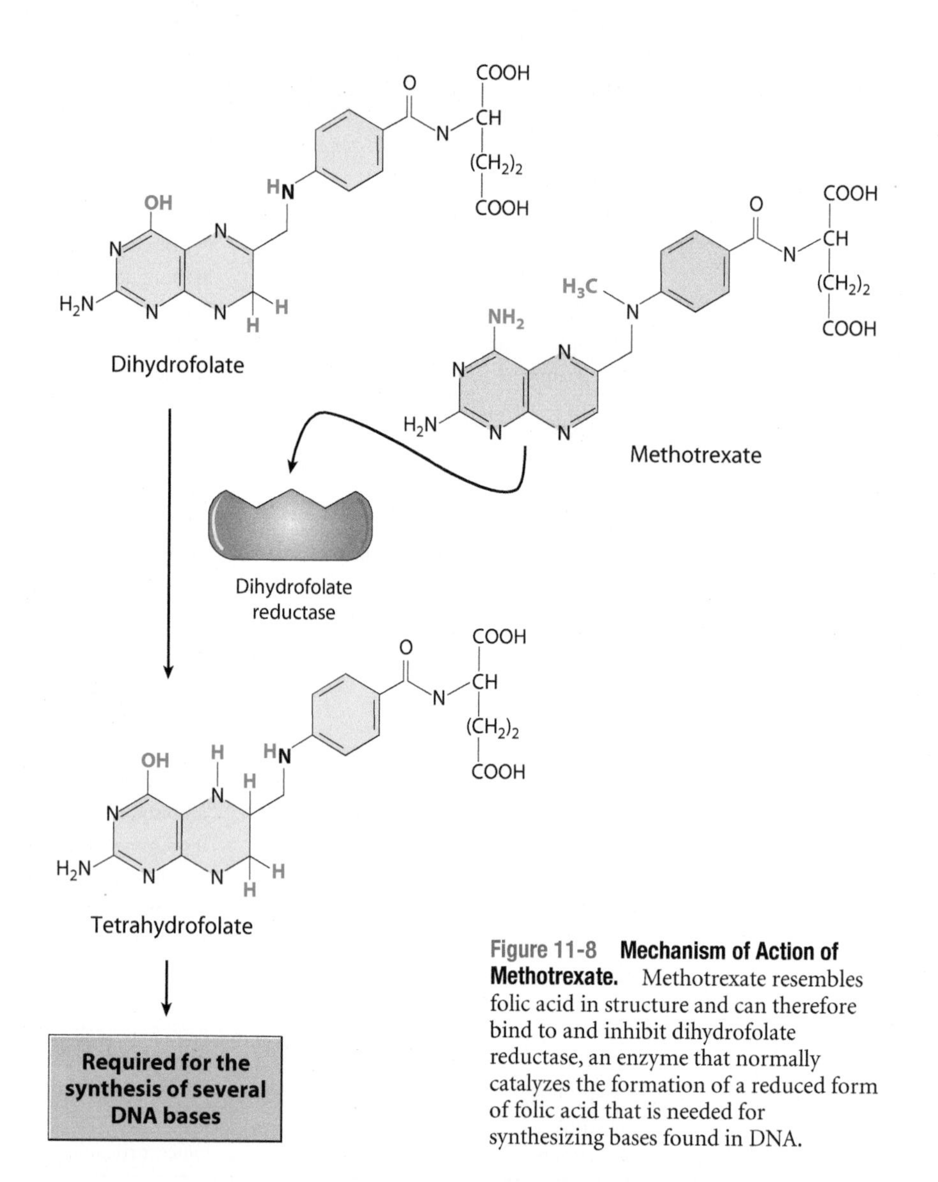

Figure 11-8 Mechanism of Action of Methotrexate. Methotrexate resembles folic acid in structure and can therefore bind to and inhibit dihydrofolate reductase, an enzyme that normally catalyzes the formation of a reduced form of folic acid that is needed for synthesizing bases found in DNA.

production of a reduced form of folic acid that is required for the synthesis of several bases found in DNA; inhibition of dihydrofolate reductase by methotrexate therefore disrupts pathways involved in DNA synthesis and repair. Shortly after its discovery, methotrexate was shown to be an effective treatment for *choriocarcinoma*, a cancer arising from cells of the placental membranes that are sometimes left behind after childbirth. Choriocarcinoma was fatal for most women who developed the disease prior to the introduction of methotrexate chemotherapy in the mid-1950s. After methotrexate began to be used, cure rates improved to almost 90%. Although its effects are not always this dramatic, methotrexate is currently used to treat a diverse spectrum of cancers, including acute leukemias and tumors of the breast, bladder, and bone.

In addition to analogs of folic acid such as methotrexate, analogs of the nitrogenous bases found in DNA are also useful for cancer chemotherapy. DNA contains two types of bases: single-ring compounds called **pyrimidines**, which include the bases cytosine (C) and thymine (T); and double-ring compounds called **purines**, which include the bases adenine (A) and guanine (G). Several analogs of pyrimidines and purines are routinely used as anticancer drugs. Examples include the pyrimidine analogs *fluorouracil* and *cytarabine* (also called *cytosine arabinoside*) and the purine analogs *mercaptopurine* and *thioguanine*. As shown in Figure 11-9, the close resemblance of these substances to normal bases found in DNA causes the analogs to bind to and thereby disrupt the activity of enzymes involved in DNA synthesis and repair. Pyrimidine and purine analogs are used mainly for treating leukemias and lymphomas, although fluorouracil is effective against a broad spectrum of other cancers as well.

Alkylating and Platinating Drugs Act by Crosslinking DNA

Alkylating agents are highly reactive organic molecules that trigger DNA damage by linking themselves directly to DNA. As we saw in Chapter 5, this ability to attack DNA molecules makes alkylating agents mutagenic as well as carcinogenic. However, alkylating agents are also employed as anticancer drugs because they kill cancer cells at higher doses, and the risk that they may cause cancer in

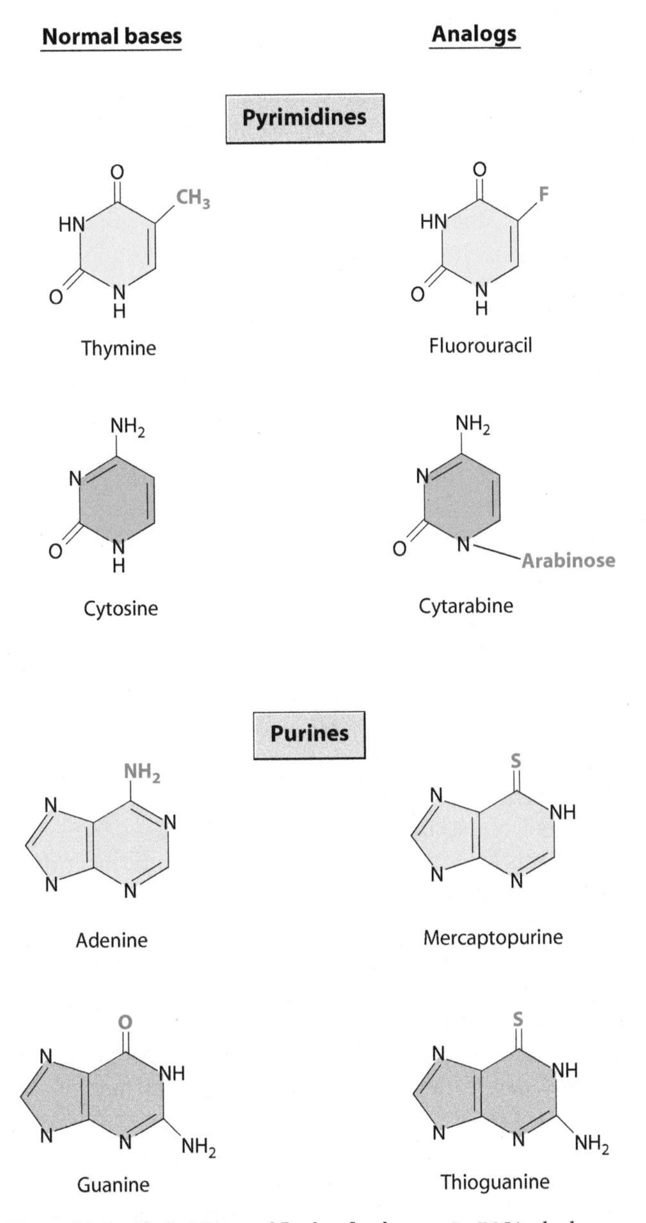

Figure 11-9 Pyrimidine and Purine Analogs. In DNA, the bases thymine (T) and cytosine (C) are pyrimidines, and adenine (A) and guanine (G) are purines. The pyrimidine analogs fluorouracil and cytarabine (cytosine arabinoside), and the purine analogs mercaptopurine and thioguanine, are shown to illustrate their close resemblance to normal bases. Red is used to highlight the chemical groups that differ between the normal bases and their corresponding analogs.

such cases is outweighed by the potential benefit of curing a cancer that already exists.

The first alkylating agent to be employed for cancer chemotherapy has an interesting history. During World War I, the German military used an oily alkylating agent called *sulfur mustard* as a chemical weapon because it vaporizes easily and causes severe blistering injuries to the skin and lungs. A more toxic version, called *nitrogen mustard*, was produced and stockpiled by both Germany and the United States during World War II. Nitrogen mustard was never employed on the battlefield, but German bombers attacked an Italian seaport in 1943 and sank a U.S. supply ship loaded with 100 tons of weapons containing the toxic chemical. Survivors pulled from the water, which had become heavily contaminated with nitrogen mustard, exhibited severe skin burns and quickly developed a variety of internal symptoms, including a dramatic drop in the number of blood lymphocytes.

Given this toxic effect on lymphocytes, scientists at Yale University decided to investigate whether nitrogen mustard would have a similar effect on cancers arising from lymphocytes. Shortly after the end of World War II, they reported that nitrogen mustard injections cause lymphocytic cancers to regress in animals and humans—the first demonstration of the potential usefulness of alkylating agents as anticancer drugs. Better alkylating agents have subsequently been developed, but nitrogen mustard (now called *mechlorethamine*) is still occasionally used to treat Hodgkin's lymphoma. Medical staff who handle the drug take precautions to avoid inhaling the vapors of this one-time chemical weapon and must be certain that it is injected cleanly into a patient's vein without contacting the skin.

Based on the initial promising results with nitrogen mustard, hundreds of other alkylating agents have been synthesized in the laboratory and tested in animals for anticancer activity. This effort has produced several drugs related to nitrogen mustard, including *cyclophosphamide*, *chlorambucil*, and *melphalan*, that are routinely used to treat cancer patients. In addition to substances related to nitrogen mustard, other alkylating agents have been developed for use as anticancer drugs, including *thiotepa* and *nitrosourea* compounds, such as *BCNU* (*bischloroethyl nitrosourea*). In general, the various alkylating agents disrupt normal DNA function by crosslinking the two strands of the DNA double helix (Figure 11-10, *top*). As a result, the two strands are unable to separate and DNA replication cannot take place, thereby preventing cell division.

Another group of DNA-crosslinking agents used in cancer chemotherapy contain the element *platinum* (see Figure 11-10, *bottom*). The ability of these substances, called **platinating agents**, to act as anticancer drugs was discovered in a roundabout manner. In some experiments performed during the 1960s that were totally unrelated to cancer biology, platinum electrodes were used to pass an electric current through a culture of bacterial cells to see how the cells react to electricity. The bacteria stopped dividing, but it was soon discovered that this response was caused not by the electricity but by an unexpected reaction involving the platinum electrodes. In essence, ammonium chloride present in the culture medium had reacted with platinum in the electrodes to form a nitrogen-containing platinum compound called *cisplatin*, which in turn inhibited bacterial cell division. The ability of cisplatin to block cell division led to successful tests on cancer cells, and the drug was approved for trials in human cancer patients in 1972. Cisplatin (trade name Platinol) is now one of the most effective agents in our arsenal of anticancer drugs, and efforts are being made to synthesize derivatives of cisplatin that might work even better.

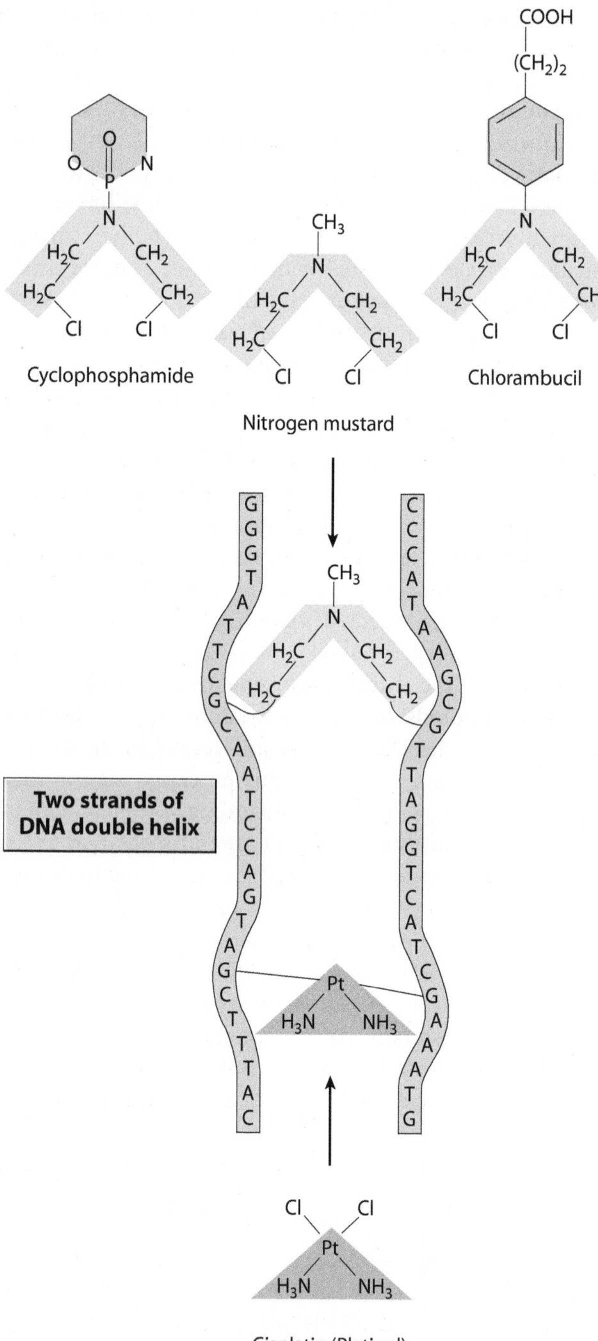

Figure 11-10 DNA Crosslinking by Alkylating and Platinating Drugs. The top of the diagram shows how nitrogen mustard and related drugs crosslink the two strands of the DNA double helix. The bottom of the diagram illustrates the comparable reaction for platinating drugs such as cisplatin (Platinol).

Antibiotics and Plant-Derived Drugs Are Two Classes of Natural Substances Used in Cancer Chemotherapy

Most of the antimetabolites and alkylating agents being used as anticancer drugs are synthetic molecules that were created in the laboratory for the purpose of treating cancer. Over the centuries, humans have also found ways of treating disease by drawing on natural substances produced by living organisms. An especially dramatic twentieth-century example was the discovery of *penicillin*, a substance produced by a fungus that turned out to be one of the first effective drugs against bacterial infections.

Penicillin is an **antibiotic**, a term that refers to any substance produced by a microorganism, or a synthetic derivative, that kills or inhibits the growth of other microorganisms or cells. Antibiotics are generally thought of as being antibacterial drugs, but some of them exhibit anticancer properties as well. One of the most fruitful sources of antibiotics for cancer chemotherapy has been a group of bacteria called *Streptomyces*. Besides producing *streptomycin*, which is an antibiotic used for treating tuberculosis and other serious bacterial infections, members of the *Streptomyces* group synthesize several antibiotics that have found their way into our arsenal of anticancer drugs, including *doxorubicin*, *daunorubicin*, *mitomycin*, and *bleomycin*. All these antibiotics target the DNA molecule, although their mechanisms of action are somewhat different. Doxorubicin and daunorubicin insert themselves into the DNA double helix and inhibit the action of *topoisomerase*, an enzyme that normally breaks and rejoins DNA strands during DNA replication to prevent excessive twisting of the double helix. In contrast, mitomycin is a DNA crosslinking agent and bleomycin triggers DNA strand breaks.

Plants are another natural source of anticancer drugs. Several of the drugs obtained from plants act as topoisomerase inhibitors; included in this category are *etoposide* and *teniposide*, derived from a substance present in the mayapple (mandrake) plant, and *topotecan* and *irinotecan*, derived from a substance present in the bark of the Chinese camptotheca tree. Another group of plant-derived drugs attack the microtubules that make up the mitotic spindle. This class of drugs includes *vinblastine* and *vincristine*, obtained from the Madagascar periwinkle plant, and *Taxol* (generic name *paclitaxel*), discovered in the bark of the Pacific yew tree. Vinblastine and vincristine block the process of microtubule assembly, whereas Taxol stabilizes microtubules and promotes the formation of abnormal microtubule bundles. In either case, the mitotic spindle is disrupted and cells cannot divide.

Hormones and Differentiating Agents Are Relatively Nontoxic Tools for Halting the Growth of Certain Cancers

One of the main problems with the drugs described thus far is that their toxic effects on DNA replication and cell division are harmful to normal cells as well as to cancer cells. When cancers arise in hormone-dependent tissues, an alternative and considerably less toxic approach can sometimes be used. This approach, known as **hormone therapy**, was pioneered in the 1940s by Charles Huggins in studies involving prostate cancer patients. Based on earlier observations in animals, Huggins believed that the proliferation of prostate cells is dependent on steroid hormones known as **androgens** (*testosterone* is one example).

In an effort to eliminate the source of androgens in men with advanced prostate cancer, he surgically removed their testicles, which produce most of the testosterone, and also treated them with the female steroid hormone, estrogen. More than half of his prostate cancer patients improved and saw their tumor growth reduced.

These early observations eventually led to the development of drugs that block the production or the actions of androgens as an alternative to removing the testicles. Androgen production is normally controlled by peptide hormones called *gonadotropins*, which are synthesized in the pituitary gland. One drug used to treat prostate cancer, named *leuprolide*, is an analog of the *gonadotropin-releasing hormone* that controls the release of these gonadotropins. By suppressing the release of gonadotropins, leuprolide inhibits androgen production by the testicles. Another group of drugs inhibit the activity of **androgen receptors**, which are receptor proteins located in prostate epithelial cells that bind incoming androgens and transmit the signal that stimulates cell division. *Flutamide* and *bicalutamide* are examples of anticancer drugs that act by blocking androgen receptors.

Similar considerations apply to breast cancers, which arise from cells whose normal proliferation is driven by steroid hormones of the **estrogen** family. For breast cancers that retain this estrogen requirement, drugs that block estrogen action may be effective cancer treatments. One widely used drug that works in this way is **tamoxifen**, a molecule that exhibits some similarities to estrogen in chemical structure (Figure 11-11). Estrogens normally exert their effects on target cells by binding to intracellular proteins called **estrogen receptors**. When tamoxifen is administered to breast cancer patients whose tumors require estrogen, it binds to estrogen receptors in place of estrogen and prevents the receptors from being activated. Another group of drugs, called *aromatase inhibitors*, inhibit one of the enzymes required for estrogen synthesis. Generally these drugs are only recommended for treating breast cancer in postmenopausal women, where they inhibit the synthesis of the small amounts of estrogen that are being produced.

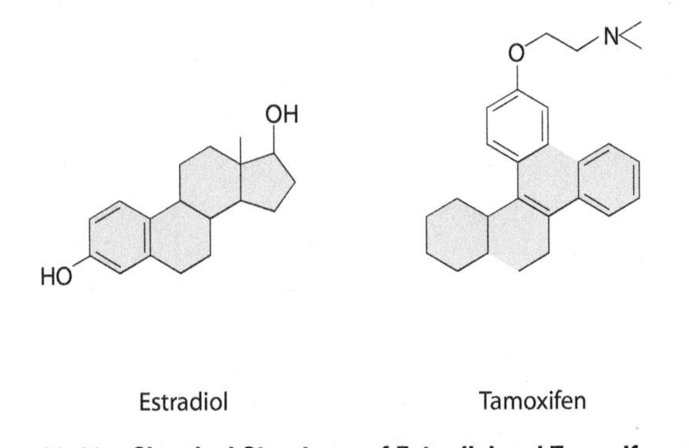

Figure 11-11 Chemical Structures of Estradiol and Tamoxifen. Tamoxifen exhibits some similarities to estrogens in its chemical structure, which is enough to allow it to bind to estrogen receptors in place of normal estrogens, such as estradiol. The binding of tamoxifen to estrogen receptors in breast cells prevents the receptors from being activated. In some other tissues, such as the uterus, tamoxifen activates estrogen receptors instead of blocking them. Color shading is used to highlight the similarities between the two molecules.

A somewhat different rationale is used when applying the principle of hormone therapy to lymphocytic cancers. The adrenal cortex produces a family of steroid hormones called **glucocorticoids**, whose properties include the ability to inhibit lymphocyte proliferation. Consequently *prednisone*, a synthetic glucocorticoid that slows down the proliferation of lymphocytes, is sometimes used in treating lymphomas and lymphocytic leukemias.

One advantage of hormone therapies is that their side effects tend to be mild because they do not destroy normal cells and because they only affect a selected group of target cells whose proliferation is controlled by the hormone in question. On the other hand, this latter property also imparts a significant limitation: Hormone-based treatments are only useful for cancers that arise in hormone-dependent tissues. And even in these tissues, cancers do not always exhibit the hormone-dependence seen in the corresponding normal cells. For example, some breast cancers lack the estrogen receptors found in normal breast cells, and some prostate cancers lack the androgen receptors found in normal prostate cells. In such cases, hormone therapies are of little value.

Another relatively nontoxic approach to cancer chemotherapy involves the use of substances called **differentiating agents**. Whereas hormone therapies are designed to restrain cell proliferation, differentiating agents promote the process by which cells acquire the specialized structural and functional traits of differentiated cells. When cells undergo differentiation, they also lose the capacity to divide (p. 5). Agents that promote cell differentiation therefore tend to decrease the overall level of cell proliferation. An example of a differentiating agent used in cancer therapy is *retinoic acid*, a form of vitamin A employed in the treatment of acute promyelocytic leukemia.

Toxic Side Effects and Drug Resistance Can Limit the Effectiveness of Chemotherapy

The ultimate goal of chemotherapy is to destroy or restrain the proliferation of cancer cells without harming normal cells. However, with the exception of hormones and differentiating agents, which are useful for only a few selected types of cancer, most chemotherapeutic drugs act by inhibiting DNA replication, damaging DNA, or blocking cell division—actions that are detrimental to normal dividing cells as well as to cancer cells. Moreover, because chemotherapeutic drugs circulate throughout the body, they encounter normal dividing cells no matter where the cells reside. For example, the hair loss that commonly accompanies chemotherapy is a toxic side effect that is triggered when circulating drugs encounter the dividing cells that line the hair follicles.

The most serious side effects of chemotherapy involve the gastrointestinal tract and the bone marrow. As with radiation therapy, damage to normal dividing cells in these tissues can lead to nausea, vomiting, diarrhea, anemia, defective blood clotting, and immune deficiency. Such side effects usually tend to be more severe with chemotherapy than with radiation because drugs cannot be easily focused on a particular region of the body to minimize toxicity to the gastrointestinal tract and bone marrow. Fortunately, some cancer cells are particularly sensitive to chemotherapy and can be destroyed without excessive toxicity to normal cells; for many cancers, however, chemotherapy may fail because the drug dosage required to kill all cancer cells would trigger overwhelmingly toxic side effects.

Another problem that can reduce the effectiveness of chemotherapy is the tendency for tumors to become resistant to the killing effects of anticancer drugs, especially after a prolonged series of treatments. Even if most of the cancer cells in a person's body are destroyed by a particular drug, a few drug-resistant cells present in the initial population could proliferate and form a new tumor that would then be completely resistant to the drug. And if drug-resistant cells are initially absent, cancers tend to be genetically unstable and may acquire mutations that impart drug resistance during the course of treatment. An illustration of this problem is provided by methotrexate, an anticancer drug that inhibits the enzyme *dihydrofolate reductase* (see Figure 11-8). In cancers that are being treated with methotrexate, the gene for dihydrofolate reductase sometimes undergoes mutation or amplification. The mutations create altered forms of dihydrofolate reductase that are no longer inhibited by methotrexate, and gene amplification leads to increased production of dihydrofolate reductase, thereby diminishing the effectiveness of methotrexate treatment. Such genetic changes, which alter the target of a drug to make it less susceptible to the drug's effects, are commonly observed in individuals receiving chemotherapy.

Given the large number of anticancer drugs available, it might seem that a simple solution would be to just switch drugs when resistance arises. Unfortunately, the situation is complicated by the fact that tumors often develop resistance to several drugs at the same time, even though only a single drug is administered. One way in which cancer cells become resistant to multiple drugs is by producing plasma membrane proteins that actively pump drugs out of the cells. These drug-pumping proteins, called **multidrug resistance transport proteins**, have a remarkably broad specificity: They export a wide range of chemically dissimilar molecules, thereby imparting resistance to a broad spectrum of drugs.

Another factor that can contribute to multidrug resistance is related to the mechanism by which anticancer drugs kill cells. Although multiple killing mechanisms appear to be involved, chemotherapeutic drugs sometime act by damaging DNA to such an extent that apoptosis is invoked to destroy the damaged cell. In such cases, the effectiveness of chemotherapy may be reduced by mutations that disable apoptosis. Mutations of this type are often present at the time of initial diagnosis, or they may arise during chemotherapy. In either case, mutations that disable apoptosis would be expected to decrease the effectiveness of any drug that kills a particular type of cancer cell primarily by triggering apoptosis.

Another possible source of drug resistance is related to the heterogeneity of tumor cell populations. A growing body of evidence suggests that in any given tumor, only a small population of cells, called **cancer stem cells**, are able to proliferate indefinitely. The existence of these cancer stem cells, which have been postulated to give rise to all the other cells found in a tumor, could help explain why treatments that cause tumors to shrink until they are undetectable may still not cure the disease. While the treatment may eliminate the bulk of the cancer cells, a few remaining cancer stem cells may be all that is needed to replenish the tumor cell population. According to this theory, existing anticancer drugs may be more effective at killing the majority of a person's tumor cells than they are at killing the rare cancer stem cells, which then regenerate the tumor after treatment is stopped. Researchers are currently exploring this idea by searching for cancer stem cells in various tumor types and testing to see whether they exhibit any unique properties that could be targeted by future anticancer drugs.

Combination Chemotherapy and Stem Cell Transplants Are Two Strategies for Improving the Effectiveness of Chemotherapy

For certain kinds of cancer, chemotherapy is successful in restoring normal life expectancies to many patients. Sometimes the chemotherapy by itself is responsible for the improved prognosis, but it is more common for chemotherapy to be used in conjunction with surgery or radiation. Despite these successes, the effectiveness of chemotherapy is often hindered by the emergence of drug resistance and by the toxic side effects that restrict the dose that can be safely administered. Additional challenges are raised by the need for delivery techniques that convey drugs to tumor sites at the proper concentration for an appropriate period of time and by the existence of heterogeneous tumor populations containing mixtures of cells that respond differently to the same drug.

One strategy for trying to improve the effectiveness of chemotherapy is to administer several drugs in combination rather than a single agent alone. Drug combinations are often named using an acronym that is derived from the initials of the drugs being used. For example, *BEP chemotherapy* (bleomycin, etoposide, and Platinol) is the name of a treatment for testicular cancer, and *CMF chemotherapy* (cyclophosphamide, methotrexate, and fluorouracil) is the name of a treatment for breast cancer.

This general approach, known as **combination chemotherapy**, is most effective with drugs that differ in their mechanisms of action. For example, consider three drugs exhibiting different side effects that limit the dose of each that can be safely administered. Combining the three drugs at their maximum tolerated doses will increase the overall tumor-killing effectiveness compared with each drug by itself, and yet the overall toxicity may remain at an acceptable level because each drug works in a different way. Another advantage of drug combinations is that cancer cells are less likely to become resistant to chemotherapy when several drugs are administered simultaneously, especially if the drugs differ in their chemical properties, cellular targets, and mechanisms of action. The enormous challenge of combination therapy is finding the most effective drug mixtures for each type of cancer, especially given the dozens of drugs that could in theory be administered in thousands of different combinations.

Another approach for improving the effectiveness of chemotherapy deals with the potential problem of bone marrow damage. Many anticancer drugs are capable of killing all cancer cells if the dose is raised high enough. The dose that can be realistically administered, however, is limited by toxicity to the bone marrow, which contains the **hematopoietic stem cells** whose proliferation gives rise to blood cells. If too many of these stem cells are destroyed during high-dose chemotherapy, blood cells will not be produced and a person cannot survive. One approach for addressing this problem is to use high-dose chemotherapy to destroy all cancer cells and then follow the treatment with **stem cell transplantation** (also called *bone marrow transplantation*) to replenish the person's hematopoietic stem cells. Under such conditions, higher drug doses can be used because the blood-forming stem cells destroyed by the chemotherapy are subsequently being replaced.

The stem cells used for transplantation can be obtained either from a cancer patient's own bone marrow or blood prior to administration of high-dose chemotherapy, or from the bone marrow or blood of a genetically compatible individual who is willing to serve as a stem cell donor. Unfortunately, each approach has its complications. Using a cancer patient's own stem cells for subsequent transplantation creates the risk of either reintroducing cancer cells or relying on stem cells that have been damaged during earlier cancer treatments. On the other hand, finding an appropriately matched donor can be difficult, and immune cells present in the donor's blood or bone marrow sometimes attack the tissues of the cancer patient, thereby creating a potentially life-threatening condition known as **graft-versus-host disease**.

An alternative is to use *umbilical cord blood* rather than bone marrow or peripheral blood as a source of stem cells for transplantation. The umbilical cord, which is normally discarded at birth, contains blood with a large number of hematopoietic stem cells. These cells elicit a lower incidence of graft-versus-host disease, do not require as close a genetic match as do adult stem cells, and are readily obtained from blood banks that store frozen umbilical cord blood taken from healthy newborns. The possible usefulness of cord blood as a source of stem cells for cancer patients is currently under investigation.

Molecular and Genetic Testing Is Beginning to Allow Cancer Treatments to Be Tailored to Individual Patients

A final approach for enhancing the effectiveness of chemotherapy involves the possibility of designing drug treatments that are personalized for each individual patient. It has been known for many years that cancer patients with tumors that are indistinguishable from one another by traditional criteria often exhibit different outcomes after receiving the same treatment. Experiments using DNA microarray technology to analyze gene activity have provided a likely explanation: Cancers of the same type exhibit different patterns of gene expression that cause them to behave differently. We saw earlier in the chapter that the Oncotype DX gene expression test, which measures the activity of 21 key genes in breast cancer cells, is able to predict which patients are most likely to have their cancers recur after surgery. In the absence of such information, doctors would usually recommend that most patients receive chemotherapy. The value of gene expression testing is that it can help identify those patients who really need chemotherapy and are likely to benefit from it.

Taking this approach one step further, analyzing cancer specimens for gene expression patterns and the presence of specific mutations may provide information about the exact type of cancer treatment that is most appropriate for each person. A striking example is provided by *Iressa*, a member of a new class of drugs that will be described later in the chapter when we cover the concept of molecular targeting. Iressa, which acts by inhibiting the receptor for epidermal growth factor (EGF), has been approved for use in the treatment of lung cancer. Tumor shrinkage occurs in only about 10% of the patients treated with Iressa, but when the drug does work, it works extremely well.

The reason Iressa is more effective in some individuals than others has been traced to the presence of a mutant form of the EGF receptor gene in the cancers of those patients who respond well to the drug. When lung cancer cells containing the mutant form of the EGF receptor are grown in laboratory culture, they are found to be much more sensitive to the growth-inhibiting effects of Iressa than are cancer cells that contain the normal form of the EGF receptor (Figure 11-12). This discovery opens the door to a personalized type of cancer therapy in which genetic testing of cancer cells is used to identify those particular patients who are most likely to benefit from treatment with Iressa.

A patient's hereditary background can also affect how he or she responds to different types of treatment. For example, inherited genes that influence steps in drug metabolism have been found to influence how well a person responds to different kinds of drugs. It is therefore hoped that a better understanding of patient-specific and tumor-specific differences in genetic makeup will eventually allow treatments to be tailor-made for each individual cancer patient.

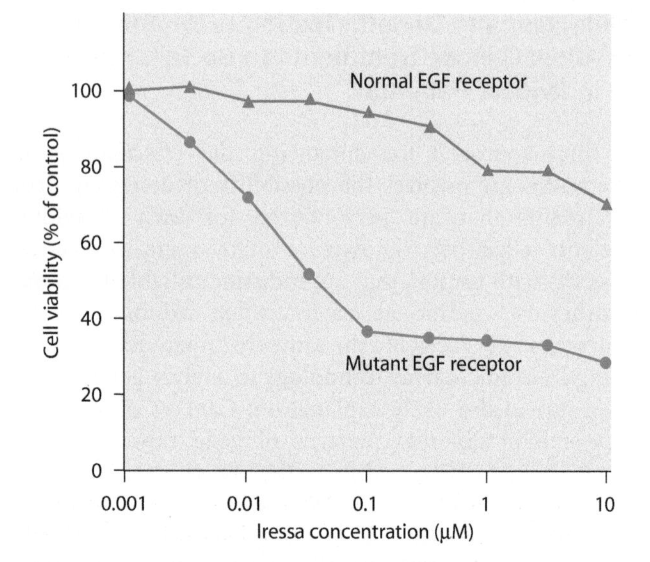

Figure 11-12 Sensitivity of Lung Cancer Cells to Iressa. Lung cancer cells with or without a mutant EGF receptor were exposed to various concentrations of Iressa in cell culture. After 72 hours of treatment, the rate of cell proliferation was measured and expressed relative to the rate in cells that had not been treated with Iressa. The data show that the presence of the mutant form of the EGF receptor makes lung cancer cells more susceptible to the growth-inhibiting effects of Iressa. [Data from J. G. Paez et al., *Science* 304 (2004): 1497 (Figure 3A).]

EMERGING TREATMENTS: IMMUNOTHERAPY AND MOLECULAR TARGETING

The use of surgery, radiation, or chemotherapy—either alone or in various combinations—can cure or significantly prolong survival times for many types of cancer, especially when the disease is diagnosed early. However, some of the more aggressive cancers, including those involving the lung, pancreas, or liver, are difficult to control in these ways, nor are current approaches very successful with cancers diagnosed in their advanced stages. In trying to find more effective ways of treating such cancers, scientists have been working to develop "magic bullets" that will selectively seek out and destroy cancer cells without damaging normal cells in the process. Although this goal presents a formidable challenge, several approaches for achieving better selectivity in targeting cancer cells are beginning to show signs of success.

Immunotherapies Exploit the Ability of the Immune System to Recognize Cancer Cells

One way of introducing better selectivity into cancer treatments is to exploit the ability of the immune system to recognize cancer cells. This general approach, called **immunotherapy**, was first proposed in the 1800s after doctors noticed that tumors occasionally regress in people who develop bacterial infections. Since infections stimulate the immune system, it was postulated that the stimulated immune cells might be attacking cancer cells as well as the invading bacteria. Efforts were therefore made to build on this idea by using live or dead bacteria to provoke the immune system of cancer patients. Some success was eventually seen with **Bacillus Calmette-Guérin (BCG)**, a bacterial strain that does not cause disease but elicits a strong immune response at the site where it is introduced into the body. One use of BCG is in the treatment of early stage bladder cancers that are localized to the bladder wall. After the cancer is surgically removed, inserting BCG into the bladder elicits a prolonged activation of immune cells that leads to lower rates of cancer recurrence.

Although this example demonstrates the potential value of stimulating the immune system, BCG must be administered directly into the bladder to provoke an immune response at the primary tumor site. With other types of cancer, especially when they have metastasized to unknown locations, it becomes necessary to stimulate an immune response against cancer cells wherever they may have traveled. For this purpose scientists have turned to molecules called **cytokines**, which are proteins produced by the body to stimulate immune responses against infectious agents.

The first cytokine found to be helpful in treating cancer was *interferon alpha*, a protein produced in response to viral infections. *Interferon alpha* is used in the treatment of several kinds of cancer, including hairy cell leukemia and Kaposi's sarcoma. *Interleukin-2* (*IL-2*) and *tumor necrosis factor* (*TNF*) are two other cytokines that are being evaluated for possible use as immune stimulators in cancer patients. IL-2 and TNF both elicit a strong antitumor response in laboratory animals, but they are extremely toxic when administered to humans. At present, TNF is still under active investigation and IL-2 is an approved treatment for advanced kidney cancer and melanoma. As we will see shortly, IL-2 is also being used experimentally to stimulate antitumor lymphocytes that are isolated from a patient's tumor site and grown in the laboratory prior to being injected back into the bloodstream.

Large Quantities of Identical Antibody Molecules Can Be Produced Using the Monoclonal Antibody Technique

BCG and cytokines are relatively nonspecific approaches to immunotherapy because they strengthen the overall activity of the immune system rather than preferentially directing an attack against cancer cells. Devising immunotherapies that act more selectively requires approaches for distinguishing cancer cells from normal cells.

The immune system sometimes recognizes cancer cells through the presence of specific antigens that cancer cells carry (see p. 38 for a discussion of *tumor-specific* and *tumor-associated antigens*). One way in which the immune system responds to antigens is by producing **antibodies**, which are soluble proteins manufactured by immune cells known as **B lymphocytes**. Antibodies circulate in the bloodstream and penetrate into extracellular fluids, where they specifically bind to the antigens that triggered the immune response. Antibody molecules recognize and bind to their corresponding antigens with extraordinary precision, making antibodies ideally suited to serving as "magic bullets" that selectively target antigens that are unique to (or preferentially concentrated in) cancer cells.

For many years, the use of antibodies for treating cancer was hampered by the lack of a reproducible method for producing large quantities of pure antibody molecules directed against the same antigen. Then in 1975, Georges Köhler and César Milstein solved the problem by devising the procedure illustrated in Figure 11-13. In this technique, animals are injected with material containing an antigen of interest, and antibody-producing lymphocytes are isolated from the animal a few weeks later. Within such a heterogeneous lymphocyte population, each lymphocyte produces a single type of antibody directed against one particular antigen. To facilitate the selection and growth of individual lymphocytes, the lymphocytes are fused with cells that divide rapidly and have an unlimited lifespan in culture. The resulting hybrid cells are then individually selected and grown to form a series of clones called *hybridomas*. The antibodies produced by hybridomas are referred to as **monoclonal antibodies** because each one is a pure antibody produced by a cloned population of lymphocytes. Hybridomas can be maintained in culture indefinitely and represent inexhaustible sources of individual antibody molecules, each directed against a different antigen.

Monoclonal Antibodies Can Be Used to Trigger Cancer Cell Destruction Either by Themselves or Linked to Radioactive Substances

The ability to obtain monoclonal antibodies in large quantities gave rise to high expectations regarding their usefulness for selectively targeting cancer cells. The basic strategy is to immunize animals with human cancer tissue

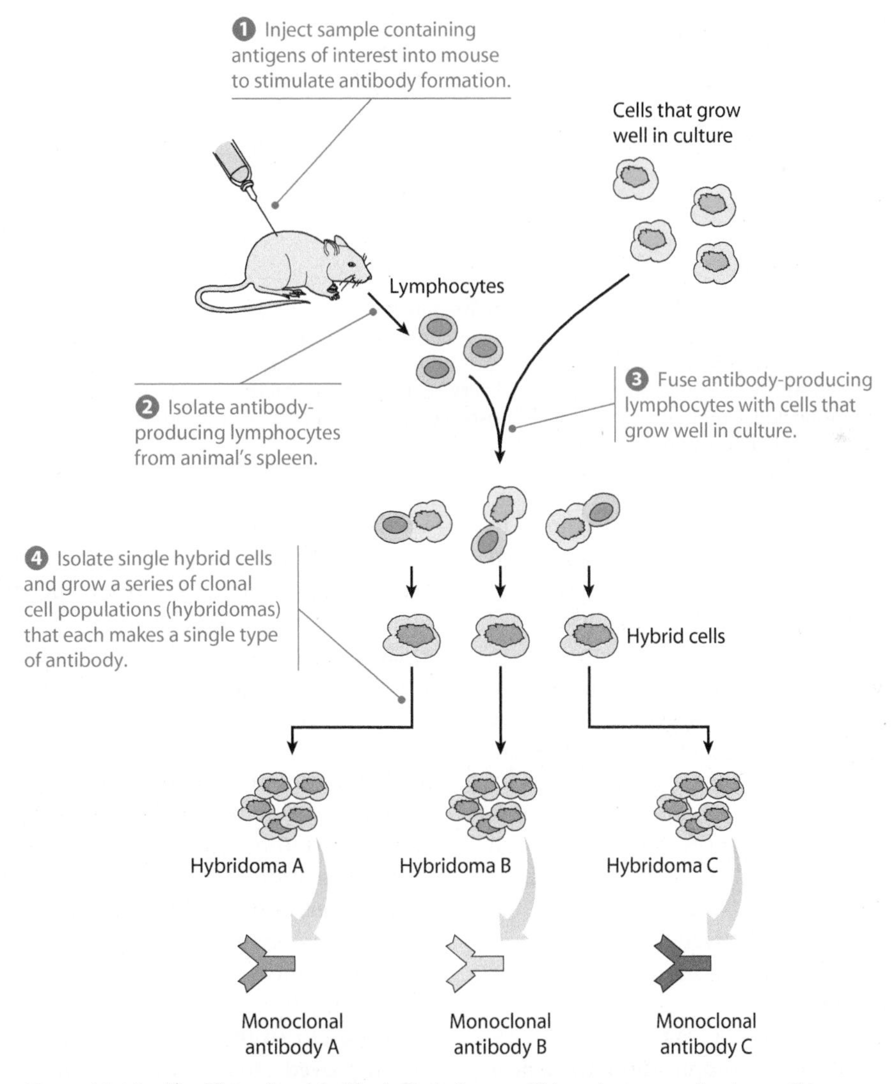

Figure 11-13 The Monoclonal Antibody Technique. This technique makes it possible to produce large amounts of pure populations of antibody molecules, each directed against a single antigen. Many of the monoclonal antibodies isolated in any given experiment will be directed against antigens to which the animal has been exposed prior to the experiment, so extensive screening is required to find a hybridoma that makes an antibody directed against the antigen of interest.

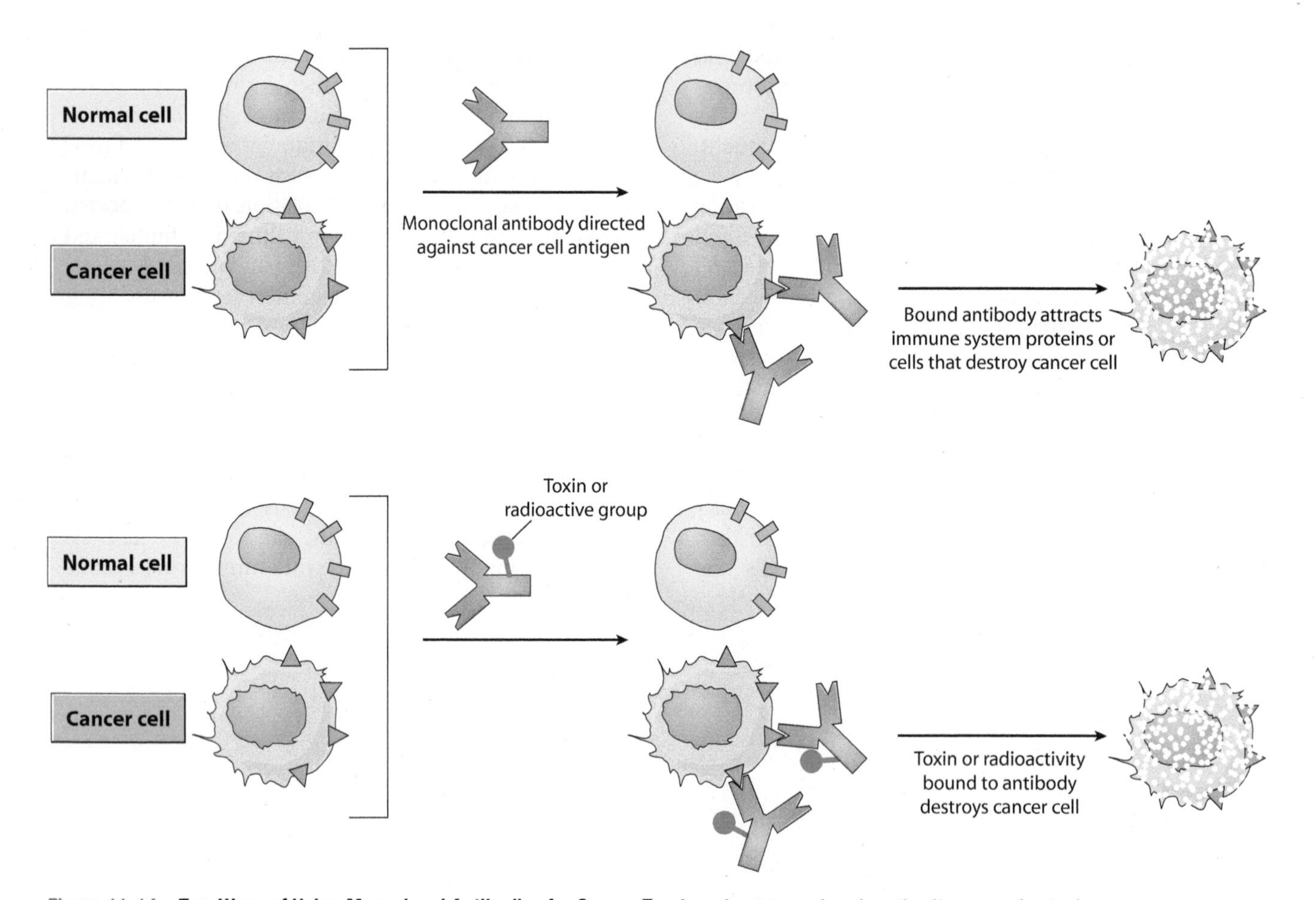

Figure 11-14 Two Ways of Using Monoclonal Antibodies for Cancer Treatment. Monoclonal antibodies can selectively target cancer cells by binding to tumor-specific antigens located on the outer cell surface. (*Top*) After monoclonal antibodies become selectively bound to cancer cells, the antibody's presence triggers an attack by other cells or proteins of the immune system. (*Bottom*) Antibodies can also be used as delivery vehicles for radioactive groups or other toxic substances. Linking them to monoclonal antibodies allows such substances to be concentrated at tumor sites without accumulating to toxic levels elsewhere in the body.

and then select those monoclonal antibodies that bind to antigens on the cancer cell surface. When they are injected into individuals with cancer, these antibody molecules would be expected to circulate throughout the body until they encounter cancer cells. The antibodies then bind to the cancer cell surface, where their presence triggers an immune attack that destroys only those cells to which the antibody is attached (Figure 11-14, *top*). Antibodies can also be used as delivery vehicles for toxic molecules by linking them to radioactive substances, chemotherapeutic drugs, or other kinds of toxic substances that are too lethal to administer alone (Figure 11-14, *bottom*). Attaching these substances to monoclonal antibodies allows the toxins or radioactivity to be selectively concentrated at tumor sites by the antibody without accumulating to toxic levels elsewhere in the body.

Although this strategy sounds simple in theory, several obstacles have slowed its application to cancer patients. One problem is that monoclonal antibodies are usually produced in mice by injecting them with human cancer tissue. The resulting antibodies are therefore recognized as foreign proteins when administered to cancer patients, who mount an immune response that inactivates the mouse antibody molecules, especially if the antibody is administered more than once. For this reason, monoclonal antibodies cannot be used for repeated treatments unless they are first made more human-like by replacing large parts of the mouse antibody molecule with corresponding sequences derived from human antibodies. A second complication encountered with monoclonal antibodies is that the cancer cell antigens they recognize may be present on certain normal cells as well. Each newly developed antibody must therefore be tested by linking it to a radioisotope and injecting it into patients to see whether the radioactivity becomes preferentially localized to sites where tumor cells are present.

The preceding issues have complicated the development of antibody-based therapies, but several successes have already been achieved. For example, the monoclonal antibodies *Rituxan*, *Zevalin*, and *Bexxar* are now among the approved treatments for non-Hodgkin's B cell lymphoma. All three antibodies target B lymphocytes for destruction by binding to the *CD20 antigen*, which is present on the surface of malignant as well as normal B lymphocytes. Although antibodies that target CD20 are toxic to normal B lymphocytes, CD20 is not present

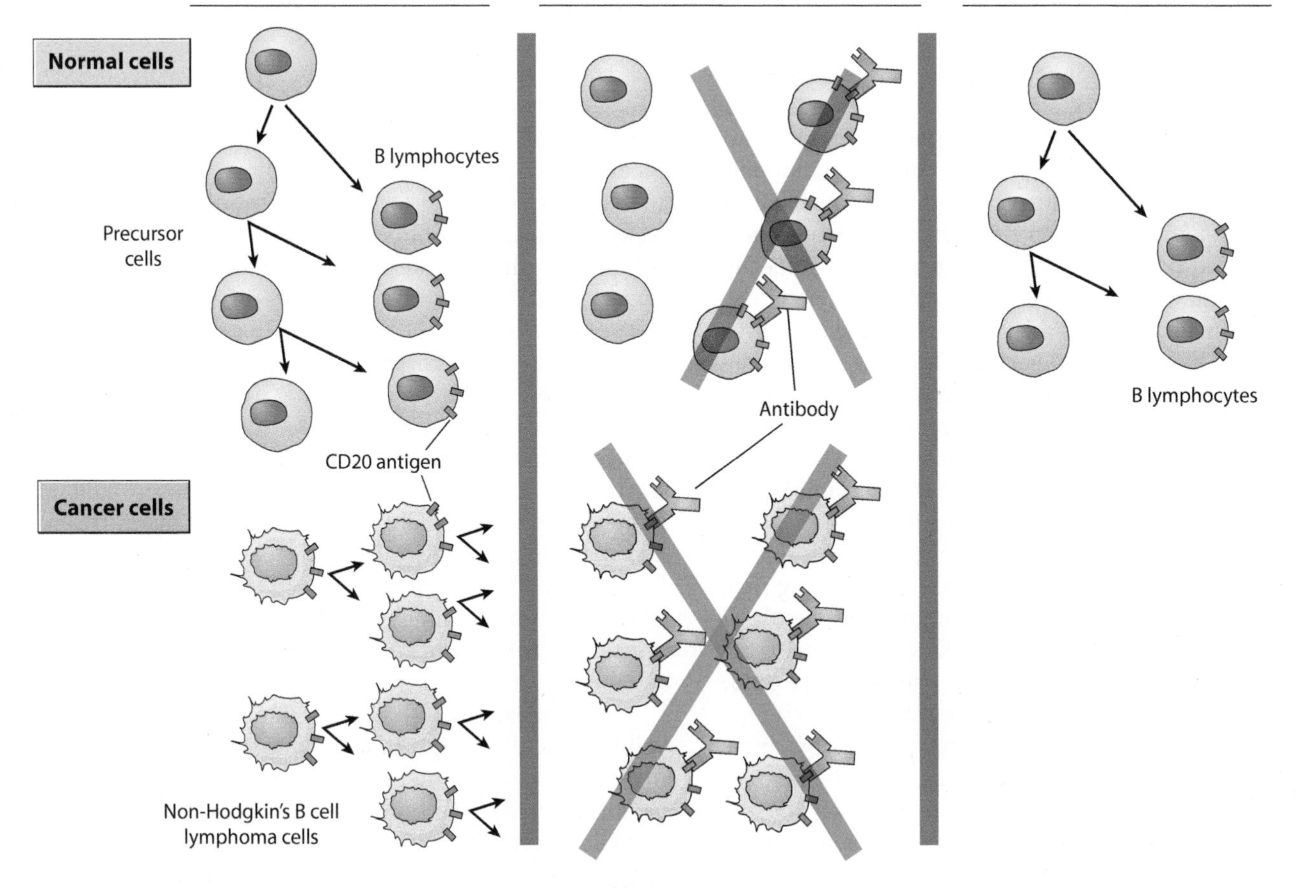

Figure 11-15 Use of Monoclonal Antibodies Directed Against CD20. When monoclonal antibodies directed against CD20 are injected into cancer patients with non-Hodgkin's B cell lymphoma, the antibody promotes the destruction of normal and malignant B lymphocytes. Unharmed precursor cells then replenish the normal B lymphocyte population.

on the precursor cells whose proliferation gives rise to B lymphocytes. These precursor cells therefore replenish the normal B lymphocyte population that is inadvertently destroyed along with malignant B lymphocytes during antibody treatment (Figure 11-15). Besides being administered by themselves, monoclonal antibodies directed against CD20 have been linked to radioactive chemicals and used to direct high doses of radiation to tumor sites, which may be more effective in killing cancer cells than the use of antibodies alone. Radioactive antibodies are also useful for determining where cancer cells are localized and for monitoring changes in tumor cell numbers in response to treatment.

The value of monoclonal antibodies is not restricted to their ability to target cancer cells for destruction. Monoclonal antibodies have also been developed that target signaling pathway components required by cancer cells for their proliferation. For example, some breast cancer patients are being treated with *Herceptin*, a monoclonal antibody that binds to and blocks a growth factor receptor. Because monoclonal antibodies are not the only tools used for targeting signaling pathway components, we will delay a discussion of this type of cancer therapy until the section on molecular targeting.

Several Types of Cancer Vaccines Are Currently Under Development

Antibodies are one of two basic mechanisms used by the immune system for attacking foreign antigens. The second mechanism, known as *cell-mediated immunity*, utilizes **cytotoxic T lymphocytes** that bind to the surface of cells exhibiting foreign antigens and kill the targeted cells by causing them to burst. This tactic is normally used to destroy cells harboring infectious agents such as viruses, bacteria, and fungi, and it also plays a role in the destruction of foreign tissue grafts and organ transplants.

The realization that cytotoxic T lymphocytes might be able to mount an attack against cancer cells first emerged in the 1940s from studies in which cancer was induced in mice by exposing them to carcinogenic chemicals or viruses. The resulting tumors were found to contain antigens whose administration to other mice immunized the animals against transplants of the same tumor. When T lymphocytes were isolated from the immunized animals, these T lymphocytes could kill tumor cells in culture and transfer tumor immunity when injected into other animals. In contrast, antibodies produced by the tumor-bearing animals were relatively ineffective at killing cancer cells or transferring immunity.

These observations have stimulated interest in the idea of developing **vaccines** that will stimulate a cancer patient's own T lymphocytes to attack cancer cells. The underlying rationale is that tumor antigens tend to be weak antigens that do not elicit a strong immune response, but an appropriate vaccine might be able to present the antigens in a way that would stimulate the immune system to become more aware of their existence. Among the candidates for vaccine antigens are the abnormal proteins that cancer cells produce as a result of genetic mutations. Since these proteins are not produced by normal cells, putting them into vaccines should stimulate an immune response that is selectively directed against cancer cells. Other proteins that are overproduced by tumors might also be useful candidates for incorporation into cancer vaccines.

It is possible to vaccinate cancer patients by simply injecting them with tumor antigens, but attempts are being made to improve vaccination efficiency by first introducing the antigens into **dendritic cells** for antigen processing. (Recall from Chapter 2 that triggering an efficient immune response requires that antigens be broken into fragments and presented to the immune system by antigen-presenting cells such as dendritic cells.) When dendritic cells obtained from cancer patients are grown in the laboratory together with tumor antigens, the dendritic cells take up the antigens, chop them into pieces, and present the resulting fragments on their cell surface in a way that activates an immune response. Experiments are currently under way to determine whether the injection of such antigen-loaded dendritic cells into patients is a feasible tactic for treating cancer.

Adoptive-Cell-Transfer Therapy Uses a Person's Own Antitumor Lymphocytes That Have Been Selected and Grown in the Laboratory

Adoptive-cell-transfer (ACT) therapy is an alternative to vaccination in which a patient's own lymphocytes are first isolated, selected, and grown in the laboratory to enhance their cancer-fighting properties prior to injecting the cells back into the body. The underlying reasoning is that individuals with cancer often possess lymphocytes that are capable of attacking tumor cells, but these lymphocytes are not produced in sufficient quantities to keep the tumor under control. ACT therapy attempts to solve this problem by removing some of these lymphocytes from the body and increasing their numbers by growing them in culture prior to reintroducing the cells into the patient.

If a person with cancer has any lymphocytes that are capable of attacking tumor cells, the most likely place to find them would be within the tumor itself. Lymphocytes that are located at the tumor site, called *tumor-infiltrating lymphocytes* (*TILs*), have therefore been used as a source of cells for ACT therapy. In one set of studies, illustrated in Figure 11-16, multiple samples of TILs were isolated from the tumors of advanced stage melanoma patients and tested for their ability to attack tumor cells. TIL samples exhibiting the greatest anti-tumor activity were then selected and grown in culture in the presence of *interleukin-2* (*IL-2*), a cytokine that stimulates the proliferation and cancer-destroying properties of the lymphocytes. Before introducing the tumor-killing lymphocytes back into the body, each cancer patient was treated with high-dose chemotherapy to destroy a large fraction of their existing lymphocytes. The tumor-killing lymphocytes were then injected back into the bloodstream and the patients were treated with IL-2 to further stimulate the proliferation of the injected cells. The net result was that tumor-killing lymphocytes became a large portion of each person's immune system, and a significant number of patients experienced tumor regressions.

ACT therapy is still an experimental procedure and will be difficult to apply to large numbers of patients, but these results suggest that cancer therapies may eventually be able to exploit the ability of lymphocytes to recognize and kill cancer cells. Several problems remain to be solved, however. First, the possibility exists that lymphocytes targeted against cancer cell antigens will mistakenly attack healthy cells possessing similar antigens. Another problem is that cancer cells can devise ways of evading immune attack (p. 39). For example, sometimes cancer cells acquire mutations that cause them to stop making the antigens being targeted by the immune system. In other cases, cancer cells become resistant to immune attack by producing molecules that either kill lymphocytes or disrupt their ability to function. Of course, the possibility that resistance will develop is not unique to immunotherapy; we have already seen that resistance arises with chemotherapy as well. For this reason, a combination of different therapeutic approaches may end up being the best approach for treating cancer.

Herceptin and Gleevec Are Anticancer Drugs That Illustrate the Concept of Molecular Targeting

Until the early 1980s, research into new cancer treatments focused largely on the development of drugs that disrupt DNA synthesis and interfere with cell division. Although

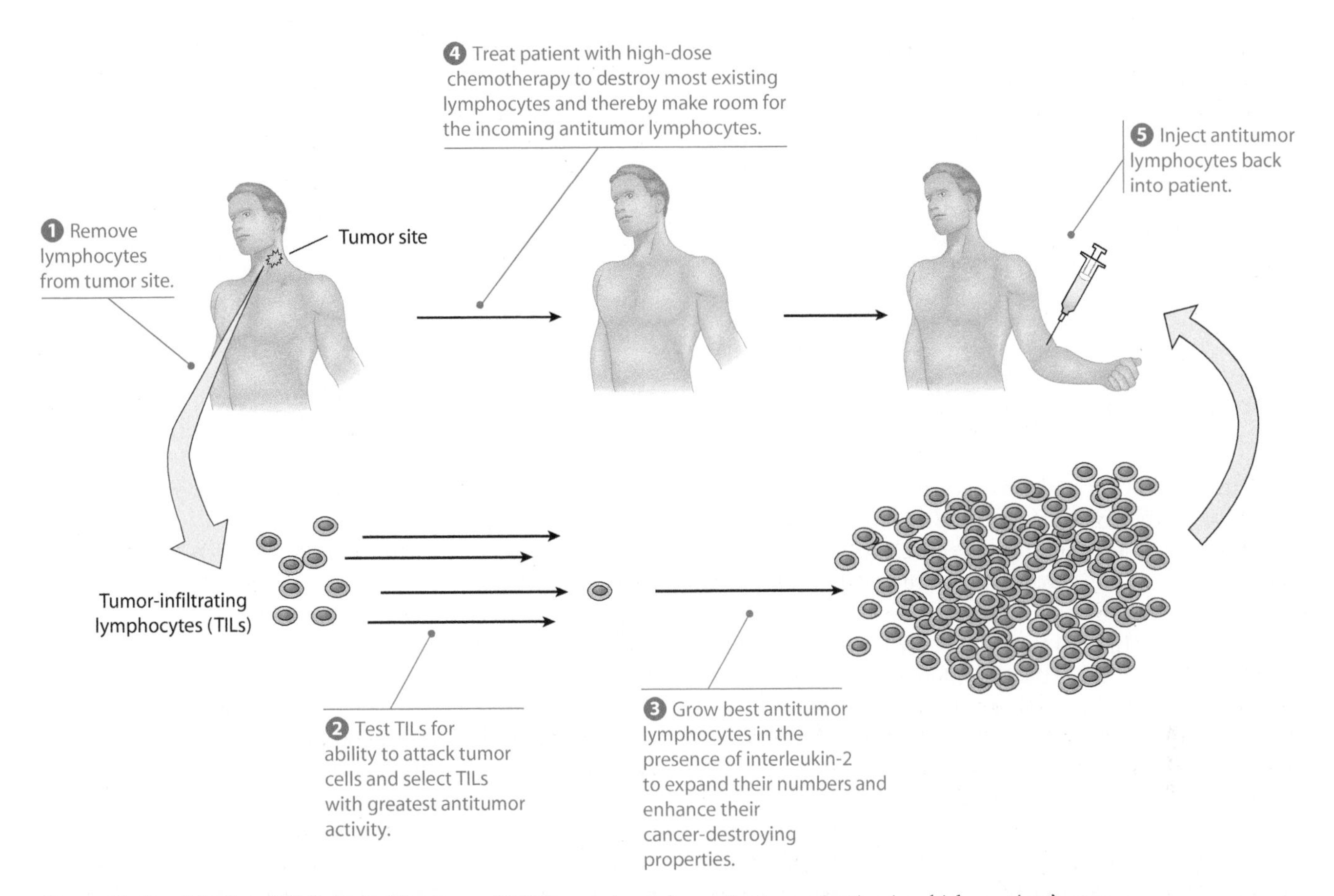

Figure 11-16 Adoptive-Cell-Transfer Therapy. ACT therapy is an alternative to vaccination in which a patient's own lymphocytes are isolated and grown in the laboratory to enhance their cancer-fighting properties prior to injecting them back into the body. Before reintroducing the antitumor lymphocytes, the patient is treated with high-dose chemotherapy to destroy most existing lymphocytes and thereby make room for the incoming antitumor lymphocytes, which become a large portion of the person's immune system.

some of the resulting drugs have turned out to be useful in treating cancer, their effectiveness is often limited by toxic effects on normal dividing cells. In the past two decades, the identification of specific genes whose mutation or altered expression can lead to cancer has opened up a new possibility—**molecular targeting**—in which drugs are designed to target those proteins that are critical to the cancerous state.

One way to pursue the goal of molecular targeting is to take advantage of the specificity of antibodies. Substantial efforts are currently being made to develop monoclonal antibodies that bind to and inactivate key proteins involved in the signaling pathways required for cancer cell proliferation. The first such antibody to be approved for use in treating cancer patients, called **Herceptin**, binds to and inactivates a cell surface growth factor receptor called the **ErbB2 receptor**, which is produced by the *ERBB2* gene (also called *HER2*). About 25% of all breast and ovarian cancers have amplified *ERBB2* genes, which produce excessive amounts of ErbB2 receptor that in turn causes hyperactive signaling. When individuals whose cancers overexpress the ErbB2 receptor are treated with Herceptin, the Herceptin antibody binds to the ErbB2 receptor and the ability of the receptor to stimulate cell proliferation is blocked, thereby slowing or stopping tumor growth.

Monoclonal antibodies are not the only way to target specific molecules for inactivation. Another approach, called **rational drug design**, involves the laboratory synthesis of *small molecule inhibitors* that are designed to bind to and inactivate specific target molecules. Unlike antibodies, these inhibitors are small enough to enter cells and affect intracellular proteins. One of the first such drugs to be developed, called **Gleevec** (generic name *imatinib*), is a small molecule that binds to and inhibits the abnormal tyrosine kinase produced by the *BCR-ABL* oncogene present in chronic myelogenous leukemias. As described in Chapter 9, *BCR-ABL* is a fusion gene generated during the chromosomal translocation that creates the Philadelphia chromosome. Because it arises from the fusion of DNA sequences derived from two different genes, *BCR-ABL* produces a structurally abnormal protein—the Bcr-Abl tyrosine kinase—that represents an ideal drug target because it is produced only by cancer cells.

Initial studies of the effectiveness of Gleevec as a treatment for chronic myelogenous leukemia were extremely encouraging. In patients with early stage disease, more than 50% had no signs of cancer six months after treatment (a response rate ten times better than had been seen before). Unfortunately, patients with late stage disease frequently develop mutations that alter the structure of the Bcr-Abl tyrosine kinase, thereby making it resistant to Gleevec. Additional small molecule inhibitors that overcome this resistance to Gleevec have been developed, but it takes many years to take each new compound through the necessary testing before it can be approved for routine medical use.

A Diverse Group of Potential Targets for Anticancer Drugs Are Currently Being Investigated

The drugs Herceptin and Gleevec illustrate two different approaches—monoclonal antibodies and small molecule inhibitors—for targeting specific proteins found in cancer cells. These two drugs are relatively recent accomplishments in the long history of cancer drug research; Herceptin was introduced in 1998 and Gleevec in 2001. As might be expected, their success has stimulated interest in developing other drugs that target molecules important to cancer cells. For example, the introduction of Gleevec in 2001 was followed in 2003 by another small molecule inhibitor called **Iressa** (generic name *gefitinib*). As mentioned earlier in the chapter, Iressa targets the receptor for epidermal growth factor and is effective in a subset of lung cancer patients whose cancer cells possess a mutant form of the EGF receptor (see Figure 11-12).

Dozens of other drugs based on the principle of molecular targeting are currently under investigation. Tyrosine kinases and growth factor receptors (the targets for Gleevec and Herceptin, respectively) are just two of many potential targets. As we saw in Chapters 9 and 10, the uncontrolled proliferation of cancer cells can be traced to disruptions in a variety of growth signaling pathways, including the Ras-MAPK, Jak-STAT, Wnt, and PI3K-Akt pathways. Any of the proteins involved in these pathways could represent a potential target for an anticancer drug. Other proteins whose activities contribute to the six hallmark traits of cancer cells (p. 195) might likewise be good candidates. Table 11-3 lists some examples of proteins in these various categories that are now being investigated as potential targets for anticancer drugs.

Despite the attractiveness of molecular targeting, many of the drugs developed after the initial successes with Herceptin and Gleevec have failed to work well when tested in cancer patients. While such disappointments may simply mean that these particular drugs are ineffective, several factors complicate the testing of anticancer drugs that could have contributed to the failures. First, targeted therapies would only be expected to work in those individuals whose cancer cells exhibit the appropriate molecular target. Since cancers of the same type often differ in their molecular properties from person to person, obtaining a molecular profile of each person's tumor might assist in identifying patients most likely to benefit from a given type of treatment.

Second, testing of new drugs is generally done in patients who also receive standard chemotherapy, which might obscure the benefits of an experimental drug. For example, in the case of tamoxifen, which targets the estrogen receptor, inferior results are obtained when tamoxifen is combined with standard chemotherapy compared with giving tamoxifen either alone or after

Table 11-3 Examples of Possible Targets for Anticancer Drugs

Target Protein	Pathway or Function	Drugs Approved*
ErbB2 receptor	Growth factor receptor	Herceptin
EGF receptor	Growth factor receptor	Iressa, Erbitux, Tarceva
FGF receptor	Growth factor receptor	
PDGF receptor	Growth factor receptor	
VEGF	Angiogenesis signaling	Avastin
Bcr-Abl kinase	Apoptosis signaling	Gleevec
Src kinase	Ras-MAPK pathway	
Raf kinase	Ras-MAPK pathway	
Ras	Ras-MAPK pathway	
Cyclin-dependent kinases	Cell cycle progression	
PI 3-kinase	PI3K-Akt pathway	
Hsp90	Stabilizes growth signaling proteins	
Mdm2	Apoptosis inhibitor	
Bcl2	Apoptosis inhibitor	
Matrix metalloproteinases	Invasion/metastasis/angiogenesis	
Proteasome	Targeted protein degradation	Velcade
Telomerase	Limitless replicative potential	

*Drugs listed in this column have already been approved for treating cancer patients.

chemotherapy. In theory, the most reliable results would be obtained by comparing a new drug given to one group of patients versus standard chemotherapy given to another group of patients. However, ethical considerations make it inappropriate to withhold standard treatment from the first group of patients if the standard treatment is known to be beneficial.

A third type of problem is related to the need for better drug delivery methods that reliably convey drugs to tumor sites at the proper concentration for an appropriate period of time. In many cases, drugs are simply degraded too quickly after entering the body and do not accumulate in tumor tissues. One way to improve drug delivery is through the use of water-soluble polymers such as *polyethylene glycol* or *N-(2-hydroxypropyl)methacrylamide.* Binding drugs to these polymers prolongs a drug's lifetime in the body and alters its pattern of distribution. The reason for the altered behavior is that the large size of drug-polymer complexes prevents them from passing out of the bloodstream and into cells as rapidly as the free drug itself. In addition, tumor blood vessels tend to be "leaky," causing drug-polymer complexes to leave the bloodstream and enter tumor tissues more readily than normal tissues.

A final problem that complicates drug testing is that clinical trials are usually carried out in late-stage cancer patients after all other treatments have failed. At this advanced stage, targeted molecular therapy may no longer be useful. For example, consider the behavior of drugs that inhibit *matrix metalloproteinases* (*MMPs*), which are attractive targets because they play important roles in angiogenesis, tissue invasion, and metastasis (see Chapter 3). Animal studies have shown that MMP inhibitors are effective antitumor agents during the early stages of cancer progression, when tumor invasion and metastasis are just beginning. Human testing, however, has been performed mainly in patients with late stage disease, when MMP inhibitors appear to be largely ineffective. This is just one of many examples of experimental anticancer drugs that have been tested in late stages of cancer progression rather than early in the disease, when they are more likely to work. Such problems are difficult to avoid for the simple reason that experimental new treatments are not likely to be tried on patients until other treatments have failed, at which point the disease may have reached an advanced stage that makes it unresponsive to targeted therapies.

Anti-angiogenic Therapy Illustrates the Difficulties Involved in Translating Laboratory Research into Human Cancer Treatments

We saw in Chapter 3 that tumor growth and metastasis depend on *angiogenesis*—that is, the growth of blood vessels that supply nutrients and oxygen to tumor cells and remove waste products. It is therefore logical to expect that angiogenesis inhibitors might be useful for treating cancer patients. Initial support for this concept of **anti-angiogenic therapy** came from the studies of Judah Folkman, who reported that treating tumor-bearing mice with the angiogenesis-inhibiting proteins *angiostatin* and *endostatin* makes tumors shrink and disappear (see Figure 3-6). When these experiments were first described in 1998 in a front page story appearing in the *New York Times*, a distinguished scientist was quoted as saying, "Judah is going to cure cancer in two years."

Needless to say, such sensational news coverage led to unrealistic expectations concerning the prospects for an immediate cancer cure. Applying the results of animal studies to human patients takes many years of testing, and humans do not always respond in the same way as animals. Dozens of angiogenesis-inhibiting drugs are therefore being evaluated in cancer patients to see if the promising results observed in animals will apply to humans. On the positive side, the early human studies showed that anti-angiogenic therapy elicits few of the harsh side effects seen with chemotherapy, and in a few cancer patients, tumors seemed to stop growing. However, some disappointment was expressed with the early results because they failed to show the quick cure for cancer that people had been led to expect.

Of course, expectations for a quick cancer cure were unrealistic, and there are many reasons why it would be premature to come to any definitive conclusions at this point regarding the effectiveness of anti-angiogenic therapy. First, the early human trials were carried out mainly on cancer patients with late stage disease, and anti-angiogenic therapy may work better at earlier stages. Second, the optimal dose for angiogenesis-inhibiting drugs may need to be tailored to each individual patient based on the concentration of angiogenesis-stimulating molecules their tumors produce. Third, angiogenesis inhibitors may work best when their concentration within the body is maintained at a relatively constant level, which is quite different from the way in which standard chemotherapy is typically administered using large intermittent doses. Finally, the effectiveness of anticancer drugs is usually measured by assessing their ability to make tumors shrink or disappear. This outcome might be an appropriate expectation for a drug that kills cancer cells, but inhibiting blood vessel growth may simply stop tumors from becoming any larger. Such a state, called *stable disease*, could represent an acceptable outcome for an anti-angiogenic drug if it allowed patients to live with cancer as a chronic but manageable disease condition, especially in view of the minimal side effects associated with the use of angiogenesis inhibitors.

The complexities raised by the preceding issues mean that it will take many years to assess the effectiveness of angiogenesis-inhibiting drugs and determine how best to use them. Nonetheless, signs of progress are already evident. In 2004, **Avastin** became the first anti-angiogenic drug to be approved for routine medical use in cancer patients. Avastin is a monoclonal antibody that binds to and inactivates the angiogenesis-stimulating growth factor, *VEGF* (p. 48). In tumors that depend on VEGF to stimulate angiogenesis, blocking VEGF with Avastin would be expected to inhibit angiogenesis and thereby inhibit tumor growth. Human clinical trials have shown that patients with metastatic colon cancer who received standard chemotherapy plus Avastin lived longer than patients who received standard chemotherapy without Avastin. These results were one of the first signs that anti-angiogenic therapy may one day become an integral component of human cancer treatment.

Engineered Viruses Are Potential Tools for Repairing or Killing Cancer Cells

Over the past two decades, the roles played by oncogenes and tumor suppressor genes in the development of cancer have become increasingly apparent. This discovery raises the possibility of attacking the disease at its root cause: defective genes. In other words, rather than trying to kill or restrain the proliferation of cancer cells, it might be possible to repair the defective genes that are responsible for the cancerous state.

The process of replacing defective genes with normal versions is called **gene therapy**. Gene therapy was initially envisioned as a treatment for genetic diseases in which a person inherits a single defective gene, such as a gene responsible for cystic fibrosis, hemophilia, or certain immune deficiencies. Curing illnesses of this type would simply require that a normal copy of the single defective gene be inserted into a person's cells under conditions that allow the inserted gene to be actively expressed. While the concept sounds simple in theory, it is difficult to transfer genes into cells efficiently under conditions that permit the transferred genes to become permanently incorporated and expressed. As a result, gene therapy had been of limited usefulness in treating genetic diseases thus far.

Applying gene therapy to cancer is even more complicated than treating an inherited genetic disease because it may be necessary to repair the defect in all cancer cells, not just some of them. Moreover, cancer cells usually exhibit defects in several genes rather than just one, although it may not be necessary to repair them all. As mentioned in Chapter 10, human cancers often exhibit defects in the p53 pathway that prevent cells from undergoing apoptosis. If this single pathway could be restored, the other abnormalities exhibited by cancer cells might trigger the p53 pathway and cause the cells to self-destruct by apoptosis. Attempts have therefore been made to repair the *p53* gene in cancers in which this gene is defective (Figure 11-17). Support for this approach has come from animal studies showing that tumor regression can be induced by injecting animals with a virus whose DNA contains a normal copy of the p53 gene. In early human testing, a similar virus injected into the tumors of lung cancer patients has been found to restore p53 production and induce disease stabilization in some patients.

An alternative to using viruses for gene therapy is to engineer them to kill cancer cells selectively. It has been known for many years that some viruses cause infected cells to rupture and die, a process called **lysis**. Attempts are therefore being made to create viruses that selectively infect and cause the lysis of cancer cells. One of the first of these viruses to be tested in humans was *ONYX-015*, an adenovirus containing a mutation designed to permit the virus to replicate only in cells with a defective p53 pathway. Since the p53 pathway is defective in a majority of human cancers, it was predicted that ONYX-015 might be a broadly useful tool for killing cancer cells. Early investigations appeared to verify the ability of ONYX-015 to replicate preferentially in cancer cells, but follow-up studies failed to confirm the dependence of viral replication on the presence of a defective p53 pathway and future development of this particular virus is uncertain.

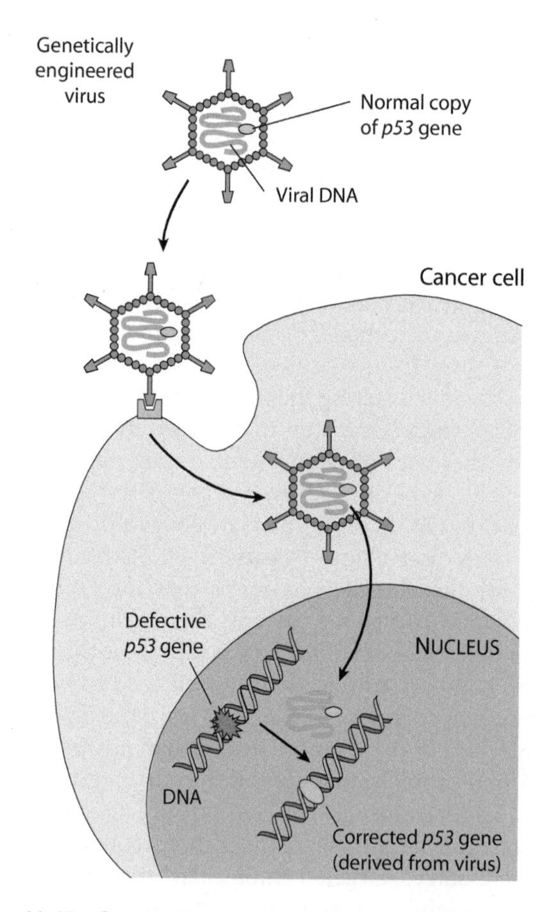

Figure 11-17 Strategy for Using Gene Therapy to Repair a Defective Cancer Cell Gene. Many human cancers exhibit defects in the *p53* gene. If these defects could be corrected, restoration of the p53 pathway might cause cancer cells to self-destruct by apoptosis. Viruses engineered to contain a normal copy of the *p53* gene have therefore been used in gene therapy experiments to infect tumors and insert the normal *p53* gene into the DNA of cancer cells.

ONYX-015, however, represents just one of many engineered viruses that are being developed to kill cancer cells without harming normal cells. Like ONYX-015, these viruses have been genetically altered to make their replication dependent either on the absence of genes that are inactive only in cancer cells or on the presence of genes that are active only in cancer cells (Figure 11-18, *left*). Another potential strategy is to modify viruses in ways that cause them to interact preferentially with cancer cells,

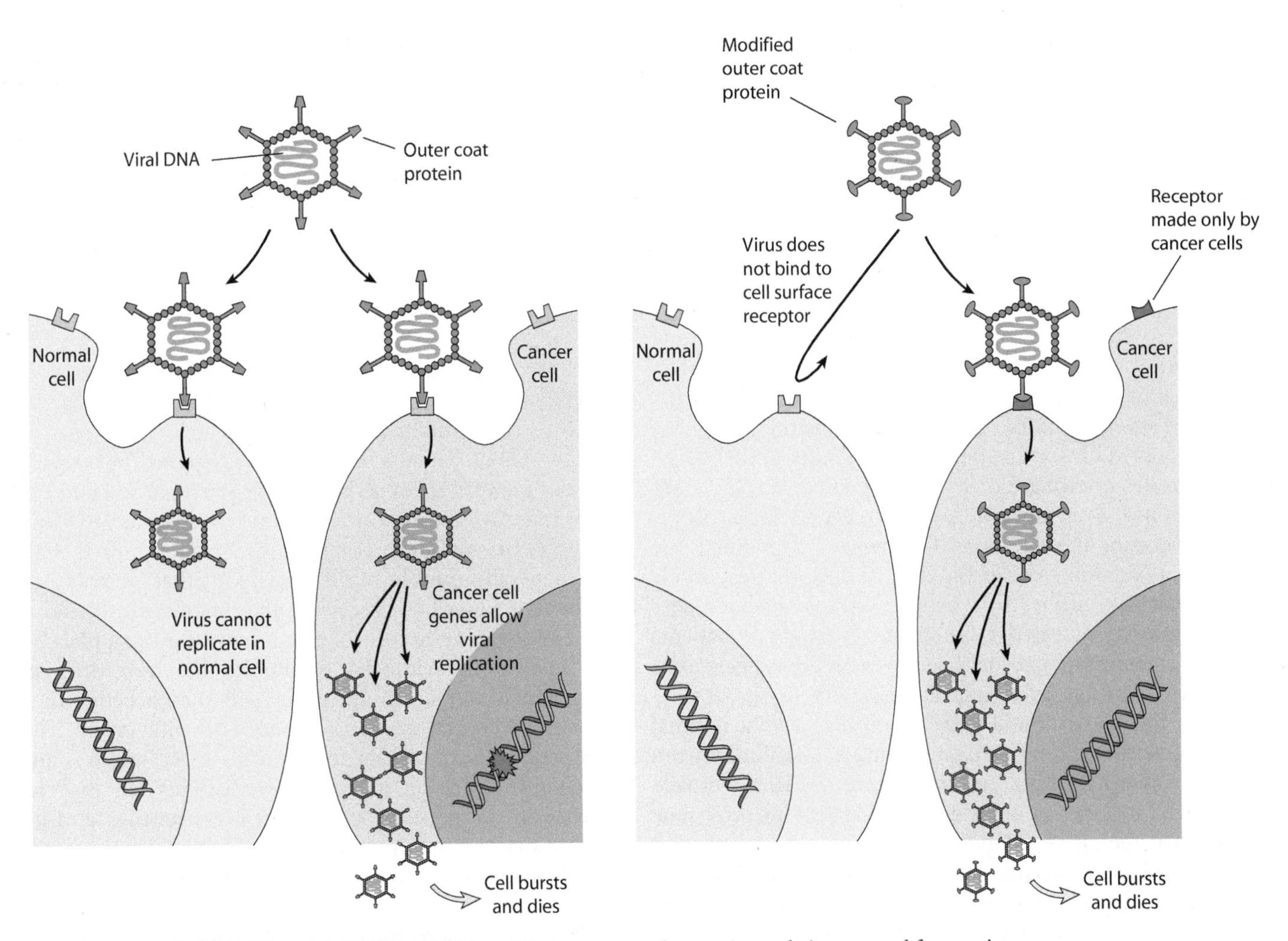

Figure 11-18 Designing Viruses to Kill Cancer Cells. Two experimental strategies are being pursued for creating viruses that might selectively kill cancer cells. (*Left*) One approach uses viruses that infect normal cells as well as cancer cells. The viruses are modified, however, to make their replication dependent either on the absence of genes that are inactive only in cancer cells or on the presence of genes that are active only in cancer cells. In either case, such viruses would be expected to replicate preferentially in (and therefore kill) cancer cells. (*Right*) Alternatively, the coat proteins of some viruses have been modified so that they bind to receptors that are only present on cancer cells. These viruses would be expected to infect cancer cells but not normal cells.

perhaps by altering viral coat proteins so that they bind to receptors present on the surface of cancer cells (see Figure 11-18, *right*). Such approaches are currently under active investigation to see whether they might be of any use in the treatment of cancer.

CLINICAL TRIALS AND OTHER APPROACHES

Before any new treatment can be incorporated into standard medical practice, it must first undergo a lengthy and painstaking evaluation process. In the early days of cancer research, identifying and evaluating new treatments was especially time consuming because anticancer drugs were often discovered through a largely random approach. For example, the National Cancer Institute established a massive screening program in the mid-1960s that systematically tested thousands of chemical compounds for possible anticancer activity. Those substances that exhibited the most promise in killing cancer cells in laboratory culture or in animal studies were eventually tested in humans, and a number of drugs now used in cancer chemotherapy were discovered in this way.

In recent years, our growing understanding of the molecular abnormalities exhibited by cancer cells has permitted more selective approaches for developing drugs that target cancer cells. Nonetheless, such drugs still require extensive testing before they can be incorporated into standard medical practice. The testing process, which is regulated in the United States by the Food and Drug Administration (FDA), requires that any drug proposed for human use first undergo **preclinical testing** in animals to demonstrate that the treatment is safe and effective. If successful, animal testing is followed by an extensive series of human tests to determine whether the drug works in humans and whether it compares favorably to existing methods of treatment.

Human Clinical Trials Involve Multiple Phases of Testing

Evaluating a new drug in humans involves a series of tests called **clinical trials**. Patients who volunteer for a clinical trial are given information regarding the nature of their disease, the potential risks and benefits of the treatment being tested, and the availability of other treatment options. Before participating, all patients must sign **informed consent** documents indicating their understanding of these conditions and providing their voluntary consent. Each trial involves several phases of testing, often requiring five to ten years to complete at a cost of several hundred million dollars, before a drug can be approved for routine medical use (Figure 11-19).

In the first phase of testing, called a *Phase I clinical trial*, a new drug is administered to several dozen people to determine the safe dose. The first few individuals are given a very low dose of the drug and monitored closely for toxic side effects. If the drug is well tolerated, the dose is gradually increased in subsequent groups of patients until an appropriate dose is determined that is likely to be effective without severe side effects.

If the drug is found to be reasonably safe, the optimal dose determined during Phase I testing is then administered to a somewhat larger group of cancer patients—usually from 25 to 100—in a *Phase II clinical trial* to determine whether the drug exhibits any effectiveness in treating cancer. Evidence of effectiveness might be the complete disappearance of a tumor (*complete response*), a tumor that gets smaller (*partial response*), or a tumor that stops growing (*stable disease*). To justify further testing, a significant percentage of the treated patients must exhibit one of these three responses. The required percentage may vary, however, depending on the type of cancer being treated and the effectiveness of currently available drugs. For example, a 10% response rate might justify continued testing of a treatment for an aggressive tumor like pancreatic cancer that tends to be resistant to most current drugs, whereas a much higher response rate would be required for low-grade lymphomas, for which several effective drugs are already available.

If a drug exhibits sufficient signs of anticancer activity in Phase II testing, its effectiveness and safety are thoroughly evaluated in a *Phase III clinical trial*. A Phase III trial is a **randomized trial** in which hundreds or thousands of patients are randomly assigned to two different groups: an experimental group that receives the new treatment and a control group that does not. To avoid possible bias in interpreting the results, randomized trials are generally **double blind**; that is, neither doctors nor patients know who is receiving the treatment and who is not. Patients in the control group may be given a **placebo** (inactive substance) that resembles the new drug in appearance so that no individual will know whether they are in the control group or the experimental group. The purpose is to control for the **placebo effect**, which is any beneficial effect on a patient's condition that may be caused by a person's expectations concerning a drug rather than by the drug itself. Placebos, however, are not used to substitute for currently existing treatments that are known to be beneficial. For example, the experimental group might receive the standard treatment along with the

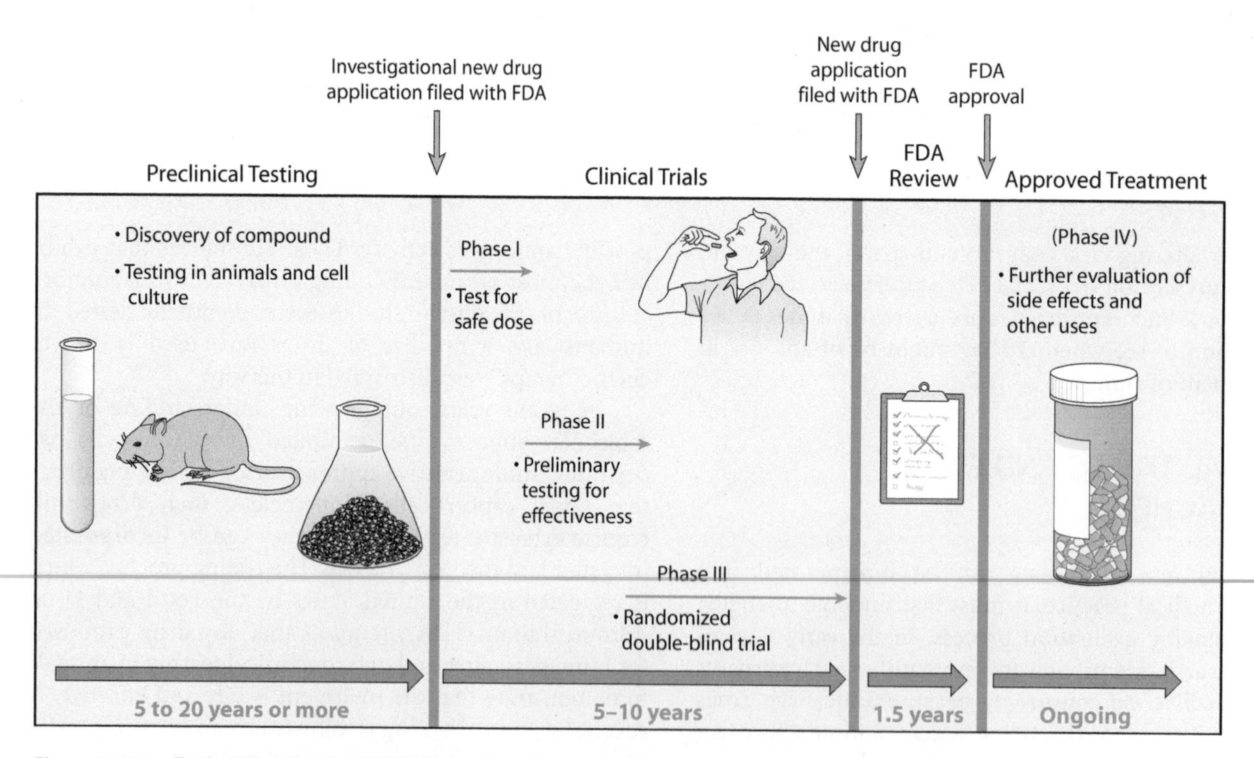

Figure 11-19 Typical Timeline for Developing a New Cancer Drug. Developing a new cancer treatment takes many years and requires numerous steps, including preclinical laboratory and animal testing, clinical trials in cancer patients, and FDA approval. [Adapted from J. A. Zivin, *Sci. Amer.* 282 (April 2000): 70.]

new drug, while the control group receives the standard treatment along with a placebo.

Based on the results of Phase III randomized trials, the FDA decides whether or not to approve a new drug as an acceptable treatment for standard medical use. After approval has been granted, further *Phase IV clinical trials* may be carried out to answer additional questions concerning the best ways to use the drug or to explore possible side effects that were not detected in earlier testing.

Complementary and Alternative Cancer Treatments Are Frequently Used by People Who Have Cancer

Prior to obtaining FDA approval, new drugs undergoing laboratory and clinical testing are referred to as **experimental treatments.** It usually takes many years to obtain enough evidence to justify incorporating an experimental treatment into standard medical practice. In addition to experimental treatments, a diverse array of unproven and largely untested cancer treatments exist that are not part of standard medical practice. These treatments can be subdivided into **complementary treatments**, which are used along with standard medical care, and **alternative treatments**, which are used as a substitute for standard medical care. Complementary and alternative treatments include herbal remedies, vitamins, special diets, and a variety of physical and psychological practices such as massage and relaxation techniques. More than half of all individuals with cancer have been reported to use one of more of these practices, often without discussing it with their doctor.

Complementary treatments are usually used to control symptoms and improve a person's quality of life while under standard medical care. In contrast, many *alternative* treatments are claimed to cure cancer. Individuals who rely solely on these alternative remedies may put themselves at considerable risk. A striking example is provided by the history of **laetrile** (also called *amygdalin* or *vitamin B17*), a natural substance extracted from apricot pits that attracted considerable attention in the 1970s when medical clinics in Mexico claimed that it cured cancer. Research in laboratory animals failed to show any anticancer effects of laetrile, so it did not meet the normal standards for human testing in the United States. However, the prominence of laetrile and its use by thousands of Americans (many of whom traveled to Mexico for treatment) led the National Cancer Institute to sponsor a human clinical trial despite the absence of supporting data from animal testing. After the trial showed laetrile to be ineffective against cancer, its popularity gradually declined. In retrospect, the lack of anticancer properties was not the only problem with laetrile. The drug also has hidden dangers because it breaks down to form cyanide, and some people treated with laetrile may have died of cyanide poisoning rather than their cancers. Another risk incurred by individuals who rely on unproven remedies such as laetrile is that they deny themselves the benefits of any proven methods that may be genuinely useful for their particular type of cancer.

Although the experience with laetrile highlights the need for caution, it does not mean that alternative remedies are always risky and without value. An intriguing example involves an herbal product called **PC-SPES** ("PC" for Prostate Cancer and "SPES" from the Latin word for "hope"). PC-SPES, which consists of extracts from eight herbs, was introduced in 1996 as a remedy for prostate cancer. Because it was being sold as a dietary supplement consisting entirely of natural ingredients, PC-SPES did not require a doctor's prescription or fall under governmental regulations for purity or effectiveness. Shortly after it was introduced, several studies reported that PC-SPES slowed the growth of prostate cancer in humans (Figure 11-20). These encouraging results made PC-SPES one of the best prospects for an alternative cancer treatment that might stand up to the scrutiny of rigorous scientific testing. However, chemical analyses of PC-SPES subsequently revealed that this supposedly "all natural" herbal mixture was contaminated with several synthetic drugs, and the manufacturer voluntarily stopped selling it. When it was discovered that several other herbal products sold by the same company were also adulterated with synthetic drugs, the company went out of business and PC-SPES is no longer available today. This cautionary tale illustrates the problems that arise when trying to evaluate the possible effectiveness of

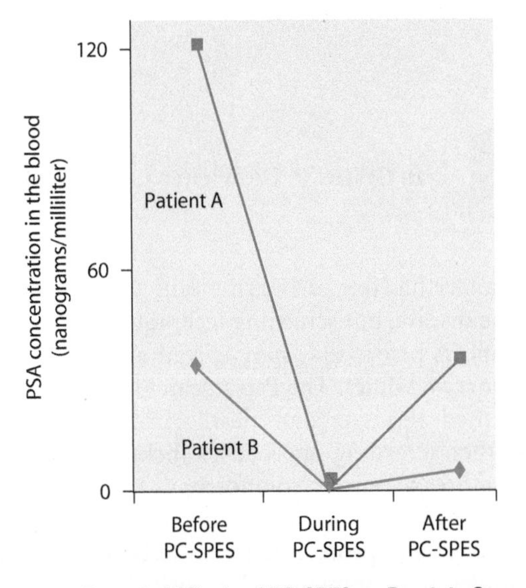

Figure 11-20 Reported Effects of PC-SPES on Prostate Cancer Patients. This graph shows the reported effects of the herbal remedy PC-SPES in two patients with prostate cancer. Patient B, but not patient A, had his prostate gland removed prior to using PC-SPES. During PC-SPES treatment, both patients exhibited a dramatic decrease in cancer growth as determined by measuring blood PSA levels. Similar effects were reported for other patients, but subsequent analyses revealed that PC-SPES was contaminated with synthetic drugs and this herbal remedy is no longer being sold. The lack of governmental regulations regarding the purity of herbal remedies makes it difficult to determine the effectiveness of alternative treatments like PC-SPES. [Based on data from R. S. DiPaola, *New England J. Med.* 339 (1998): 785 (Table 3).]

herbal remedies, which are not subject to the kinds of strict governmental regulations for purity, composition, and effectiveness that apply to the drugs manufactured by pharmaceutical companies.

Psychological Factors Are Not a Significant Cause of Cancer but May Influence the Course of the Disease

Cancer patients who have been treated with placebos in clinical trials sometimes exhibit improvement in symptoms such as pain and poor appetite. Since placebos contain no active ingredients, this phenomenon raises the question of the relationship between psychological factors and cancer.

Numerous investigations into the role of psychological factors have been carried out over the years, with special attention paid to stress and depression because both can trigger changes in the immune system. Early studies revealed that cancer patients are more likely than other individuals to be depressed and anxious, which was initially interpreted to mean that psychological stress can cause cancer. A more straightforward interpretation, however, is that depression and anxiety are triggered by the discovery that a person has cancer, occurring after the disease arises rather than being the underlying cause.

Large prospective studies in which psychological traits are measured in healthy individuals who are then followed into the future to see who develops cancer have generally failed to support the idea that psychological factors are a significant cause of cancer. Some studies have documented an increased rate of cancer deaths among people who have recently experienced a highly stressful or depressing event, such as the loss of a spouse or other close family member. However, cancer typically takes ten or more years to develop, so these cancer deaths are likely to involve tumors that were already growing in the body at the time of the psychological disturbance (even if the disease had not yet been diagnosed). The overall body of evidence therefore suggests that psychological factors are not a significant cause of cancer, but they may influence the course of the disease after it has already begun.

If that is true, it raises the question of whether psychological interventions might be beneficial for cancer patients. In 1989, a widely publicized study reported that women with advanced breast cancer who participated in cancer support groups lived about 18 months longer than women who did not. Although these findings were widely accepted at the time, subsequent studies have failed to confirm the conclusion that cancer patients participating in support groups live longer than nonparticipating patients. Support group participation is, however, consistently associated with improvements in patients' awareness about their illness and reductions in anxiety and distress. Whether support groups or other types of psychological intervention can extend survival for certain individuals remains an open question.

Summary of Main Concepts

- Cancer has few early symptoms that are specific for the disease, but screening techniques can detect some cancers before they have spread and thereby prevent cancer fatalities. The Pap smear, for example, has prevented thousands of deaths from cervical cancer. Other screening approaches include mammography for breast cancer; colonoscopy, X-ray imaging, and FOBT for colon cancer; the PSA test for prostate cancer; and proteomic blood tests that are currently under development. False positives, false negatives, and overdiagnosis are potential problems that can limit the effectiveness of cancer screening techniques.

- Abnormal results from cancer screening tests require a follow-up biopsy, which is the only reliable way to diagnose most cancers. If a biopsy detects a cancer before invasion and metastasis have occurred, the disease can usually be cured by surgery alone. When surgery is not sufficient or practical, radiation or chemotherapy or both are employed to attack cancer cells that remain in the body.

- Radiation therapy kills cancer cells by triggering apoptosis or mitotic death. To minimize damage to normal tissues, radiation treatments are focused on the tumor region and the total radiation dose is divided into a series of smaller dose treatments. Brachytherapy involves the use of tiny sources of radiation that can be inserted directly into or close to a tumor.

- Chemotherapy uses drugs that circulate through the bloodstream to reach cancer cells wherever they may reside. Most chemotherapeutic drugs act by disrupting DNA structure, interfering with DNA replication, or blocking cell division. The main classes of drugs that work in this way are (1) antimetabolites that block pathways required for DNA synthesis by substituting for normal molecules involved in these pathways, (2) alkylating and platinating drugs that crosslink the DNA double helix, (3) antibiotics that bind to DNA or inhibit topoisomerases required for DNA replication, and (4) plant-derived drugs that target topoisomerases or the mitotic spindle. Hormone therapy, which is a relatively nontoxic approach for treating hormone-dependent cancers, is a fifth group of drugs used for chemotherapy.

- Most chemotherapeutic drugs are toxic to normal dividing cells and their effectiveness is therefore limited by side effects such as nausea, vomiting, diarrhea, anemia, defective blood clotting, and immune deficiency. Another problem is the tendency of cancer cells to develop drug resistance, which can be created by membrane proteins that pump drugs out of cells, mutations that alter a drug's target to make it less susceptible to the drug, or disruptions in the pathways by which drug-induced damage would otherwise cause cells to self-destruct by apoptosis. Approaches for improving the effectiveness of chemotherapy include the use of multiple drugs in combination, stem cell transplants to replace the blood-forming cells that are destroyed by chemotherapy, and drug treatments that are tailored to individual patients.
- Immunotherapies exploit the ability of the immune system to recognize and attack cancer cells. BCG and cytokines are relatively nonspecific approaches that stimulate the overall activity of the immune system rather than targeting cancer cells specifically. Monoclonal antibodies are a more selective tool because they bind to and target specific cancer cell components. Vaccines and adoptive-cell-transfer therapy are being explored as possible ways of stimulating a patient's own antitumor lymphocytes.
- Molecular targeting involves the use of drugs that target specific proteins known to be required for the proliferation or spread of cancer cells. Among the first such agents to receive FDA approval were Herceptin, a monoclonal antibody that targets the ErbB2 receptor produced in elevated amounts by some breast cancers, and Gleevec, a small molecule that inhibits the Bcr-Abl tyrosine kinase associated with chronic myelogenous leukemia. Monoclonal antibodies and small molecule inhibitors that target a variety of other proteins are currently being developed and evaluated.
- Anti-angiogenic therapies, which target the blood vessels needed by growing tumors, are receiving considerable attention because of the success seen with this approach in animal studies. Although expectations regarding a quick cure for cancer were unrealistic, dozens of anti-angiogenic drugs are currently being evaluated in humans and progress is slowly being made in determining how such drugs may best be used.
- Engineered viruses are being investigated as possible tools for cancer therapy, either as vehicles for repairing defective genes or as agents that selectively infect and destroy cancer cells.
- Developing new cancer therapies requires many years of preclinical laboratory and animal testing, followed by clinical trials in cancer patients, before a new treatment is approved by the FDA for routine medical use. To eliminate possible bias or the placebo effect, human clinical testing includes randomized double-blind trials in which neither doctors nor patients know who is receiving the new treatment and who is receiving a placebo.
- Psychological factors are not a significant cause of cancer, but they may influence the course of the disease after it has begun.

Key Terms for Self-Testing

Cancer Screening and Diagnosis

Pap smear (p. 202)
biopsy (p. 202)
mammography (p. 202)
colonoscopy (p. 203)
fecal occult blood test (FOBT) (p. 203)
PSA test (p. 203)
proteomic analysis (p. 204)
false negative (p. 205)
false positive (p. 205)
overdiagnosis (p. 205)
imaging techniques (p. 206)
computed tomography scan (CT scan) (p. 206)
magnetic resonance imaging (MRI) (p. 206)
ultrasound imaging (p. 206)
basal lamina (p. 206)
carcinoma *in situ* (p. 206)
tumor staging (p. 206)
tumor grading (p. 207)

Surgery, Radiation, and Chemotherapy

mastectomy (p. 208)
radiation therapy (p. 209)
mitotic death (p. 209)
brachytherapy (p. 210)
hyperthermia (p. 210)
chemotherapy (p. 210)
antimetabolite (p. 211)
analog (p. 211)
pyrimidine (p. 212)
purine (p. 212)

alkylating agent (p. 212)
platinating agent (p. 213)
antibiotic (p. 214)
hormone therapy (p. 214)
androgen (p. 214)
androgen receptor (p. 215)
estrogen (p. 215)
tamoxifen (p. 215)
estrogen receptor (p. 215)
glucocorticoid (p. 215)
differentiating agent (p. 215)
multidrug resistance transport protein (p. 216)
cancer stem cells (p. 216)
combination chemotherapy (p. 216)
hematopoietic stem cell (p. 217)
stem cell transplantation (p. 217)
graft-versus-host disease (p. 217)

Emerging Treatments: Immunotherapy and Molecular Targeting

immunotherapy (p. 218)
Bacillus Calmette-Guérin (BCG) (p. 218)
cytokine (p. 218)
antibody (p. 218)
B lymphocyte (p. 218)
monoclonal antibody (p. 219)
cytotoxic T lymphocyte (p. 221)
vaccine (p. 222)
dendritic cell (p. 222)
adoptive-cell-transfer (ACT) therapy (p. 222)
molecular targeting (p. 223)
Herceptin (p. 223)
ErbB2 receptor (p. 223)
rational drug design (p. 223)
Gleevec (p. 223)
Iressa (p. 224)
anti-angiogenic therapy (p. 225)
Avastin (p. 225)
gene therapy (p. 226)
lysis (p. 226)

Clinical Trials and Other Approaches

preclinical testing (p. 227)
clinical trial (p. 228)
informed consent (p. 228)
randomized trial (p. 228)
double blind (p. 228)
placebo (p. 228)
placebo effect (p. 228)
experimental treatment (p. 229)
complementary treatment (p. 229)
alternative treatment (p. 229)
laetrile (p. 229)
PC-SPES (p. 229)

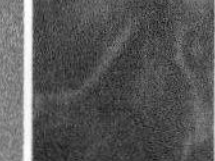

Suggested Reading

Cancer Screening and Diagnosis

Bok, R. A., and E. J. Small. Bloodborne biomolecular markers in prostate cancer development and progression. *Nature Reviews Cancer* 2 (2002): 918.

Davies, R. J., R. Miller, and N. Coleman. Colorectal cancer screening: prospects for molecular stool analysis. *Nature Reviews Cancer* 5 (2005): 199.

Etzioni, R., et al. Overdiagnosis due to prostate-specific antigen screening: Lessons from U.S. prostate cancer incidence trends. *J. Natl. Cancer. Inst.* 94 (2002): 981.

Etzioni, R., et al. The case for early detection. *Nature Reviews Cancer* 3 (2003): 243.

Hanks, G. E., and P. T. Scardino. Does screening for prostate cancer make sense? *Sci. Amer.* 275 (September 1996): 114.

Nyström, L., et al. Long-term effects of mammography screening: updated overview of the Swedish randomized trials. *Lancet* 359 (2002): 909.

Paik, S., et al. A multigene assay to predict recurrence of tamoxifen-treated, node-negative breast cancer. *New England J. Med.* 351 (2004): 2817.

Petricoin, E. F., et al. Use of proteomic patterns in serum to identify ovarian cancer. *Lancet* 359 (2002): 572.

Ransohoff, D. F. Bias as a threat to the validity of cancer molecular-marker research. *Nature Reviews Cancer* 5 (2005): 142.

Schoen, R. E. The case for population-based screening for colorectal cancer. *Nature Reviews Cancer* 2 (2002): 65.

Traverso, B. A., et al. Detection of *APC* mutations in fecal DNA from patients with colorectal tumors. *New England J. Med.* 346 (2002): 311.

Wulfkuhle, J. D., L. A. Liotta, and E. F. Petricoin. Proteomic applications for the early detection of cancer. *Nature Reviews Cancer* 3 (2003): 267.

Surgery, Radiation, and Chemotherapy

Barker, J. N., and J. E. Wagner. Umbilical-cord blood transplantation for the treatment of cancer. *Nature Reviews Cancer* 3 (2003): 526.

Bernier, J., E. J. Hall, and A. Giaccia. Radiation oncology: a century of achievements. *Nature Reviews Cancer* 4 (2004): 737.

Brown, J. M., and L. D. Attardi. The role of apoptosis in cancer development and treatment response. *Nature Reviews Cancer* 5 (2005): 231.

Chabner, B. A., and T. G. Roberts Jr. Chemotherapy and the war on cancer. *Nature Reviews Cancer* 5 (2005): 65.

Dean, M., T. Fojo, and S. Bates. Tumour stem cells and drug resistance. *Nature Reviews Cancer* 5 (2005): 275.

Gottesman, M. M. Mechanisms of cancer drug resistance. *Annu. Rev. Med.* 53 (2002): 615.

Grem, J. L., and B. Keith. Mechanisms of Action of Cancer Chemotherapeutic Agents: Antimetabolites. In *The Cancer Handbook* (M. Alison, ed.). London: Nature Publishing Group, 2002, Chapter 84A.

Gudkov, A. V., and E. A. Komarova. The role of p53 in determining sensitivity to radiotherapy. *Nature Reviews Cancer* 3 (2003): 117.

Johnstone, R. W., A. A. Ruefli, and S. W. Lowe. Apoptosis: A link between cancer genetics and chemotherapy. *Cell* 108 (2002): 153.

Jordan, M. A., and L. Wilson. Microtubules as a target for anticancer drugs. *Nature Reviews Cancer* 4 (2004): 253.

Jordan, V. C. Tamoxifen: A most unlikely pioneering medicine. *Nature Reviews Drug Discovery* 2 (2003): 205.

Lichter, A. S. Radiation Therapy. In *Clinical Oncology* (M. D. Abeloff, ed.). New York: Churchill Livingstone, 2000, Chapter 18.

McBride, W. H., G. J. Dougherty, and L. Milas. Molecular Mechanisms of Radiotherapy. In *The Cancer Handbook* (M. Alison, ed.). London: Nature Publishing Group, 2002, Chapter 86.

Minna, J. D., et al. A bull's eye for targeted lung cancer therapy. *Science* 304 (2004): 1458.

Moses, M. A., H. Brem, and R. Langer. Advancing the field of drug delivery: Taking aim at cancer. *Cancer Cell* 4 (2003): 337.

Neidle, S., and D. E. Thurston. Chemical approaches to the discovery and development of cancer therapies. *Nature Reviews Cancer* 5 (2005): 285.

Niederhuber, J. E. Surgical Therapy. In *Clinical Oncology* (M. D. Abeloff, ed.). New York: Churchill Livingstone, 2000, Chapter 19.

Siddik, Z. H. Mechanisms of Action of Cancer Chemotherapeutic Agents: DNA-interactive Alkylating Agents and Antitumour Platinum-based Drugs. In *The Cancer Handbook* (M. Alison, ed.). London: Nature Publishing Group, 2002, Chapter 84B.

Sparreboom, A., K. Nooter, and J. Verweij. Mechanisms of Action of Cancer Chemotherapeutic Agents: Antitumor Antibiotics. In *The Cancer Handbook* (M. Alison, ed.). London: Nature Publishing Group, 2002, Chapter 84E.

Emerging Treatments: Immunotherapy and Molecular Targeting

Allen, T. M. Ligand-targeted therapeutics in anticancer therapy. *Nature Reviews Cancer* 2 (2002): 750.

Banchereau, J., and A. K. Palucka. Dendritic cells as therapeutic vaccines against cancer. *Nature Reviews Immunology* 5 (2005): 296.

Blattman, J. N., and P. D. Greenberg. Cancer immunotherapy: A treatment for the masses. *Science* 305 (2004): 200.

Capdeville, R., et al. Glivec (STI571, imatinib), a rationally developed, targeted anticancer drug. *Nature Reviews Drug Discovery* 1 (2002): 493.

Carter, P. Improving the efficacy of antibody-based cancer therapies. *Nature Reviews Cancer* 1 (2001): 118.

Dudley, M. E., and S. A. Rosenberg. Adoptive-cell-transfer therapy for the treatment of patients with cancer. *Nature Reviews Cancer* 3 (2003): 666.

Dudley, M. E., et al. Cancer regression and autoimmunity in patients after clonal repopulation with antitumor lymphocytes. *Science* 298 (2002): 850.

Gilboa, E. The promise of cancer vaccines. *Nature Reviews Cancer* 4 (2004): 401.

Klein, S., F. McCormick, and A. Levitzki. Killing time for cancer cells. *Nature Reviews Cancer* 5 (2005): 573.

Lake, R. A., and B. W. S. Robinson. Immunotherapy and chemotherapy—a practical partnership. *Nature Reviews Cancer* 5 (2005): 397.

Nettelbeck, D. M., and D. T. Curiel. Tumor-busting viruses. *Sci. Amer.* 289 (October 2003): 136.

O'Hare, T., et al. AMN107: Tightening the grip of imatinib. *Cancer Cell* 7 (2005): 117.

Zou, W. Immunosuppressive networks in the tumour environment and their therapeutic relevance. *Nature Reviews Cancer* 5 (2005): 263.

Clinical Trials and Other Approaches

Chvetzoff, G., and I. F. Tannock. Placebo effects in oncology. *J. Natl. Cancer Inst.* 95 (2003): 19.

Goodwin, P. J., et al. The effect of group psychosocial support on survival in metastatic breast cancer. *New England J. Med.* 345 (2001): 1719.

Nelson, P. S., and B. Montgomery. Unconventional therapy for prostate cancer: good, bad or questionable? *Nature Reviews Cancer* 3 (2003): 845.

Richardson, M. A. Complementary/alternative medicine use in a comprehensive cancer center and the implications for oncology. *J. Clin. Oncol.* 18 (2000): 2505.

Sampson, W. Controversies in cancer and the mind: effects of psychosocial support. *Seminars Oncology* 29 (2002): 595.

White, J. PC-SPES—A lesson for future dietary supplement research. *J. Natl. Cancer Inst.* 94 (2002): 1261.

Zivin, J. A. Understanding clinical trials. *Sci. Amer.* 282 (April 2000): 69.

12 Preventing Cancer

Cancer research is motivated to a large extent by the belief that a better understanding of the disease will lead to better approaches for treatment and prevention. For example, we just saw in Chapter 11 how our increasing knowledge of the role played by specific genes and proteins in cancer development has prompted a variety of new treatment strategies. Likewise, knowing the causes of cancer provides crucial information as to how cancer can be prevented, and as the old saying goes, "An ounce of prevention is worth a pound of cure."

One approach for preventing cancer is based on the discovery that many types of cancer are caused by known environmental agents and behaviors; avoiding these causes of cancer will therefore decrease a person's risk of developing the disease. A second approach for preventing cancer is based on the realization that cancers arise through a multistep process that usually unfolds over a period of several decades rather than occurring as a single discrete event. So even if a person has already been exposed to agents that "cause" cancer, there are opportunities to intervene and block one of the subsequent steps required for the progression to malignancy. Current statistics suggest that if people adopt both prevention approaches—that is, reduce their exposure to cancer-causing agents and take steps that help protect against cancer development after such agents have been encountered—more than half of all cancer deaths could be prevented. ■

AVOIDING THE CAUSES OF CANCER

When people talk about cancer prevention, they are really referring to steps a person can take to minimize their *risk* of developing the disease. The most obvious actions that can be taken are those that reduce exposure to the known causes of cancer. In the first part of this chapter, we will examine those agents whose avoidance would have the greatest impact in decreasing cancer incidence and deaths.

Not Smoking Is the Most Effective Way to Reduce the Risk of Developing a Fatal Cancer

Cigarette smoking is the number one cause of preventable cancer deaths. Tobacco smoke contains dozens of carcinogenic chemicals (see Table 4-2), and the evidence implicating these chemicals in the causation of cancer was thoroughly discussed in Chapter 4. If people did not smoke cigarettes, roughly one of every three cancer deaths would be prevented.

Progress has been made over the past several decades in lowering smoking rates in the United States, but these gains have been offset by a growing epidemic of cigarette smoking elsewhere in the world. Even in the United States, close to 50 million people smoke cigarettes and smoking is the leading cause of preventable death, killing more people than AIDS, car accidents, murder, alcohol, illegal drugs, and suicides combined. Figure 12-1 illustrates the trends in smoking rates in the United States since 1900. In the first half of the twentieth century, cigarette consumption increased from a yearly average of several dozen cigarettes per person in 1900 to more than 4000 per person in 1963. Then in 1964, the U.S. government published the first *Surgeon General's Report on Smoking and Health*. Based on an analysis of more than 7000 scientific investigations involving a variety of experimental and epidemiological approaches, the report concluded that cigarette smoking causes lung cancer. At that point, smoking rates in the United States stopped growing and slowly began to decline.

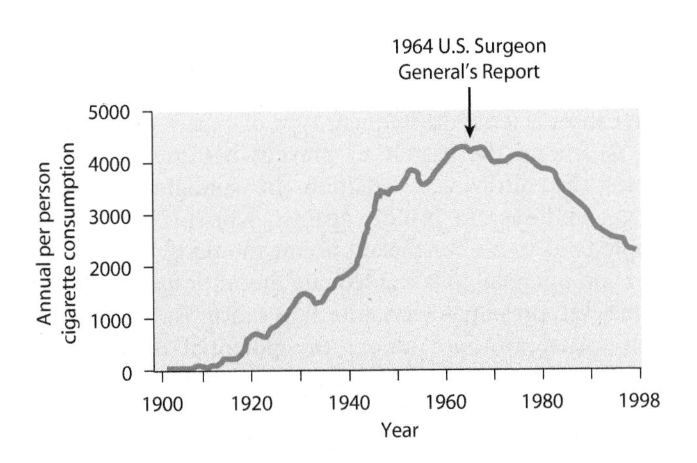

Figure 12-1 Cigarette Smoking Trends in the United States. During the first half of the twentieth century, cigarette smoking increased from an average of several dozen cigarettes per person in 1900 to more than 4000 per person in 1963. After publication of the *Surgeon General's Report on Smoking and Health* in 1964, cigarette use began a gradual decline. [Based on S. S. Hecht in *The Cancer Handbook* (M. Alison, ed., London: Nature Publishing Group, 2002), Chapter 29 (Figure 2).]

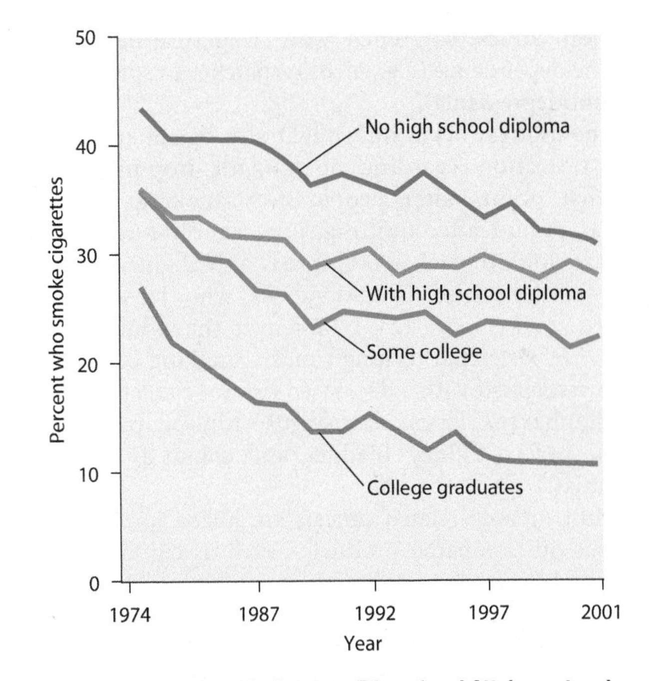

Figure 12-2 Relationship Between Educational Attainment and Smoking Behavior. This graph shows the percentage of adults over age 25 who smoke cigarettes in the United States as a function of their overall level of education. Smoking rates among individuals who never graduated from high school are roughly three times higher than smoking rates seen in those who graduated from college. [Data from *Cancer Prevention & Early Detection Facts & Figures 2004* (Atlanta, GA: American Cancer Society, 2004), p. 12.]

Improved education has played an important role in stimulating this decline. In general, smoking behavior is inversely related to a person's overall level of education: Statistics show that roughly 30% of those without a high school diploma smoke cigarettes compared with about 10% of college graduates (Figure 12-2). A variety of governmental actions have also contributed to the decline in smoking rates. Included in this category are laws requiring warning labels on cigarette packages, which inform people about the health hazards of tobacco, and high taxation of tobacco products and restrictions on smoking in public places, both of which tend to discourage smoking behaviors.

One obstacle to further progress is the difficulty that people encounter when they try to quit smoking. About half of all adult smokers express a desire to stop smoking and make at least one attempt to quit annually, yet fewer than 15% can refrain from smoking for more than 30 days. The addictive power of the nicotine in tobacco is clearly a major impediment. Although some people overcome their nicotine addiction and quit smoking through sheer willpower, success rates are improved by supportive social arrangements, such as smoking cessation clinics, combined with drug treatments for nicotine dependence (e.g., nicotine patches or sprays and even antidepressants).

The motivation to quit might also be enhanced by better education regarding the dramatic drop in cancer risk that occurs after people stop smoking. In the 20-year period after quitting, lung cancer rates drop about tenfold to reach a level that is only slightly higher than that observed in individuals who have never smoked at all (Figure 12-3). Moreover, the reduction in risk is not restricted to lung cancer. Smoking cessation is also associated with a decreased risk for cancers of the mouth, pharynx, larynx, esophagus, stomach, pancreas, uterine cervix, kidney, bladder, and colon, as well as leukemias.

Most tobacco-related cancers are linked to cigarette use, but other tobacco products—such as cigars, pipes, and smokeless tobacco—also cause cancer. Smoking cigars or a pipe triggers most of the same cancers as cigarettes, although the risk of lung cancer is less than with cigarettes because cigar or pipe smoke is not inhaled as deeply into the lungs. Secondhand smoke can also cause cancer, although the risk is small compared with that associated with the direct use of tobacco products. Smoke is not even required for tobacco to exert its carcinogenic effects. Smokeless tobacco, which is chewed rather than smoked, causes numerous cancers of the mouth and throat, as well as some in the esophagus, stomach, and pancreas.

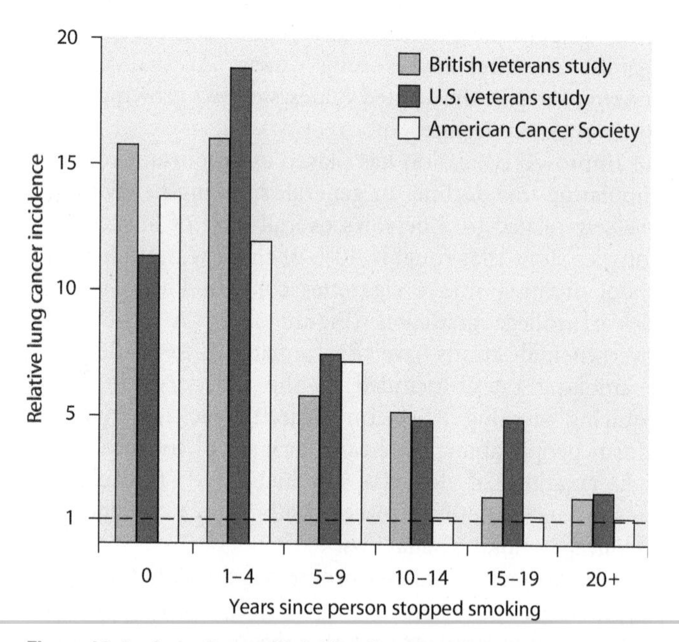

Figure 12-3 Lung Cancer Risk After Smoking Cessation. The data in this histogram come from three independent studies of lung cancer rates in people who have stopped smoking. Lung cancer rates are expressed as the ratio of cancer rates in ex-smokers to cancer rates in lifelong nonsmokers. The dashed line represents a value of "1", which would indicate no increase in cancer risk relative to that of lifelong nonsmokers. [Based on data from S. S. Hecht in *The Cancer Handbook* (M. Alison, ed., London: Nature Publishing Group, 2002), Chapter 29 (Table 5).]

Alcohol Consumption Is Associated with Cancer Risks, but Also Cardiovascular Benefits When Used in Moderate Amounts

It was pointed out in Chapter 4 that alcohol consumption is associated with an increased risk for several types of cancer, including those of the mouth, pharynx, larynx, esophagus, stomach, and liver. Alcohol also interacts synergistically with tobacco to create cancer risks that are significantly greater than the sum of the effects produced by each acting alone (see Figure 4-10). Limiting one's consumption of alcoholic beverages can therefore be an effective way of decreasing cancer risk.

Recommendations concerning alcohol consumption are complicated by the fact that unlike tobacco, which has no known health benefits, drinking moderate amounts of alcohol has been linked to about a 25% reduction in the risk of coronary heart disease. The observation that the incidence of heart disease is relatively low in France despite high rates of saturated fat intake and smoking has led to the claim that red wine is particularly beneficial. However, epidemiological studies have failed to show a reliable difference in the heart benefits of drinking wine versus any other kind of alcohol.

Although alcohol's beneficial effects on the heart have been fairly well established, these benefits are only seen at moderate levels of drinking—that is, no more than one to two drinks per day. Alcohol consumption in excess of these levels has no discernable health benefits and creates a variety of health hazards, including an increased risk of cancer. Anyone considering modest levels of alcohol consumption to exploit its beneficial effects on the heart might want to consider their own medical and family history to determine whether they have any predispositions to cardiovascular disease (where alcohol may be helpful) or cancer (where alcohol is more likely to be harmful).

Protecting Against the Ultraviolet Radiation in Sunlight Reduces the Risk of Skin Cancer

Skin cancer is the most frequent type of cancer worldwide but is among the easiest to prevent because its main cause—the ultraviolet radiation in sunlight—is well known and easy to protect against. Nonetheless, many people tend to be nonchalant about the dangers of sunlight and often fail to take adequate precautions to protect themselves, presumably because skin cancer is rarely fatal. Such casual attitudes ignore the potential dangers of melanoma, a type of tumor whose tendency to metastasize makes it the most lethal form of skin cancer. Melanomas represent only 5% of all skin cancers, but the 5% figure corresponds to enough cancer cases to place melanoma among the ten most common types of cancer.

Surveys have revealed that only one-third of Americans routinely take the three precautions that are most effective in preventing skin cancer: (1) restricting exposure to the sun, especially during midday when its ultraviolet radiation is the strongest; (2) wearing protective clothing; and (3) using sufficient amounts of sunscreen lotion. As described in Chapter 6, modern sunscreen lotions typically contain a mixture of ingredients for blocking and absorbing both UVB and UVA radiation (p. 109). The strength of any given sunscreen is expressed by its **SPF** (**sun-protection factor**), a number that reflects how much time it takes for skin treated with sunscreen to burn compared with unprotected skin. In other words, an SPF of 15 means that skin covered with sunscreen lotion will take 15 times longer to burn than unprotected skin. If it would normally take your skin 20 minutes to burn under a given set of sunlight conditions, proper use of a sunscreen lotion with an SPF of 15 would allow an exposure of $20 \times 15 = 300$ minutes (5 hours) before burning. However, people typically apply sunscreen in amounts that achieve only about half of the protection suggested by the SPF value. Thus in actual practice, a sunscreen labeled with an SPF of 30 needs to be used to achieve an SPF of 15.

It has been estimated that worldwide death rates from melanoma could be reduced by at least 50% if people restrict their exposure to intense sunlight, wear protective clothing, and properly use sunscreens that achieve an SPF of at least 15. In Australia, which has the highest incidence and death rates for skin cancer in the world, significant public education efforts have been made to try to influence sun-related attitudes and behaviors. Three years after the launching of one prevention program in the Australian state of Victoria, the rate of sunburn dropped by 35% and people reported an increase in the wearing of hats and the use of sunscreen. Such results indicate that exposure to melanoma risk factors can be reduced fairly rapidly in response to health promotion campaigns.

Cancer Risks Created by Ionizing Radiation Tend to Be Small for Most Individuals

We saw in Chapter 6 that *ionizing radiation* is a high-energy type of radiation that removes electrons from molecules, thereby generating reactive ions that trigger DNA damage. Although large-dose exposures to ionizing radiation can cause many types of cancer, the doses encountered by the average person tend to be relatively small and account for less than 5% of all cancers.

Most of the ionizing radiation encountered on a regular basis—other than from the radioactive polonium in tobacco smoke—comes from natural background sources that cannot be readily avoided (see Figure 6-12). The largest contributor to background radiation is *radon*, a radioactive gas emitted from underground rock formations in the earth's crust. Increased rates of lung cancer have been detected in underground mine workers exposed to high concentrations of radon for long periods of time, but comparable exposures are rarely experienced by the general public. Nevertheless, radon gas can seep into homes and accumulate in significant amounts if the ventilation is inadequate, especially in regions of the country where large mounts of radon are emitted from the earth's surface. It is therefore advisable to test for radon levels in the home and, if they are exceptionally high, install an improved ventilation system in the basement to minimize the amount of radon that accumulates. Inexpensive radon testing kits suitable for this purpose are readily available at hardware stores or through public health agencies.

Natural background radiation accounts for about 80% of the ionizing radiation encountered by the average person. Of the remaining 20%, medical X-rays make the largest contribution. The health benefits of medical X-rays almost always outweigh the small risks involved, as long as X-rays are used only when a clear medical necessity exists. A striking example of the dangers associated with unnecessary medical X-rays occurred in the mid-1900s, when X-ray treatments were employed for clearing up facial acne in adolescent children. Individuals who received these treatments later developed thyroid cancer at significantly elevated rates. Caution is also appropriate when using a high-dose X-ray procedure known as a *computed tomography scan* (*CT scan*) with young children. CT scans deliver up to 100 times more radiation than standard medical X-rays, and growing children are more sensitive than adults to the hazards of ionizing radiation.

In general, the average citizen tends to overestimate the risks posed by ionizing radiation. For comparison purposes, it is worth pointing out that among the Japanese residents of Hiroshima and Nagasaki who survived the immediate effects of the atomic explosions in 1945, only about 1% have died of radiation-induced cancers despite receiving radiation doses that were enormously greater than most people will encounter in their entire lifetimes. Another interesting comparison involves perceptions regarding the dangers of the radiation associated with nuclear power. Living near a nuclear power plant normally exposes people to a radiation dose that is less than 1% of the background radiation we all receive from natural sources. Yet when individuals are asked to rank their perceived risk of dying from various activities, nuclear power tends to be ranked ahead of much riskier activities, such as smoking cigarettes. In reality, the risk of dying from a radiation-induced cancer is quite small compared to most of the other risks associated with everyday life (see Table 6-4).

Avoiding Exposure to Certain Infectious Agents Can Reduce Risks for Several Types of Cancer

Although cancer is not usually perceived by the general public as being an infectious disease, we saw in Chapter 7 that roughly 15% of all cancers worldwide are linked to viral, bacterial, or parasitic infections. People can therefore lower their risk of developing cancer by avoiding behaviors that expose them to the relevant infectious agents.

Human papillomavirus (HPV) is responsible for the largest number of virally linked cancers, mainly cervical cancer and, to a lesser extent, cancer of the penis. HPV is spread predominantly through sexual activity, so the risk of becoming infected can be reduced by avoiding sexual contact with multiple partners. Using a condom also decreases the risk of HPV-induced cancers, but condoms are not fully protective because they fail to prevent the spread of HPV caused by contact with infected areas not covered by the condom.

The hepatitis B and hepatitis C viruses are responsible for most cases of liver cancer, a common form of cancer in Southeast Asia, China, and Africa. The hepatitis B virus is transmitted by exchange of bodily fluids, such as blood or semen, and can be easily spread through sexual activity. The proper use of latex condoms may help reduce the transmission of hepatitis B, although the efficacy of the protection is unknown. The hepatitis C virus is difficult to transmit by mechanisms other than direct contact with contaminated blood, so the main route for acquiring hepatitis C is through the sharing of dirty needles by intravenous drug users. Hepatitis viruses can also be transmitted by blood transfusions involving contaminated blood, but the blood supply used for medical purposes is now routinely screened for both the hepatitis B and hepatitis C viruses.

The only bacterium to be clearly linked to a human cancer is *H. pylori*, the main cause of stomach cancer. Roughly half the world's population has been infected with *H. pylori*, and stomach cancer is therefore one of the top cancer killers globally. Unlike the situation in many countries, the prevalence of stomach cancer in the United States has declined substantially over the past several decades. One factor contributing to this marked decline is the widespread use of antibiotics, which has diminished the rate of *H. pylori* infections and thereby reduced the chances that people will come in contact with infected individuals. Public health conditions also play an important role because transmission of *H. pylori* from person to person is facilitated by poor sanitation and crowded living environments, which are less prevalent in the United States than in most developing countries (Figure 12-4).

Minimizing Exposure to Cancer-Causing Chemicals and Drugs Can Reduce Cancer Risks

More than two hundred chemical substances are known to cause—or are reasonably anticipated to cause—cancer in humans (see Appendix B). Chapter 4 described how these carcinogens have been identified through the combined use of epidemiological evidence, animal testing, and the Ames test. Chapter 5 then discussed the types of chemicals involved and the mechanisms by which they cause cancer.

In practice, many of the chemicals that pose a significant cancer risk for humans were not identified until large numbers of cancers had arisen in workers exposed to these substances on a regular basis, especially in the rubber, chemical, plastic, mining, fuel, and dye industries. Although some of the responsible chemicals have now been banned from the workplace, others remain in use. Workers must therefore take appropriate precautions, such as wearing masks and protective clothing, working in properly ventilated environments, and in some cases breathing through a respirator. As a result of the protective measures that have been implemented, many of the occupational cancers that were once prevalent in the United States have declined in frequency, and workplace exposure to carcinogenic chemicals now accounts for less than 5% of all fatal cancers.

Chemical carcinogens are usually encountered in high concentration only in industrial workplaces or in tobacco smoke, but small amounts escape into the environment and contaminate the air we breathe, the water we drink, and the food we eat. It has therefore become fashionable to blame chemical pollution of the environment for creating a cancer epidemic. In fact, the data do not generally support this conclusion (p. 83). Those chemical carcinogens that do pollute the environment tend to be present in concentrations that are thousands of times lower than typical workplace exposures, and most evidence suggests that the cancer threat they pose for the average person is quite small.

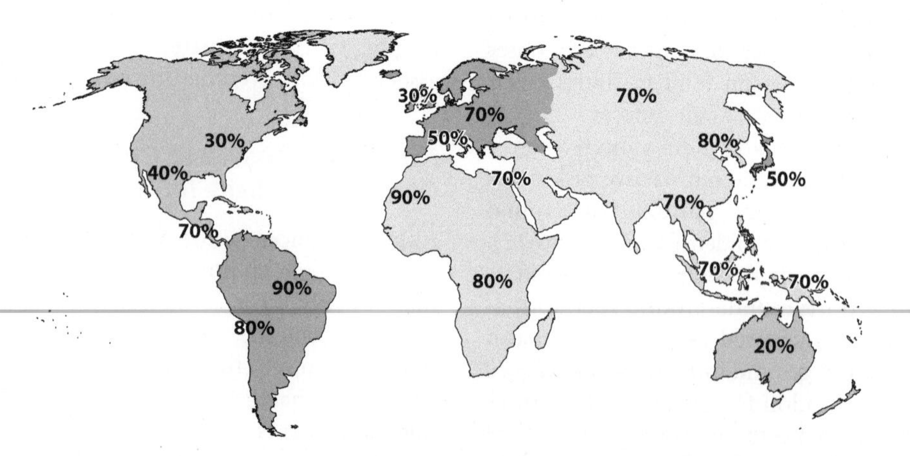

Figure 12-4 Worldwide Distribution of *H. pylori* Infections. Infections with *H. pylori*, which are linked to stomach cancer, occur most frequently in those areas of the world that have lower standards of living and where poor sanitation and crowded living conditions are more common. [Data from The Helicobacter Foundation (http://www.helico.com).]

Of course, not everyone is an "average person." For example, people who live near a site where toxic chemicals have been stored or released may be exposed to carcinogen concentrations that are thousands of times higher than those encountered by the general public. Similarly, people who live in older buildings constructed with asbestos-containing materials may have higher-than-normal exposures to asbestos, especially if the construction materials are deteriorating. Each person therefore needs to assess his or her own living situation to determine whether it involves any unusual contact with carcinogenic chemicals.

People also encounter high concentrations of specific chemicals when they take prescription drugs to treat chronic health problems. If a single drug is used for a protracted period of time, exposure to the substance in question will be high and it is important to know whether any cancer risks are involved. Table 5-2 lists a number of prescription drugs that are already known to cause cancer. Because of the cancer hazard, most of these drugs are only prescribed for serious medical conditions in which the potential benefits of the drug in question far outweigh the risk that cancer might arise.

One problem in assessing the cancer hazards associated with prescription drugs is that many years of human use may be required before the risks become apparent. Chapter 5 described such a situation with diethylstilbestrol (DES), a synthetic estrogen that was prescribed to pregnant women for 30 years before its carcinogenic properties were discovered (p. 88). A more recent illustration is provided by **hormone replacement therapy** (**HRT**), a treatment sometimes prescribed for postmenopausal women because their lowered estrogen production can lead to several health problems, including increased risks for heart disease and osteoporosis (bone loss). In the 1950s and 1960s, doctors routinely prescribed estrogen pills to postmenopausal women to prevent these problems. Then it was reported in the mid-1970s that estrogen administration creates an increased risk for uterine (endometrial) cancer, and its use began to decline. When subsequent studies showed that the uterine cancer risk can be reduced or eliminated by combining estrogen with a *progestin* (a substance exhibiting progesterone-like activity), estrogen-progestin became the predominant form of HRT.

Although this newer type of HRT was initially thought to be safe, a study of 16,000 postmenopausal women sponsored by the Women's Health Initiative was prematurely halted in 2002 when results began to suggest that the harmful effects associated with estrogen-progestin treatment outweigh the potential benefits. One of the main problems to be reported was an increased incidence of breast cancer, but elevated rates for strokes, heart attacks, and blood clots were noticed as well. While the study also revealed that estrogen-progestin treatment lowers the risk of osteoporosis and colon cancer, these benefits were not considered to be sufficient to outweigh the other risks (Figure 12-5). In 2003, a Swedish trial of HRT in postmenopausal women with a previous history of breast cancer was also halted, in this case because of an unacceptably high recurrence rate for breast cancer.

Reduced Consumption of Saturated Fat, Red Meat, Total Calories, and Dietary Carcinogens May Decrease Cancer Risk

Of the thousands of chemicals we routinely encounter on a daily basis, most that enter our bodies are deliberately ingested because they are natural components of food. People who differ in the foods they eat also tend to differ in the cancers they develop, suggesting that diet has an important influence on cancer risks. Unfortunately, the effects of individual foods are difficult to pinpoint because the eating habits of different people vary in numerous ways, making it hard to assess the impact of each ingredient. The situation is further complicated by the fact that in any given diet, the carcinogenic effects of a substance found in one type of food may be blocked by the protective action of another substance present in the same diet. In other words, cancer risk is related both to foods that increase cancer risk (whose intake should be reduced) and to foods that decrease cancer risk (whose intake should be increased).

It has been widely recommended that one component of the diet whose consumption should be reduced is animal

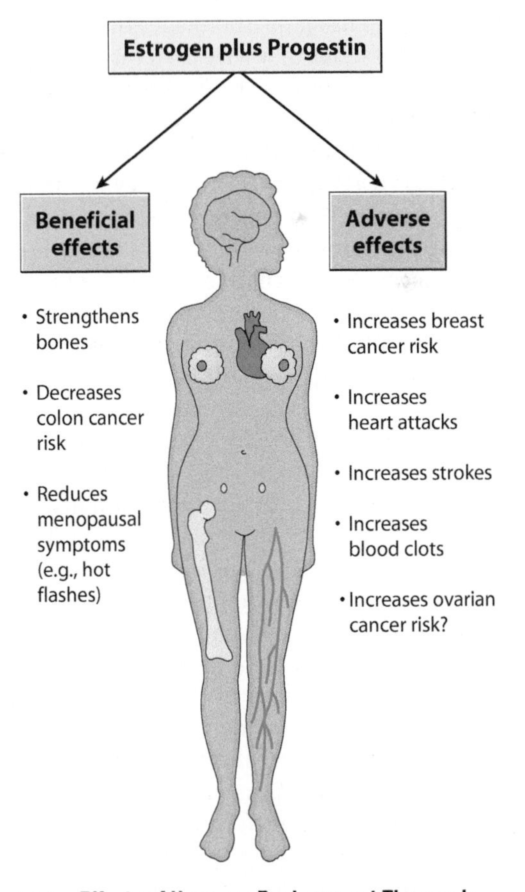

Figure 12-5 Effects of Hormone Replacement Therapy in Postmenopausal Women. Large prospective studies suggest that the harmful effects associated with estrogen-progestin treatment outweigh the potential benefits.

fat. Unlike the fats found in fruits and vegetables, which are largely unsaturated, animal fats are rich in saturated fat. (The term *saturated* refers to a fat molecule in which all carbon atoms are bound to the maximum number of hydrogen atoms.) Diets rich in saturated fat in general, and red meat in particular, have been linked in numerous epidemiological studies to an elevated risk for several types of cancer, including colon, prostate, and breast cancers.

The main problem with this evidence is that most of the early epidemiological studies involved a retrospective approach in which people who had already developed cancer were asked to recall their dietary habits from many years earlier. Because retrospective studies are susceptible to numerous sources of bias, several recent studies have used the more powerful prospective approach in which large groups of individuals without cancer are monitored into the future to see who develops the disease. Participants are repeatedly asked about their eating habits along the way, which provides more reliable information than can be obtained by asking people for their recollections many years later. Large studies of this type have continued to support the conclusion that eating red meat is a risk factor for colon cancer, although the relationship between saturated fat intake and cancer risk is less certain. Part of the reason for this uncertainty may be that even large prospective studies are hampered by imprecise methods for assessing what people really eat. (See page 75 for a discussion of the inaccuracies inherent to using questionnaires that ask people to report the kinds of foods they have been eating.)

The apparent linkage seen in some studies between saturated fat intake and cancer risk may arise in part because fats have a higher calorie content than other nutrients. Two reasons exist for believing that extra calories are a cancer risk. First, animal studies have shown that reducing the number of calories in the diet leads to a decrease in cancer rates, both for spontaneous tumors and for tumors induced by carcinogenic chemicals or radiation (Figure 12-6). Second, people who eat high-calorie diets are in danger of becoming overweight, especially if they do not exercise regularly, and we will see later in the chapter that obesity is a risk factor for many types of cancer.

In addition to reducing red meat, saturated fat, and calorie intake, several other approaches exist for decreasing cancer hazards in the diet. One is to reduce the use of high-temperature cooking techniques, such as open flame grilling or deep frying, which generate carcinogenic polycyclic aromatic hydrocarbons and aromatic amines that were not initially present in the foods being cooked (p. 75). Food preservation techniques are another potential source of dietary carcinogens that can be avoided. For example, decreased consumption of foods that are smoked, cured, pickled, or heavily salted has been linked to a lower risk for stomach cancer. Taking simple measures to reduce pesticide contamination, such as washing or peeling fruits and vegetables, is also a reasonable precaution, although the potential hazards of pesticide contamination are probably overemphasized (p. 73).

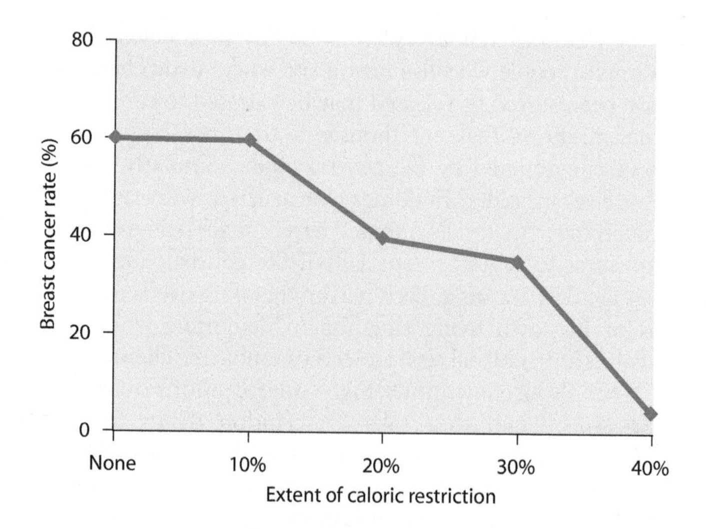

Figure 12-6 Influence of Calorie Intake on Cancer Risk. Rats were fed diets in which the calorie content was restricted to varying degrees relative to what would have been ingested if food were made freely available in unlimited amounts. The carcinogen dimethylbenz[*a*]anthracene (DMBA) was administered to all animals and breast cancer rates were then measured. Note the dramatic drop in cancer rates observed in animals whose calorie intake was restricted. Reduced cancer incidence after caloric restriction has also been observed for spontaneous tumors. [Based on data from D. Kritchevsky, *Toxicological Sci.* 52 Suppl. (1999): 13 (Table 4).]

Reducing Exposure to Carcinogens Involves a Combination of Personal and Governmental Actions

There are two basic ways in which exposure to carcinogens can be decreased. One is for individuals to take personal actions like those we have been discussing—that is, reducing exposure to tobacco smoke, alcohol, sunlight, ionizing radiation, cancer-causing infectious agents, and chemical carcinogens encountered in the workplace, drugs, and the diet. (Of course, people need to be educated about these dangers before they can be expected to take action.) The second approach is for the government to create laws and regulations that reduce or eliminate the carcinogenic hazards to which people are exposed.

In practice, the two approaches tend to complement each other. A good illustration of this principle is provided by the history of cigarette smoking in the United States. As we saw earlier in the chapter, smoking rates increased dramatically in the twentieth century until 1964, the year in which the first *Surgeon General's Report on Smoking and Health* was published (see Figure 12-1). This report concluded that cigarette smoking causes lung cancer and stated, "Cigarette smoking is a health hazard of sufficient importance in the United States to warrant appropriate remedial action." As a consequence, Congress passed the Federal Cigarette Labeling and Advertising Act of 1965 and the Public Health Cigarette Smoking Act of 1969. These laws required health warnings on cigarette packages, restricted cigarette advertising, and called for periodic government reports on the health consequences of smoking. Over the ensuing decades a growing body of federal and local regulations imposed further controls, including restrictions on smoking in public places, airplanes, office buildings, restaurants, and the workplace.

Despite the severe health hazards involved, there has been no attempt to ban cigarettes outright, an approach that would clearly be impractical. Government and public health organizations have instead worked to educate people about the dangers of smoking and to promote a nonsmoking cultural environment that encourages individuals to make their own personal decision not to smoke. This strategy has been fairly successful: Smoking rates have gradually declined over the past 40 years and people are giving up smoking in increasing numbers. In fact, nearly half of all living American adults who ever smoked have now quit, and smoking rates among men have fallen by roughly 50% since 1964.

OSHA and the FDA Are Governmental Agencies That Regulate Carcinogens in the Workplace and Food Supply

Government involvement in protecting the public from carcinogenic hazards is not restricted to tobacco products. Another area of interest is the industrial workplace, where employees may be exposed to high doses of chemical carcinogens for prolonged periods of time. In the first half of the twentieth century, many cancers were caused by such occupational exposures before the risks were fully appreciated. Finally, in 1970 an act of Congress created the **Occupational Safety and Health Administration** (**OSHA**) to formulate and enforce regulations that protect the safety and health of workers. OSHA regulations have eliminated some of the most dangerous carcinogens from the workplace and have created requirements that limit worker exposure to others. As a consequence, occupational cancers that were once prevalent in the United States are declining in frequency and now account for less than 5% of all cancer deaths each year.

Regulation of carcinogens in the food supply has had a more complicated history. In the United States, food additives and contaminants that might cause cancer are regulated by the **Food and Drug Administration** (**FDA**), a federal agency whose jurisdiction includes most foods other than meat. A law passed in 1938 empowered the FDA to prohibit the marketing of any food containing a substance that may render it injurious to health. In 1958, Congress passed the **Delaney Amendment**, which specifically directed the FDA to ban any food additive or contaminant if it was found to cause cancer in animals at any dose. The Delaney Amendment, which expired in 1996, was referred to as a *zero-risk* standard because it allowed for no acceptance of even the tiniest amount of an additive or contaminant in food once it has been shown to exhibit any carcinogenic properties in animals.

Using the authority provided by the preceding laws, the FDA has banned a number of food additives and contaminants suspected of being human carcinogens. Among the prohibited substances are several food-coloring agents (e.g., *Red No. 2* and *Red No. 4*), artificial sweeteners (e.g., *cyclamate*), and packaging materials that contaminate food (e.g., *acrylonitrile* used in making plastic beverage bottles). When the FDA attempted to ban the artificial sweetener *saccharin*, however, the difficulties inherent to its approach for assessing carcinogenic hazards became evident.

Saccharin is an artificial sweetener with a long history of use in foods and beverages. After several studies showed that saccharin causes bladder cancer in laboratory animals, the FDA proposed in 1977 to prohibit its use based on the requirements of the Delaney Amendment. At that time, saccharin was widely used in diet foods—millions of pounds were consumed annually in soft drinks alone—and a public outcry ensued. In response, Congress passed a law overriding the FDA ban and saccharin remained legal. However, the law did require a warning label on saccharin-containing foods and beverages indicating that saccharin causes cancer in animals.

In 1981, the government used this animal evidence as justification for placing saccharin on its list of substances that are "reasonably anticipated" to be human carcinogens. Twenty years later, the government reversed its position and saccharin was removed from the list. The reason for the reversal was a series of studies showing that the original animal data, derived mainly from rats, were not likely to be relevant to humans. The newer experiments revealed that saccharin causes bladder cancer in rats by binding to proteins in the urine, thereby triggering the formation of calcium phosphate-containing crystals that irritate the bladder wall and lead to cancer. Formation of these crystals requires the presence of high concentrations of calcium phosphate and protein in the urine, conditions unique to the rat bladder that do not exist in humans.

Regulatory Standards for Environmental Pesticides and Pollutants Are Established Using Quantitative Assessments of Risks and Benefits

The history of government efforts to regulate saccharin shows how difficult it can be to make decisions about food additives or contaminants that act as weak carcinogens in animals but may or may not cause cancer in humans. In the case of pesticides that contaminate the food supply, the situation is further complicated by the fact that different standards have been applied to raw and processed foods. For many years pesticide contaminants detected in processed foods were considered to be additives covered by the Delaney Amendment, which imposed a standard of **zero tolerance** for any substance that causes cancer in animals. If the same pesticide happened to be detected in raw foods, the risk was analyzed quantitatively based on all available animal and human evidence so that an **acceptable tolerance** level could be established based on typical dietary intake.

Because of the confusion created by this inconsistent approach to pesticide regulation, Congress passed the Food Quality Protection Act of 1996 to eliminate the distinction between standards for raw and processed foods. All pesticide residues are now regulated under the standard of "reasonable certainty of no harm," which means that acceptable tolerance levels are established based on an assessment of what is thought to be safe. Unlike the zero-tolerance standard of the Delaney Amendment, which does not discriminate between potent and weak carcinogens, the newer type of analysis establishes different tolerance levels for carcinogens of differing potency. Moreover, benefits as well as risks associated with the use of each pesticide are considered when establishing such standards.

The **Environmental Protection Agency** (**EPA**) uses a similar approach when creating regulatory standards for carcinogens that contaminate the environment. An instructive example is provided by the environmental carcinogen *arsenic*, which is present in trace amounts in the drinking water of many wells and municipal water systems. Before 2001, the EPA regulatory standard for arsenic in drinking water had been set at 50 parts per billion (ppb). When epidemiological studies revealed that people who regularly drink water containing that much arsenic add about 1% to their lifetime risk of dying of cancer, the EPA proposed to lower the arsenic standard from 50 to 10 ppb. In an ideal world, any proposal to reduce the threat from an environmental carcinogen would be immediately accepted, but in the real world of limited financial resources, the cost of each regulatory decision needs to be balanced against the benefits to be gained. It was estimated that the public and private spending required to lower the arsenic levels in drinking water from 50 to 10 ppb would range between $600,000 and $6,000,000 per avoided cancer death. Although the EPA eventually decided that reducing arsenic levels was worth the expenditure, there is a point at which the cost of reducing the concentration of a given environmental carcinogen to a lower level becomes so great as to make the proposed effort impractical.

Since it is impossible to create a completely risk-free world in which all carcinogens have been removed, society needs to find ways of establishing priorities as to which carcinogenic hazards require the most attention. When establishing such priorities, two variables are particularly relevant. First, how potent is each carcinogen? Second, how much are we exposed to? These two variables have been combined into a single measure called the **HERP value**, where HERP stands for Human Exposure and Rodent Potency. The HERP value for any given agent is calculated by determining the typical human lifetime dose of the substance in question and dividing it by the dose administered to mice or rats that causes half the animals to develop tumors within their lifetime. The HERP value is thus a measure of potential hazard that takes into account both human exposure and carcinogenic potency in animals. In other words, powerful carcinogens with large human exposures exhibit the highest HERP values, weak carcinogens with small human exposures have the lowest HERP values, and potent carcinogens with small human exposures, or weak carcinogens with high human exposures, exhibit intermediate HERP values.

HERP calculations are not precise indicators of human hazard because of the uncertainties involved in extrapolating data on carcinogenic potency from animals to humans. Nonetheless, comparing the HERP values of various carcinogens with one another can provide insights as to which ones are likely to be most hazardous (Table 12-1). Such analyses have revealed that workplace

Table 12-1 Ranking Some Possible Carcinogenic Hazards Using HERP Values

HERP Value*	Source of Human Exposure	Carcinogen
140	Occupational exposure	Ethylene dibromide
4.0	Occupational exposure	Formaldehyde
0.1	Apple (1)	Caffeic acid
0.1	Mushroom (1)	Hydrazines
0.03	Peanut butter (1 sandwich)	Aflatoxin
0.008	Swimming pool (1 hour)	Chloroform
0.005	Coffee (1 cup)	Furfural
0.003	Home air (14 hr/day)	Benzene
0.002	Pesticide contamination of food	DDT (before 1972 ban)
0.001	Celery (1 stalk)	8-Methoxypsoralen
0.001	Chlorinated tap water (1 liter)	Chloroform
0.0004	Pesticide contamination of food	EBD (before 1984 ban)
0.000001	Pesticide contamination of food	Lindane
0.0000004	Pesticide contamination of food	PCNB
0.000000008	Pesticide contamination of food	Folpet
0.000000006	Pesticide contamination of food	Captan

Data from L. S. Gold. *Science* 258 (1992): 261.
Abbreviations: DDT = dichlorodiphenyltrichloroethane; EDB = ethylene dibromide; PCNB = pentachloronitrobenzene.
*The HERP value is a combined measure that reflects human exposure to each carcinogen as well as its carcinogenic potency when tested in animals.

exposures to chemical carcinogens tend to have high HERP values and therefore deserve considerable attention. In contrast, the pesticides and pollutants that contaminate our environment, water, and food supply generally exhibit HERP values that are quite low when compared with the background of naturally occurring carcinogens encountered in a typical diet.

BLOCKING THE DEVELOPMENT OF CANCER

The first part of this chapter has focused on preventing cancer by avoiding exposure to cancer-inducing agents, such as tobacco smoke, alcohol, sunlight, ionizing radiation, certain infectious agents, and carcinogenic chemicals encountered in the workplace, drugs, and food. Actions that are taken to avoid these agents will significantly reduce cancer risk, although it is impossible to shield ourselves completely from all carcinogens.

The second, complementary approach to cancer prevention involves actions people can take to protect against cancer development even though they have undoubtedly been exposed to some carcinogens. The combination of both approaches to cancer prevention is nicely illustrated by decisions we make regarding our diets. We have already seen that consumption of some foods should be *decreased* to avoid the cancer-causing agents they may contain. We will begin the second part of the chapter by considering those components of the diet whose consumption should be *increased* because they may help prevent the development of cancer after cancer-causing agents have been encountered.

The Proposed Role of Fruits and Vegetables in Protecting Against Cancer Is Based Largely on Retrospective Studies

During the past twenty years, the possible effectiveness of fruits and vegetables in reducing cancer risk has been investigated in more than 200 epidemiological studies. These investigations have consistently (but not universally) supported the conclusion that people who eat more fruits and vegetables have lower cancer rates than people who eat fewer fruits and vegetables. The association is strongest for cancers involving gastrointestinal and respiratory organs, such as stomach, colon, and lung cancers, and is weak for hormone-related cancers, such as breast and prostate cancers.

Unfortunately, interpreting the significance of the epidemiological evidence is complicated by two factors: (1) The observed reductions in cancer rates are generally small, and (2) most of the early studies involved a retrospective approach in which people with cancer were asked to recall their previous dietary habits. As pointed out in Chapter 4, retrospective studies are susceptible to several sources of bias and are generally less reliable than prospective studies that monitor healthy people into the future to see who develops cancer. When large prospective studies have been carried out, they generally detect little or no connection between overall fruit and vegetable consumption and cancer incidence.

This inconsistency in the epidemiological evidence has created some uncertainty regarding the precise role played by fruits and vegetables in protecting against cancer. Part of the problem may be that "fruits and vegetables" is an imprecise label that includes dozens of different food items, only some of which may be useful in reducing cancer risk. Because of the difficulty in comparing conclusions from different studies in which the exact food items are not clearly identified, we need to focus on individual fruits or vegetables—and the molecules they contain—to determine what roles, if any, these foods play in protecting against cancer.

Vitamins, Minerals, and Dietary Fiber Are Being Investigated as Possible Sources of Protection Against Cancer

A variety of natural substances present in fruits and vegetables have been investigated as possible sources of anticancer activity. Early research focused mainly on vitamins and minerals, which are known to be essential components of a healthy diet. *Vitamins A, B6, folic acid, B12, C, D,* and *E* and the minerals *calcium, iron, zinc,* and *selenium* are among the vitamins and minerals that have been suggested to play a protective role against cancer. Dietary intake of each of these substances has been linked to decreased cancer rates in some reports; other studies, however, have failed to confirm the results, and there is currently insufficient high-quality data to justify strong conclusions.

Further progress requires a better understanding of how vitamins and minerals exert their postulated anticancer effects. One property shared by several of these substances is their ability to act as **antioxidants**—that is, inhibitors of oxidation reactions. We saw in Chapters 5 and 6 that carcinogenic chemicals and radiation sometimes cause DNA damage through oxidation reactions involving **free radicals**, which are atoms or molecules possessing an unpaired electron. A free radical is a highly unstable substance that quickly attacks another nearby molecule, taking an electron that it needs to gain stability (Figure 12-7). Upon losing an electron, the second molecule also becomes a free radical and removes an electron from a third molecule, starting a chain reaction of oxidation reactions (*oxidation* refers to the process of losing an electron). An uncontrolled cascade of free radical reactions can create considerable damage within a cell, including DNA mutations. Antioxidant vitamins halt this scenario by acting as *free radical scavengers* that donate their own electrons to free radicals without becoming unstable in the process. For example, ascorbic acid (vitamin C) readily donates electrons to free radicals, undergoing oxidation to dehydroascorbic acid in the process. Dehydroascorbic acid is then converted back to

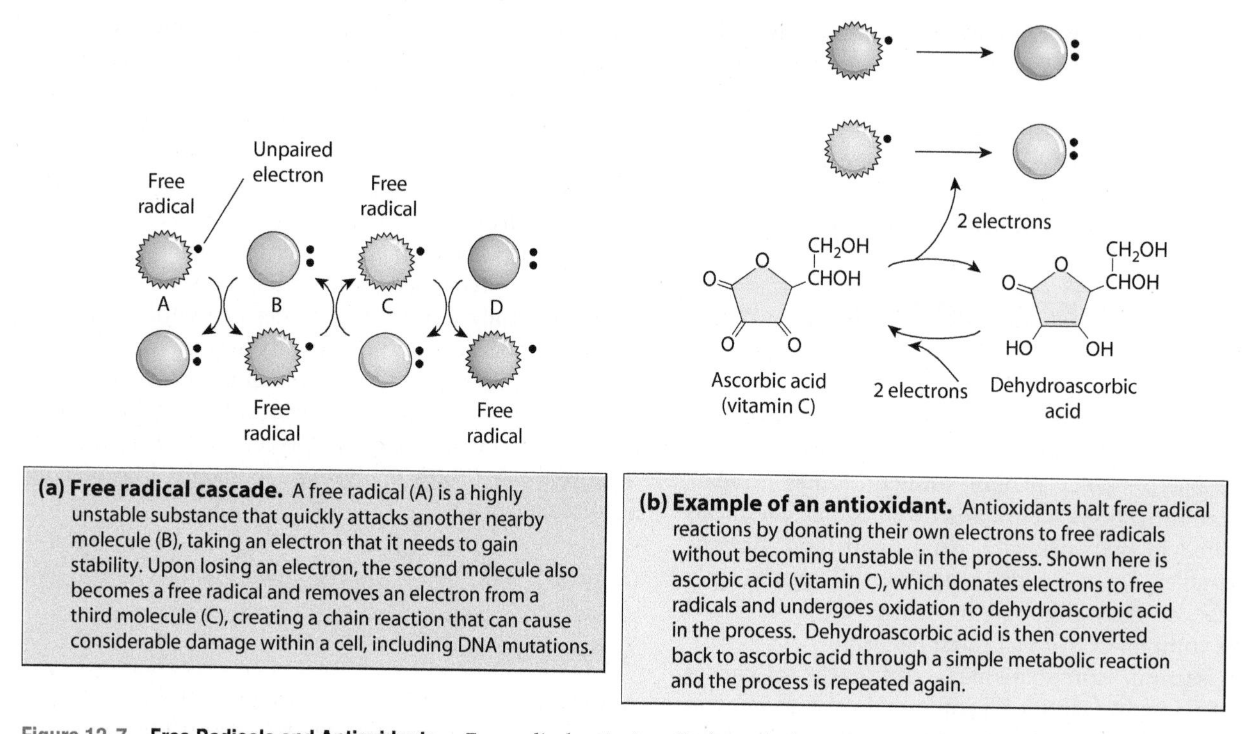

(a) Free radical cascade. A free radical (A) is a highly unstable substance that quickly attacks another nearby molecule (B), taking an electron that it needs to gain stability. Upon losing an electron, the second molecule also becomes a free radical and removes an electron from a third molecule (C), creating a chain reaction that can cause considerable damage within a cell, including DNA mutations.

(b) Example of an antioxidant. Antioxidants halt free radical reactions by donating their own electrons to free radicals without becoming unstable in the process. Shown here is ascorbic acid (vitamin C), which donates electrons to free radicals and undergoes oxidation to dehydroascorbic acid in the process. Dehydroascorbic acid is then converted back to ascorbic acid through a simple metabolic reaction and the process is repeated again.

Figure 12-7 Free Radicals and Antioxidants. Free radicals arise in cells through the action of toxic chemicals, radiation, and even as part of normal metabolic reactions and cellular activities. (**a**) Once a free radical has been created, it starts a chain reaction of damaging oxidation reactions. (**b**) Antioxidants, such as vitamin C, are free radical scavengers that can halt these chain reactions.

ascorbic acid through a simple metabolic reaction and the process is repeated again.

The ability of antioxidant vitamins to function as free radical scavengers has led to the proposal that they might be able to inhibit carcinogenesis by preventing oxidative DNA damage. The principal antioxidant vitamins found in fruits and vegetables are vitamins A, C, and E. Many retrospective studies have indicated that people who report high intake of these vitamins—either from food or through vitamin supplements—exhibit decreased cancer rates. The effects are generally small, however, and could easily be caused by other unmeasured differences in lifestyle or environment. The only way to provide convincing proof would be a randomized trial in which participants are randomly assigned to receive either a vitamin supplement or a placebo.

During the mid-1980s, a randomized trial of vitamins A and E was initiated in Finland to see whether these vitamins can reduce lung cancer rates, as was suggested by earlier retrospective evidence. The Finnish trial involved more than 29,000 male smokers who were randomly assigned to receive vitamin A (in the form of *beta-carotene*), vitamin E, both, or neither (placebo). Over a period of seven years, the trial failed to detect any protective effect of either of the two vitamins on lung cancer rates. In fact, a statistically significant *increase* in lung cancer cases was actually observed in the men who received vitamin A supplements (Figure 12-8). A second independent study carried out in the United States

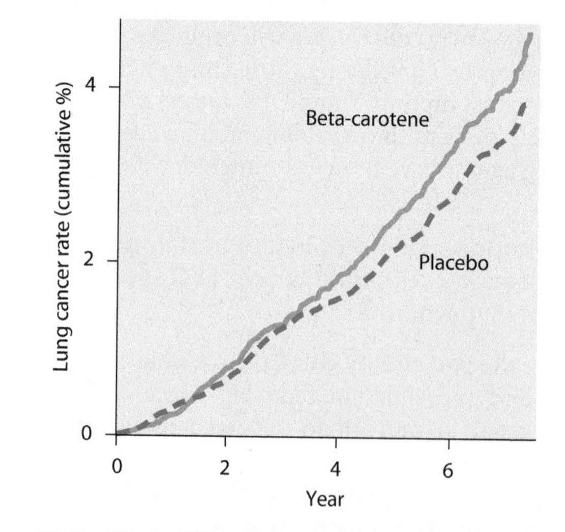

Figure 12-8 Effect of Vitamin A on Lung Cancer Rates in Male Smokers. This graph shows the results of a randomized, placebo-controlled trial carried out in Finland to determine whether supplementing the diet with beta-carotene (a form of vitamin A) can reduce lung cancer rates. After seven years, a statistically significant increase in lung cancer rates was observed in the men who received beta-carotene, which is the opposite of what had been expected. [Data from The Alpha-Tocopherol, Beta Carotene Cancer Prevention Study Group, *New England J. Med.* 330 (1994): 1029 (Figure 1).]

confirmed the ability of vitamin A to increase lung cancer rates. Such results cast serious doubt on the usefulness of vitamin A for cancer protection and reveal the possible risks associated with using vitamins in high doses (the amounts of vitamin A employed in these studies were five to ten times higher than the normally recommended dietary intake).

The preceding results do not rule out the possibility that certain vitamins might exhibit some useful anticancer properties when consumed directly from a well-balanced diet or obtained from pills designed to deliver normally recommended vitamin doses rather than high doses. One vitamin thought to protect against cancer in normal doses is *folic acid*. The typical dietary intake of folic acid in the United States is less than optimal, and low intake of folic acid can lead to DNA damage (folic acid normally provides the methyl group needed for synthesizing the base thymine in DNA). The possibility that low intake might create an increased cancer risk is supported by reports of increased colon cancer rates in people whose diets are deficient in folic acid. In one long-term prospective study, a 75% reduction in colon cancer rates was seen in individuals who take multivitamin pills containing folic acid, although the effect did not become evident until the vitamins were used for 15 years (Figure 12-9). An obvious problem in such studies is that multivitamin pills contain a variety of different vitamins and minerals, and it is possible that ingredients other than (or in addition to) folic acid were responsible for the reduction in cancer rates. Vitamin D, for example, is thought to exert some protective effects against colon cancer and might have contributed to the observed reduction in cancer rates. The ideal test would involve a randomized trial of folic acid or vitamin D versus placebo pills, but it is generally impractical to run trials for the long duration required in this particular case (15 years).

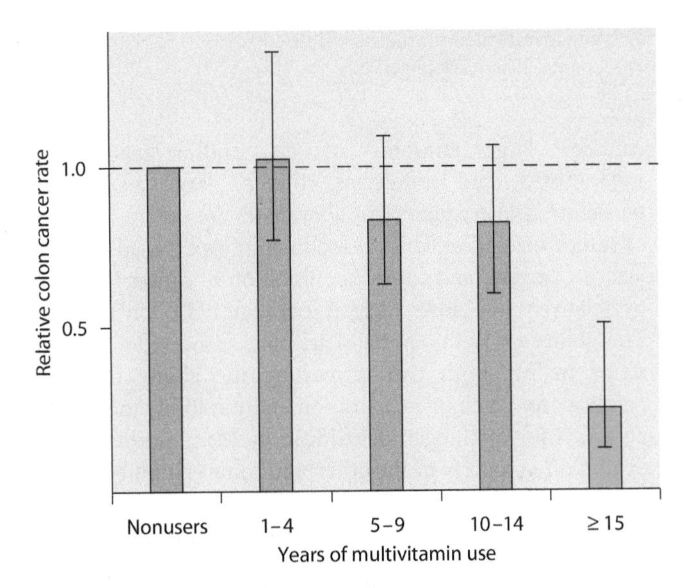

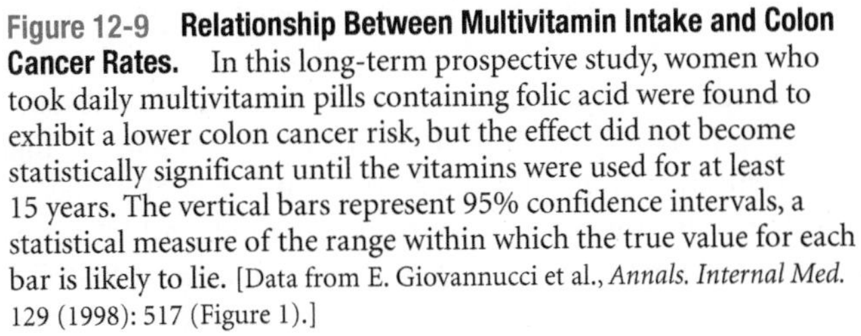
Figure 12-9 Relationship Between Multivitamin Intake and Colon Cancer Rates. In this long-term prospective study, women who took daily multivitamin pills containing folic acid were found to exhibit a lower colon cancer risk, but the effect did not become statistically significant until the vitamins were used for at least 15 years. The vertical bars represent 95% confidence intervals, a statistical measure of the range within which the true value for each bar is likely to lie. [Data from E. Giovannucci et al., *Annals. Internal Med.* 129 (1998): 517 (Figure 1).]

Vitamins are not the only constituents of fruits and vegetables for which it has been difficult to obtain reliable evidence. A similar situation exists with **dietary fiber**, the indigestible portion of plant foods that is composed mainly of complex polysaccharides. Early retrospective studies suggested that colon cancer rates are decreased in people who consume large amounts of dietary fiber, but most of these studies included little information regarding other components of the diet and did not permit the effects of fiber to be clearly distinguished from the effects of other components of plant foods. The retrospective nature of the studies also raised the possibility of numerous sources of bias, including the problem of asking individuals with cancer to recall their dietary habits from many years earlier. Large prospective studies, which tend to be more reliable, have generally found little relationship between dietary fiber intake and colon cancer protection. It has been suggested, however, that the populations being studied do not vary sufficiently in fiber intake to detect an effect. One large prospective study of European populations, which tend to exhibit substantial variability in fiber intake, has reported that doubling the average fiber intake is associated with a 40% reduction in colorectal cancer incidence.

Fruits and Vegetables Contain Dozens of Phytochemicals That Exhibit Possible Cancer-Fighting Properties

Although early investigations focused largely on vitamins and minerals, fruits and vegetables contain many other substances that might play roles in blocking or slowing cancer development. The collective term **phytochemicals** is commonly used when referring to this large group of plant-derived molecules. Technically speaking, any chemical produced by a plant is a "phytochemical"; the term is usually restricted, however, to plant chemicals that are thought to have health-related effects but are not essential nutrients like proteins, carbohydrates, fats, minerals, and vitamins.

Phytochemicals are a structurally diverse group of molecules that exhibit a wide range of properties with potential relevance for fighting cancer. Among the most relevant properties are the abilities to inhibit carcinogen activation, stimulate carcinogen detoxification, inhibit carcinogen binding to DNA, inhibit free radical and other DNA-damaging oxidative pathways, inhibit cell proliferation, inhibit oncogene expression, induce apoptosis, induce differentiation, inhibit angiogenesis, and inhibit invasion and metastasis. Laboratory studies have identified dozens of phytochemicals that might help protect against cancer through one or more of these mechanisms. A few representative examples are illustrated in Figure 12-10 and briefly described below.

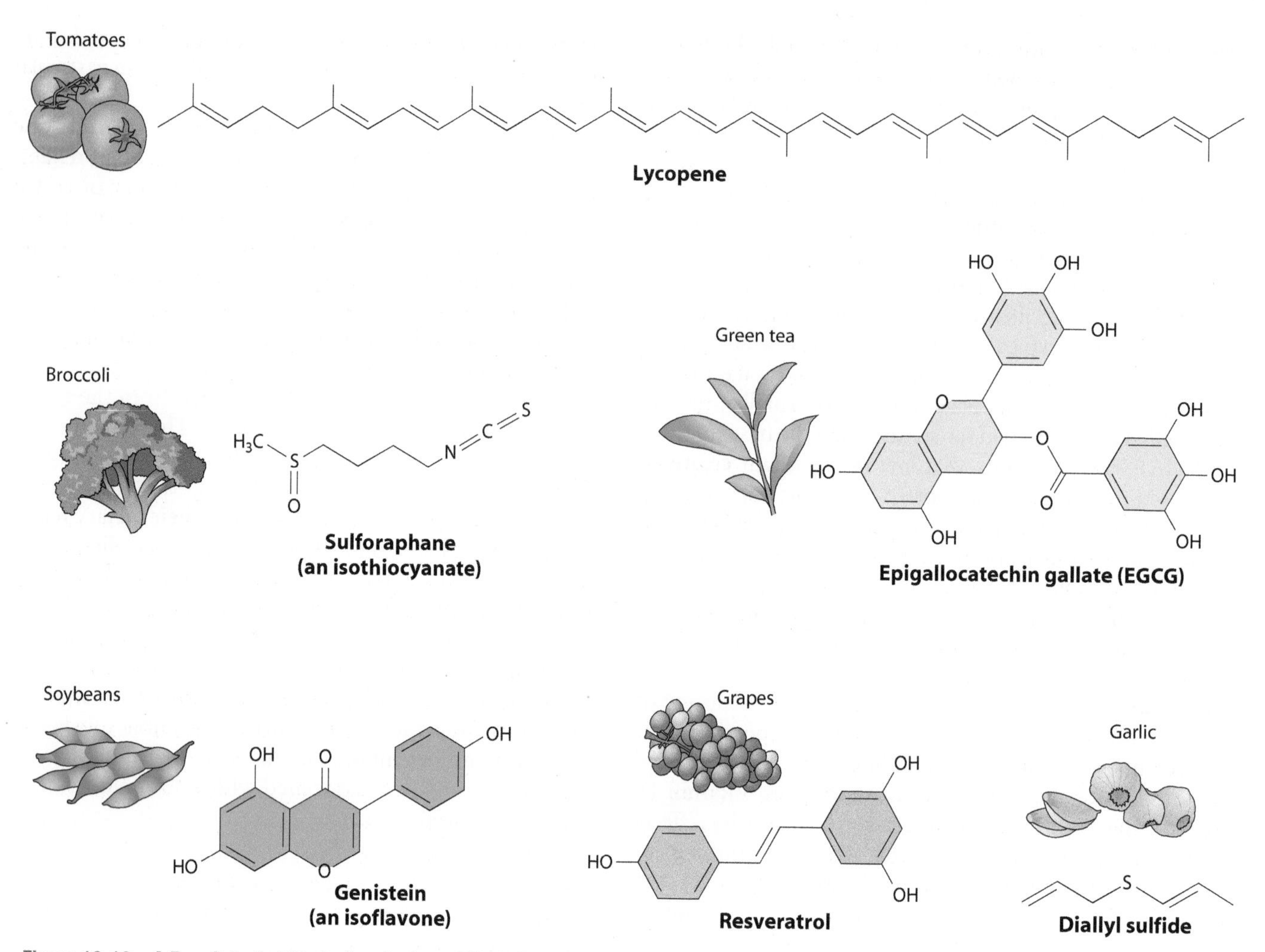

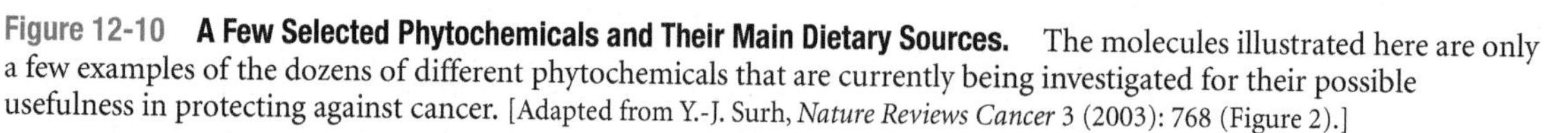

Figure 12-10 A Few Selected Phytochemicals and Their Main Dietary Sources. The molecules illustrated here are only a few examples of the dozens of different phytochemicals that are currently being investigated for their possible usefulness in protecting against cancer. [Adapted from Y.-J. Surh, *Nature Reviews Cancer* 3 (2003): 768 (Figure 2).]

1. Lycopene. Lycopene is a natural red pigment that gives tomatoes and other fruits and vegetables their reddish colors. The chemical structure of lycopene is related to that of beta-carotene (a vitamin A precursor), and, like beta-carotene, lycopene is a potent antioxidant. Epidemiological studies have linked dietary intake of lycopene, derived mainly from tomatoes and tomato-based sauces, to a reduced incidence of several cancers, especially those of the prostate, lung, and stomach. In some cases, lycopene levels have been directly measured in blood samples to verify dietary intake. Such measurements have shown that individuals with higher concentrations of lycopene in their blood exhibit lower rates of cancer than people with lower concentrations. Lycopene's antioxidant properties are thought to be responsible for the lowering of cancer risk, but lycopene also exerts other effects on metabolism, gene activity, and cell signaling that might be involved.

2. Isothiocyanates. Isothiocyanates are a diverse group of compounds, characterized by the presence of a —N=C=S group, that occur in especially high concentration in the vegetable family that includes broccoli, brussels sprouts, cabbage, cauliflower, and watercress. Broccoli, for example, contains large amounts of an isothiocyanate called *sulforaphane*, and watercress contains large amounts of *phenethyl isothiocyanate*. Isothiocyanates are potent inhibitors of tumor formation when tested in animals exposed to chemical carcinogens, and complete inhibition of cancer formation by relatively low doses of isothiocyanates is often observed. The ability of isothiocyanates to block cancer development stems mainly from two properties: the ability to inhibit pathways for carcinogen activation and the ability to stimulate pathways for carcinogen detoxification. Many isothiocyanates exhibit one property or the other, and some exhibit both.

3. Epigallocatechin gallate. Numerous animal and laboratory experiments, along with some human epidemiological studies, have indicated a possible benefit of green tea in protecting against certain cancers, especially those of the esophagus, stomach, colon, and bladder. The substance *epigallocatechin gallate* (*EGCG*), which is present in high concentration in green tea, exhibits a number of properties that may help explain how green tea could protect against cancer. Among these properties are the abilities of EGCG to act as an antioxidant, inhibit carcinogen activation, stimulate carcinogen detoxification, inhibit cell proliferation, induce cell cycle arrest, induce apoptosis, and inhibit angiogenesis. It remains to be determined, however, if any of these activities are exerted to a significant extent by the small amounts of EGCG a person typically ingests when drinking green tea.

4. Isoflavones. Breast and prostate cancer rates are relatively low in Japan and China, where soybean-derived foods represent a significant portion of the diet. A group of soybean compounds known as *isoflavones* have been proposed to play a role in these reduced cancer rates. One isoflavone, called *genistein*, exhibits weak estrogen-like effects on the body and is therefore referred to as a *phytoestrogen* ("plant estrogen"). The growth of breast cancer cells in culture is inhibited by high concentrations of genistein, suggesting that this isoflavone can act as an antiestrogen. However, at lower concentrations that may be closer to levels seen in humans who eat soy products, genistein has been reported to stimulate breast cell proliferation. Given this potential to either inhibit or stimulate breast cell proliferation, a precautionary approach to consuming soy to prevent breast cancer is advisable until the role of soy isoflavones is better understood.

5. Resveratrol. Resveratrol, an antioxidant produced by a variety of plants, is present in especially high concentration in the skin of red grapes and in red wines made from these grapes. It has been proposed that resveratrol is responsible for the reported linkage between red wine consumption and lower death rates from cardiovascular disease and certain cancers. In addition to acting as an antioxidant, resveratrol exhibits a number of other properties that might be relevant to cancer prevention. For example, resveratrol can promote carcinogen detoxification, inhibit DNA mutation, inhibit cell proliferation, and induce apoptosis. Resveratrol also inhibits the synthesis and metabolic activity of *cyclooxygenases*, which are enzymes whose role in promoting inflammation and stimulating tumor cell growth will be described shortly.

6. Sulfides. For more than 3000 years, folklore has advocated garlic as a remedy for a variety of health problems. Recent epidemiological studies have provided some support for these beliefs by linking high levels of garlic consumption with reduced cancer rates, especially for cancers of the stomach, colon, and prostate. Garlic contains a variety of *organic sulfides* (compounds containing sulfur joined to carbon by single bonds) that might be relevant to the postulated benefits of garlic. For example, a compound called *diallyl sulfide*, which is present in high concentration in garlic, has been shown to inhibit the development of colon, esophageal, and lung cancers in animals that have been exposed to chemical carcinogens. Organic sulfides exhibit a number of properties of potential relevance to cancer protection, including the ability to inhibit carcinogen activation, stimulate DNA repair, inhibit cell proliferation, and induce apoptosis.

The preceding list includes only a few of the numerous phytochemicals that are currently being investigated. The reported existence of cancer-preventing properties in these molecules is certainly encouraging, but a note of caution is appropriate because most investigations are still in their early stages. For comparison purposes, it is worth pointing out that the initial body of evidence regarding the postulated role of vitamin A in preventing cancer was also very encouraging, but randomized human trials eventually revealed that the early claims were unjustified. Since the cancer-fighting properties of phytochemicals have not been evaluated in randomized human trials, it would be premature to draw any definitive conclusions regarding their usefulness in preventing cancer.

Aspirin and Other Anti-inflammatory Drugs May Help Prevent Certain Types of Cancer

Thus far, we have been focused on inhibiting cancer development through the use of substances present in food, either by eating specific foods or by ingesting some of the chemicals they contain, such as vitamins, minerals, and phytochemicals. This practice of using specific chemical substances for protecting against cancer is called **chemoprevention**. Besides natural substances found in foods, a variety of synthetic drugs are also being explored for possible use in cancer chemoprevention efforts.

One drug known to be helpful in protecting against cancer is **aspirin**. The synthesis of aspirin was first reported in the late 1890s by Felix Hoffmann, a German chemist who was searching for a way to relieve his father's arthritis pain. Motivated by folklore concerning the pain-relieving and anti-inflammatory properties of willow bark, Hoffman synthesized aspirin from an ingredient in willow bark called salicylic acid. The ability of aspirin (acetylsalicylic acid) to reduce pain, fever, and inflammation with minimal short-term toxicity has made it the world's most widely used medication. Aspirin works by inhibiting the enzyme **cyclooxygenase (COX)**, which catalyzes the production of chemical messengers called **prostaglandins**. Some prostaglandins act on blood vessels, nerves, and cells of the immune system to trigger tissue **inflammation**, which is characterized by swelling, redness, pain, and heat. By inhibiting cyclooxygenase, aspirin reduces prostaglandin production and thereby diminishes these symptoms.

During the 1990s, a series of retrospective and prospective epidemiological studies revealed that people who use aspirin on a regular basis exhibit up to a 50% reduction in colon cancer rates. Lower rates for several other types of cancer were reported as well, including cancers of the prostate, lung, mouth, throat, and esophagus. Such results suggest that aspirin might be a simple and effective tool for cancer chemoprevention. Unfortunately, chronic use of aspirin has drawbacks because the drug also causes stomach irritation, bleeding, and ulcers. The reason for these side effects is that cyclooxygenase, the enzyme inhibited by aspirin, exists in multiple forms with differing functions. One form, called *COX-1*, is expressed in many cell types and plays a variety of normal roles, such as protecting the lining of the stomach against acid irritation. Inhibition of COX-1 by aspirin interferes with this protective function, thereby triggering the adverse effects of aspirin on the stomach.

A second form of cyclooxygenase, called *COX-2*, is also inhibited by aspirin. Normal tissues contain relatively little COX-2 but production of the enzyme is enhanced at sites of tissue damage, where it synthesizes prostaglandins that trigger tissue inflammation. COX-2 is also produced by colon cancers and other epithelial malignancies, where it may contribute to tumor development by synthesizing a prostaglandin known as *prostaglandin E2 (PGE2)*. PGE2 exhibits several properties that could help cancer cells proliferate, survive, and evade the immune system. First, PGE2 inhibits the activity of cytotoxic T lymphocytes, which might otherwise attack tumor cells. PGE2 also stimulates angiogenesis, which is required for tumor growth beyond a tiny size, and it inhibits apoptosis in cancer cells. Each of these properties could play a role in cancer development and progression.

The discovery that COX-2 produces prostaglandins that trigger tissue inflammation and facilitate cancer development suggests that aspirin's ability to decrease cancer risk stems from its inhibitory effects on COX-2. Aspirin's adverse effects, such as stomach bleeding and ulcers, are linked to its inhibition of COX-1. To obtain the beneficial effects of aspirin without the adverse side effects, several new anti-inflammatory drugs were introduced in the late 1990s that selectively inhibit COX-2 rather than inhibiting both COX-1 and COX-2 like aspirin does. These selective *COX-2 inhibitors*, sold under the trade names *Vioxx*, *Celebrex*, and *Bextra*, were initially introduced to provide relief from chronic arthritis pain without causing the stomach problems seen with aspirin. Preliminary evidence suggested that COX-2 inhibitors might also be useful for reducing colon cancer risk, and a large clinical trial was initiated in early 2000. Unfortunately, the trial was halted in late 2004 when it was discovered that COX-2 inhibitors double the risk of heart attack and stroke, and the future of these drugs is now uncertain.

Several other anti-inflammatory drugs in addition to aspirin—for example, acetaminophen (Tylenol) and ibuprofen—have also been reported to exhibit some protective effects against cancer, although the data are not nearly as extensive as for aspirin. The possible usefulness of these and other anti-inflammatory drugs for reducing cancer risk is likely to be an area of active future investigation.

Hormone-Blocking Drugs Are Potentially Useful for Preventing Hormone-Dependent Cancers

We saw in Chapter 11 that cancers arising in hormone-dependent tissues, such as the breast and prostate, are sometimes treated with drugs that interfere with the actions of the required hormones. For example, **tamoxifen** blocks the estrogen receptors of breast cells and inhibits the ability of estrogen to drive the proliferation of these cells, thereby making the drug useful for treating estrogen-dependent breast cancers.

Tamoxifen may also be helpful for preventing breast cancer in women at high risk for the disease. The rationale for this approach is related to events that normally control the proliferation of breast cells. During each menstrual cycle, estrogen triggers the proliferation of epithelial cells that line the milk glands in the breast. If pregnancy does not occur, estrogen levels decline at the end of the menstrual cycle and those breast cells that have proliferated in that month deteriorate and die. For the average woman, the result is hundreds of cycles of cell division and death repeated over a span of roughly 40 years, from puberty to menopause. These repeated cycles of estrogen-induced cell division increase the risk of developing cancer in two ways: (1) estrogen stimulates the division of any cells that may have already acquired DNA mutations, thereby increasing the number of mutant cells that might progress to malignancy; and (2) repeated cycles of estrogen-induced proliferation increase the chances of new mutations arising as a result of errors in DNA replication.

Because estrogen-driven proliferation of breast cells increases the risk of cancer, using tamoxifen to block estrogen action would be expected to lower the cancer risk. To test this hypothesis, the National Cancer Institute sponsored a study during the 1990s that involved more than 13,000 healthy women considered to be at high risk for breast cancer based on either their family or medical history. Half the women were given tamoxifen and the other half were given a placebo. Over a five-year period, the women receiving tamoxifen experienced a roughly 50% reduction in new cases of breast cancer (Figure 12-11).

If a reduction in breast cancer risk were the only effect of tamoxifen, the drug would be widely recommended for breast cancer prevention. Unfortunately, tamoxifen has a side effect that limits its usefulness. This side effect arises because tamoxifen does not block estrogen receptors in every tissue as it does in the breast. In the uterus, tamoxifen mimics the action of estrogen when it binds to estrogen receptors, thereby stimulating cell proliferation and increasing the risk of uterine endometrial cancer (see Figure 12-11, *right*). Tamoxifen is therefore unsuitable as a cancer prevention drug in women with no obvious susceptibility to breast cancer because of the elevated risk of developing uterine cancer.

Drugs like tamoxifen, which block estrogen action in some tissues but act like an estrogen in others, are called **selective estrogen receptor modulators (SERMs)** because they selectively stimulate or inhibit the estrogen receptors of different target tissues. Scientists have been working to develop other SERMs that might exhibit the beneficial properties of tamoxifen on the breast without its harmful effects on the uterus. One example is **raloxifene**, a drug that is useful for preventing osteoporosis (bone loss) in postmenopausal women. Raloxifene functions like

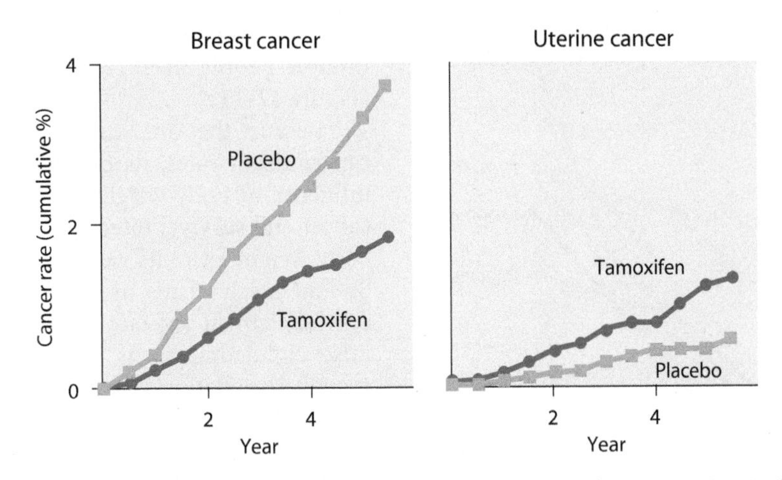

Figure 12-11 Effectiveness of Tamoxifen in Cancer Chemoprevention. A randomized trial was carried out involving more than 13,000 healthy women at high risk for breast cancer. Women receiving tamoxifen experienced a 50% reduction in new cases of invasive breast cancer, but the rate of uterine (endometrial) cancer more than doubled. This adverse side effect involving uterine cancer is not relevant to women who have had their uterus removed for other medical reasons. [Data from B. Fisher et al., *J. Natl. Cancer Inst.* 90 (1998): 1371 (Figures 2 and 5).]

estrogen in bone, acting to maintain bone strength and increase bone density. In contrast, raloxifene blocks estrogen action in the breast and uterus and has been shown to reduce the incidence of breast and uterine cancers in animal studies. The National Cancer Institute is therefore sponsoring a human trial called the *Study of Tamoxifen and Raloxifene* (*STAR*) to directly compare the effects of the two drugs in preventing cancer.

Prostate cancer is another type of cancer arising in a hormone-dependent tissue that might be preventable using drugs that block hormone action. Since the proliferation of prostate cells depends on steroid hormones called *androgens* (testosterone is one example), prevention efforts have focused on drugs that inhibit androgen activity. One such drug, **finasteride**, is an enzyme inhibitor that blocks the production of *dihydrotestosterone*, the active form of testosterone in the prostate. In a randomized trial involving more than 18,000 men, daily use of finasteride was found to reduce the incidence of new cases of prostate cancer by 25% compared with the placebo group.

While these results suggest that hormonal prevention of prostate cancer might be an attainable goal, there are significant concerns about the use of finasteride for this purpose. The most serious problem is the discovery that men receiving finasteride in the randomized trial developed a higher proportion of high-grade cancers (37% of their tumors were high grade) than did the men who received the placebo (22% of their tumors were high grade). So even though men receiving finasteride developed fewer cancers, the tumors that did develop were more aggressive on average than those observed in the control group. One possible explanation for this unexpected result is that the androgen-deficient environment created by finasteride treatment may favor the development of tumors that are more aggressive because they do not depend on androgens for their growth. Whatever the correct explanation turns out to be, concerns about the high incidence of aggressive cancers makes it inappropriate to recommend finasteride for routine use as a cancer prevention agent.

A Healthy Body Weight and Regular Physical Exercise Can Reduce Cancer Risks

It has been known for many years that individuals who are overweight have a higher-than-normal incidence of several types cancer, mainly uterine, breast, and colon cancers. Only recently has it become apparent that the impact of being overweight is much broader than this. In 2003, the American Cancer Society published a study in which body weight and cancer death rates were tracked in more than 900,000 people for 16 years. The data revealed that obesity is linked to increased death rates for cancers of the uterus, breast, colon, esophagus, pancreas, kidney, gallbladder, ovary, liver, and prostate, as well as cancers of blood cells such as multiple myeloma and non-Hodgkin's lymphoma.

To determine who was overweight, these studies estimated the amount of body fat each person carries using a formula based on height and weight called the **Body Mass Index** (**BMI**):

$$\text{BMI} = \frac{\text{Weight in kilograms}}{(\text{Height in meter})^2} \quad \text{or}$$

$$\text{BMI} = \frac{\text{Weight in pounds}}{(\text{Height in inches})^2} \times 703$$

Although this formula tends to overestimate body fat in muscular individuals, people are generally considered to be *overweight* if their BMI is 25 or higher and *obese* if it is 30 or higher. When the BMI was used to assess body

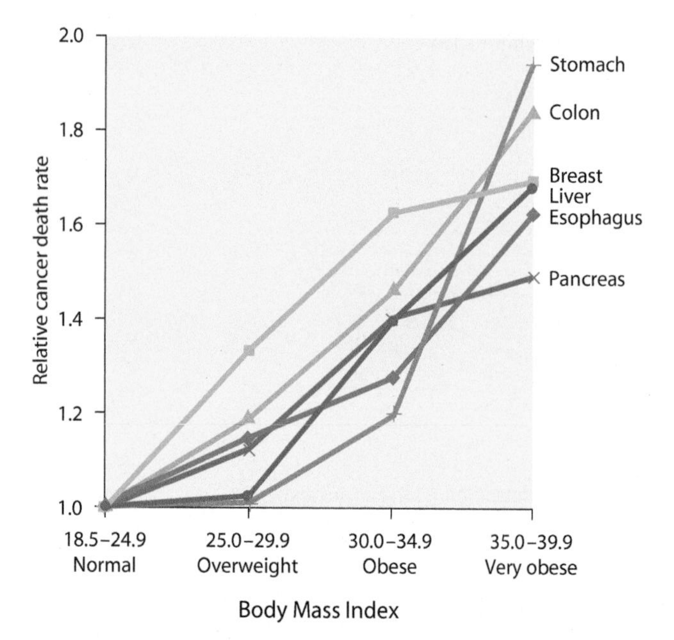

Figure 12-12 Effect of Body Weight on Cancer Risks. A direct relationship exists between the amount of excess weight a person carries and a progressively increased risk of dying from various types of cancer. [Based on data from E. E. Calle et al., *New England J. Med.* 348 (2003): 1625 (Tables 1 and 2).]

weight in the American Cancer Society study, a direct relationship was detected between the amount of excess weight a person carries and a progressively greater risk of dying of cancer (Figure 12-12). The linkage between cancer and excess body weight is especially troubling because a growing epidemic of obesity has been occurring in the United States since the early 1980s. By the year 2000, roughly 60% of U.S. adults were categorized as being overweight or obese, putting more than 100 million people at increased risk for dying of cancer (Figure 12-13).

Because the American Cancer Society study measured cancer death rates, its conclusions reflect the combined influence of body weight on both the risk of developing cancer and survival rates after diagnosis. How does excess body weight exert its negative influence on these events? Several mechanisms are thought to be involved, but a central element appears to be the tendency of obesity to raise circulating levels of insulin, estrogen, and other growth-stimulating hormones. Such molecules can contribute to the promotion phase of carcinogenesis by stimulating cell proliferation and may also stimulate tumor growth after cancer has arisen.

There are two ways of avoiding the accumulation of excess body weight that leads to increased cancer risk. One is to eat a diet that does not contain too many calories, and the other is to engage in regular physical exercise. Exercise plays an especially important role because its effects on cancer risk are not limited to maintaining a healthy body weight. When individuals of the same exact weight are compared with one another, those who are physically active still have lower cancer rates than those who are inactive. One possible explanation is that individuals who exercise regularly tend to have more muscle tissue and less body fat than individuals of the same weight who do not exercise, and reduced levels of body fat are associated with hormonal and metabolic changes that may reduce cancer risks. Exercise-induced stimulation of immune function and antioxidant pathways has also been reported, but the relevance of these changes (if any) to the development of cancer remains to be determined.

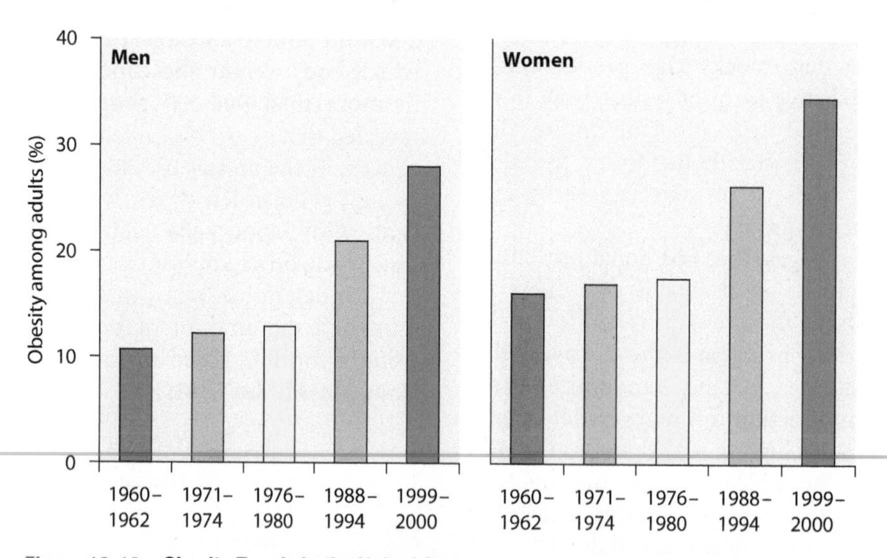

Figure 12-13 Obesity Trends in the United States. A growing epidemic of obesity has occurred since 1980 in both men and women. These bar graphs show that by the year 2000, approximately 30% of all adults were categorized as being obese (BMI = 30 or higher). Another 30% were categorized as being overweight (data not shown), putting 60% of the adult population at increased risk for dying of cancer. [Data from E. E. Calle and R. Kaaks, *Nature Reviews Cancer* 4 (2004): 579 (Figure 1).]

Vaccines and Antibiotics Are Useful for Preventing Cancers Caused by Viruses and Bacteria

The importance of avoiding exposure to cancer-causing infectious agents—such as human papillomavirus, hepatitis B virus, hepatitis C virus, and the bacterium *H. pylori*—was discussed earlier in this chapter. Complete avoidance, however, is extremely difficult because of the widespread prevalence of these infectious agents. It is therefore important to pursue approaches for preventing cancer from arising in people who have already been exposed.

One strategy is the development of vaccines that immunize against cancer-causing infectious agents. In the case of the hepatitis B virus, effective vaccines have been available since 1982. Many countries now include hepatitis B vaccination in their routine childhood immunization programs, and reduced liver cancer rates have already been reported in immunized individuals. Developing a vaccine for hepatitis C has taken longer because the hepatitis C virus exhibits greater genetic variability, but candidate vaccines are currently being developed and evaluated.

Vaccination against human papillomavirus (HPV), the main cause of cervical cancer, is also being actively pursued. Creating a vaccine that protects against HPV is not a straightforward task because more than 100 different types of HPV exist and roughly a dozen have been implicated in the development of cancer. HPV 16 was the first to be pursued as a vaccine target because it is the most common strain of HPV in invasive cervical cancers, occurring in more than 50% of such tumors (see Figure 7-3). In 2002, results from the first successful trial of an HPV 16 vaccine were reported in a study involving more than 2300 women. The vaccine was found to be 100% effective in preventing HPV 16 infection and cervical cancer over a period of two years. Preliminary testing of another vaccine, which includes protection against HPV 18 as well as HPV 16, has also been successful, and it is anticipated that these vaccines will soon meet FDA requirements for general use.

Finally, *H. pylori* is one of the most common disease-causing microbes worldwide, infecting roughly two-thirds of the global population. Efforts to prevent stomach cancer caused by chronic infection with this bacterium are being pursued in two different ways. First, a vaccine designed to block persistent infection with *H. pylori* has been developed and is currently undergoing human testing. In addition, the fact that *H. pylori* is a bacterium rather than a virus makes it susceptible to treatment with antibiotics. When antibiotics are administered to infected individuals early in the course of infection to rid them of *H. pylori*, the risk of stomach cancer is dramatically reduced.

Optimal Reductions in Cancer Risk Require a Combination of Prevention Strategies

Progress in finding ways to prevent cancer has been greater than people may generally recognize. Part of the reason for this lack of recognition might be that no single piece of advice can be given for preventing all cancers. Avoiding tobacco smoke certainly has a major impact for anyone who takes such action, reducing the risk of lung cancer by almost 90% and reducing the overall risk of dying of cancer by roughly one-third. But two-thirds of all cancer fatalities—in other words, a majority of cancer deaths—cannot be prevented by this action alone.

When it comes to reducing the incidence of other fatal cancers, the actions that need to be taken require a combination of strategies. A good illustration is provided by colon cancer, the second leading cause of cancer deaths in the United States. A strong hereditary predisposition explains only a small fraction of colon cancer cases; lifestyle factors play a major role in the rest. Mounting evidence suggests that six of the most important lifestyle risk factors for colon cancer are obesity, physical inactivity, alcohol consumption, cigarette smoking, red meat consumption, and low folic acid intake. Researchers have estimated that the majority of all colon cancers could be prevented if people adopt healthy behaviors regarding this constellation of risk factors. Periodic screening for colon cancer after age 50, which allows removal of benign polyps before they can become malignant, would reduce the cancer incidence even further.

The inescapable conclusion is simple but perhaps surprising to many individuals: Most of the cancer deaths caused by the top two cancer killers in the United States (lung cancer and colon cancer) can already be prevented if people are appropriately educated and take the required actions. Such a conclusion is cause for considerable optimism about future prospects for reducing the health burden that cancer places on society.

Summary of Main Concepts

- The two main strategies for cancer prevention involve minimizing exposure to agents that cause cancer and adopting behaviors that protect against the development of cancer after any such agents have been encountered. If all the appropriate actions discussed in this chapter were taken, more than half of all cancer deaths could be prevented.
- Choosing not to smoke or use tobacco products is the single most effective step that can be taken to reduce

the risk of developing a fatal cancer. Avoiding the excessive consumption of alcohol, especially in combination with smoking, is another effective way of reducing cancer risk.

- Since most skin cancers are not fatal, people often fail to take adequate precautions to protect themselves from the ultraviolet radiation in sunlight. Such attitudes ignore the dangers of melanoma, the most lethal type of skin cancer. Actions for protecting against skin cancer include restricting exposure to intense sunlight, wearing protective clothing, and using a sunscreen lotion that achieves an SPF of at least 15.
- The dose of ionizing radiation experienced by the average person is relatively small and comes largely from natural background sources. Reasonable precautions for reducing cancer risks from ionizing radiation include checking for excessive radon accumulation in homes and buildings, and avoiding unnecessary medical X-rays.
- Reducing exposure to cancer-causing viruses and bacteria, such as human papillomavirus, hepatitis B virus, hepatitis C virus, and *H. pylori*, would lead to significant reductions in cancer rates worldwide. Cancer risks can also be reduced by minimizing exposure to carcinogenic chemicals, especially in the workplace where chemical concentrations are high or in individuals taking chronic medications that may increase cancer risk.
- Diets rich in animal fat in general, and red meat in particular, have been linked in retrospective studies to an elevated risk for several cancers. Prospective studies, on the other hand, have failed to provide consistent support for the role of dietary fat in cancer risk. Fats are high in calories and a strong relationship exists between calorie intake and cancer risk, so dietary fat may contribute indirectly to cancer risk by introducing excess calories.
- Government actions have helped reduce cancer rates by educating the public regarding specific hazards, such as tobacco smoke, and by creating regulatory agencies such as the FDA, EPA, and OSHA, which create and enforce regulations concerning carcinogens found in food, the environment, and the workplace.
- Eating fruits and vegetables may provide some protection against cancer, but the data are inconsistent and uncertainty exists concerning the identity of the critical molecules. Retrospective studies have pointed to possible protective roles of several vitamins, minerals, and dietary fiber, but prospective studies, which are less susceptible to bias, have often failed to confirm these findings. Convincing proof requires randomized human trials of individual dietary components, which are difficult to perform for the long periods of time required for definitive results. In the case of vitamin A, randomized trials have revealed that large doses of the vitamin may actually increase cancer risk.
- Many phytochemicals found in fruits and vegetables exhibit properties that might be relevant to preventing cancer. Included in this category are the abilities to inhibit carcinogen activation, stimulate carcinogen detoxification, inhibit carcinogen binding to DNA, inhibit free radical and other oxidative pathways that damage DNA, inhibit cell proliferation, inhibit oncogene expression, induce apoptosis, induce differentiation, inhibit angiogenesis, and inhibit invasion and metastasis.
- In addition to phytochemicals, certain drugs may also be useful as cancer chemoprevention agents. For example, aspirin lowers colon cancer risk through its inhibitory effects on COX-2, and tamoxifen reduces breast cancer risk by blocking the estrogen receptor of breast cells.
- Limiting the number of calories in the diet, along with regular exercise, reduces the cancer risks associated with being overweight and physically inactive.
- Cancers triggered by exposure to infectious agents can be prevented by immunizing people with the appropriate vaccines. Antibiotics are useful for stopping persistent infections with *H. pylori*, which might otherwise cause stomach cancer to arise.

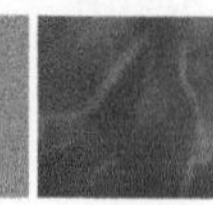

Key Terms for Self-Testing

Avoiding the Causes of Cancer

SPF (sun-protection factor) (p. 237)
hormone replacement therapy (HRT) (p. 239)
Occupational Safety and Health Administration (OSHA) (p. 241)
Food and Drug Administration (FDA) (p. 241)
Delaney Amendment (p. 241)
zero tolerance (p. 241)
acceptable tolerance (p. 241)
Environmental Protection Agency (EPA) (p. 242)
HERP value (p. 242)

Blocking the Development of Cancer

antioxidant (p. 243)
free radical (p. 243)
dietary fiber (p. 245)

phytochemical (p. 245)
chemoprevention (p. 247)
aspirin (p. 247)
cyclooxygenase (COX) (p. 247)
prostaglandin (p. 247)
inflammation (p. 247)
tamoxifen (p. 248)
selective estrogen receptor modulators (SERMs) (p. 248)
raloxifene (p. 248)
finasteride (p. 249)
Body Mass Index (BMI) (p. 249)

Suggested Reading

Avoiding the Causes of Cancer

Ames, B. N., R. Magaw, and L. S. Gold. Ranking possible carcinogenic hazards. *Science* 236 (1987): 271.

Chao, A., et al. Meat consumption and risk of colorectal cancer. *JAMA* 293 (2005): 172.

De Jong, S. A., et al. Thyroid carcinoma and hyperparathyroidism after radiation therapy for adolescent acne vulgaris. *Surgery* 110 (1991): 691.

Ellwein, L. B., and S. M. Cohen. The health risks of saccharin revisited. *Crit. Rev. Toxicol.* 20 (1990): 311.

Engstrom, P. F., et al. Prevention of Tobacco-Related Cancers. In *Holland-Frei Cancer Medicine*, 5th ed. (R. C. Bast et al., eds.). Lewiston, NY: Decker, 2000, Chapter 25.

Hill, D., et al. Changes in sun-related attitudes and behaviours, and reduced sunburn prevalence in a population at high risk of melanoma. *European J. Cancer Prevention* 2 (1993): 447.

Holmberg, L., and H. Anderson. HABITS (hormonal replacement therapy after breast cancer—is it safe?), a randomized comparison: trial stopped. *Lancet* 363 (2004): 453.

Hursting, S. D., and F. W. Kari. The anti-carcinogenic effects of dietary restriction: mechanisms and future directions. *Mutation Res.* 443 (1999): 235.

Keefe, K. A., and F. L. Meyskens Jr. Cancer Prevention. In *Clinical Oncology* (M. D. Abeloff, ed.). New York: Churchill Livingstone, 2000, Chapter 15.

Kushi, L., and E. Giovannucci. Dietary fat and cancer. *Amer. J. Med.* 113, No. 9, Suppl. 2 (2002): 63S.

Writing Group for the Women's Health Initiative Investigation. Risks and benefits of estrogen plus progestin in healthy postmenopausal women. *JAMA* 288 (2002): 321.

Blocking the Development of Cancer

Alpha-Tocopherol, Beta Carotene Cancer Prevention Study Group. The effect of vitamin E and beta carotene on the incidence of lung cancer and other cancers in male smokers. *New England J. Med.* 330 (1994): 1029.

Bingham, S., and E. Riboli. Diet and cancer—The European Prospective Investigation into Cancer and Nutrition. *Nature Reviews Cancer* 4 (2004): 206.

Calle, E. E., and R. Kaaks. Overweight, obesity and cancer: Epidemiological evidence and proposed mechanisms. *Nature Reviews Cancer* 4 (2004): 579.

Calle, E. E., et al. Overweight, obesity, and mortality from cancer in a prospectively studied cohort of U.S. adults. *New England J. Med.* 348 (2003): 1625.

Cohen, J. High hopes and dilemmas for a cervical cancer vaccine. *Science* 308 (2005): 618.

Colditz, G. A., and S. E. Hankinson. The Nurses' Health Study: lifestyle and health among women. *Nature Reviews Cancer* 5 (2005): 388.

Fisher, B., et al. Tamoxifen for prevention of breast cancer: Report of the National Surgical Adjuvant Breast and Bowel Project P-1 Study. *J. Natl. Cancer Inst.* 90 (1998): 1371.

Flood, A., and A. Schatzkin. Colorectal cancer: Does it matter if you eat your fruits and vegetables? *J. Natl. Cancer Inst.* 92 (2000): 1706.

Greenwald, P. Chemoprevention of cancer. *Sci. Amer.* 275 (September 1996): 96.

Houghton, M., and S. Abrignani. Prospects for a vaccine against the hepatitis C virus. *Nature* 436 (2005): 961.

Jang, M. Cancer chemopreventive activity of resveratrol, a natural product derived from grapes. *Science* 275 (1997): 218.

Koutsky, L. A., et al. A controlled trial of a human papillomavirus type 16 vaccine. *New England J. Med.* 347 (2002): 1645.

Kucuk, O. Cancer chemoprevention. *Cancer Metastasis Rev.* 21 (2002): 189.

Lee, I.-M. Physical activity and cancer prevention—data from epidemiological studies. *Med. Sci. Sports Exercise* 35 (2003): 1823.

Michels, K. B., et al. Prospective study of fruit and vegetable consumption and incidence of colon and rectal cancers. *J. Natl. Cancer Inst.* 92 (2000): 1740.

Platz, E. A., et al. Proportion of colon cancer risk that might be preventable in a cohort of middle-aged US men. *Cancer Causes Control* 11 (2000): 579.

Scardino, P. T. The prevention of prostate cancer—the dilemma continues. *New England J. Med.* 349 (2003): 297.

Surh, Y.-J. Cancer chemoprevention with dietary phytochemicals. *Nature Reviews Cancer* 3 (2003): 768.

Willett, W. C. Diet and cancer. *Oncologist* 5 (2000): 393.

Willett, W. C. Diet and cancer: An evolving picture. *JAMA* 293 (2005): 233.

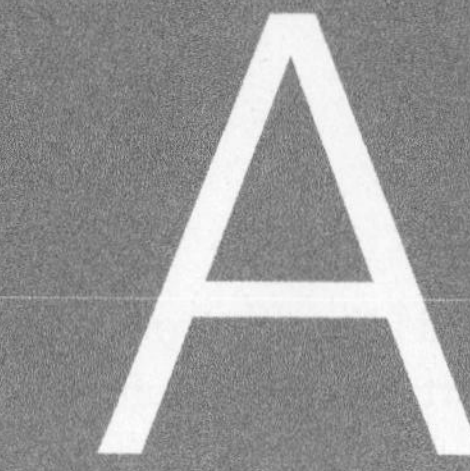

Appendix A: Some Features of the Main Types of Cancer

The chapters in this book are organized around a set of general principles that apply to cancer biology as a field of study. However, more than 100 different kinds of cancer can be distinguished from one another based on the tissue and cell type involved. When people talk about cancer, they often have a special interest in the features exhibited by a particular form of the disease. This appendix is therefore designed to provide a brief overview of the main characteristics exhibited by the most common types of cancer. More detailed information about individual cancers can be found at websites sponsored by the National Cancer Institute (http://www.cancer.gov/cancertopics) and the American Cancer Society (http://www.cancer.org).

BLADDER CANCER

Main Types. Bladder cancer arises from the epithelial cells that line the inner surface of the bladder. It is more frequent in men than in women and occurs in three main types. The most common, accounting for roughly 90% of all bladder cancers, is a *transitional cell carcinoma* that develops from the main cell type lining the inner wall of the bladder. These cells are said to be "transitional" because they change shape from cubical when the bladder is empty to flattened when the bladder is full. The other types of bladder cancer are *squamous cell carcinoma* and *adenocarcinoma*, which account for roughly 8% and 2% of the total cases, respectively. Although all three types of bladder cancer are capable of metastasizing, transitional cell carcinoma often does not.

Risk Factors. Cigarette smoking is the most common risk factor for bladder cancer. Smokers are twice as likely to develop bladder cancer as nonsmokers, and smoking has been linked to roughly 50% of bladder cancer deaths in men and almost 30% of bladder cancers deaths in women. Bladder cancer risk is also increased by exposure to chemicals used in the clothing dye, textile, leather, rubber, and print industries. These chemicals, as well as the chemicals in tobacco smoke, become concentrated in the urine and cause damage to the cells lining the bladder wall. The industrial use of 2-naphthylamine as a starting material for textile dyes in the early part of the twentieth century led to the first reported epidemic of occupational bladder cancer (p. 81). Chronic bladder infections have also been linked to increased bladder cancer risk.

Symptoms and Diagnosis. Blood in the urine and painful or frequent urination are possible symptoms of bladder cancer, especially if the symptoms do not clear up after treatment for possible bladder infections. Bladder cancer is diagnosed by microscopic examination of cells in the urine and through *cystoscopy*, a procedure in which a doctor inserts a thin tube through the urethra and into the bladder to obtain a small biopsy specimen for microscopic examination.

Treatment and Prognosis. Treatment of bladder cancer depends on the stage of the disease at the time of diagnosis. Surgical removal of the tumor followed by chemotherapy is the most common treatment for early stage disease. Superficial cancers may be treated by administering immunotherapy (p. 218) or chemotherapy directly into the bladder. Later stage cancers require complete removal of the bladder and, in some cases, surrounding tissues, usually accompanied by chemotherapy, radiation therapy, or both. Five-year survival rates approach 95% for early stage disease with no invasion or metastases, but fall to less than 10% for late stage disease in which the cancer has already metastasized to distant sites.

BRAIN AND NEURAL CANCERS

Main Types. Brain cancers are relatively rare in adults but are the most frequent type of solid tumor in children, accounting for roughly 25% of all childhood cancers. Most brain cancers arise from supporting *glial cells* rather than from neurons. These glial tumors, or *gliomas*, can develop from several different types of glial cells. The most common type of glioma develops from star-shaped glial cells called astrocytes and is therefore called an *astrocytoma*. Brain cancers originating from cell types other than glial cells are extremely rare in adults but are fairly common in young children. The main nonglial brain cancer, called a *medulloblastoma*, arises from a primitive type of neural cell and is usually located in the cerebellum. Medulloblastomas account for about 30% of all childhood brain cancers and usually affect children between the ages of 3 and 8. Another childhood neural cancer, called a *neuroblastoma*, is derived from embryonic nerve cells. Neuroblastomas are most commonly located in the abdomen and tend to appear before 2 years of age. They are the most frequently encountered solid tumor of early childhood.

Risk Factors. The causes of brain cancer have not been well established, but risks are known to be increased in individuals who have had high-dose exposure to ionizing radiation or to certain chemicals such as formaldehyde, vinyl chloride, and acrylonitrile. Brain cancer risk is also increased in individuals whose immune systems have been weakened by infection with HIV. Nonetheless, the large majority of brain cancers are not associated with any known risk factors. In recent years, the question of whether the microwave radiation emitted by cell phones can cause brain cancer has stimulated considerable interest. Thus far, epidemiological and laboratory studies have failed to detect a consistent relationship between cell phone usage and brain cancer rates, but there has been insufficient time to determine whether any long-term hazards exist (p. 116). For the moment, the low-energy nature of the radiation emitted by cell phones combined with the failure to detect any consistent carcinogenic effects suggest that if any cancer risk exists at all, it will probably turn out to be rather small.

Symptoms and Diagnosis. Brain cancers rarely spread to other parts of the body, so symptoms are usually caused by pressure within the skull created by the growing tumor. Headaches are often the first symptom, followed by neurological symptoms such as loss of coordination, dizziness, blurred vision, seizures, personality changes, and partial paralysis. The brain of persons exhibiting such neurological symptoms can be examined either by a computed tomography scan (CT scan) or by magnetic resonance imaging (MRI) to determine whether an abnormal tissue mass is present. If imaging techniques reveal the presence of a mass within the brain, surgery is generally performed to remove and diagnose the tissue mass.

Treatment and Prognosis. The initial treatment for brain cancer is usually surgical removal of the primary tumor. For some tumor types, surgery alone may be sufficient. If all the tumor cannot be removed or if it is spreading rapidly, surgery is often followed by radiation treatment. Chemotherapy is also used in some cases, although its effectiveness is limited by the existence of a *blood-brain barrier* that prevents many drugs from passing from the bloodstream into the brain. Survival rates depend on the type of cancer and whether it can all be removed surgically. Overall five-year survival rates for brain cancer are only about 30%, but the outlook is significantly better for selected tumor types. For example, about 50% of the children treated for medulloblastoma survive more than five years.

BREAST CANCER

Main Types. Breast cancer is the most common type of cancer in women other than skin cancer and is the second leading cause of cancer deaths among women after lung cancer. Breast cancer typically arises from the epithelial cells of the milk glands. The milk glands of each breast are divided into 15 to 20 lobes, each possessing a duct that runs separately into the nipple. The most frequent breast cancers are *ductal carcinomas* derived from the epithelial cells lining these ducts; a smaller number are *lobular carcinomas* that develop from the lobes of the milk glands. Although the vast majority of breast cancers occur in women, men also have small amounts of breast duct tissue and these ductal cells can develop into carcinomas.

Risk Factors. Gender and age are important risk factors for breast cancer; rates are about one hundred times higher in women than in men, and about 80% of all cases are diagnosed in women over 50 years of age. Roughly 10% of all breast cancers are linked to a family history of the disease, associated mainly with inherited mutations in the *BRCA1* or *BRCA2* genes (p. 150). Women who have had extensive exposure to ionizing radiation in the chest area incur an increased risk for breast cancer, and reproductive and hormonal history play a role as well. Women who started menstruating at an early age or go through menopause at a late age are at slightly increased risk, as are women who have had no children or who had their first child after age 30. Women who used older types of oral contraceptives, which contained higher doses of estrogen, exhibit a slight elevation in breast cancer risk. Estrogen replacement therapy after menopause has also been linked to small increases in breast cancer risk. Use of alcohol, obesity, and lack of physical exercise are additional risk factors. Some reports have suggested that environmental pollution involving pesticides that mimic the action of estrogens can promote the development of breast cancer, although current evidence does not show a consistent link (p. 85). Claims about risk associated with antiperspirants, induced abortion, or breast implants are not supported by most of the currently available evidence. For women considered to be at high risk for breast cancer based on either family or medical history, tamoxifen has been shown to decrease the risk by roughly 50% (p. 248).

Symptoms and Diagnosis. The first symptom of breast cancer is usually a lump detected during breast self-examination. Unexplained breast swelling, thickening, skin irritation or dimpling, tenderness, nipple pain, and discharge other than breast milk are other possible signs. Mammography is capable of detecting small breast cancers before they can be felt by a woman or by her physician. It is therefore currently being recommended that women over age 40 begin regular mammography to screen for breast cancer (p. 202). When physical examination or mammography indicates the presence of an abnormal mass, a biopsy is done to determine whether cancer is present.

Treatment and Prognosis. Surgical removal of the primary tumor is usually the initial treatment for breast cancer, although chemotherapy is sometimes administered before surgery to shrink tumors and decrease the amount of surgery required. The extent of the surgery depends on the size and location of the tumor. For many years, the standard operation was a *mastectomy* to remove the entire breast, with or without the underlying chest muscles. Today most breast cancers are treated with limited surgery (*partial mastectomy* or *lumpectomy*), which removes only the tumor and a small amount of surrounding normal tissue. Some of the lymph nodes under the armpit may also be removed to determine whether the tumor has begun to spread. Surgery is often followed by radiation therapy to the breast to destroy any cancer cells that may remain in the area. The risk of distant metastases can be lowered by chemotherapy. When estrogen receptors are present (about two-thirds of breast cancers), antiestrogens such as tamoxifen (p. 215) may be administered to block the action of estrogen on breast cells. A new approach for assessing the risk of recurrence involves testing of gene expression profiles (p. 207). If the risk of recurrence is high, combination chemotherapy with one of several possible drug combinations is usually employed. Breast cancers that have high levels of the growth factor receptor ErbB2 may be treated with Herceptin (p. 223), a monoclonal antibody that binds to and blocks the receptor. Death rates from breast cancer have been slowly declining in recent years, probably because of the growing use of mammography and the introduction of therapy with tamoxifen and various drug combinations. The five-year survival rate for early stage breast cancer, with no invasion or metastases, is greater than 95%, but falls to roughly 80% for disease with regional spreading and 25% for late stage disease in which the cancer has already metastasized to other parts of the body.

COLORECTAL CANCER

Main Types. Colorectal cancer, which may arise anywhere along the length of the colon or rectum, is the second leading cause of cancer deaths in the United States (after lung cancer) for men and women combined. Colorectal cancers frequently begin as polyps that are benign outgrowths (adenomas) emerging from the epithelial lining of the colon or rectum. More than 95% are *adenocarcinomas*, but *lymphomas*, *carcinoid tumors* (derived from hormone-producing cells), and *gastrointestinal stromal tumors* occasionally occur as well. Discussions regarding the properties and behavior of colorectal cancers almost always refer to adenocarcinomas.

Risk Factors. The risk of developing colorectal cancer is strongly affected by age, with rates increasing dramatically after 50 years of age. Because the lining of the colon is continually exposed to the food we consume, diet is thought to play a significant role in the development of colorectal cancer. Evidence points to diets that are high in saturated animal fat (especially from red meat) and calories as being likely risk factors, especially in individuals who are overweight or physically inactive. Diets low in fruits and vegetables, folic acid, or vitamin D are also linked to increased risk, as is a history of smoking cigarettes and alcohol consumption. Individuals with a family history of colorectal cancer are generally at higher risk for the disease, and risks are dramatically increased for individuals who inherit the gene mutations responsible for either *familial adenomatous polyposis* (p. 144) or *hereditary nonpolyposis colon cancer* (p. 150). A previous history of colon polyps or *ulcerative colitis* (inflammation of the colon) also puts individuals at higher risk. The regular use of aspirin and related anti-inflammatory drugs may reduce the risk of colorectal cancer.

Symptoms and Diagnosis. Colorectal cancers usually produce no symptoms in their early stages. As a tumor grows larger, common symptoms include blood in the stool, changes in bowel habits such as diarrhea or constipation, and abdominal pain or discomfort. Routine screening for colon cancer using colonoscopy, X-ray imaging, or the fecal occult blood test is recommended beginning at age 50 (or earlier for others at high risk). Definitive diagnosis of colorectal cancer requires a biopsy, which can be performed at the time of colonoscopy.

Treatment and Prognosis. Surgery is the main treatment for colorectal cancer. Removing a malignant polyp during colonoscopy may provide a cure if the tumor is still localized. For tumors that have begun to invade, a portion of the colon may need to be removed. Later stage cancers require removal of an extensive stretch of colon and surrounding tissues, followed by chemotherapy and sometimes radiation therapy. Five-year survival rates for early stage colorectal cancer, with no invasion or metastases, are about 90%, but fall to 65% for tumors with regional spread and 10% for late stage disease in which the cancer has metastasized to distant sites.

ESOPHAGEAL CANCER

Main Types. Esophageal cancers develop from the epithelial lining of the esophagus and are subdivided into two major types, *squamous cell carcinomas* and *adenocarcinomas*. The entire length of the esophagus is normally lined by squamous cells (thin, flat cells), so squamous cell carcinomas can arise anywhere in the esophagus. Adenocarcinomas composed of glandular cells resembling those found in the stomach typically appear in the lower third of the esophagus. Glandular cells are not a normal component of the esophageal epithelium but they sometimes replace the squamous cells in individuals with chronic *acid reflux*, a condition in which acid escapes from the stomach and backs up into the esophagus. When glandular cells replace the squamous cell lining any farther up than a few centimeters above the esophageal-stomach junction, the condition is referred to as *Barrett's esophagus* and puts people at an elevated risk for developing esophageal cancer.

Risk Factors. Esophageal cancer is roughly three times more frequent in men than in women and exhibits marked geographic differences in incidence, being especially prevalent in certain regions of China, Iran, and Africa. Smoking cigarettes or using smokeless tobacco is a major risk factor for esophageal cancer, as is the chronic heavy use of alcohol. Combining the two habits (tobacco and alcohol) is especially dangerous, leading to esophageal cancer rates that may be 30 times higher than normal. Long-term irritation of the esophagus, such as that caused by acid reflux from the stomach, increases a person's risk of developing esophageal cancer. Other sources of irritation to the esophagus include swallowing caustic or very hot liquids and chronic breathing of certain types of toxic chemical fumes. For example, occupational exposure to the chemical perchloroethylene, which is used in the dry cleaning industry, has been linked to an increased risk for esophageal cancer.

Symptoms and Diagnosis. The most common symptom of esophageal cancer is difficulty in swallowing solid foods. As the tumor grows, even the swallowing of liquids may become difficult and painful as the pathway to the stomach becomes blocked. Other possible symptoms include unexplained episodes of choking or painful spasms after eating, vomiting or coughing up blood, hoarseness, chronic cough, and pain in the throat, back, or chest. Weight loss is frequently associated with esophageal cancer because the difficulty in swallowing causes people to eat less. Individuals exhibiting any of the preceding symptoms should visit a doctor for medical examination to determine whether cancer is present. The first test to be done is usually a *barium swallow*, in which X-ray pictures are taken of the esophagus after an individual has swallowed a barium-containing liquid that coats the esophagus to help increase the contrast of the X-ray picture. The other common test is *endoscopy*, in which a doctor inserts a thin, flexible tube containing a light and a video camera down a person's throat and into the esophagus. This procedure permits visual examination of the entire length of the esophagus and allows a biopsy specimen to be removed for microscopic examination to determine whether cancer is present.

Treatment and Prognosis. The most common treatment for esophageal cancer is surgical removal of the primary tumor, along with all or a portion of the esophagus, nearby lymph nodes, and other surrounding tissues. The remaining, healthy part of the esophagus is then connected to the stomach, either directly or using a piece of intestine to bridge the gap, thereby allowing the patient to swallow and eat. If the cancer cannot be removed surgically because of extensive spread, chemotherapy and radiation therapy may be administered in combination to control the disease. Laser therapy, which employs high-energy light to destroy tumor cells, can also be used in the treatment of esophageal cancer. In a special type of laser therapy called *photodynamic therapy*, patients are injected with a harmless chemical that accumulates in cancer cells and is converted into a toxic chemical by exposing the cells to high-intensity light. Esophageal cancer is one of the more difficult cancers to cure, especially after it has metastasized to other regions of the body. Five-year survival rates for early stage tumors localized to the esophagus are only about 30%, fall to less than 15% for tumors with regional spread, and are less than 5% for late stage disease in which the cancer has metastasized to distant sites.

KIDNEY CANCER

Main Types. In adults, about 85% of all kidney cancers are *renal cell carcinomas* arising from the epithelial lining of the kidney tubules. The remainder are *transitional cell carcinomas* (a cancer that also occurs in the bladder) arising in the renal pelvis, the region of the kidney that collects urine and passes it into the ureter. Adult kidney cancers are about twice as common in men than in women. *Wilms tumor*, which occurs mainly in children under five years of age and accounts for about 5% of all childhood cancers, is a cancer of embryonic kidney cells that behaves differently from kidney cancer in adults.

Risk Factors. The risk of developing kidney cancer is roughly doubled in cigarette smokers. Risk is also increased by occupational exposure to petroleum products, coke ovens in steel plants, asbestos, heavy metals, and cadmium in products such as batteries, paints, and welding materials. An increased incidence of kidney cancer has been linked to heavy use of nonprescription pain medications containing *phenacetin*, which were once popular but are no longer sold in the United States. Some studies have shown a link between obesity, high-fat diets, and kidney cancer rates. Kidney patients on long-term dialysis to treat chronic kidney failure have an increased risk of developing kidney cancer. The inherited syndrome *von Hippel–Lindau disease* puts individuals at increased risk for renal cell carcinoma (as well as for tumors in several other organs).

Symptoms and Diagnosis. Kidney cancer has few symptoms in its early stages, but blood usually appears in the urine as the tumor grows in size. Enough blood may be present to be visible during urination, or it may be present in microscopic amounts that are only detectable when a urine sample is examined during a routine medical exam. A pain in the side or back that won't go away or a lump in the kidney area is another sign that kidney cancer may be present. When any of the preceding symptoms persist without an obvious explanation, imaging techniques that may include X-rays, MRI, and ultrasound are used to take pictures of the kidneys and surrounding tissues. If such tests reveal an abnormal tissue mass, a thin needle can be inserted into the kidney to obtain a biopsy specimen for microscopic examination.

Treatment and Prognosis. The most common treatment for kidney cancer is surgical removal of the entire kidney, often along with the adrenal gland and surrounding tissue as well. The remaining kidney is usually capable of performing the work of both kidneys. If the tumor is small and localized, it may be possible to remove just the region of the kidney that contains the tumor. Chemotherapy is usually of limited effectiveness in treating kidney cancer, but if the tumor cannot be removed entirely or signs of spread are evident, surgery is often followed by radiation treatment and immunotherapy with the cytokines interleukin-2 and interferon alpha. Five-year survival rates for adult kidney cancers with no invasion or metastases are about 90%, but fall to 60% for tumors with regional spread and 10% for late stage disease in which the cancer has metastasized to distant sites.

LEUKEMIAS

Main Types. *Leukemias* are cancers that develop from blood-forming cells in the bone marrow. Rather than arising as solid primary tumors, these cancers are characterized by the accumulation of abnormal, immature white blood cells in the bloodstream (leukemia literally means "white blood"). The various leukemias are classified on the basis of the cell type involved and whether the disease is *acute* (rapidly growing cells that are highly abnormal) or *chronic* (slower-growing cells that are less abnormal). Leukemias in children are almost always acute, whereas adults are susceptible to both acute and chronic leukemias. Leukemias develop from two different types of white blood cells called lymphoid cells and myeloid cells. The main leukemias arising from lymphoid cells are *acute lymphocytic leukemia (ALL)* and *chronic lymphocytic leukemia (CLL)*, and those that arise from myeloid cells are *acute myelogenous leukemia (AML)* and *chronic myelogenous leukemia (CML)*. Leukemias are the most common cancers to arise in children, accounting for about 30% of all cancers occurring in individuals up to 14 years of age. Although they account for a much smaller fraction of the cancers arising in adults, leukemias are about ten times more frequent in adults than in children.

Risk Factors. The risk of developing leukemia is increased in individuals who have been exposed to high doses of ionizing radiation. Risk can also be increased by smoking cigarettes or by chronic exposure to certain chemicals, such as benzene and formaldehyde. Infection with human T-lymphotropic virus I (HTLV-I) causes a rare form of chronic lymphocytic leukemia. Cancer patients who have received certain types of chemotherapeutic drugs, such as alkylating agents, are at risk for developing leukemia many years later. Increased risk for leukemia is also associated with several genetic diseases, especially those involving abnormal chromosomes or genetic instability. Included in these categories are *ataxia telangiectasia*, *Bloom syndrome*, and *Fanconi anemia* (p. 151).

Symptoms and Diagnosis. Various symptoms are associated with leukemia, depending on the number of cancer cells present in the bloodstream and where they accumulate. Common symptoms include fever, night sweats, fatigue, headache, easy bruising or bleeding, bone or joint pain, abdominal swelling, frequent infections, swollen lymph nodes, and weight loss. Such symptoms can also be caused by less serious conditions, but if they persist, a blood test should be performed to look for an excessive number of white blood cells. A definitive diagnosis of leukemia usually requires a bone marrow biopsy.

Treatment and Prognosis. The cancer cells occurring in individuals with leukemia spread through the bone marrow and bloodstream to multiple sites, so surgery is rarely useful. Chemotherapy with various drugs, either alone or in combinations, is the most common treatment, but radiation is sometimes used as well. The molecular-targeting drug Gleevec (p. 223) is especially designed for treating chronic myelogenous leukemia. Stem cell transplants in combination with high-dose chemotherapy may be useful for treating certain leukemias. Five-year survival rates vary with the type of leukemia, ranging from roughly 75% for chronic lymphocytic leukemia to 20% for acute myelogenous leukemia. Survival rates for acute lymphocytic leukemia have improved significantly over the past several decades, increasing from 60% to 90% for children and from 40% to 65% for adults.

LIVER CANCER

Main Types. Cancer originating in the liver is rare in North America and Europe but is quite common in Africa and Asia. About 90% of primary liver cancers are *hepatocellular carcinomas* that arise from *hepatocytes*, the main cell type of the liver. Most of the remaining cancers are *cholangiocarcinomas* that originate in the bile ducts. Because a large volume of blood flows through the liver, it is common for cancer cells to be carried to the liver from other parts of the body, such as the colon and stomach. Metastatic tumors that develop from these cancer cells that have spread to the liver from other sites are named according to their site of origin and are not properly referred to as liver cancers.

Risk Factors. As described in Chapter 7, the most common cause of liver cancer globally is chronic infection with hepatitis B virus (HBV) or hepatitis C virus (HCV). HBV is usually transmitted by exchange of body fluids, such as semen or blood. The main route for acquiring HCV is through the sharing of dirty needles by intravenous drug users. Infants may become infected from a mother who carries the virus. A vaccine against HBV infection has been available since 1982, and vaccines for HCV are currently under development. Ingestion of food contaminated with *aflatoxins* (p. 74) is another cause of liver cancer, especially in tropical and subtropical regions where the fungus that produces aflatoxin frequently contaminates stored nuts, grains, beans, corn, and rice. In the United States, cirrhosis of the liver triggered by excessive alcohol drinking is the main cause of liver cancer. Some cases of liver cancer have also been linked to exposure to certain carcinogenic chemicals, such as vinyl chloride and arsenic. The risk of developing liver cancer is increased in individuals who are either obese or whose immune systems have been weakened by infection with HIV. Parasitic infections with liver flukes that contaminate raw or undercooked fish are associated with cancer of the bile ducts in regions of Asia where liver flukes are common.

Symptoms and Diagnosis. Liver cancer usually causes no obvious symptoms in its early stages, but symptoms may eventually include abdominal pain or swelling, weight loss and fatigue, and signs of jaundice such as yellowing of the skin and eyes or darkly colored urine. A person exhibiting such symptoms should have blood tests performed to check for problems with liver function. One blood test detects alpha-fetoprotein, an embryonic protein whose presence in high concentrations in the blood can be a sign of liver cancer. Other blood tests are used to assess whether the liver is functioning properly. If abnormalities are detected, detailed pictures of the liver can be taken using X-ray, MRI, or ultrasound imaging techniques. When such tests indicate that liver cancer may be present, a thin needle is inserted into the liver to obtain a biopsy specimen for microscopic examination.

Treatment and Prognosis. Early stage liver cancers are usually treated by a surgical operation in which part of the liver is removed. Complete removal of the liver may be required for some patients, but this procedure requires transplantation of a new, healthy liver. A liver transplant is an option only if the disease has not spread outside the liver and only if a suitable donated liver can be found. Treatment for advanced liver cancer may involve chemotherapy, radiation therapy, or both, but liver cancer is difficult to control with most available treatments. Even for early stage liver cancers with no signs of invasion or metastasis, five-year survival rates are only about 15%, and the value falls to less than 5% for late stage disease in which the cancer has already spread to other parts of the body.

LUNG CANCER

Main Types. Lung cancer is the leading cause of cancer death in the United States, accounting for about 30% of all cancer fatalities. Cancers arising in the lungs are divided into two major groups called *small cell* and *non-small cell* lung cancers. Small cell lung cancer accounts for about 20% of all lung cancers and is the most aggressive form of the disease. Non-small cell lung cancers, which are the most common form of lung cancer, tend to grow and spread more slowly than small cell lung cancers. Three main types of non-small cell lung cancer can be distinguished from one another based on the cell type involved: *squamous cell carcinoma*, *adenocarcinoma*, and *large cell carcinoma*.

Risk Factors. The number one risk factor for lung cancer is cigarette smoking, which accounts for about 90% of all lung cancers. Cigar and pipe smokers are also at increased risk for lung cancer, as are individuals who live with someone who smokes. Some lung cancers have been linked to exposure to high concentrations of radon or asbestos, although many of the affected individuals also smoke cigarettes. Lung cancer risk is increased by chronic exposure to certain workplace chemicals, such as arsenic, beryllium, cadmium, chloromethyl esters, coal products, nickel chromates, and vinyl chloride.

Symptoms and Diagnosis. Symptoms from lung cancer do not usually develop until the disease has begun to spread. Some possible early symptoms include chronic cough, chest pain, difficulty breathing, spitting up blood, wheezing, and hoarseness. Although they can be caused by diseases other than lung cancer, any such symptoms should be followed up by appropriate medical testing to determine whether cancer is present. Such testing usually involves chest X-rays or other imaging procedures to take pictures of the lungs. If any unusual tissue masses are detected, microscopic examination of cells obtained from a deep-cough sample of mucus in the lungs may be useful in determining whether it is lung cancer. Definitive diagnosis requires microscopic examination of a biopsy specimen, which can be obtained by inserting a thin tube (*bronchoscope*) down through the windpipe and into the lungs, or by inserting a needle through, or making a surgical incision in, the chest wall.

Treatment and Prognosis. Treatments for lung cancer depend on the type of cancer involved, the size of the tumor, and the extent to which it has spread. Non-small cell lung cancers are usually treated by surgical removal of the tumor. Depending on the size of the tumor, anywhere from a small portion of the lung to an entire lung is removed. Chemotherapy and radiation therapy are then administered to destroy any cancer cells that may remain. Some patients with advanced non-small cell lung cancer respond well to Iressa (p. 224), a drug that targets the receptor for epidermal growth factor. Surgery is less suitable for treating small cell lung cancers because they tend to spread quickly through the body. Chemotherapy is therefore the standard treatment for most patients with small cell lung cancer. Many drugs are active against small cell lung cancers, which often respond well to chemotherapy. Because lung cancers commonly metastasize to the brain, patients are sometimes given radiation treatments to the brain, even though cancer is not found there, to try to prevent tumors from arising. Early stage lung cancers with no invasion or metastases at the time of diagnosis exhibit five-year survival rates of about 50%, but the value falls to less than 5% for late stage disease in which cancer has spread to other parts of the body. Unfortunately, most lung cancers have already begun to spread by the time they are diagnosed, and so the overall five-year survival is only about 15%.

LYMPHOMAS

Main Types. *Lymphoma* is a general term for cancers arising from cells of the lymphatic system. Lymphatic tissue is distributed throughout the body and lymphomas can therefore occur almost anywhere. The main cells of the lymphatic system are B lymphocytes and T lymphocytes. *B lymphocytes* protect the body from invading infectious agents, such as bacteria and viruses, by developing into plasma cells that produce immune proteins called antibodies. *T lymphocytes* are involved in destroying bacteria and foreign or abnormal cells directly; they also facilitate the function of other immune system cells. Lymphomas are subdivided into two main categories: *non-Hodgkin's lymphoma* and *Hodgkin's disease*. Non-Hodgkin's lymphomas are more common and arise from either B lymphocytes (about 85% of cases) or T lymphocytes (about 15% of cases). Hodgkin's disease is a less common type of lymphoma involving a unique class of abnormal B lymphocytes called *Reed-Sternberg cells*. These cells are much larger than normal lymphocytes and look different from the cells of non-Hodgkin's lymphomas or other cancers.

Risk Factors. The risk of developing lymphoma is increased in individuals with weakened immune systems, as occurs in people with AIDS or in organ transplant recipients treated with immunosuppressive drugs. Infection with Epstein-Barr virus (EBV) or human T-lymphotropic virus I (HTLV-I) increases the risk for specific types of lymphoma. An increased susceptibility to lymphomas is also associated with certain genetic diseases. Obesity and high-dose exposure to ionizing radiation or certain toxic chemicals have been linked to an increased incidence of non-Hodgkin's lymphoma.

Symptoms and Diagnosis. The most common symptom of lymphoma is a painless swelling of lymph nodes in the neck, underarm, or groin. Other possible indications include recurrent fever, night sweats, itchy skin, and weight loss or fatigue. Similar symptoms can also be caused by less serious conditions, such as the flu, but if the symptoms persist, a doctor should be consulted. A definitive diagnosis requires that swollen lymph nodes be biopsied to look for the presence of cancer cells. If lymphoma is detected, imaging procedures such as MRI, CT scanning, and ultrasound may be employed to assess the extent and spread of the disease.

Treatment and Prognosis. Lymphomas are usually treated with chemotherapy, radiation therapy, or both. Certain types of lymphoma, most notably non-Hodgkin's B-cell lymphomas, can also be treated using monoclonal antibodies directed against cell surface antigens (p. 220). When standard treatments have failed, some patients may receive high-dose chemotherapy to destroy all cancer cells followed by a stem cell transplant to replace the blood-forming cells damaged by chemotherapy (p. 217). Five-year survival rates for Hodgkin's disease are quite high, approaching 95% for early stage disease and dropping to about 75% for disease that has already spread. Non-Hodgkin's lymphomas have a five-year survival rate of roughly 75% for early stage disease and 30% for later stage, more aggressive disease.

MOUTH, THROAT, AND LARYNGEAL CANCERS

Main Types. A variety of different cancers arise in the mouth and throat area, a region of the body that includes the lips, the lining of the cheeks and mouth, the teeth and gums, the tongue, and the pharynx (throat) and larynx (voice box). The mouth and throat region contains several types of tissue, each containing multiple cell types that can become malignant. The surface of each of these tissues, however, is lined largely by squamous cells (thin, flat cells). *Squamous cell carcinomas* are therefore the most prevalent type of cancer, accounting for more than 90% of all cancers of the mouth and throat.

Risk Factors. The use of tobacco products—including cigarettes, cigars, pipes, and smokeless tobacco—accounts for about 90% of all cancers of the mouth, throat, and larynx. People who drink alcohol are also more likely to develop these cancers than people who do not drink. The combination of alcohol and tobacco is especially dangerous; people who smoke and drink alcohol are up to one hundred times more likely to develop laryngeal cancer than are people who neither smoke nor drink alcohol. The risk of laryngeal cancer is also increased by occupational exposure to wood dust, paint fumes, asbestos, sulfuric acid mist, nickel, and certain chemicals encountered in the metalworking, petroleum, plastics, and textile industries.

Symptoms and Diagnosis. Common symptoms of oral cancers include a chronic sore, lump, pain, bleeding, or numbness in the mouth or throat, pain or difficulty in chewing and swallowing, and chronic cough or earache. Laryngeal cancers that form on the vocal cords are often detected at an early stage because even small tumors in this area can cause hoarseness or other voice changes. Individuals exhibiting the preceding symptoms should be examined by a doctor to determine whether cancer is present. A thorough exam of the mouth and throat is usually performed using special fiber-optic devices (flexible, lighted, narrow tubes inserted through the mouth or nose), which allow direct examination of the back of the throat, base of the tongue, and larynx. If an abnormal area is seen, a biopsy specimen is taken for microscopic examination to determine whether cancer is present.

Treatment and Prognosis. Oral, throat, and laryngeal cancers are treated using surgery, radiation therapy, and chemotherapy, either alone or in appropriate combinations, depending on the type, location, and stage of the cancer. Smaller cancers of the larynx can often be removed without significant effect on the ability to speak, but larger cancers require complete removal of the larynx, which leaves an individual unable to talk using vocal cords. There are several options for restoring the ability to speak, including learning to force air through the esophagus (*esophageal speech*), surgical procedures that place a valve between the trachea and the esophagus (*tracheoesophageal puncture*), and the use of mechanical devices held against the neck or placed in the mouth. Five-year survival rates for early, localized cancers of the oral cavity, throat, and larynx are relatively high (about 80%), but fall to roughly 25% for late stage disease in which the cancer has already spread to other parts of the body.

MULTIPLE MYELOMA

Main Types. Multiple myeloma is a cancer involving immune system cells called *plasma cells*, which are specialized for the purpose of producing antibodies. Plasma cells develop from B lymphocytes as part of the normal immune response and reside mainly in the bone marrow.

When malignant plasma cells form a single tumor, the tumor is called a *plasmacytoma*. More commonly, multiple plasma cell tumors occur throughout the bone marrow and the disease is then referred to as *multiple myeloma.*

Risk Factors. The risk of developing multiple myeloma is increased by high-dose exposure to ionizing radiation or to certain chemicals used in the petroleum industry. Hereditary background sometimes plays a role, and being overweight or obese can also increase a person's risk. Most cases of multiple myeloma, however, cannot be linked to any clearly identifiable risk factors.

Symptoms and Diagnosis. The most common symptoms of multiple myeloma include bone pain, typically in the back or ribs, broken bones, weakness, fatigue, weight loss, and recurrent infections. If such symptoms suggest the possibility of multiple myeloma, the blood or urine can be tested for the presence of abnormal antibody molecules that are produced by myeloma cells. X-ray imaging of the bones and a bone marrow biopsy are usually performed to confirm the diagnosis of multiple myeloma.

Treatment and Prognosis. Because multiple myeloma tends to spread widely through the bone marrow, surgery is rarely useful. Chemotherapy is the most common treatment, but radiation is sometimes used as well. Multiple myeloma is relatively difficult to cure. Five-year survival rates for early stage disease are about 50%, and the value falls to 25% or less for late stage disease.

OVARIAN CANCER

Main Types. Ovaries, which are the female reproductive organs, consist of three main cell types: *germ cells* that produce eggs, *stromal cells* that produce the steroid hormones estrogen and progesterone, and *epithelial cells* that cover the surface of the ovary. Most ovarian cancers (~ 90%) are *adenocarcinomas* arising from the surface epithelial cells, but stromal and germ cells occasionally give rise to cancers as well. The most common germ cell tumors are benign growths called *teratomas*, which occur mainly in women of reproductive age. Teratomas are unusual tumors consisting of a variety of different cell types, including skin, bone, nervous tissue, and teeth. Ovarian germ cells can also give rise to rare cancers called *dysgerminomas*, which do not tend to grow or spread very rapidly.

Risk Factors. Most cases of ovarian cancer occur in postmenopausal women over the age of 50, but the disease affects younger women as well. An increased risk for ovarian cancer can be inherited from either the mother's or the father's side of the family. The risk is especially high if two or more first-degree relatives (mother, daughter, or sister) have had ovarian cancer. Inherited mutations in the *BRCA1* and *BRCA2* genes (p. 150), which increase breast cancer risk, are responsible for about 10% of ovarian cancers. Ovarian cancer risk is also increased in individuals with *hereditary nonpolyposis colon cancer* (p. 150), an inherited syndrome whose main effect is on colon cancer susceptibility. Women who started menstruating before age 12 or who never had children are at somewhat increased risk. Women with fewer children are more likely to develop ovarian cancer than women with more children. The use of fertility drugs that stimulate ovulation may slightly increase ovarian cancer risk. Some studies also suggest an increased risk in overweight women and in women who use hormone replacement therapy after menopause.

Symptoms and Diagnosis. Early symptoms of ovarian cancer may include bloating, increased abdominal size, fatigue, bladder problems such as a frequent need to urinate, abdominal or pelvic pain, back or leg pain, unusual vaginal bleeding, and digestive problems such as indigestion, cramps, or a constant urge to defecate. None of these symptoms is specific for ovarian cancer, but they need to be evaluated by a doctor to determine their cause. A pelvic exam is normally performed to determine whether an abnormal mass can be felt. Imaging techniques such as CT scanning, MRI, and ultrasound are employed to confirm the existence of such a tissue mass. Tests for *CA125*, a tumor antigen often found in elevated amounts in the bloodstream of women with ovarian cancer, may also be performed. If cancer is suspected based on the results of the preceding tests, definitive diagnosis requires microscopic examination of a biopsy specimen removed through a surgical incision made in the lower abdomen. Screening tests for ovarian cancer based on specific patterns of proteins in the blood are currently under development (p. 204).

Treatment and Prognosis. The initial treatment for ovarian cancer is surgical removal of one or both ovaries. For women who are not concerned about having children in the future, the fallopian tubes and uterus may be removed as well. If the disease has already begun to spread, surgery is followed by chemotherapy or radiation therapy. Five-year survival rates for early stage cancers localized to the ovary are very high, usually about 95%. Relatively few cases are detected that early, however, and less than 50% of women with the disease survive more than five years. For ovarian cancers that have metastasized to distant locations before they are detected, the five-year survival rate is only 30%.

PANCREATIC CANCER

Main Types. The pancreas consists of two main cell types: *exocrine* cells that produce digestive enzymes and *endocrine* cells that produce insulin and other hormones. Cancers of endocrine cells are quite rare and tend to have a better outcome than cancers of exocrine cells. About 95% of all pancreatic cancers are *adenocarcinomas* of the exocrine pancreas. They usually develop in the ducts of the pancreas but sometimes arise directly in the glands that produce the pancreatic enzymes.

Risk Factors. The risk of developing pancreatic cancer is strongly related to age, with more than 70% of all cases being diagnosed in individuals older than 65. Pancreatic cancer rates are elevated up to threefold in people who smoke cigarettes (or cigars), a habit that is thought to be the direct cause of about 30% of all pancreatic cancers. This finding may explain why the incidence of pancreatic cancer increased during most of the twentieth century (like lung cancer) and has recently begun to decline now that smoking rates in the United States are falling. A person's risk of developing pancreatic cancer triples if their mother, father, sister, or brother had the disease, making inheritance a factor in roughly 10% of all pancreatic cancers. A family history of colon or ovarian cancer may also increase the risk of pancreatic cancer. Diabetes and chronic inflammatory conditions of the pancreas have been linked to increased pancreatic cancer rates, and some studies suggest that workplace exposure to certain chemicals, a diet high in fat, and obesity may also play a role.

Symptoms and Diagnosis. Pancreatic cancer usually causes no symptoms in its early stages, but the growing cancer may eventually trigger pain in the upper abdomen or back, yellowing of the skin and eyes (from jaundice), dark urine, and digestive problems such as nausea, vomiting, indigestion, and weight loss. None of these symptoms is specific for pancreatic cancer, but a doctor needs to evaluate their cause. A physical exam will be performed to look for abdominal masses or fluid accumulation. Regional lymph nodes and the liver are examined for enlargement or swelling that might indicate the spread of a cancer. The blood may be tested for *CA 19-9* and *CEA* (carcinoembryonic antigen), which are proteins that are sometimes present in elevated amounts in individuals with pancreatic cancer. Imaging techniques such as CT scanning, MRI, and ultrasound are used to take pictures of the pancreas to see whether an abnormal tissue mass is present. If cancer is suspected based on the preceding tests, a definitive diagnosis is made through microscopic examination of a tissue specimen. The most common biopsy procedure is *fine needle aspiration biopsy*, in which a doctor inserts a thin needle through the skin and into the pancreas, using CT or ultrasound images to ensure that the needle is located in the tumor. Tiny pieces of tissue are then removed through the needle for microscopic examination.

Treatment and Prognosis. Depending on the location and spread of the primary tumor, the initial treatment for pancreatic cancer is partial or complete removal of the pancreas, possibly accompanied by removal of surrounding tissues such as part of the stomach, small intestine, bile duct, gallbladder, spleen, and adjacent lymph nodes. Radiation therapy may be employed afterward to destroy any cancer cells that remain in the area after surgery. Chemotherapy is used for cancers that have already metastasized, but pancreatic cancer is one of the most difficult cancers to cure. The five-year survival rate for early stage, localized pancreatic cancers is only about 15%, and the value falls to less than 5% for late stage disease in which the tumor has already metastasized to distant sites.

PROSTATE CANCER

Main Types. Prostate cancer is the most common type of cancer in men other than skin cancer and is the second leading cause of cancer deaths among men after lung cancer. The prostate gland is composed of several cell types, but 99% of all prostate cancers are *adenocarcinomas* arising from the glandular cells that produce seminal fluid, a milky liquid that nourishes sperm. Prostate cancers occasionally grow and spread quickly, making them potentially life threatening, but most develop slowly, create few problems, and often remain undetected. Autopsies performed on men who died of other causes have revealed that many older men had a prostate cancer of which they were unaware.

Risk Factors. The risk of developing prostate cancer is strongly related to age. In the United States, more than 70% of all cases are diagnosed in individuals older than 65, and the average age of patients at the time of diagnosis is 70. The disease is more common in African American men and less common in Asian men than in white men. Risk is elevated in men whose father or brother has had the disease. Some evidence suggests that a diet high in animal fat is a risk factor for prostate cancer, and the risk of dying of the disease is increased in men who are overweight.

Symptoms and Diagnosis. Prostate cancer often has no symptoms in its early stages but may eventually cause problems such as frequent urination, difficulty in starting or holding back urination, burning or painful urination, weak flow of urine, difficulty in having an erection, painful ejaculation, blood in the urine or semen, or chronic pain or stiffness in the lower back, hips, or upper thighs. For men over 50 years of age, many doctors recommend periodic screening for prostate cancer using the PSA blood test (p. 203) and a rectal exam in which the doctor feels the prostate gland by inserting a lubricated, gloved finger into the rectum. Introduction of PSA blood testing in 1988 led to a dramatic increase in the number of prostate cancers detected. When screening results or symptoms suggest the possible presence of prostate cancer, a biopsy is performed to obtain tissue for microscopic examination. The procedure involves inserting a narrow needle through the wall of the rectum and into the prostate gland. Up to a dozen or more tissue samples may be removed to ensure that samples are obtained from different regions of the prostate.

Treatment and Prognosis. Many options are available for treating prostate cancers. With small, localized tumors that are growing slowly, doctors may suggest "watchful waiting" for older men or men with other health problems because in such cases, the side effects of surgery, radiation therapy, or chemotherapy may outweigh the possible benefits. Surgery in which the doctor removes all or part of the prostate is a common treatment for prostate cancer. In some cases, nerve damage that leads to incontinence and impotence is an unavoidable consequence of prostate surgery. Radiation therapy, either alone or after surgery, is another common treatment choice. Radiation may be directed at the tumor from an external machine, or tiny radioactive pellets may be inserted directly into the tumor site (*brachytherapy*). For prostate cancers that have already begun to spread, doctors often prescribe drugs that interfere with the production or action of testosterone, the steroid hormone that stimulates the proliferation of prostate cells. If this approach does not work (or eventually stops working), other types of chemotherapeutic drugs may be employed. In general, prostate cancer is slow to develop and the vast majority of men are diagnosed before it has metastasized. Five-year survival rates for such nonmetastatic prostate cancers are close to 100%, but fall to 35% for late stage disease in which the cancer has already spread to other parts of the body at the time of diagnosis.

SARCOMAS

Main Types. Sarcomas are malignant tumors of supporting tissues that can occur almost anywhere in the body. *Soft tissue sarcomas* include tumors of fat, muscle, fibrous tissue, blood vessels, tendons, and cartilage. These tumors are grouped together because they share certain traits, produce similar symptoms, and are often treated in similar ways. Malignant tumors of bone are also sarcomas, but they behave and are treated somewhat differently. The most common type of bone cancer is *osteosarcoma*, which develops from growing bone cells. Another form of bone cancer, called *Ewing's sarcoma*, arises from immature nerve tissue in bone marrow. Osteosarcoma and Ewing's sarcoma occur most frequently in children and adolescents. As a group, sarcomas account for only about 1% of all cancers.

Risk Factors. The risk of developing soft tissue sarcomas is increased by certain inherited conditions, such as *neurofibromatosis*, *Gardner's syndrome*, *Li-Fraumeni syndrome* (p. 145), and *familial retinoblastoma* (p. 142). All four of these diseases are associated with inheritance of a mutant tumor suppressor gene. Some studies suggest that workers exposed to certain herbicides and wood preservatives have an increased risk of developing soft tissue sarcomas. Occupational exposure to vinyl chloride, which is used in the manufacturing of certain plastics, has been linked to a rare blood vessel tumor (angiosarcoma) of the liver. High-dose exposure to ionizing radiation can also cause sarcomas, although it appears to account for only a small percentage of these tumors. A common source of radiation exposure in individuals who develop sarcomas is from radiation given to treat other tumors. Exposure to Kaposi's sarcoma-associated herpesvirus (KSHV) has been implicated in the development of Kaposi's sarcoma, a rare cancer of blood vessels in the skin and mucous membranes that occurs predominantly in people with AIDS (p. 125). Bone cancers are most common in children and young adults, sometimes arising in individuals who have had radiation or chemotherapy treatments for other conditions.

Symptoms and Diagnosis. Because soft tissues are generally quite flexible, sarcomas often grow rather large, pushing aside normal tissue, before they cause any problems. The first symptom is usually a painless lump or swelling, which may eventually cause pain or soreness as it presses against nearby nerves and tissues. Pain is the most common symptom of bone cancer, which can also weaken bones and lead to fractures. If the preceding signs indicate the possible presence of a tumor, a biopsy is performed to obtain a tissue specimen for microscopic examination.

Treatment and Prognosis. The most common treatment for soft tissue sarcomas is surgical removal of the tumor. Depending on the size and location of the primary tumor, amputation of all or part of an arm or leg may be required. It is often possible, however, to perform limb-sparing surgery, which involves removing as much of the tumor as possible followed by radiation, chemotherapy, or both to destroy the remaining cancer cells. Chemotherapy may also be used if the cancer has spread to other regions of the body. The five-year survival rate for sarcomas is about 90% if the cancer is found before it has spread, but the value falls to less than 30% for sarcomas that have already metastasized at the time of diagnosis.

SKIN CANCER

Main Types. Skin cancers are by far the most common type of cancer. More than one million new cases of skin cancer are diagnosed in the United States each year, accounting for roughly half of all new cancers. Skin cancers are divided into two general categories, *nonmelanoma* and *melanoma*. There are several types of nonmelanoma skin cancer, but two predominate: *basal cell carcinoma* and *squamous cell carcinoma*. Basal cell carcinomas arise in the lowermost layer of the epithelium of the skin and account for about 75% of all skin cancers. Squamous cell carcinomas develop from cells in the middle layers of the epithelium and account for about 20% of all skin cancers. Neither of these cancers invades or metastasizes very frequently. In contrast, *melanomas* arise from pigmented cells (melanocytes) and are far more dangerous because they often metastasize before the tumor has been noticed.

Melanomas account for only 5% of all skin cancers but are responsible for the vast majority of skin cancer fatalities (see Figure 6-2).

Risk Factors. As was described in Chapter 6, ultraviolet (UV) radiation from the sun is the main cause of skin cancer. The risk is greatest for people who have fair skin that burns easily. Artificial sources of UV radiation, such as sunlamps and tanning booths, can also cause skin cancer. Nonmelanoma skin cancers occur mainly in individuals who are exposed to the sun on a regular basis, and tumors tend to appear on areas of the body that receive the most sunlight, such as the face, arms, and hands. In contrast, melanomas often arise in areas of the body that are not routinely exposed to the sun, such as the legs and back, and tend to occur in people who work indoors but have intense periodic exposures to sunlight on weekends or vacations, or who had intense sunburn episodes when they were young. Roughly 10% of all melanomas are linked to a family history of the disease. The risk of developing skin cancers can be reduced by avoiding exposure to intense sunlight and by using protective clothing and sunscreen lotions when strong sunlight cannot be avoided. Although sunlight exposure is the primary risk factor for skin cancer, depressed immune function (as occurs in AIDS) and occupational exposure to certain types of chemicals and radiation have also been linked to increased risk.

Symptoms and Diagnosis. The most common sign of skin cancer is a new growth, blemish, or bump on the skin, especially if it is growing in size, or a sore that does not heal. For pigmented growths, a convenient guide for distinguishing a normal mole (benign tumor of pigmented cells) from a melanoma is the "ABCD" rule, which consists of the following four danger signs: (1) *Asymmetry* of shape in which one side of a pigmented growth does not look the same as the other; (2) *Border* irregularities in which the edges of the pigmented growth are ragged, notched, or blurred; (3) *Color* variations within a pigmented growth involving various shades of tan, brown, or black, as well as white, gray, red, pink, or blue; and (4) *Diameter* of a pigmented growth that exceeds 5 millimeters (roughly the size of a pencil eraser). Melanomas vary in appearance and may exhibit only one or two of the ABCD features. Any suspicious growth on the skin should be examined by a doctor and, if necessary, surgically removed and examined under the microscope to determine whether cancer is present.

Treatment and Prognosis. Most skin cancers are easily cured by removing them surgically. Often the cancer will be completely removed at the time of the initial biopsy and no further treatment is needed. If there are signs of spread, however, surgery to remove surrounding tissue and, in some cases, chemotherapy, radiation therapy, immunotherapy, or a combination of these treatments may be appropriate. Basal cell carcinomas metastasize in less than one person per thousand and squamous cell carcinomas metastasize in only one person in every twenty, so cure rates for these two types of skin cancer are very high (about 99%). Melanomas metastasize more frequently and have a somewhat lower cure rate (roughly 90% for all stages combined). Early diagnosis of melanoma is especially important because the five-year survival rate falls to less than 15% for melanomas that have already metastasized at the time of diagnosis.

STOMACH CANCER

Main Types. More than 90% of stomach cancers are *adenocarcinomas* arising from the glandular epithelial cells that form the inner lining of the stomach. A rare form of stomach cancer known as a *gastrointestinal stromal tumor* develops from cells in the stomach wall that are involved in controlling motility of the stomach muscle. Another rare type of stomach cancer, called a *carcinoid tumor*, arises from hormone-producing cells of the stomach. Carcinoid tumors do not usually spread to other organs.

Risk Factors. Infection with the bacterium *H. pylori* is a major cause of stomach cancer (p. 126). The incidence of stomach cancer in the United States has declined dramatically since the 1930s (see Figure 1-2), perhaps because the frequent use of antibiotics to treat childhood infections has led to a decrease in the prevalence of *H. pylori*. High rates of stomach cancer are common in countries where large amounts of smoked, cured, salted, and pickled foods are consumed. Damage to the stomach lining caused by such preservatives, along with less frequent use of antibiotics, may help explain why stomach cancer is more common in certain parts of the world, such as Japan, Korea, and Latin America, than it is in the United States. Stomach cancer rates are roughly doubled in cigarette smokers, and some studies suggest that drinking alcohol may also contribute to increased risk. A small increase in stomach cancer risk is experienced by people who inherit certain cancer susceptibility syndromes, such as *hereditary nonpolyposis colorectal cancer* (p. 150), *familial adenomatous polyposis* (p. 145), and *familial breast cancer* (p. 151). People are also more likely to develop stomach cancer if several close blood relatives have had the disease.

Symptoms and Diagnosis. Stomach cancer rarely produces symptoms in its early stages but may eventually cause problems such as chronic abdominal pain, swelling, or discomfort, a sense of fullness after eating a small meal, heartburn, indigestion, nausea, and vomiting with or without blood. Individuals exhibiting such symptoms should visit a doctor for medical examination to determine the cause. One common test is an *upper GI series*, an imaging procedure in which X-ray pictures are taken of the upper digestive system after a person has swallowed a barium-containing liquid that coats the stomach and makes it easier to see tumors. The other main test is *endoscopy*, in which a thin, flexible tube containing a light and video camera is inserted down a person's throat and into the stomach. This instrument allows a doctor to view the lining of the esophagus, stomach, and first part of the small intestine. If abnormalities are noted, a biopsy can be taken for microscopic examination to determine whether cancer is present.

Treatment and Prognosis. The most common treatment for stomach cancer is surgery to remove all or part of the stomach along with some of the surrounding tissue. After partial removal of the stomach, the remaining portion is connected to the esophagus or small intestine. If the stomach is completely removed, the esophagus is connected directly to the small intestine. Changes in diet and eating habits—including use of vitamin supplements and eating smaller, more frequent meals—are usually necessary after stomach surgery. Surgery is commonly followed by chemotherapy, radiation therapy, or both, aimed at destroying any cancer cells that may not have been removed during surgery. The five-year survival rate for stomach cancer is nearly 60% when the disease is diagnosed early, but falls to less than 5% for late stage disease in which the cancer has spread to other parts of the body. The overall five-year survival rate for stomach cancer is only about 20% because the disease is often detected at an advanced stage.

TESTICULAR CANCER

Main Types. Cancer of the testes is relatively rare, accounting for only 1% of all cancers in men. Most testicular cancers fall into one of two general categories: *seminoma* and *nonseminoma*. Seminomas arise from sperm-producing cells and account for roughly 30% of all testicular cancers. Nonseminomas are a more diverse group of cancers that include four main tumor types: *choriocarcinoma*, *embryonal carcinoma*, *teratocarcinoma*, and *yolk sac carcinoma*. Despite structural differences in the cell types involved, the various types of nonseminoma tend to behave in similar ways.

Risk Factors. Nonseminomas occur most frequently between the ages of 15 and 40, whereas seminomas develop later in life, usually in men between their late 30s and late 50s. Men with a history of abnormal testicular development, including undescended testicles that did not move down into the scrotum, exhibit an increased risk for testicular cancer. Risk is also increased by certain chromosomal abnormalities and by a family history of testicular cancer. Data from early studies suggested a slightly increased risk in men who have had a vasectomy, a surgical procedure in which the sperm ducts are tied off in individuals who do not wish to have more children. Recent studies have failed to confirm the existence of such a risk, however.

Symptoms and Diagnosis. Testicular cancers tend to be discovered relatively early because men will experience symptoms such as a painless lump, swelling, or enlargement in a testicle, a feeling of heaviness or fluid in the scrotum, pain in the testicle or scrotum, or a dull ache in the lower abdomen, back, or groin. If such symptoms persist, a doctor may first suggest the use of ultrasound imaging of the testes. Ultrasound pictures can determine whether a mass is present and are helpful in ruling out noncancerous conditions, such as swelling caused by infection. Blood tests may be performed for proteins such as *alpha-fetoprotein*, *human chorionic gonadotropin*, and *lactate dehydrogenase*, which are commonly elevated in individuals with testicular cancer. If cancer is suspected, biopsy is the only way to make a definitive diagnosis. In most cases, the entire affected testicle is removed through an incision in the groin and the tissue is then examined with a microscope.

Treatment and Prognosis. When cancer is diagnosed in a testicle that has already been removed, the need for additional treatment is determined by both the type and stage of the disease. Seminoma cells are particularly sensitive to radiation-induced killing, so radiation therapy to the groin area is commonly recommended to destroy any seminoma cells that may remain behind after surgery. Chemotherapy is generally used after surgery for nonseminomas because these tumors are less responsive to radiation treatment. Cure rates for testicular cancer are very high. Five-year survival rates are 95% for all stages combined and reach 99% if the disease is diagnosed while still localized to the testes. Even when distant metastases are present at the time of diagnosis, five-year survival rates are close to 75%.

THYROID CANCER

Main Types. The thyroid gland consists of two main cell types: *follicular cells* that take up iodine from the blood and make thyroid hormone, and *C cells* that produce the hormone calcitonin. About 90% of thyroid cancers are either *papillary* or *follicular carcinomas*, both of which are derived from follicular cells. *Medullary thyroid carcinomas* develop from C cells and account for 5% to 10% of thyroid cancers. *Anaplastic thyroid cancer* is the rarest type of thyroid cancer, accounting for only 1% to 2% of all cases. It arises from follicular cells, but the cancer cells are highly abnormal and poorly differentiated.

Risk Factors. Papillary and follicular thyroid cancers are most prevalent in countries where the normal diet is low in iodine. In the United States and many other developed countries, this risk is minimized because iodine is added to table salt and other foods. High-dose exposure to ionizing radiation is the most clearly established risk factor for papillary and follicular thyroid cancers. Between the 1920s and 1950s, some doctors used high-dose X-rays to treat children with superficial skin conditions of the head and neck, such as ringworm and acne, or for enlarged tonsils. Individuals treated in this way eventually developed thyroid cancers at much higher rates than normal. Radioactive fallout from atomic weapons and nuclear power accidents contains a radioactive form of iodine that becomes concentrated in the thyroid gland and increases the risk of thyroid cancer. Some inherited conditions also increase the risk of developing the disease. For example, almost all individuals who inherit certain mutations in the *RET* gene (p. 152) eventually develop medullary thyroid cancer.

Symptoms and Diagnosis. Thyroid cancers tend to cause few symptoms in their early stages but as they continue to grow, symptoms may include a lump in the front of the neck, pain in the throat or neck, hoarseness or difficulty speaking, swollen lymph nodes in the neck, or difficulty in swallowing or breathing. If a person develops symptoms that suggest the possible existence of thyroid cancer, a doctor may evaluate thyroid function by testing for the concentration of thyroid-related hormones and calcium in the blood. Ultrasound or other imaging techniques are used to take pictures of the thyroid gland, which can reveal whether an abnormal tissue mass is present. If these tests suggest the possibility of thyroid cancer, a biopsy is performed by inserting a thin needle into the thyroid to obtain a biopsy specimen for microscopic examination.

Treatment and Prognosis. Surgery to remove all or part of the thyroid gland is the most common treatment for thyroid cancer. After surgery, patients are given thyroid hormone pills to compensate for loss of the natural hormone produced by the thyroid gland. Thyroid hormone pills also inhibit production of *thyroid stimulating hormone*, a hormone produced by the pituitary gland that stimulates the growth of thyroid cells. Patients with papillary or follicular cancers are often treated after surgery with *radioactive iodine*, usually in liquid or capsule form, to destroy cancer cells that may have spread in the neck area or elsewhere in the body. Because thyroid cells absorb most of the iodine circulating in the blood, the radioactive iodine becomes concentrated in these cells and delivers a toxic dose of radiation with minimum damage to other body tissues. Cancers that have spread beyond the thyroid gland and do not respond to radioactive iodine therapy may be treated with external radiation. Chemotherapy is generally not very effective with thyroid cancer, although it is sometimes useful if the disease no longer responds to other treatments. In general, most thyroid cancers grow rather slowly and cure rates are high. Five-year survival rates are 95% for all stages combined and reach 99% if the disease is diagnosed while still localized to the thyroid gland. When distant metastases are present at the time of diagnosis, five-year survival rates fall to roughly 60%.

UTERINE ENDOMETRIAL AND CERVICAL CANCERS

Main Types. The uterus is a hollow, pear-shaped organ consisting of an upper main body called the *corpus* and a lower narrower region called the *cervix*, which forms a canal that opens into the vagina. Most cancers arising in the corpus are derived from the inner lining of the uterus, called the endometrium, and are referred to as *endometrial cancers*. Nearly all endometrial cancers are *adenocarcinomas*. Beneath the endometrium is a layer of muscle that can give rise to sarcomas, a fundamentally different type of cancer described in the section of this appendix that covers "Sarcomas". Cancers arising in the cervix are called *cervical cancers*. There are two main types of cervical cancers: *squamous cell carcinomas*, which account for 80% to 90% of cervical cancers, and *adenocarcinomas*, which account for the remaining 10% to 20%.

Risk Factors. The risk factors for endometrial and cervical cancers differ from each other. Many of the risk factors for endometrial cancer are based on exposure to estrogens, which stimulate endometrial cell proliferation. For example, an increased risk for endometrial cancer has been observed in women who use estrogen to control symptoms of menopause. Obesity, which increases estrogen levels by storing estrogen in fatty tissues, increases the risk of endometrial cancer severalfold. The drug tamoxifen, which acts like an estrogen in the uterus, also increases the risk of endometrial cancer. In some cases susceptibility to endometrial cancer can be hereditary, especially in individuals who inherit the colon cancer susceptibility syndrome known as *hereditary nonpolyposis colon cancer* (p. 150).

Unlike endometrial cancer, the risk of developing cervical cancer is related to sexual behavior. Cervical cancer rates are elevated in women who have had many sexual partners, in women who start sexual activity at an early age, and in women having sexual relationships with men who have had many sexual partners. The increased risk associated with these sexual behaviors has been linked to sexual transmission of certain high-risk types of human papillomavirus (HPV). The most prevalent member of this group, HPV 16, is detected in roughly half of all cervical cancers, followed in frequency by HPV 18, HPV 45, HPV 31, and a small group of others (see Figure 7-3). Smoking cigarettes is another risk factor for cervical cancer, roughly doubling a woman's chance of developing the disease. Women whose mothers were given diethylstilbestrol during pregnancy to prevent miscarriage, a common practice between 1940 and 1970, are also at increased risk (p. 88). Some reports suggest that women whose immune systems have been weakened by either HIV infection or through the use of immunosuppressive drugs are also more likely to develop cervical cancer. For example, in organ transplant patients treated with immunosuppressive drugs to decrease the risk of immune rejection of the transplanted organ, cervical cancer rates are almost ten times higher than normal. It has also been reported that overweight women are at increased risk of dying of cervical cancer.

Symptoms and Diagnosis. Abnormal vaginal bleeding or discharge is the most common symptom of uterine cancers. Other possible symptoms include difficult or painful urination, pain during intercourse, or pain in the pelvic region. Screening with the Pap smear technique (p. 202), sometimes accompanied by testing for HPV, is very effective for the early detection of cervical cancer but does not detect most endometrial cancers. The increased prevalence of Pap screening in recent years has allowed cervical cancer to be detected more frequently in its preinvasive stages. If a Pap smear shows abnormal cells, a doctor will usually perform a test called a *colposcopy*, which employs an instrument with magnifying lenses to view the cervix. If abnormal areas are seen, a biopsy is performed. In cases in which endometrial cancer is suspected, a sample of endometrial tissue is obtained through a thin, flexible tube inserted into the uterus through the cervix. Microscopic examination of biopsy specimens is the only way to obtain a definitive diagnosis of endometrial or cervical cancer.

Treatment and Prognosis. Treatment for uterine cancer begins with surgery to remove the tumor tissue. For precancerous or early stage cancers located on the surface of the cervix, the doctor may use cryosurgery (freezing), cauterization (burning), or laser surgery to destroy the abnormal area without harming nearby healthy tissue. If the tumor is larger and invasive but has not spread beyond the cervix, an operation may be performed to remove the tumor while leaving the rest of the uterus and the ovaries intact. Endometrial cancers or cervical cancers that have spread beyond the cervix are treated by removing the entire uterus (*hysterectomy*), sometimes along with both fallopian tubes and both ovaries. Surgery is usually followed by radiation therapy aimed at destroying any cancer cells that might have been left behind after surgery. Chemotherapy, hormonal therapy, or both may also be administered. The five-year survival rate for uterine cancers exceeds 90% when the disease is diagnosed before it has spread, but falls to less than 25% for cancers that have already metastasized to other parts of the body.

Appendix B

Known and Suspected Human Carcinogens

Part A. Known to Be a Human Carcinogen

The following is a list of carcinogens for which the data from animal studies have been supplemented with enough human data to clearly establish a cancer risk for humans.

Aflatoxins
Alcoholic beverage consumption
4-Aminobiphenyl
Arsenic compounds, inorganic
Asbestos
Azathioprine
Benzene
Benzidine and dyes metabolized to benzidine
Beryllium and beryllium compounds
1,3-Butadiene
1,4-Butanediol dimethylsulfonate (Myleran®)
Cadmium and cadmium compounds
Chlorambucil
1-(2-Chloroethyl)-3-(4-methylcyclohexyl)-1-nitrosourea (MeCCNU)
bis(Chloromethyl) ether and technical-grade chloromethyl methyl ether
Chromium hexavalent compounds
Coal tars and coal tar pitches
Coke oven emissions
Cyclophosphamide
Cyclosporin A
Diethylstilbestrol
Erionite
Estrogens, steroidal
Ethylene oxide
Hepatitis B virus
Hepatitis C virus
Human papillomaviruses: some genital-mucosal types
Melphalan
Methoxsalen with ultraviolet A therapy (PUVA)
Mineral oils (untreated and mildly treated)
Mustard gas (sulfur mustard)
2-Naphthylamine
Neutrons
Nickel compounds
Phenacetin, analgesic mixtures containing
Radon
Silica, crystalline (respirable size)
Soots
Sulfuric acid, strong inorganic acid mists containing
Tamoxifen
2,3,7,8-Tetrachlorodibenzo-p-dioxin (TCDD); "dioxin"
Thiotepa
Thorium dioxide
Tobacco related exposures (tobacco smoking, smokeless tobacco, environmental tobacco smoke)
Ultraviolet radiation (solar radiation, broad spectrum ultraviolet radiation, sunlamps or sunbeds)
Vinyl chloride
Wood dust
X-radiation and gamma radiation

Part B. Reasonably Anticipated to Be a Human Carcinogen

The following is a list of carcinogens for which the potential cancer hazard to humans has been extrapolated largely from animal studies.

Acetaldehyde
2-Acetylaminofluorene
Acrylamide
Acrylonitrile
Adriamycin® (doxorubicin hydrochloride)
2-Aminoanthraquinone
o-Aminoazotoluene
1-Amino-2,4-dibromoanthraquinone
1-Amino-2-methylanthraquinone
2-Amino-3,4-dimethylimidazo[4,5-*f*]quinoline (MeIQ)
2-Amino-3,8-dimethylimidazo[4,5-*f*]quinoxaline (MeIQx)
2-Amino-3-methylimidazo[4,5-*f*]quinoline (IQ)
2-Amino-1-methyl-6-phenylimidazo[4,5-*b*]pyridine (PhIP)
Amitrole
o-Anisidine hydrochloride
Azacitidine (5-azacytidine®, 5-azaC)
Benz[*a*]anthracene (polycyclic aromatic hydrocarbon)
Benzo[*b*]fluoranthene (polycyclic aromatic hydrocarbon)
Benzo[*j*]fluoranthene (polycyclic aromatic hydrocarbon)
Benzo[*k*]fluoranthene (polycyclic aromatic hydrocarbon)
Benzo[*a*]pyrene (polycyclic aromatic hydrocarbon)
Benzotrichloride
Bromodichloromethane
2,2-bis-(bromoethyl)-1,3-propanediol (technical grade)
Butylated hydroxyanisole (BHA)
Carbon tetrachloride
Ceramic fibers (respirable size)
Chloramphenicol
Chlorendic acid
Chlorinated paraffins (C_{12}, 60% chlorine)
1-(2-Chloroethyl)-3-cyclohexyl-1-nitrosourea
bis(Chloroethyl) nitrosourea
Chloroform
3-Chloro-2-methylpropene
4-Chloro-*o*-phenylenediamine
Chloroprene
p-Chloro-*o*-toluidine and *p*-chloro-*o*-toluidine hydrochloride
Chlorozotocin
C.I. Basic Red 9 monohydrochloride
Cisplatin
Cobalt sulfate
p-Cresidine
Cupferron
Dacarbazine
Danthron (1,8-dihydroxyanthraquinone)
2,4-Diaminoanisole sulfate
2,4-Diaminotoluene
Diazoaminobenzene
Dibenz[*a,h*]acridine (polycyclic aromatic hydrocarbon)
Dibenz[*a,j*]acridine (polycyclic aromatic hydrocarbon)
Dibenz[*a,h*]anthracene (polycyclic aromatic hydrocarbon)
7H-Dibenzo[*c,g*]carbazole (polycyclic aromatic hydrocarbon)
Dibenzo[*a,e*]pyrene (polycyclic aromatic hydrocarbon)
Dibenzo[*a,h*]pyrene (polycyclic aromatic hydrocarbon)
Dibenzo[*a,i*]pyrene (polycyclic aromatic hydrocarbon)
Dibenzo[*a,l*]pyrene (polycyclic aromatic hydrocarbon)
1,2-Dibromo-3-chloropropane
1,2-Dibromoethane (ethylene dibromide)
2,3-Dibromo-1-propanol
tris(2,3-Dibromopropyl) phosphate
1,4-Dichlorobenzene
3,3′-Dichlorobenzidine and 3,3′-dichlorobenzidine dihydrochloride
Dichlorodiphenyltrichloroethane (DDT)
1,2-Dichloroethane (ethylene dichloride)
Dichloromethane (methylene chloride)
1,3-Dichloropropene (technical grade)
Diepoxybutane
Diesel exhaust particulates
Diethyl sulfate
Diglycidyl resorcinol ether
3,3′-Dimethoxybenzidine and dyes metabolized to 3,3′-dimethoxybenzidine
4-Dimethylaminoazobenzene
3,3′-Dimethylbenzidine and dyes metabolized to 3,3′-dimethylbenzidine
Dimethylcarbamoyl chloride
1,1-Dimethylhydrazine
Dimethyl sulfate
Dimethylvinyl chloride
1,6-Dinitropyrene
1,8-Dinitropyrene
1,4-Dioxane
Disperse Blue 1
Epichlorohydrin
Ethylene thiourea
di(2-Ethylhexyl) phthalate
Ethyl methanesulfonate
Formaldehyde (gas)
Furan
Glass wool (respirable size)
Glycidol
Hexachlorobenzene
Hexachlorocyclohexane isomers
Hexachloroethane
Hexamethylphosphoramide

Hydrazine and hydrazine sulfate
Hydrazobenzene
Indeno[1,2,3][*c,d*]pyrene (polycyclic aromatic hydrocarbon)
Iron dextran complex
Isoprene
Kepone® (chlordecone)
Lead and lead compounds
Lindane
2-Methylaziridine (propylenimine)
5-Methylchrysene (polycyclic aromatic hydrocarbon)
4,4′-Methylenebis(2-chloroaniline)
4-4′-Methylenebis(*N,N*-dimethyl)benzenamine
4,4′-Methylenedianiline and its dihydrochloride salt
Methyleugenol
Methyl methanesulfonate
N-Methyl-*N*′-nitro-*N*-nitrosoguanidine
Metronidazole
Michler's Ketone [4,4′-(dimethylamino)benzophenone]
Mirex
Naphthalene
Nickel, metallic
Nitrilotriacetic acid
o-Nitroanisole
Nitrobenzene
6-Nitrochrysene
Nitrofen (2,4-dichlorophenyl-*p*-nitrophenyl ether)
Nitrogen mustard hydrochloride
Nitromethane
2-Nitropropane
1-Nitropyrene
4-Nitropyrene
N-Nitrosodi-*n*-butylamine
N-Nitrosodiethanolamine
N-Nitrosodiethylamine
N-Nitrosodimethylamine
N-Nitrosodi-*n*-propylamine
N-Nitroso-*N*-ethylurea
4-(*N*-Nitrosomethylamino)-1-(3-pyridyl)-1-butanone
N-Nitroso-*N*-methylurea
N-Nitrosomethylvinylamine
N-Nitrosomorpholine
N-Nitrosonornicotine
N-Nitrosopiperidine
N-Nitrosopyrrolidine
N-Nitrososarcosine
Norethisterone
Ochratoxin A
4,4′-Oxydianiline
Oxymetholone
Phenacetin
Phenazopyridine hydrochloride
Phenolphthalein
Phenoxybenzamine hydrochloride
Phenytoin
Polybrominated biphenyls (PBBs)
Polychlorinated biphenyls (PCBs)
Polycyclic aromatic hydrocarbons
Procarbazine hydrochloride
Progesterone
1,3-Propane sultone
β-Propiolactone
Propylene oxide
Propylthiouracil
Reserpine
Safrole
Selenium sulfide
Streptozotocin
Styrene-7,8-oxide
Sulfallate
Tetrachloroethylene (perchloroethylene)
Tetrafluoroethylene
Tetranitromethane
Thioacetamide
4,4′-Thiodianaline
Thiourea
Toluene diisocyanate
o-Toluidine and *o*-toluidine hydrochloride
Toxaphene
Trichloroethylene
2,4,6-Trichlorophenol
1,2,3-Trichloropropane
Ultraviolet A radiation
Ultraviolet B radiation
Ultraviolet C radiation
Urethane
Vinyl bromide
4-Vinyl-1-cyclohexene diepoxide
Vinyl fluoride

From *Report on Carcinogens, Eleventh Edition*; U.S. Department of Health and Human Services, Public Health Service, National Toxicology Program, 2005.

Glossary

A

A: see *adenine.*

Abl kinase: tyrosine kinase that functions in the cell nucleus as part of a signaling pathway that causes cells with damaged DNA to self-destruct by apoptosis. (p. 171)

acceptable tolerance: regulatory standard based on "reasonable certainty of no harm," which means that a quantitative assessment of all available animal and human data is used to determine what level of exposure to possible carcinogens is thought to be safe. (p. 241)

acquired immunodeficiency syndrome: see *AIDS.*

ACT therapy: see *adoptive-cell-transfer therapy.*

adenine (A): nitrogen-containing base present in DNA and RNA that forms a complementary base pair with thymine (T) or uracil (U) by hydrogen bonding. (p. 13)

adoptive-cell-transfer (ACT) therapy: procedure in which a cancer patient's own lymphocytes are isolated, selected, and grown in the laboratory to enhance their cancer-fighting properties prior to injecting the cells back into the body. (p. 222)

aflatoxin: potent carcinogen produced by the mold *Aspergillus,* which grows on grains and nuts stored under humid conditions. (p. 74)

AIDS (acquired immunodeficiency syndrome): disease in which the immune system is weakened by infection with the human immunodeficiency virus (HIV); associated with an increased risk of virus-induced cancers, especially Kaposi's sarcoma. (p. 125)

Akt: protein kinase involved in the PI3K-Akt pathway; catalyzes the phosphorylation of several target proteins that suppress apoptosis and inhibit cell cycle arrest. (p. 183)

alkylating agents: highly reactive organic molecules that trigger DNA damage by linking themselves (or a reactive chemical group) directly to DNA; used in cancer chemotherapy, but many are also carcinogenic. (pp. 89, 212)

allele: any of the alternative forms of a given gene, differing from each other in base sequence. (p. 141)

alpha particles: type of nuclear radiation composed of positively charged particles containing two neutrons plus two protons. (p. 110)

alternative treatments: a diverse array of largely untested treatments for cancer (or other diseases) that are used as an alternative to standard medical care. (p. 229)

Ames test: screening test for potential carcinogens that assesses whether a substance causes mutations in bacteria. (p. 66)

analog: derivative of a chemical compound that closely resembles the parent compound in structure. (p. 211)

anaphase: stage of mitosis when the chromosomes move toward opposite poles of the mitotic spindle. (p. 189)

anaphase-promoting complex: group of proteins whose activity is required for initiating chromosome movement toward the spindle poles; activates separase, an enzyme that breaks down the cohesin proteins that otherwise hold the duplicated chromosomes together and prevent them from moving toward opposite spindle poles. (p. 189)

anaplastic: poorly differentiated and abnormal in appearance. (p. 11)

anchorage-dependent: trait exhibited by the cells of normal tissues, which must be attached to a solid surface such as the extracellular matrix before they can proliferate. (p. 19)

anchorage-independent: trait exhibited by cancer cells, which grow well not just when they are attached to a solid surface, but also when they are freely suspended in a liquid or semisolid medium. (p. 19)

androgen receptor: intracellular protein to which androgens bind in target cells, thereby leading to changes in cell behavior such as an activation of cell proliferation. (p. 215)

androgens: steroid hormones, such as testosterone, that are produced mainly in the testes and exert their effects by binding to androgen receptors in target tissues; they stimulate cell proliferation in the prostate gland and exert a variety of effects on other tissues. (p. 214)

aneuploid: containing an abnormal number of chromosomes. (pp. 35, 189)

angiogenesis: growth of new blood vessels. (p. 45)

angiostatin: protein that inhibits the growth of blood vessels; a fragment of the protein plasminogen. (p. 49)

anoikis: cell death triggered by lack of contact with the extracellular matrix. (p. 19)

anti-angiogenic therapy: treatment of cancer patients with drugs that inhibit angiogenesis. (p. 225)

antibiotic: natural substance produced by a microorganism, or a synthetic derivative, that kills or inhibits the growth of other microorganisms or cells. (p. 214)

antibody: class of proteins produced by B lymphocytes that bind with extraordinary specificity to substances, referred to as antigens, that provoke an immune response. (pp. 37, 218)

anticarcinogen: agent that protects against the development of cancer. (p. 74)

antigen: any foreign or abnormal substance capable of triggering an immune response. (p. 37)

antigen-presenting cell: cell that degrades and processes antigens and displays the resulting antigen fragments on its surface to activate an immune response by lymphocytes. (p. 37)

antimetabolites: molecules that resemble substances involved in cellular metabolism and that interact with enzymes in place of the normal substance, thereby disrupting metabolic pathways; used in cancer chemotherapy. (p. 211)

antioxidant: molecule that inhibits oxidation reactions, often by reacting with free radicals. (p. 243)

***APC* gene:** tumor suppressor gene, frequently mutated in colon cancers, that codes for a protein involved in the Wnt pathway. (pp. 145, 182)

apoptosis: cell suicide mediated by a group of protein-degrading enzymes called caspases; involves a programmed series of events that leads to the dismantling of the internal contents of the cell. (p. 28)

apoptotic bodies: cell fragments produced as a cell is dismantled by the process of apoptosis. (p. 29)

ARF protein: protein produced by the *CDKN2A* gene that binds to and promotes the degradation of Mdm2, which is the protein that normally targets the p53 protein for destruction. (p. 186)

aromatic amines: organic molecules that possess an amino group attached to a carbon backbone containing one or more benzene rings; many are carcinogenic. (p. 89)

asbestos: mineral composed of fine fibers that are woven into materials that exhibit insulating and fire-retarding properties; potent carcinogen that causes lung cancer and mesothelioma. (p. 82)

aspirin: drug used for alleviating pain, fever, and inflammation that has been found to reduce the risk of colon cancer and perhaps several other types of cancer as well. (p. 247)

ataxia telangiectasia: inherited syndrome characterized by loss of coordination, dilation of small blood vessels, immune system deficiencies, and roughly a 40% risk of developing cancer; caused by inheritance of two mutant alleles of the *ATM* tumor suppressor gene, which codes for a protein kinase involved in the DNA damage response. (p. 151)

***ATM* gene:** tumor suppressor gene coding for the ATM protein kinase, which plays a central role in DNA damage response pathways; inheriting two mutant alleles causes ataxia telangiectasia. (p. 151)

ATM kinase: protein kinase activated by damaged DNA; plays a central role in the DNA damage response by catalyzing the phosphorylation of p53 and several other target proteins. (p. 179)

autophosphorylation: phosphorylation of a receptor molecule by a receptor molecule of the same type. (p. 168)

Avastin: monoclonal antibody used in the treatment of colon cancer; inhibits angiogenesis by binding to and inactivating the angiogenesis growth factor, VEGF. (p. 225)

B

B lymphocyte: class of immune cells specialized for producing antibodies. (pp. 37, 218)

Bacillus Calmette-Guérin (BCG): bacterial strain that does not cause disease but elicits a strong immune response at the site where it is introduced into the body; sometimes used in immunotherapy. (p. 218)

basal cell: relatively undifferentiated type of epithelial cell whose proliferation gives rise to cells that are more specialized; division of basal cells in the skin produces more basal cells plus cells that migrate toward the outer layer of the skin, gradually differentiating into flattened squamous cells. (p. 5)

basal cell carcinoma: cancer of basal cells; accounts for about 75% of all skin cancers but few deaths because this type of cancer rarely metastasizes. (p. 104)

basal lamina: thin dense layer of protein-containing material that forms a barrier between epithelial cell layers and underlying tissues. (p. 206)

BCG: see *Bacillus Calmette-Guérin.*

Bcl2: protein that inhibits apoptosis. (p. 172)

***BCR-ABL*:** oncogene created by a reciprocal translocation between chromosomes 9 and 22 that leads to the fusion of portions of the *BCR* and *ABL* genes; codes for an abnormal intracellular tyrosine kinase that contributes to the development of chronic myelogenous leukemia and is targeted by the anticancer drug, Gleevec. (p. 162)

benign tumor: tumor that grows only locally, unable to invade neighboring tissues or spread to other parts of the body. (p. 5)

BERT (Background Equivalent Radiation Time): unit of measurement that converts a given dose of ionizing radiation into the amount of time it would take a person to receive that same dose from natural sources of background radiation. (p. 115)

beta particles: type of nuclear radiation composed of negatively charged electrons. (p. 110)

bias: distortion of data that causes an observed experimental result to deviate from the true value; caused by the failure to account for some influencing factor. (p. 62)

biopsy: cutting out a small sample of tissue for microscopic examination to determine whether cancer is present. (pp. 10, 202)

blood fluke: tiny flatworm that causes inflammation of the blood vessels of the intestine or bladder (schistosomiasis), occasionally leading to bladder cancer. (p. 127)

Bloom syndrome: inherited syndrome characterized by short stature, sun-induced facial rashes, immunodeficiency, decreased fertility, and an elevated risk of developing cancer before age 20; caused by inheritance of two mutant alleles of the *BLM* tumor suppressor gene, which codes for a DNA helicase involved in DNA repair. (p. 151)

BMI: see *Body Mass Index.*

Body Mass Index (BMI): formula for calculating body fat based on a person's height and weight. (p. 249)

brachytherapy: type of cancer treatment that uses a radiation source, such as tiny pellets, that can be inserted directly within (or close to) the tumor. (p. 210)

***BRAF*:** oncogene that codes for mutant forms of the Raf protein; occurs in roughly two-thirds of human melanomas and at a lower frequency in a variety of other cancers. (p. 170)

***BRCA1* and *BRCA2* genes:** tumor suppressor genes in which inheritance of a single mutant copy creates a high risk for breast and ovarian cancer; code for proteins involved in repairing double-strand DNA breaks. (pp. 150, 188)

Bub proteins: family of proteins that, along with Mad proteins, transmit a "wait" signal that inhibits the anaphase-promoting complex from initiating chromosome movement before all chromosomes are properly attached to the mitotic spindle. (p. 190)

Burkitt's lymphoma: a lymphocytic cancer associated with infection by Epstein-Barr virus along with a chromosome translocation in which the *MYC* gene is activated by moving it from chromosome 8 to 14. (p. 120)

C

C: see *cytosine.*

cancer: uncontrolled, growing mass of cells that is capable of invading neighboring tissues and spreading via body fluids, especially the bloodstream, to other parts of the body; also called a *malignant tumor*. (p. 5)

cancer stem cells: small subpopulation of cancer cells within a tumor that are the only cells capable of unlimited proliferation; give rise to all other cells of the tumor. (p. 216)

carcinogen: any cancer-causing agent. (p. 65)

carcinogen activation: conversion of precarcinogens into carcinogens by enzymatic reactions that occur in the liver. (p. 91)

carcinoma: a malignant tumor (cancer) arising from the epithelial cells that cover external and internal body surfaces. (p. 6)

carcinoma *in situ*: an epithelial cancer that has not yet invaded through the underlying basal lamina. (p. 206)

caretaker: class of tumor suppressor genes that are involved in DNA repair or chromosome sorting; loss-of-function mutations in such genes contribute to genetic instability. (pp. 148, 186)

caspase: any of a family of proteases that degrade other cellular proteins as part of the process of apoptosis. (p. 30)

Cdk: see *cyclin-dependent kinase.*

Cdk inhibitor: a protein that restrains cell proliferation by inhibiting the activity of Cdk-cyclin complexes that are required for progression through the cell cycle. (p. 181)

***CDKN2A* gene:** tumor suppressor gene that codes for two different proteins: the p16 protein, which is required for proper Rb signaling, and the ARF protein, which is required for proper p53 signaling. (p. 185)

cell-cell adhesion protein: class of proteins located at the outer cell surface that cause cells to adhere to one another. (p. 51)

cell cycle: series of stages, known as G1, S, G2, and M phase, that are involved in preparing for and carrying out cell division. (p. 23)

cell differentiation: process by which cells acquire the specialized properties that distinguish different types of cells from each other. (p. 5)

cell fusion: technique in which cells are artificially induced to fuse together by exposing them to inactivated viruses or chemical treatments; initially creates a cell in which the nuclei from the original two cells share the same cytoplasm, but cell division subsequently creates a hybrid cell that has a single nucleus containing chromosomes derived from both of the original cells. (p. 176)

cellular oncogene: oncogene that arises from a normal cellular proto-oncogene by point mutation, gene amplification, chromosomal translocation, insertional mutagenesis, or DNA rearrangement. (p. 159) Also see *viral oncogene.*

centrosome: small structure located adjacent to the nucleus that functions as an organizing center for microtubules; centrosomes are duplicated prior to cell division and the two centrosomes organize the assembly of the microtubules of the mitotic spindle, with each centrosome ending up at one end of the spindle. (p. 190)

CFC: see *chlorofluorocarbon.*

checkpoint: pathway that monitors conditions within the cell and transiently halts the cell cycle if conditions are not suitable for continuing. (p. 26) Also see DNA *damage checkpoint, DNA replication checkpoint,* and *spindle checkpoint.*

chemoprevention: use of chemical substances for the purpose of protecting against the development of cancer. (p. 247)

chemotherapy: using drugs to treat cancer. (p. 210)

chlorofluorocarbon (CFC): group of chemical compounds, used as refrigerants and for other industrial purposes, that are responsible for partially destroying the earth's ozone layer. (p. 106)

chromosomal translocation: process in which a piece of one chromosome is broken off and moved to another chromosome. (p. 162)

clinical trials: procedures in which drugs or other treatments proposed for human use are first evaluated in humans for safety, dose, and effectiveness; consist of Phase I, Phase II, and Phase III trials. (p. 228)

clone: population of identical cells produced from the reproduction of a single parental cell; also applied to genes or organisms. (pp. 19, 53, 98)

codon: sequence of three nucleotides in an mRNA molecule that serves as a coding unit for an amino acid (or a start or stop signal) during protein synthesis. (p. 13)

colonoscopy: examination of the colon using a flexible, fiber-optic tube that is inserted through the anus. (p. 203)

combination chemotherapy: treating cancer with several drugs in combination rather than a single agent alone. (p. 216)

complementary base pairing: in DNA and RNA, the ability of the base G to form a hydrogen-bonded base pair with C, and the base A to form a hydrogen-bonded base pair with T or U (pp. 13, 30)

complementary treatments: a diverse array of largely untested treatments for cancer (or other diseases) that are not part of standard medical practice, but are used along with standard medical care. (p. 229)

complete carcinogen: an agent that can trigger both the initiation and promotion stages of carcinogenesis. (p. 99)

computed tomography scan (CT scan): imaging technique in which X-ray pictures taken from many different angles are assembled by a computer to create a series of detailed cross-sectional images. (p. 206)

confounding: confusion introduced by a variable that influences the risk of developing cancer and is linked in some way to the factor being investigated. (p. 63)

constitutively active: protein or gene that remains active at all times rather than being regulated by another molecule. (pp. 133, 168)

COX: see *cyclooxygenase.*

CTL: see *cytotoxic T lymphocyte.*

CT scan: see *computed tomography scan.*

cyclin: any of a group of proteins that activate the cyclin-dependent kinases (Cdks) that are involved in regulating progression through the cell cycle. (p. 24)

cyclin-dependent kinase (Cdk): any of several protein kinases that are activated by different cyclins and that control progression through the cell cycle by phosphorylating various target proteins. (p. 23)

cyclooxygenase (COX): enzyme existing in multiple forms that catalyzes the production of prostaglandins. (p. 247)

cytochrome P450: family of enzymes that oxidize ingested foreign chemicals, such as drugs and pollutants, with the aim of making molecules less toxic and easier to excrete; reactions catalyzed by cytochrome P450 sometimes convert substances inadvertently into carcinogens. (p. 91)

cytokine: any of numerous proteins produced by the body to stimulate an immune response to foreign infectious agents. (p. 218)

cytokinesis: division of the cytoplasm during the M phase of the cell cycle. (p. 23)

cytosine (C): nitrogen-containing base present in DNA and RNA that forms a complementary base pair with guanine (G) by hydrogen bonding. (p. 13)

cytotoxic T lymphocyte (CTL): class of immune cells specialized for attacking foreign or abnormal cells, or cells that have been infected with an invading bacterium or virus. (pp. 37, 56, 221)

D

Delaney Amendment: a law in existence from 1958 to 1996 that directed the FDA to ban any food additive or contaminant if it was found to cause cancer in animal studies at any dose. (p. 241)

deletion: chromosomal abnormality involving the loss of nucleotides from DNA; can range in size from a single nucleotide to large stretches of DNA containing many genes. (p. 163)

dendritic cells: antigen-presenting cells that break antigens into fragments and present the fragments on the cell surface, where the fragments activate lymphocytes to generate an immune response against those antigens. (p. 222)

density-dependent inhibition of growth: tendency of cell division to stop when cells growing in culture reach a high population density. (p. 18)

detection bias: type of bias that arises when equivalent procedures are not used to assess cancer rates in different populations that are being investigated. (p. 62)

dietary fiber: indigestible portion of plant foods that is composed mainly of complex polysaccharides. (p. 245)

differentiating agents: substances that promote the process by which cells acquire the specialized characteristics of differentiated cells. (p. 215)

differentiation: see *cell differentiation.*

diploid: containing two sets of chromosomes and therefore two copies of each gene; can describe a cell, nucleus, or organism composed of such cells. (pp. 35, 141)

DMs: see *double minutes.*

DNA adduct: complex formed by the covalent linkage of a chemical carcinogen to DNA. (p. 92)

DNA damage checkpoint: mechanism that monitors for DNA damage and halts the cell cycle at various points, including late G1, S, and late G2, if damage is detected. (p. 27)

DNA damage response: network of cellular pathways invoked as a protective response to assaults on DNA integrity. (p. 151)

DNA methylation: addition of methyl groups to nucleotides in DNA; leads to epigenetic silencing of gene transcription when methyl groups are attached to the base C at sites where it is located adjacent to the base G. (p. 193)

DNA microarray: tiny chip of glass or plastic that has been spotted at fixed locations with thousands of different single-stranded DNA fragments; used for monitoring gene expression. (p. 191)

DNA polymerase: any of a group of enzymes involved in DNA replication and repair that catalyze the addition of successive nucleotides to the 3′end of a growing DNA strand, using an existing DNA strand as a template. (p. 32)

DNA rearrangement: change in DNA sequence organization caused by deletion, insertion, transposition, or inversion. (p. 163)

DNA replication checkpoint: mechanism that monitors the state of DNA replication to ensure that DNA synthesis is completed prior to permitting the cell to exit from G2 and begin mitosis. (p. 27)

dominant: pattern of inheritance in which a given gene (allele) determines how the trait will appear in a organism, whether present in heterozygous or homozygous form. (p. 141)

dominant negative mutation: a loss-of-function mutation in one copy of a gene that inactivates the gene's protein product, even in the presence of another normal copy of the same gene; seen in some genes that produce proteins consisting of more than one copy of the same protein chain, where a single mutant chain can disrupt the function of the protein even though the other protein chains are normal. (p. 181)

dose-response relationship: condition in which the rate of a particular disease increases or decreases in direct relation to the amount of exposure to the agent being investigated. (p. 64)

double blind: testing procedure in which neither doctors nor patients know who is receiving the treatment and who is not. (p. 228)

double helix: two intertwined helical chains of a DNA molecule, held together by complementary base-pairing between the bases A and T, and between the bases C and G. (p. 30)

double minutes (DMs): independent, chromosome-like bodies that are much smaller than typical chromosomes, often appearing as spherical, paired structures; contain amplified DNA consisting of several dozen to several hundred copies of one or more genes. (p. 160)

dysplasia: abnormal tissue growth in which cell and tissue organization is disrupted; may be an early stage in cancer development. (p. 5)

E

E2F transcription factor: protein that, when not bound to the Rb protein, activates the transcription of genes coding for proteins required for DNA replication and entrance into S phase of the cell cycle. (p. 178)

E6 oncoprotein: protein produced by an oncogene present in human papillomavirus (HPV) that binds to a cell's normal p53 protein, thereby targeting it for destruction. (pp. 135, 181)

E7 oncoprotein: protein produced by an oncogene present in human papillomavirus (HPV) that binds to and inactivates a cell's normal Rb protein, thereby interfering with the ability of the Rb protein to restrain cell proliferation. (pp. 136, 178)

EBV: see *Epstein-Barr virus.*

E-cadherin: cell-cell adhesion protein that binds epithelial cells to one another. (p. 51)

EGF: see *epidermal growth factor.*

electromagnetic radiation: waves of electric and magnetic fields propagated through space at the speed of light. (p. 105)

electrophilic: describes a compound with electron-deficient atoms that readily reacts with substances possessing atoms that are electron-rich. (p. 92)

ELF (Extremely Low Frequency): electric and magnetic fields, emitted by high-voltage power lines, that represent the low-energy end of the radiofrequency region of the electromagnetic spectrum. (p. 117)

endostatin: protein (fragment) that inhibits the growth of blood vessels; a fragment of the protein collagen. (p. 49)

endothelial cell: type of cell that lines the internal surface of blood and lymphatic vessels. (p. 45)

Environmental Protection Agency (EPA): U.S. government agency whose responsibilities include creating regulatory standards for carcinogens that contaminate the environment. (p. 242)

EPA: see *Environmental Protection Agency.*

epidemiology: branch of medical science that investigates the frequency and distribution of diseases in human populations. (p. 61)

epidermal growth factor (EGF): protein that stimulates the growth and division of a wide variety of epithelial cell types. (pp. 22, 165)

epigenetic change: alteration in cellular properties brought about by a change in gene expression rather than by gene mutation. (pp. 99, 191)

episome: DNA that can persist and replicate inside a cell independent of the chromosomal DNA. (p. 129)

epithelial cell: type of cell that forms the covering layers over external and internal body surfaces. (p. 6)

epoxide: three-membered ring consisting of an oxygen atom covalently bonded to two carbon atoms; the two carbons are electron-deficient and therefore tend to react with atoms that are electron-rich. (p. 92)

Epstein-Barr virus (EBV): virus associated with Burkitt's lymphoma, nasopharyngeal carcinoma, and the noncancerous condition known as infectious mononucleosis. (p. 121)

***ERBB2* gene:** gene that is amplified in about 25% of all breast and ovarian cancers; codes for the ErbB2 growth factor receptor, which is targeted by the anticancer drug Herceptin. (p. 161)

ErbB2 receptor: growth factor receptor present in excessive amounts in certain breast and ovarian cancers; targeted by Herceptin, a monoclonal antibody used in treating breast cancer. (p. 223)

estrogen receptor: intracellular protein to which estrogens bind in target cells, thereby leading to changes in cell behavior such as an activation of cell proliferation. (p. 215)

estrogens: steroid hormones produced by the ovaries that exert their effects by binding to estrogen receptors in target tissues; they stimulate cell proliferation in the breast and uterus, and exert a variety of effects on other tissues as well. (p. 215)

excision repair: DNA repair mechanism that removes and replaces abnormal nucleotides. (pp. 32, 150)

experimental treatment: any new drug or other treatment that is undergoing clinical testing but has not yet been approved for use in standard medical practice. (p. 229)

experimenter bias: type of bias in which the expectations of a person carrying out an investigation interfere with objective evaluation of the data. (p. 62)

extracellular matrix: insoluble meshwork of protein fibers and polysaccharides that fills the spaces between neighboring cells. (p. 19)

F

false negative: situation in which a person who has cancer obtains a negative test result that fails to detect the presence of the disease. (p. 205)

false positive: situation in which a person who does not have cancer obtains a positive test result that incorrectly suggests that the disease may be present. (p. 205)

familial adenomatous polyposis: inherited condition in which numerous polyps develop in the colon, eventually leading to cancer; caused by inheritance of a single defective or missing copy of the *APC* gene. (p. 145)

familial (hereditary) cancer: cancers that arise as a result of an inherited mutation that creates a predisposition to developing cancer. (p. 141)

Fanconi anemia: inherited syndrome characterized by an inability of the bone marrow to produce a sufficient number of blood cells, accompanied by skeletal malformations, organ deformities, reduced fertility, and marked predisposition to developing leukemias and squamous cell carcinomas; caused by inheritance of two mutant alleles of any of at least 11 different tumor suppressor genes coding for proteins involved in DNA damage response pathways. (p. 152)

FDA: see *Food and Drug Administration.*

fecal occult blood test (FOBT): test to check for tiny amounts of blood in the feces, which may be a sign of colon cancer. (p. 203)

FGF: see *fibroblast growth factor.*

fiber: see *dietary fiber.*

fibroblast growth factor (FGF): signaling protein that stimulates the growth of new blood vessels. (p. 48)

finasteride: inhibitor of testosterone production that reduces the overall risk of prostate cancer but results in a high proportion of aggressive cancers, thereby limiting its usefulness as a cancer prevention agent. (p. 249)

five-year survival rate: percentage of people who are still alive five years after diagnosis of a disease. (p. 9)

FOBT: see *fecal occult blood test.*

Food and Drug Administration (FDA): U.S. government agency responsible for regulating the marketing of food, drugs, medical devices, and cosmetics. (p. 241)

free radicals: atoms or molecules with an unpaired electron, which makes them extremely reactive. (pp. 113, 243)

fusion gene: a gene containing sequences derived from two different genes spliced together. (p. 162)

fusion protein: a protein containing amino acid sequences encoded by portions of two different genes that have been fused together. (p. 162)

G

G: see *guanine.*

G protein: class of protein molecules whose activity is regulated by binding to the small nucleotides GTP and GDP. (p. 169)

G0 phase (G zero): offshoot of the G1 phase of the cell cycle in which cells have exited from the cell cycle and are no longer proliferating. (p. 23)

G1 phase: stage of the cell cycle between the end of the previous cell division and the onset of chromosomal DNA replication. (p. 22)

G2 phase: stage of the cell cycle between the completion of chromosomal DNA replication and the onset of cell division. (p. 22)

gamma rays: type of nuclear radiation consisting of electromagnetic waves with a wavelength shorter than that of X-rays. (p. 110)

gap junction: structure involved in cell-cell communication that joins cells together in a way that allows small molecules to pass directly from one cell to another; constructed from a protein called connexin. (p. 40)

gatekeeper: class of tumor suppressor genes that are directly involved in restraining cell proliferation; loss-of-function mutations in such genes can lead to excessive cell proliferation and tumor formation. (pp. 148, 186)

GEF: see *guanine-nucleotide exchange factor.*

gene: nucleotide base sequence in DNA (or RNA in some viruses) that codes for a functional product, usually a protein chain. (p. 13)

gene amplification: mechanism for creating extra copies of individual genes by selectively replicating specific DNA sequences numerous times in succession, thereby creating dozens, hundreds, or even thousands of copies of the same stretch of DNA. (p. 160)

gene cloning: set of laboratory procedures for generating multiple copies of a specific DNA (gene) sequence. (p. 159)

gene therapy: treating genetic diseases by inserting normal copies of genes into the cells of people who have defective, disease-causing genes. (pp. 137, 226)

genetic instability: trait of cancer cells in which abnormally high mutation rates are caused by defects in DNA repair or chromosome sorting mechanisms. (pp. 35, 187)

genetic testing: laboratory analysis that is performed to determine whether a person's DNA carries any hereditary mutations that confer susceptibility to cancer or other diseases. (p. 154)

genome: total genetic information of a virus, cell, or organism. (pp. 129, 148)

genotoxic: class of carcinogens that act by causing DNA damage. (p. 95)

Gleevec: small molecule drug used in treating chronic myelogenous leukemia; binds to and inhibits the abnormal tyrosine kinase produced by the *BCR-ABL* oncogene. (p. 223)

glucocorticoids: group of steroid hormones, including cortisone, that are produced by the adrenal cortex and have a variety of metabolic and anti-inflammatory properties, including the ability to inhibit lymphocyte proliferation. (p. 215)

grading: see *tumor grading.*

graft-versus-host disease: a potentially life-threatening condition that can occur during stem cell transplantation when immune cells in the donated stem cell population attack the tissues of the person receiving the transplanted cells. (p. 217)

gray (Gy): unit of measurement for indicating how much energy is absorbed when a given type of radiation interacts with biological tissue; replaces an older historical unit called the *rad*, which corresponded to 0.01 Gy. (p. 111)

growth factor: class of extracellular signaling proteins that stimulate (or in some cases inhibit) the proliferation of particular cell types by binding to specific receptor proteins located on the outer cell surface. (pp. 22, 165)

guanine (G): nitrogen-containing base present in DNA and RNA that forms a complementary base pair with cytosine (C) by hydrogen bonding. (p. 13)

guanine-nucleotide exchange factor (GEF): a protein that triggers the release of GDP from the Ras protein, thereby permitting Ras to acquire a molecule of GTP. (p. 170)

Gy: see *gray.*

H

HBV: see *hepatitis B virus.*

HCV: see *hepatitis C virus.*

***Helicobacter pylori* (*H. pylori*):** bacterium that causes stomach inflammation and ulcers, which can eventually lead to cancer. (p. 126)

hematopoietic stem cell: unspecialized cell type whose normal proliferation gives rise to more hematopoietic stem cells as well as to cells that differentiate into the various types of blood cells. (p. 217)

hepatitis: inflammation of the liver. (p. 124)

hepatitis B virus (HBV): DNA virus that causes hepatitis and liver cancer; transmitted by exchange of bodily fluids, such as semen or blood. (p. 124)

hepatitis C virus (HCV): RNA virus that causes hepatitis and liver cancer; transmitted by direct contact with contaminated blood. (p. 125)

Herceptin: monoclonal antibody used in the treatment of breast cancer; binds to and inactivates a cell surface growth factor receptor known as ErbB2 (p. 223)

hereditary nonpolyposis colon cancer (HNPCC): inherited susceptibility to colon cancer caused by inherited defects in DNA mismatch repair. (p. 150)

HERP value: quantitative measure of the potential hazard posed by a given carcinogen, based on typical human exposure multiplied by its carcinogenic potency when tested in animals; an acronym for Human Exposure Rodent Potency. (p. 242)

heterozygous: having two different alleles for a given gene. (p. 141) Also see *homozygous.*

HIV: see *human immunodeficiency virus.*

HNPCC: see *hereditary nonpolyposis colon cancer.*

Hodgkin's disease: type of lymphoma characterized by the presence of Reed-Sternberg cells, which possess two nuclei and have a uniquely distinctive appearance. (p. 122)

homogeneously staining region (HSRs): chromosome regions that stain homogeneously rather than exhibiting the alternating pattern of light and dark staining bands that is typical of normal chromosomes; contain amplified DNA consisting of several dozen to several hundred copies of one or more genes. (p. 160)

homologous chromosomes: two copies of a specific chromosome, one derived from each parent; homologous chromosomes contain the same set of genes (identical or slightly different copies) arranged in the same order. (p. 141)

homologous recombination: method for fixing double-strand breaks in DNA in which an intact copy of a chromosomal DNA molecule serves as a template for guiding the repair of the broken DNA; also used for the exchange of genetic information between two DNA molecules exhibiting extensive sequence similarity. (pp. 34, 188)

homozygous: having two identical alleles for a given gene. (p. 141) Also see *heterozygous.*

hormetic model: U-shaped dose-response relationship in which cancer rates decline at very low doses of a carcinogen and then begin to rise as the dose is further increased. (p. 86)

hormone replacement therapy (HRT): use of estrogen combined with progestin for treating the symptoms and long-term health effects of menopause. (p. 239)

hormone therapy: cancer treatments that involve the use of hormones or drugs that interfere with hormone synthesis or action. (p. 214)

HPV: see *human papillomavirus.*

HRT: see *hormone replacement therapy.*

HSRs: see *homogeneously staining regions.*

HTLV-I: see *human T-cell lymphotropic virus-I.*

human immunodeficiency virus (HIV): retrovirus that causes AIDS (acquired immunodeficiency syndrome). (p. 125)

human papillomavirus (HPV): virus linked to cervical cancer; possesses oncogenes that produce oncoproteins that interfere with the functioning of a cell's Rb and p53 proteins. (p. 122)

human T-cell lymphotropic virus-I (HTLV-I): virus associated with adult T-cell leukemia/lymphoma, an aggressive type of cancer that is rare in the United States but prevalent in certain parts of Japan, Africa, and the Caribbean. (p. 125)

hybrid cell: cell whose nucleus contains chromosomes derived from two different cells that were joined together by the technique of cell fusion. (p. 176)

hyperplasia: tissue growth based on an increase in the number of cells, but cell organization is normal. (p. 4)

hyperthermia: raising the temperature of body tissues a few degrees to sensitize cancer cells to the killing effects of radiation used in cancer treatment. (p. 210)

hypertrophy: tissue growth based on an increase in cell size. (p. 4)

I

imaging techniques: techniques for taking pictures of the inside of the body, including conventional X-ray procedures, computed tomography (CT scan), magnetic resonance imaging (MRI), and ultrasound imaging. (p. 206)

immune surveillance theory: theory postulating that immune rejection of cancer cells helps protect people against the development of cancer. (p. 38)

immunosuppressive drug: a drug that inhibits the immune system; given to organ transplant recipients to prevent immune rejection of a transplanted organ, such as a heart or kidney. (p. 88)

immunotherapy: treatment of a disease, such as cancer, by stimulating the immune system or administering antibodies made by the immune system. (p. 218)

incomplete carcinogen: an agent that can function in the initiation or promotion stage of carcinogenesis, but not both. (p. 99)

infectious mononucleosis: nonmalignant proliferation of lymphocytes that produces temporary flu-like symptoms; caused by infection with Epstein-Barr virus. (p. 122)

inflammation: response of body tissues to injury, irritation, or infection that is characterized by swelling, redness, pain, and heat. (p. 247)

informed consent: process in which individuals who want to participate in a clinical trial are given information regarding risks, benefits, and other options, and then sign a document indicating their understanding of these conditions and their voluntary consent. (p. 228)

initiation (stage of carcinogenesis): conversion of a cell to a precancerous state by agents that cause DNA mutation. (p. 94)

inorganic substance: class of compounds that do not contain carbon and hydrogen. (p. 91)

insertion: chromosomal abnormality involving the addition of nucleotides to a DNA molecule; can range in size from a single nucleotide to large stretches of DNA containing many genes. (p. 163)

insertional mutagenesis: change in gene activity or structure caused by the chromosomal integration of DNA derived from another source, usually a virus. (pp. 133, 164)

integrase: enzyme that inserts viral genes into a cell's chromosomal DNA. (p. 129)

interphase: portion of the cell cycle situated between successive cell divisions (M phases); composed of G1, S, and G2 phases. (p. 23)

invasion: direct spread of cancer cells into neighboring tissues. (p. 51)

inversion: chromosomal abnormality in which a DNA segment is excised and then reinserted backwards in the same location. (p. 163)

ionizing radiation: high-energy forms of radiation that remove electrons from molecules, thereby generating highly reactive ions that cause DNA damage; includes X-rays and radiation emitted by radioactive elements. (p. 109)

Iressa: monoclonal antibody used in the treatment of lung cancer; binds to and inactivates the epidermal growth factor receptor. (p. 224)

J

Jak kinase: tyrosine kinase that catalyzes the phosphorylation of cytoplasmic proteins called STATs; phosphorylated STATs trigger changes in gene expression that can stimulate cell proliferation. (p. 171)

Jak-STAT pathway: signaling pathway in which receptors that do not possess their own tyrosine kinase activity stimulate the activity of an independent tyrosine kinase, Jak, which in turn catalyzes the phosphorylation of cytoplasmic proteins called STATs; phosphorylated STATs trigger changes in gene expression that can stimulate cell proliferation. (p. 169)

K

Kaposi's sarcoma: cancer arising from blood vessels in the skin; generally quite rare, but rates are increased 100-fold in people infected with HIV. (p. 125)

Kaposi's sarcoma-associated herpesvirus (KSHV): sexually transmitted DNA virus that has been linked to Kaposi's sarcoma. (p. 125)

KSHV: see *Kaposi's sarcoma-associated herpesvirus.*

L

laetrile: a natural substance extracted from apricot pits that was claimed to be a cure for cancer but failed to have any useful effects when evaluated in randomized clinical trials. (p. 229)

latent (virus): condition in which a virus remains "hidden" inside a cell with no new virus particles being produced or released. (p. 129)

lead time bias: tendency for survival rates to appear to be improved by early diagnosis. (p. 10)

lectin: carbohydrate-binding protein possessing two or more carbohydrate-binding sites, thereby allowing a single lectin molecule to link two cells together by binding to carbohydrate groups exposed on the surface of each cell. (p. 40)

leukemia: cancer of blood or lymphatic origin in which the cancer cells reside mainly in the bloodstream rather than growing as solid masses of tissue. (p. 7)

Li-Fraumeni syndrome: susceptibility to a broad range of cancers caused by inheritance of a single missing or defective copy of the *p53* gene. (p. 145)

linear model: a linear dose-response relationship exhibiting no threshold. (p. 86)

liver fluke: parasitic flatworm that causes inflammation of the bile ducts; chronic infections may eventually lead to bile duct cancer (cholangiocarcinoma). (p. 127)

long terminal repeats (LTRs): DNA sequences located at the ends of the genome of retroviruses; involved in integrating the viral DNA into the host chromosomal DNA and in activating the transcription of viral as well as nearby host genes. (p. 134)

loss of heterozygosity: process by which one more genes initially present in a heterozygous state are converted to the homozygous state. (p. 177)

LTRs: see *long terminal repeats.*

lymphocyte: class of white blood cells involved in immune responses; B lymphocytes produce antibodies, whereas cytotoxic T lymphocytes kill targeted cells directly. (p. 37)

lymphoma: cancer of blood or lymphatic origin that grows mainly as solid masses of tissue. (p. 7)

lysis: cell death through rupturing of the plasma membrane. (p. 226)

M

M phase: stage of the cell cycle when the nucleus and the rest of the cell divide. (p. 22)

Mad proteins: family of proteins that, along with Bub proteins, transmit a "wait" signal that inhibits the anaphase-promoting complex from initiating chromosome movement before all chromosomes are properly attached to the mitotic spindle. (p. 190)

magnetic resonance imaging (MRI): imaging technique in which radio waves and magnetic fields are used instead of X-rays to create a series of detailed cross-sectional images of the body. (p. 206)

major histocompatibility complex (MHC): type of cell surface protein recognized by the immune system as part of the process of inducing an immune response. (pp. 37, 56)

malignant tumor: tumor that can invade neighboring tissues and spread through the bloodstream to other parts of the body; also called a *cancer.* (p. 5)

mammography: screening technique for breast cancer that uses low-dose X-rays to create detailed pictures that reveal the internal tissues of the breast. (p. 202)

MAP kinase (MAPK): intracellular protein kinase that plays a key role in the Ras-MAPK pathway, entering the nucleus and phosphorylating several different transcription factors. (p. 170)

MAPK: see *MAP kinase.*

mastectomy: treatment for breast cancer that involves surgical removal of the breast. (p. 208)

matrix metalloproteinase (MMP): family of enzymes that degrade proteins found in the extracellular matrix; referred to as metalloproteinases because they require zinc or calcium atoms to function properly. (p. 48)

maximum tolerated dose (MTD): highest dose of a suspected carcinogen that can be administered to animals without causing serious weight loss or signs of immediate life-threatening toxicity. (p. 86)

Mdm2: protein that interacts with p53 and links it to ubiquitin, thereby targeting the p53 protein for destruction by proteasomes. (p. 179)

melanin: family of brown pigments synthesized by melanocytes; responsible for skin color and the suntan that occurs in response to sunlight. (p. 109)

melanoma: a cancer arising from pigmented cells, usually in the skin. (p. 104)
mesothelioma: rare form of cancer derived from the mesothelial cells that cover the interior surfaces of the chest and abdominal cavities; usually caused by exposure to asbestos. (p. 82)
messenger RNA (mRNA): RNA molecule whose base sequence codes for the amino acid sequence of a protein chain. (p. 13)
metaphase: stage during mitosis when the chromosomes become lined up at the center of the mitotic spindle just before being parceled out to the two newly forming cells. (p. 189)
metastasis: spread of tumor cells from one part of the body to another via the bloodstream or other body fluids. (pp. 5, 51)
metastasis promoting gene: class of genes coding for proteins that are required for or activate events associated with invasion and metastasis. (p. 57)
metastasis suppressor gene: class of genes coding for proteins that inhibit or block events associated with invasion and metastasis. (p. 58)
MHC: see *major histocompatibility complex.*
mismatch repair: DNA repair mechanism that detects and corrects base pairs that are improperly hydrogen bonded. (pp. 33, 150)
mitosis: division of the nucleus during the M phase of the cell cycle. (p. 23)
mitotic death: cell death caused by high-dose radiation exposure that causes chromosomal damage so severe that it prevents cells from progressing through mitosis. (p. 209)
mitotic index: percentage of cells in a population that are undergoing mitosis at any given moment. (p. 11)
MMP: see *matrix metalloproteinase.*
molecular targeting: developing drugs that specifically target proteins known to be critical to cancer cells. (p. 223)
monoclonal antibody: purified antibody directed against a single antigen; obtained using a laboratory technique for producing and selecting cloned populations of antibody-producing cells called hybridomas. (p. 219)
MRI: see *magnetic resonance imaging.*
mRNA: see *messenger RNA.*
MTD: see *maximum tolerated dose.*
multidrug resistance transport proteins: family of plasma membrane proteins that actively pump a broad spectrum of drugs out of cells. (p. 216)
multiple endocrine neoplasia type II: hereditary condition associated with the development of both benign and malignant tumors of endocrine glands; caused by inheritance of a single mutant copy of the *RET* proto-oncogene—in other words, a *RET* oncogene. (p. 152)
mutagen: chemical or physical agent that is capable of inducing mutations. (pp. 31, 66)
mutation: change in the base sequence of a DNA molecule. (p. 30)
***MYC* gene:** normal human proto-oncogene that codes for the Myc protein, a transcription factor that activates the transcription of genes required for cell proliferation. (pp. 133, 161) Also see *v-myc gene.*
Myc protein: transcription factor that activates the transcription of genes required for cell proliferation. (p. 134)

N

nasopharyngeal carcinoma: cancer of the epithelial lining of the nasal passages and throat caused by infection with the Epstein-Barr virus; frequent in Southeast Asia but rare elsewhere in the world. (p. 122)
natural products: substances produced by living organisms. (p. 91)
neoplasia: abnormal growth process in which cells proliferate in an uncontrolled, relatively autonomous fashion. (p. 5)
neoplasm: growing mass of tissue created by an abnormal process in which cells proliferate in an uncontrolled, relatively autonomous fashion, leading to a continual increase in the number of dividing cells; may be either benign or malignant; also called a *tumor.* (p. 5)
NF-kappa B (NF-κB): transcription factor activated in tissues where inflammation is occurring; activates the transcription of genes that produce proteins that stimulate cell division and make cells resistant to apoptosis. (p. 128)
N-nitroso compounds: organic chemicals that contain a nitroso group (N=O) joined to a nitrogen atom; include nitrosamines and nitrosoureas, which are potent carcinogens. (p. 89)
nonhomologous end-joining: mechanism for repairing double-strand DNA breaks that uses proteins that bind to the ends of the two broken DNA fragments and join them together. (p. 34)
nonreceptor tyrosine kinase: intracellular tyrosine kinase that does not possess a receptor site. (p. 171)
nuclear radiation: radiation emitted by radioactive elements; includes electromagnetic radiation (gamma rays) and particulate radiation (alpha and beta particles). (p. 110)
nucleotide: molecule consisting of a nitrogen-containing base (A, G, C, T, or U) linked to a five-carbon sugar (ribose or deoxyribose) that is attached to a phosphate group; building block of DNA and RNA. (p. 13)

O

Occupational Safety and Health Administration (OSHA): U.S. government agency that formulates and enforces regulations designed to protect the safety and health of workers. (p. 241)
oncogene: any gene whose presence can lead to cancer; some oncogenes are introduced by viruses, but in human cancers, most arise by mutation from normal cellular genes called *proto-oncogenes.* (pp. 12, 129, 152, 158)
oncogenic virus: virus that can cause cancer. (pp. 77, 120)
oncoprotein: protein that contributes to the development of cancer; produced by an oncogene. (p. 133)
organic chemical: any carbon-containing compound. (p. 89)
OSHA: see *Occupational Safety and Health Administration.*
overdiagnosis: detection of cancers that would not otherwise have been a health hazard. (p. 205)
ozone: gas composed of three oxygen atoms that is present in the upper atmosphere, where it absorbs much of the harmful ultraviolet radiation emitted by the sun. (p. 105)

P

***p* value:** probability that an observed difference between two measurements would appear by chance, even if there were in reality no such difference at all; *p* values of 0.05 or less are generally required before it can be concluded that an observed difference is statistically significant. (p. 62)
p15 protein: Cdk inhibitor that halts progression through the cell cycle by inhibiting several different Cdk-cyclins. (p. 185)
p16 protein: Cdk inhibitor produced by the *CDKN2A* gene that suppresses the activity of the Cdk-cyclin complex that normally phosphorylates the Rb protein. (p. 186)
p21 protein: Cdk inhibitor that halts progression through the cell cycle by inhibiting several different Cdk-cyclins. (p. 181)
***p53* gene:** tumor suppressor gene that codes for the p53 protein, a transcription factor involved in preventing genetically damaged cells from proliferating; most frequently mutated gene in human cancers; also called the *TP53* gene in humans. (pp. 106, 146, 179)
p53 protein: molecule that accumulates in the presence of damaged DNA and activates genes whose products halt the cell cycle and trigger apoptosis. (pp. 27, 179)

Pap smear: screening technique for early detection of cervical cancer in which cells obtained from a sample of vaginal secretions are examined with a microscope. (p. 202)

pathogen: any disease-producing agent. (p. 120)

pathologist: doctor who specializes in diagnosing disease by examining tissues and cells with a microscope. (p. 10)

PC-SPES: herbal remedy for prostate cancer that appeared to be effective in preliminary human testing, but was taken off the market when it was found to be contaminated with synthetic drugs. (p. 229)

PDGF: see *platelet-derived growth factor.*

penetrance: frequency with which a given dominant or homozygous recessive allele yields the expected trait within a population. (p. 141)

Philadelphia chromosome: fragment of chromosome 22 found in the cancer cells of about 90% of people with chronic myelogenous leukemia; generated by a reciprocal exchange of DNA (translocation) between chromosomes 9 and 22. (pp. 35, 162)

phorbol ester: class of chemical compounds found in croton oil that function as tumor promoters. (p. 96)

phosphatidylinositol 3-kinase (PI 3-kinase or PI3K): enzyme that catalyzes the addition of a phosphate group to the plasma membrane lipid PIP_2, thereby converting it into PIP_3. (p. 183)

phosphorylation: addition of a phosphate group. (p. 23)

phytochemical: term applied to plant chemicals that are thought to have health-related effects but are not essential nutrients like proteins, carbohydrates, fats, minerals, and vitamins. (p. 245)

PI 3-kinase: see *phosphatidylinositol 3-kinase.*

PI3K: see *phosphatidylinositol 3-kinase.*

PI3K-Akt pathway: one of the signaling pathways activated by the binding of growth factors to cell surface receptors; involves the activation PI 3-kinase, which converts PIP_2 into PIP_3 molecules that stimulate the phosphorylation and activation of Akt protein kinase; this pathway is inhibited by PTEN, an enzyme that removes a phosphate group from PIP_3 and thereby abolishes its ability to activate Akt. (p. 183)

PIP_2: abbreviation for the plasma membrane lipid phosphatidylinositol-4,5-bisphosphate); can be converted by addition of a phosphate group to PIP_3, a component of the PI3K-Akt pathway. (p. 183)

PIP_3: abbreviation for the plasma membrane lipid phosphatidylinositol-3,4,5-trisphosphate; component of the PI3K-Akt pathway. (p. 183)

placebo: an inactive substance that resembles a drug in appearance; given to patients in the control group in a randomized double-blind trial. (p. 228)

placebo effect: any beneficial effect on a patient's condition that is caused by a person's expectations concerning a drug rather than by the drug itself. (p. 228)

plasma membrane: membrane composed of lipids and proteins that defines the outer boundary of the cell and regulates the flow of materials into and out of the cell; also called the cell membrane. (p. 22)

plasmin: protease produced from the precursor plasminogen by the action of plasminogen activator. (p. 52)

plasminogen activator: protease that converts plasminogen into plasmin. (p. 52)

platelet-derived growth factor (PDGF): protein produced by blood platelets that stimulates the proliferation of connective tissue and smooth muscle cells. (pp. 22, 165)

platinating agents: highly reactive platinum-containing compounds, such as cisplatin, that are used in cancer chemotherapy; crosslink DNA by forming chemical bonds between the platinum atom and DNA bases. (p. 213)

point mutation: mutation involving a single nucleotide. (p. 160)

polonium: radioactive metal produced by the radioactive decay of radon gas; forms tiny particles that may be inhaled and become lodged in a person's lungs, where the radioactive decay of polonium produces alpha particles that can cause lung cancer. (p. 111)

polycyclic aromatic hydrocarbons: diverse group of compounds constructed from multiple, fused benzene rings; many are carcinogenic. (p. 89)

polycyclic hydrocarbons: see *polycyclic aromatic hydrocarbons.*

polyp: any mass of tissue that arises from the wall of a hollow organ and protrudes into the lumen. (p. 10)

post hoc fallacy: drawing the incorrect conclusion that two events exhibit a cause-and-effect relationship just because they are associated with each other. (p. 64)

precarcinogen: any substance that is capable of causing cancer only after it has been metabolically activated. (p. 91)

preclinical testing: process in which a drug proposed for human use is first tested in animals. (p. 227)

primary tumor: tumor at the site where cancer initially arose. (p. 8)

promotion (stage of carcinogenesis): gradual process by which cells previously exposed to an initiating carcinogen are subsequently converted into cancer cells by agents that stimulate cell proliferation. (p. 94)

prospective study: epidemiological approach that monitors people into the future to see who will develop a particular disease. (p. 63)

prostaglandin: group of hormone-like, fat-soluble molecules that trigger inflammation and influence a variety of other functions. (p. 247)

protease: class of enzymes that degrade protein molecules. (p. 52)

proteasome: multiprotein complex that catalyzes the degradation of proteins linked to ubiquitin. (p. 181)

protein kinase: any of numerous enzymes that catalyze protein phosphorylation—that is, the addition of phosphate groups to protein molecules. (pp. 23, 167)

protein kinase C: protein kinase involved in a cellular signaling pathway that influences cell proliferation; normally activated by diacylglycerol, but also activated by tumor promoting phorbol esters. (p. 96)

protein phosphatase: any of numerous enzymes that catalyze protein dephosphorylation—that is, the removal of phosphate groups from protein molecules. (p. 24)

proteomic analysis: technique that permits rapid examination of the protein makeup of a tiny sample, such as a drop of blood, creating a snapshot of thousands of proteins at once. (p. 204)

proto-oncogene: normal cellular gene that can be converted into an oncogene by point mutation, gene amplification, chromosomal translocation, local DNA rearrangement, or insertional mutagenesis. (pp. 130, 152, 159)

provirus: a DNA copy of the genetic information of an RNA virus that has been integrated into the chromosomal DNA of a host cell. (p. 129)

PSA test: screening technique for early detection of prostate cancer that measures how much prostate-specific antigen (PSA) is present in the blood. (p. 203)

PTEN: enzyme that removes a phosphate group from PIP_3, thereby converting it to PIP_2; inhibitor of the PI3K-Akt pathway. (p. 184)

publication bias: type of bias arising from the common practice of scientific journals not to publish studies in which investigators have failed to detect some kind of relationship. (p. 62)

Puma: protein involved in triggering cell death by apoptosis; expression of the gene coding for Puma is activated by p53. (p. 181)

purine: two-ringed nitrogen-containing molecule; parent compound of the bases adenine (A) and guanine (G). (p. 212)
pyrimidine: single-ringed nitrogen-containing molecule; parent compound of the bases cytosine (C), thymine (T), and uracil (U). (p. 212)
pyrimidine dimer: mutation in which two adjacent pyrimidine bases (CC, CT, TC, or TT) in DNA are joined together by covalent bonds; pyrimidine dimers distort DNA structure, thereby causing incorrect nucleotides to be inserted during DNA replication. (pp. 31, 106)

R

Rad50 exonuclease complex: multiprotein complex involved in the first phase of the pathway for repairing double-strand DNA breaks by homologous recombination; removes nucleotides from one strand of the double helix to expose a single-stranded segment on the opposite strand. (p. 188)
Rad51 repair complex: multiprotein complex involved in the second phase of the pathway for repairing double-strand DNA breaks by homologous recombination; catalyzes a "strand invasion" reaction in which the exposed single-stranded DNA segment at the end of the broken DNA molecule displaces one of the two strands of the intact DNA molecule being used as a template. (p. 188)
radiation: energy traveling through space; exists in varying forms characterized by differences in wavelength and energy content. (pp. 76, 103)
radiation therapy: form of cancer treatment that uses high-energy X-rays or other types of ionizing radiation to kill cancer cells. (p. 209)
radioactivity: spontaneous decay of an unstable atomic nucleus accompanied by the emission of alpha, beta, or gamma radiation. (p. 110)
radiofrequency (RF) waves: electromagnetic radiation with a wavelength longer than that of microwaves. (p. 116)
radium: radioactive element that resembles calcium in some of its chemical properties. (p. 112)
radon: radioactive gas produced from the spontaneous breakdown of radium found in underground rock formations; can accumulate in poorly ventilated buildings and cause lung cancer. (p. 111)
Raf kinase: intracellular protein kinase that is activated by Ras as part of the Ras-MAPK pathway. (p. 170)
raloxifene: drug that inhibits cell proliferation in the breast and uterus by blocking estrogen receptors, but acts like estrogen in bone, acting to maintain bone strength and increase bone density. (p. 248)
randomized trial: experiment in which people are randomly assigned to different groups that receive different doses (or no dose at all) of a drug or treatment. (pp. 65, 228)
Ras: GTP-binding protein associated with the inner surface of the plasma membrane; central component of the Ras-MAPK pathway, a signaling mechanism for stimulating cell proliferation. (p. 169)
Ras-MAPK pathway: signaling pathway that plays a central role in stimulating cell proliferation in response to the binding of growth factors to cell surface receptors; activated receptors transmit a signal to the GTP-binding protein Ras; activated Ras triggers a cascade of intracellular protein phosphorylation reactions that lead to activation of MAP kinase (MAPK); activated MAPK then enters the nucleus and phosphorylates transcription factors that activate the transcription of genes required for cell proliferation; many oncogenes code for hyperactive versions or excessive quantities of proteins involved in this pathway. (p. 165)
***RAS* oncogene:** altered form of a *RAS* proto-oncogene, usually generated by point mutation, that codes for a mutant Ras protein that stimulates excessive cell proliferation and can lead to the development of cancer. (p. 159)
***RAS* proto-oncogene:** normal cellular gene coding for the Ras protein, which is involved in the Ras-MAPK pathway that activates cell proliferation. (p. 159)
rational drug design: laboratory synthesis of small molecule inhibitors that are designed to bind to and inactivate specific target molecules. (p. 223)
***RB* gene:** tumor suppressor gene coding for the Rb protein. (pp. 143, 177)
Rb protein: protein whose phosphorylation controls passage through the restriction point of the cell cycle. (pp. 26, 178)
RBE: see *relative biological effectiveness.*
reading frame: any of the three possible ways of reading a sequence of bases in messenger RNA as a series of triplets, depending on the starting point. (p. 185)
recall bias: differences in the ability to recall past events caused by the presence of a disease. (p. 62)
receptor: protein that contains a binding site for a specific signaling molecule. (pp. 22, 167)
receptor tyrosine kinase: a receptor whose activation causes it to catalyze the phosphorylation of the amino acid tyrosine in target proteins, usually other receptor molecules of the same type. (p. 167)
recessive: pattern of inheritance in which a given gene (allele) determines how the trait will appear in a organism only when the gene is present in the homozygous form; masked by a dominant allele when heterozygous. (p. 141)
relative biological effectiveness (RBE): correction factor that is multiplied by the absorbed dose of radiation (measured in grays) to obtain the biologically equivalent dose (measured in sieverts). (p. 111)
restriction point: control point near the end of G1 phase of the cell cycle where the cycle can be halted until conditions are suitable for progression into S phase; regulated to a large extent by the presence or absence of extracellular growth factors. (p. 23)
***RET* gene:** proto-oncogene that codes for a growth factor receptor present on the surface of endocrine cells; inheritance of a single mutant copy is responsible for multiple endocrine neoplasia type II. (p. 152)
retinoblastoma: rare cancer of young children that arises in the light-absorbing retinal cells located at the back of the eye; about 40% of cases involve a hereditary predisposition created by inheritance of a single defective or deleted copy of the *RB* gene. (p. 142)
retrospective study: epidemiological approach that assesses the past exposures of people who have already developed a disease. (p. 63)
retrovirus: any RNA virus that uses reverse transcriptase to make a DNA copy of its RNA. (p. 129)
reverse transcriptase: enzyme that uses an RNA template to synthesize a complementary molecule of double-stranded DNA. (p. 129)
RF: see *radiofrequency waves.*
risk factor: any agent or condition that increases the risk of developing cancer. (p. 100)
Rous sarcoma virus: virus that causes sarcomas in chickens; first cancer virus to be discovered. (p. 77)

S

S phase: stage of the cell cycle when the chromosomal DNA is replicated. (p. 22)

sarcoma: any cancer arising from a supporting tissue, such as bone, cartilage, fat, connective tissue, or muscle. (p. 6)
selection bias: type of bias that arises when people nonrandomly self-select for participation in a research study. (p. 62)
selective estrogen receptor modulator (SERM): class of drugs that selectively stimulate or inhibit the estrogen receptors of different target tissues. (p. 248)
serine/threonine kinases: family of protein kinases that catalyze the phosphorylation of the amino acids serine and threonine in target proteins; play a prominent role in signaling pathways. (p. 170)
SERM: see *selective estrogen receptor modulator*.
sievert (Sv): unit of measurement for the biologically equivalent dose of radiation, which is a reflection of the relative damage caused when a given type of radiation interacts with biological tissue; replaces an older historical unit called the *rem*, which corresponded to 0.01 Sv. (p. 111)
Smads: class of proteins involved in the signaling pathway triggered by transforming growth factor β (TGFβ); upon activation, Smads enter the nucleus and regulate gene expression. (p. 185)
SPF: see *sun-protection factor*.
spindle checkpoint: mechanism that halts mitosis at the junction between metaphase and anaphase if chromosomes are not properly attached to the spindle. (pp. 27, 189)
sporadic (nonhereditary) cancer: cancer that arises in the absence of any inherited mutation creating a predisposition to developing the disease. (p. 141)
squamous cell: epithelial cell exhibiting a thin, flattened shape; found in the outer layer of the skin and in several other types of epithelial tissue. (p. 5)
squamous cell carcinoma: cancer of squamous cells; accounts for about 20% of all skin cancers and occurs in some other epithelial tissues as well. (p. 104)
***SRC* gene:** normal human proto-oncogene that is related to the v-*src* oncogene of the Rous sarcoma virus; codes for the Src protein, a normal tyrosine kinase involved in growth signaling pathways. (p. 130)
Src kinase: tyrosine kinase produced by the *SRC* proto-oncogene; an abnormal version of this protein is produced by the v-*src* oncogene of the Rous sarcoma virus. (pp. 133, 171).
staging: see *tumor staging*.
statistically significant: likely to be a genuine result rather than a random fluctuation; generally requires a *p* value of 0.05 or less. (p. 62)
STATs: components of the Jak-STAT signaling pathway; phosphorylated STATs trigger changes in gene expression that can stimulate cell proliferation. (p. 169)
stem cell transplantation: technique in which bone marrow, peripheral blood, or umbilical cord blood is transplanted from one individual to another, or removed from and transplanted to the same individual, so as to replenish hematopoietic stem cells that are required for producing new blood cells. (p. 217)
sunburn: reddening and peeling of the skin observed after intense sunlight exposure; caused by apoptosis triggered by p53 in response to sunlight-induced DNA damage. (p. 108)
sun-protection factor (SPF): a number that indicates how much time it takes for skin treated with sunscreen to burn compared with unprotected skin. (p. 237)
sunscreen: substance that protects the skin against sunburn by absorbing or blocking the penetration of ultraviolet radiation. (p. 109)
Sv: see *sievert*.
SV40 (simian virus 40): DNA virus derived from monkey cells that causes cancer in animals and that contaminated early batches of polio vaccine. (p. 126)
synergistic: acting together in such a way that two agents in combination produce an effect that is greater than the sum of the effects produced by each acting alone. (p. 72)

T

T: see *thymine*.
tamoxifen: drug that exhibits some structural similarities to estrogens; inhibits cell proliferation in the breast by blocking estrogen receptors, but stimulates cell proliferation in the uterus by activating estrogen receptors; used for both the treatment and prevention of breast cancer. (pp. 215, 248)
telomerase: enzyme that adds new copies of the telomeric repeat sequence to the ends of existing DNA molecules. (p. 22)
telomere: end of a linear chromosomal DNA molecule; contains the same short base sequence repeated over and over again. (p. 21)
TGFβ: see *transforming growth factor β*.
TGFβ-Smad pathway: signaling pathway in which the binding of transforming growth factor β (TGFβ) to its cell surface receptor leads to the phosphorylation of Smad proteins that move into the nucleus and activate the expression of genes that inhibit cell proliferation; two key genes produce the Cdk-inhibiting proteins p15 and p21. (p. 184)
threshold: dose that must be exceeded before cancer rates begin to rise. (p. 86)
threshold model: dose-response relationship in which there is no cancer risk at low doses and a linear dose-response relationship exists after a threshold dose is exceeded. (p. 86)
thrombospondin: signaling protein that inhibits the growth of blood vessels. (p. 49)
thymine (T): nitrogen-containing base present in DNA and RNA that forms a complementary base pair with adenine (A) by hydrogen bonding. (p. 13)
TNM system: system for assessing tumor stage that considers three criteria: "T" for Tumor size, "N" for lymph Node involvement, and "M" for Metastasis. (p. 11)
***TP53* gene:** see *p53 gene*.
transcription: process by which RNA is synthesized using one strand of DNA as a template. (p. 13)
transcription factor: protein that binds to DNA and contributes to the activation of a specific set of genes. (p. 171)
transfection: transfer of foreign DNA into cells under artificial conditions. (pp. 36, 159)
transforming growth factor β (TGFβ): growth factor that can either stimulate or inhibit cell proliferation, depending on the target cell type; especially relevant for tumor development because it is a potent inhibitor of epithelial cell proliferation and roughly 90% of human cancers are of epithelial origin. (p. 184)
translation: process by which the base sequence of an mRNA molecule guides the sequence of amino acids incorporated into a protein chain; occurs on ribosomes. (p. 13)
translesion synthesis: DNA replication across regions where the DNA template is damaged. (p. 32)
translocation: see *chromosomal translocation*.
transposition: chromosomal abnormality in which a DNA segment is moved from one location to another. (p. 163)
***TRK* oncogene:** oncogene created by a DNA inversion in which one end of the *TPM3* gene (which codes for tropomyosin) becomes

fused to the opposite end of the *NTRK1* gene (which codes for a growth factor receptor); produces a fusion protein that functions as a constitutively active receptor that continually stimulates cell proliferation, regardless of whether an appropriate growth factor is present. (p. 163)

tumor: growing mass of tissue created by an abnormal process in which cells proliferate in an uncontrolled, relatively autonomous fashion, leading to a continual increase in the number of dividing cells; may be either benign or malignant; also called a *neoplasm*. (p. 5)

tumor angiogenesis: process by which cancer cells stimulate the development of a blood supply. (p. 47)

tumor dormancy: formation of microscopic tumor masses that remain dormant for prolonged periods of time. (p. 50)

tumor grading: assignment of numerical grades to tumors based on differences in their microscopic appearance; higher-grade cancers tend to grow and spread more aggressively and be less responsive to therapy than lower-grade cancers. (pp. 11, 207)

tumor-host interaction: interactions between tumor cells and surrounding normal cells that influence tumor cell proliferation, invasion, and metastasis. (p. 56)

tumor progression: gradual changes in tumor properties observed over time as cancer cells acquire more aberrant traits and become increasingly aggressive. (pp. 55, 98)

tumor staging: estimate of how far cancer has progressed based on a tumor's size and extent of spread at the time of diagnosis. (pp. 11, 206) Also see *TNM system*.

tumor suppressor gene: gene whose loss or inactivation by deletion or mutation can lead to cancer. (pp. 12, 144, 158, 175) Also see *gatekeeper* and *caretaker*.

two-hit model: requirement that both alleles of the same gene undergo mutation before cancer will arise; applies to tumor suppressor genes. (p. 142)

tyrosine kinases: family of protein kinases that catalyze the phosphorylation of the amino acid tyrosine in target proteins; play a prominent role in signaling pathways. (p. 133)

U

U: see *uracil*.

ubiquitin: small protein that is linked to other proteins as a way of marking the targeted protein for degradation by proteasomes. (p. 179)

ultrasound imaging: technique in which sound waves and their echoes are used to produce a picture of internal body structures. (p. 206)

ultraviolet radiation (UV): electromagnetic radiation with a wavelength between that of visible light and X-rays; shortest wavelength component of sunlight. (p. 105)

uracil (U): nitrogen-containing base present in RNA that forms a complementary base pair with adenine (A) by hydrogen bonding. (p. 13)

UV: see *ultraviolet radiation*.

UVA: portion of the UV spectrum whose wavelength falls between 315 and 400 nm; the predominant form of ultraviolet radiation to reach the earth. (p. 105)

UVB: portion of the UV spectrum whose wavelength falls between 280 and 315 nm; main cause of skin cancer. (p. 105)

UVC: portion of the UV spectrum whose wavelength falls between 100 and 280 nm; can cause severe burns but is absorbed by the upper layers of the atmosphere before reaching the earth. (p. 106)

V

vaccine: preparation containing antigens that stimulates an immune response toward those antigens. (p. 222)

vascular endothelial growth factor (VEGF): signaling protein that plays a central role in stimulating the growth of new blood vessels. (p. 48)

VEGF: see *vascular endothelial growth factor*.

VEGF receptor: protein located on the outer surface of endothelial cells that contains a binding site for VEGF; binding of VEGF activates the receptor and thereby stimulates cell proliferation. (p. 48)

viral oncogene: cancer-causing gene found within the genome of a virus. (p. 159) Also see *cellular oncogene*.

virus: tiny particle, composed of DNA or RNA plus a protein coat, that is incapable of reproducing independently; invades and infects cells, where it may reside in a latent form or where it may redirect the host cell's synthetic machinery toward the production of more virus. (p. 129)

v-*myc* gene: oncogene of the avian myelocytomatosis virus; codes for an abnormal version of the Myc transcription factor. (p. 134)

v-Src (protein): abnormal tyrosine kinase produced by the v-*src* oncogene of the Rous sarcoma virus. (p. 131)

v-*src* gene: oncogene of the Rous sarcoma virus; codes for an abnormal tyrosine kinase. (p. 129)

W

Wnt pathway: signaling pathway that plays a prominent role in controlling cell proliferation and differentiation during embryonic development; abnormalities in this pathway occur in some cancers. (p. 182)

X

xeroderma pigmentosum: inherited susceptibility to cancer (mainly skin cancer) caused by inherited defects in DNA excision repair or translesion synthesis of DNA. (p. 148)

X-rays: electromagnetic radiation with a wavelength shorter than that of ultraviolet radiation. (p. 109)

Z

zero tolerance: regulatory standard that assumes that there is no safe dose for any carcinogen and therefore requires the banning of any food additive or contaminant if it causes cancer in animal studies at any dose. (p. 241)

Index

Note: An *f* following a page number indicates a Figure and a *t* indicates a Table. Page numbers in bold indicate pages on which Key Terms are defined.

A

B

C

D

E

F

G

H

I

J

K

L

M

N

O

P

T

U

V

W

X

Y

Z